D1312352

Nelson
BIOLOGY

Nelson
BIOLOGY

BOB RITTER
University of Alberta/Austin O'Brien
High School
Edmonton, Alberta

RICHARD F. COOMBS
Science Coordinator
Roman Catholic School Board
for St. John's
St. John's, Newfoundland

DR. R. BRUCE DRYSDALE
Coordinator of Instruction:
Science/Health
Red Deer Public School District #104
Red Deer, Alberta

DR. GRANT A. GARDNER
Associate Professor
Department of Biology
Memorial University of Newfoundland
St. John's, Newfoundland

DAVE T. LUNN
Science Department Head
Henry Wise Wood High School
Calgary, Alberta

Nelson Canada

Published in 1993 by
Nelson Canada,
A Division of Thomson Canada Limited
1120 Birchmount Road, Scarborough, Ontario, M1K 5G4

This book is printed on acid-free paper. The choice of paper reflects Nelson Canada's goal of using, within the publishing process, the available resources, technology, and suppliers that are as environment friendly as possible.

All student investigations in this textbook have been designed to be as safe as possible, and have been reviewed by professionals specifically for that purpose. As well, appropriate warnings concerning potential safety hazards are included where applicable to particular investigations. However, responsibility for safety remains with the student, the classroom teacher, the school principal, and the school board.

ISBN 0-17-603870-1

Canadian Cataloguing in Publication Data

Main entry under title:

Nelson biology

Includes index.
ISBN 0-17-603870-1

1. Biology. I. Ritter, Robert John, 1950-

QH308.7.N45 1993   574   C93-093226-9

Cover photos: G.V. Faint/The Image Bank; Inset, Leonard Lee Rue III/Animals Animals/Earthscenes. The cover shows the Columbia Icefield glacier, located along the British Columbia–Alberta border. The inset shows a mother polar bear and her cub. **6:** Left, Dr. Bryan Eyden/Science Photo Library; Right, Illustrated by Dave Mazierski; **7:** Left, Illustrated by Dave Mazierski; Right, CNRI/Science Photo Library; **8:** Left, Monfred Kage/Peter Arnold Inc.; Right, Illustrated by Dave Mazierski; **9:** Joseph Drivas/The Image Bank; **10:** Jim Brandenburg/Minden Pictures; **11:** Left, Norbert Wu/Oxford Scientific Films; Right, Dale Wilson/First Light; **12:** Left, Howard Sochurek/Masterfile; Right, Illustrated by Dave Mazierski; **13:** Science Source/Photo Researchers; **14:** Left, Dr. Tony Brian & David Parker/Masterfile; Top left, Nuridsany et Perennou, The National Audubon Society Collection/Photo Researchers; Bottom right, Illustrated by Dave Mazierski; **15:** Mark Tomalty/Masterfile; **Unit 1** Main photo: Giant trees in a rain forest on Vancouver Island; Sidebar: Microscopic image of B-lymphocyte cells; **Unit 2** Main photo: A capillary (green) densely packed with red blood cells; Sidebar: The beginning of a blood clot. The slim threads are fibrinogen fibers; **Unit 3** Main photo: Skydivers join together to form a complex display while falling toward the earth; Sidebar: A laser is used to correct an eye disorder; **Unit 4** Main photo: This spectacular photograph is a compilation of thousands of separate images from the Tiros-N series of meteorological satellites; Sidebar: A glacier stream; **Unit 5** Main photo: A river otter; Sidebar: Cherry blossoms; **Unit 6** Main photo: Birth of a baby; Sidebar: Sunrise over the Atlantic Ocean; **Unit 7** Main photo: Stickleback eggs close to hatching; Sidebar: Seeds of a bean plant; **Unit 8** Main photo: A school of tropical fish; Sidebar: Penguins on the Antarctic continent.

Executive Editor:	*Lynn Fisher*	**Photo Research:**	*Sandra Mark, Ann Ludbrook,*
Project Manager:	*Ruta Demery*		*Leonard Lessin*
Project Coordinator:	*Jennifer Dewey*	**Photo Editor:**	*Stephen Cowie*
Senior Supervising Editor:	*Susan Green*	**Design and Art Direction:**	*John Robb*
Copy Editor:	*James Leahy*	**Assistant Art Director:**	*Liz Nyman*

Printed and Bound in the United States of America

3 4 5 6 7 8 9 AG 2 1 0 9 8 7 6

CONTENTS

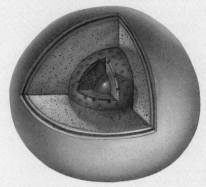

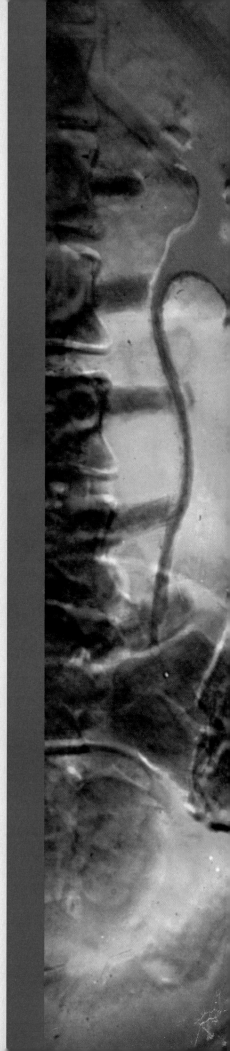

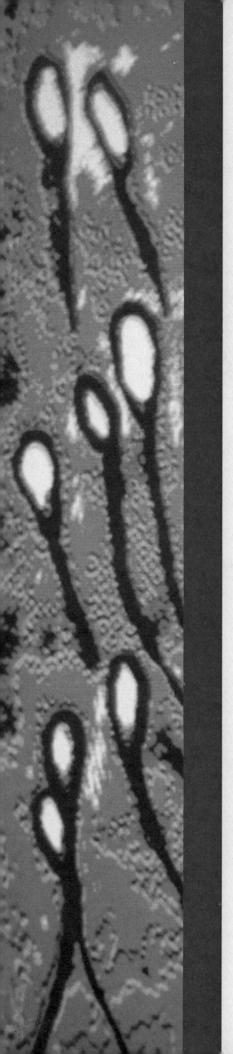

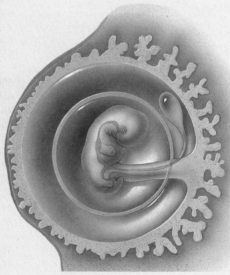

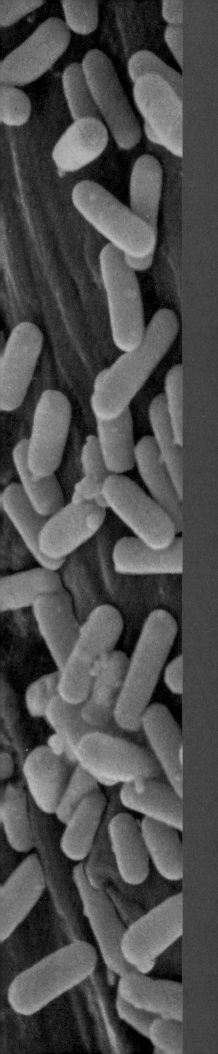

A number of science-related social issues are introduced throughout this textbook. At the end of each chapter, you are given an opportunity to debate one of these issues with other members of your class. An understanding of the issues requires an understanding of scientific principles, and the potentials and limitations of technological applications. Background information is provided in each Social Issue section as well as throughout the chapter. However, it is recommended that you do further research before the decision-making and debating begins. Collect a number of articles that present different viewpoints on the issue. In addition, you may want to speak with scientists, economists, or other professionals on the topic as you are formulating an opinion.

Read the point-counterpoint arguments presented in the debates. Note that these are not designed to provide a summary, nor are they meant to be comprehensive arguments. Rather, the arguments outline the complexity of the issues and act as a springboard for further research. Once you have completed your research, take an opportunity to reflect on your findings and discuss your viewpoint with others both inside and outside the classroom.

If you are not accustomed to viewing science and technology from a social context, you may expect science to provide you with unequivocal answers.

It is easy to fall into the trap of reducing any controversial issue to a "good-versus-bad" argument. The key to developing an understanding of why experts occasionally arrive at different conclusions is grounded in an understanding of the nature of science. Science does not provide absolute truths but presents an approach for interpreting nature. While research answers some questions, it raises many others.

When preparing for your debate it is also important to be aware of the limitations of scientific and technological problem-solving. Different social, moral, religious, and economic considerations must be recognized and valued. Controversy arises when people with competing world views draw divergent conclusions about the impact of science and technology on society. The debate format will enable you to recognize the difference between rejecting an idea and rejecting the person with the idea. The uniqueness of Canadian society stems, in part, from a plurality of beliefs and values.

The authors hope that these conflict-resolution scenarios will provide you with an opportunity to think critically, to reason, to argue logically, to devise answers that are supported by evidence, to reflect on your thinking, and to listen to others. The group debates will encourage you to share your concerns and will foster an appreciation of divergent world views.

Concept maps are used to link ideas. Although a variety of concept maps are presented throughout this textbook, the most meaningful ones are those you construct yourself. Research indicates that students who use concept maps as a study tool are often more successful on exams. Concept maps help you connect the facts you have learned in a way that is personally meaningful.

Your concept maps may be somewhat sparse at first—it is often difficult to identify the relationships between things you have studied. However, if you persist and go back to your concept maps before a test or exam, you will find that they provide an excellent summary as well as a strategy for understanding the material covered.

You may find the following points useful when using concept maps.

- Construct a concept map after every lesson. Because concept maps for most lessons do not need to be detailed, this process should only take you a few min-

utes. Check to see that all concepts presented in the lesson have been included in the concept map. If one seems to be missing, ask the teacher where it fits in.

- After every chapter, review your lesson concept maps and make a detailed chapter concept map linking the concepts learned in each lesson. You may want to look at concept maps made by other people in your class. Have they included any relationships that you might have missed?

- To prepare for unit tests, look for links between chapter concept maps. To accomplish this major review, you will need several sheets of paper. You may wish to use string or thread to make the interconnections.

The following concept map shows some of the relationships that exist in an ecosystem. How many relationships can you identify?

Relationships in Ecosystems

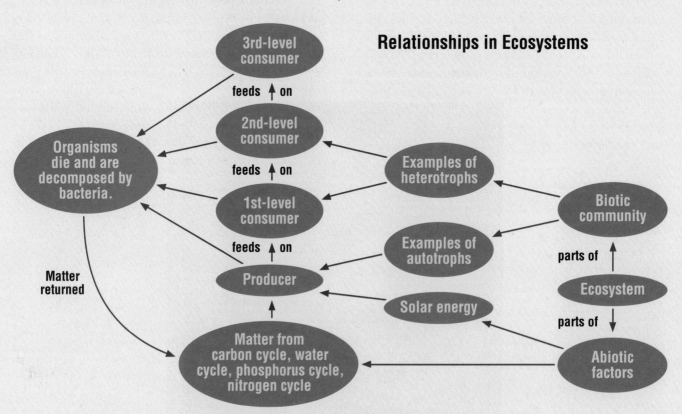

*T*he laboratory is a dynamic and exciting setting for learning biology; however, it is also a setting that can be potentially dangerous. The laboratory exercises in this textbook have been designed to minimize any hazards. However, to prevent placing yourself or other students in a harmful situation, it is important that you follow all safety procedures.

Safety Equipment

Your instructor will acquaint you with the safety equipment found in your laboratory. Locate and familiarize yourself with the use of the following safety items:

- Eyewash bottle or eye bath
- Chemical shower
- Fire extinguisher
- Fire blanket
- First-aid kit

Safety Symbols

The following symbols are used throughout the textbook to alert you to the use of specific safety equipment and procedures. You should recognize these symbols and know what they represent.

 Safety goggles: You must wear eye protection during any lab that involves flames, chemicals, or the possibility of broken glass or other small particles. Since safety goggles are shatter resistant and protect the side of the eyes, you must wear them even if you wear glasses.

 Laboratory apron: Wear a laboratory apron when handling any materials that could damage your skin or clothing.

 Wash your hands: Always wash your hands and forearms thoroughly with soap after any experiment.

General Safety Rules

1 Before you begin a lab, clear all unnecessary items from your work area and remove them to a designated place. Ensure that your hair, clothing, or jewelry will not interfere with your work. Maintain a clean and tidy work area during the laboratory. **2** Read all the directions for an experiment carefully several times before you begin your laboratory work. Before beginning any step, make sure that you understand what to do. Review any safety procedures that are specific to the exercise. Follow the steps exactly as they are written or modified by your instructor. Do not experiment on your own unless your instructor has approved your procedures and you have received permission to proceed. **3** Report all injuries, accidents, and spills, no matter how minor, to your instructor immediately. **4** Notify your instructor of any piece of equipment that is broken or does not work. Do not attempt to fix any broken or defective equipment on your own. **5** Before using any glassware, inspect it for chips or cracks. If it is chipped or cracked, do not use it. If glassware breaks, notify your instructor and dispose of it in the proper waste container using a safe procedure. Never force glass tubing into a rubber stopper; use gloves and a turning motion and a lubricant to help you. When heating a test tube, always point the open end away from yourself and others and move the tube back and forth through the flame. Use a clamp or tongs when handling glassware that has been heated. Before putting glassware away, clean it thoroughly as instructed by your teacher. **6** When you smell something, it should be done by a wafting of the fumes. Do not put your nose directly over the substance. **7** Do not taste or touch any material unless you are told to do so by your teacher. Do not eat or chew gum in the laboratory. **8** Caution must be exercised when dissecting an organism. **9** Disconnect electrical equipment by removing plugs from sockets and not pulling on the cord. Also remember that it is hazardous to have water or wet hands near an electrical source. **10** In microbiology labs, it is standard practice to assume that all microorganisms are potentially hazardous. **11** Follow carefully all directions given regarding handling, sterilization, and disposal of cultures and associated materials. **12** Dispose of any unused or waste materials, specimens, chemicals, etc., as instructed by your teacher. **13** When you have completed your laboratory exercise, clean up your work area, and return all equipment to its proper place and wash your hands thoroughly before leaving.

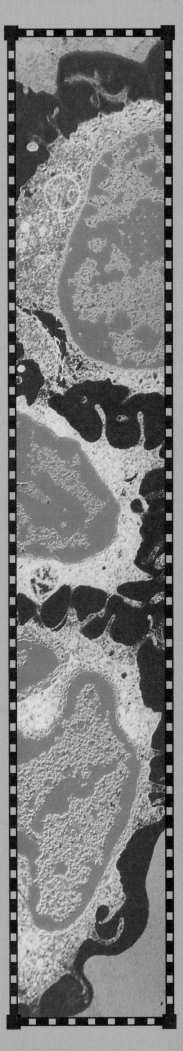

Exchange of Energy and Matter in Cells

∎

One
Development of the Cell Theory

Two
Chemistry of Life

Three
Energy within the Cell

U N I T

*D*evelopment of the Cell Theory

IMPORTANCE OF THE CELL THEORY

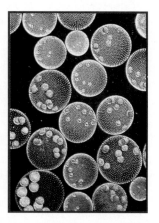

The cell is the basis of life. It has been estimated that the human body is made up of about one hundred trillion cells. Different cells in the body are specialized to perform various tasks. Muscle cells, for example, are capable of rapid contraction. If they did not function properly, movement would not be possible. Nerve cells transmit electrochemical messages. If they did not work properly, a person would not be aware of his or her environment, or be able to respond to changes in it. Vision, hearing, taste, smell, and touch all depend on nerves. The transport of oxygen, defense against disease, and communication all depend on specialized body cells. Cells are not exclusive to humans; they are also found in plants, bread mold, and pond scum. In fact, all life forms are composed of at least one cell.

Most of the cells that you encounter every day are invisible to the naked eye. The idea that you are not alone becomes evident if you examine the cells that co-exist with you. A small, rod-shaped bacterium, called *Escherichia coli*, lives in your gut. This microbe, or its

Figure 1.1

A head louse clings to a human hair. The head louse inhabits the hair of the head, gluing its eggs to the individual hairs.

descendants, supplies you with important vitamins and will likely be with you until the end of your life. A survey of your skin may reveal different types of fungi. Athlete's foot is an example of a fungus. Your skin also harbors a sphere-shaped bacterium called *Staphylococcus epidermis*. You might also see tiny mites in your eyebrows, plant spores and pollen that have found their way into your lungs, and a host of microbes and microscopic animal eggs that enter your body with the food you consume. Your body is a walking ecosystem.

BIOLOGY CLIP
It has been estimated that one-half of the dust in a furnace filter is actually dead skin cells.

Cells outside your body also play an important role in your life. The yogurt that you ate for lunch is a living community of bacteria. A bacterium called *Lactobacillus bulgaricus* causes the milk in the yogurt to sour, while a second bacterium, *Streptococcus lactis*, enhances the taste. The processing of cheese, beer, and wine depends on cells. Even the tanning of leather could not be accomplished if it were not for cells. Is it any wonder that an introduction to biology can begin with a study of the cell?

In spite of their varied size, shape, and appearance, cells have several things in common. All cells digest nutrients, excrete wastes, synthesize needed chemicals, and reproduce. Cells define both life and death.

ABIOGENESIS

Biology, like other forms of science, progresses by observation. Unfortunately, observations of nature are often flawed by interpretations that attempt to explain what was observed. This is due, in part, to the fact that the observer is actually part of nature. Therefore, the observer is subject to many of the same factors as the objects being observed. In an attempt to interpret natural events, scientists often propose explanations. An explanation is also called a **hypothesis.** The hypotheses proposed by early scientists were almost never tested by experiment. Often one unsubstantiated hypothesis became the basis for another, and scientists moved further and further from the truth.

Early scientists noticed that ponds dried up during a long period of drought and that no living fish were found in the mud. When rain began to fall in the spring and the pond filled with water, observers noticed that the pond was teeming with frogs and fish. Some concluded that the frogs and fish must have fallen to earth during the rainstorm. Incredible as this explanation may seem now, it seemed logical to many people in earlier times. The fact that nobody had ever been hit by a frog or a fish during a rainstorm did not seem to have occurred to anyone!

Aristotle, the great philosopher of ancient Greece, did not accept the hypothesis of fish and frogs falling from the sky. He proposed that fish and frogs came from the mud. He also believed that flies came from rotting meat, because he had always observed flies on decayed meat. Aristotle was so persuasive in his arguments that scientists accepted his theory of **abiogenesis** for nearly 2000 years. Abiogenesis is the theory that proposes that nonliving things can be transformed into living things spontaneously. The theory is sometimes referred to as "spontaneous generation." A mere 300 years ago, a Belgian doctor, Jean van Helmont, concluded that mice could be created from grains of wheat and a dirty shirt. Van Helmont had placed grains of

*A **hypothesis** is a possible solution to a problem or an explanation of an observed phenomenon.*

Abiogenesis *is a theory that states that nonliving things can be transformed into living things.*

wheat and a dirty shirt in a container, and within 21 days mice appeared. According to van Helmont, the sweat in the shirt caused the wheat to ferment. The fermenting wheat bubbled and was eventually transformed into mice.

In 1668, Francesco Redi, an Italian physician, conducted an experiment to test the hypothesis that rotting meat can be transformed into flies. Prior to Redi's work, science was based on logical analysis rather than on experimentation. Redi placed bits of snake, eel, fish, and veal in four different jars. He repeated the same steps in four other jars, but sealed the second set of jars. The open set of jars was designated the **experimental** group, while the closed set was designated the **control** group. What do you think happened next? After a period of time, Redi noticed that maggots were crawling all over the meat in the open jars. Apparently, flies had been attracted to the meat and began laying eggs on the food supply. The eggs hatched into maggots, which began feeding on the meat. The maggots then became flies and the cycle continued. Redi concluded that flies come from other flies, not from rotting meat. However, Redi's critics replied that the sealed jars were different from the control set, because no fresh air circulated around the meat. Air, claimed the critics, is the "active ingredient" that causes spontaneous generation. Fresh air must circulate around the meat in order for flies to appear.

Once again, Redi turned to experimentation for his answer. He repeated the experiment, but this time placed fine, meshed wire over the opening of the experimental set of jars. As Redi had predicted, flies were not found inside the experimental jars, despite the fact that air circulated around the meat. Once again, Redi proved that rotting meat cannot be transformed into flies.

Experimental variables are designed to test a hypothesis. Experimental groups test a single variable at a time.

Controls are standards used to verify a scientific experiment. Controls are often conducted as parallel experiments.

A WINDOW ON THE INVISIBLE WORLD

Inventions in one field of science often contribute to advancements in others. Three thousand years ago, when the artisans of Egypt and Mesopotamia first combined silicon dioxide, boric oxide, and aluminum oxide, they produced a supercooled liquid that would forever change the world. The miracle substance, glass, became the material of lenses. In the early 1200s a famous scientist named Roger Bacon described how crystal lenses might help improve the vision of the elderly. By the late 1200s, an Italian scientist, Salvino degli Amati, provided the world with spectacles. Lenses were now fashioned by craftsmen, and a new branch of physics, called optics, began to explain the movement of light.

The two greatest inventions of the 1600s, the telescope and the microscope, changed the way in which people understood and explained the universe. The telescope allowed scientists to see a far greater universe than was previously known, while the microscope opened the doors to a hidden world. The boundaries of the universe were no longer limited by what people could see with the naked eye.

No one scientist is responsible for the development of the microscope. Like many other inventions, the development of the microscope was an ongoing process that involved technological advances in glass making and lens polishing, along with refinements to existing models. One of the first great contributors to the development of the microscope was the Dutch naturalist, Anton van Leeuwenhoek (1632–1723). A draper by profession, van Leeuwenhoek became an expert lens maker, and proved to be a remarkably observant scientist. Although simple by today's standards, his microscope was the first to reveal microorganisms and human blood cells. Van Leeuwenhoek's techniques, however, remain shrouded in mystery. Modern scientists are still trying to explain how he was able to record and describe the invisible world so accurately. A private man, van Leeuwenhoek did not sell or display his most effective microscopes. It was only after his death, when his daughter made over 200 of his finest microscopes available to the scientific community, that his true genius was recognized.

Early microscopes were named after two great astronomers, Galileo and Kepler, as an acknowledgment of their pioneering work with telescope lenses. While the telescope was designed to magnify distant objects—to bring large objects near—the microscope took advantage of a similar construction to magnify near, but tiny objects. In fact, the microscope is so similar to the telescope that it is often described as bringing small objects near. Although this may seem an odd way to describe magnification under a microscope, consider microscopic objects as being so distant from the human eye that they are invisible. The microscope brings tiny objects closer to the eye, making them appear larger. To focus on a human blood cell, you would have to place it within one millimeter of your eye. However, the human eye cannot focus on objects that close. The microscope brings the image close to the eye through a series of lenses.

Magnification and Resolution

The most important feature of a microscope is not its magnification, but its ability to distinguish fine detail. This is referred to as *resolving power*. The resolving power of a human eye is about 0.1 mm. This means that the eye can distinguish one dot from another as long as the dots are separated by a distance of at least 0.1 mm.

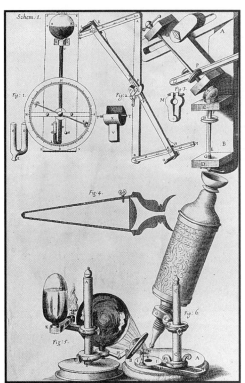

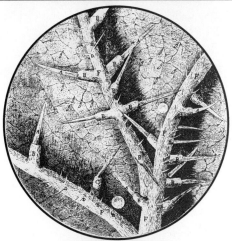

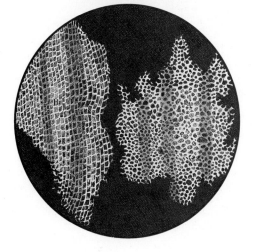

Figure 1.2

The impact of the microscope was so great that most scholars of the 17th century proudly displayed their microscopes in full view. In 1637, the famous mathematician and philosopher, René Descartes, published drawings of specimens he had observed under a microscope. Like many new inventions, the microscope at first was greeted with suspicion. Descartes was alarmed when his fellow townspeople referred to a prepared slide of a flea as "a devil shut up in a glass."

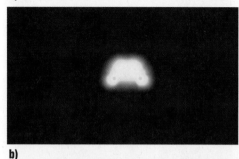

a)

b)

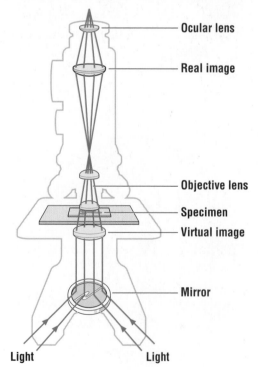

c)

Figure 1.3

Demonstration of magnification with resolution. (a) Car headlights as they appear at a distance. (b) Magnification without resolution. The eye and camera fail to detect a space between the headlights. (c) Magnification with resolution. Each headlight appears as a distinct object.

Figure 1.4

The compound light microscope uses a system of two lenses.

— Ocular lens

— Real image

— Objective lens

— Specimen

— Virtual image

— Mirror

Light   Light

The microscope employs two lenses that bring the specimen closer to the eye. The image is magnified once by the *ocular lens*, which is near the observer's eye, and again by an *objective lens*, which rests above the specimen. Should the objective lens provide a magnification of 4× (four times) and the ocular lens an additional 10× magnification, the image would be enlarged 40×. Likewise, an objective of 40×, along with a 10× ocular magnification, would increase the size of the image 400×.

Microscope Safety

1 Always carry the microscope with two hands. One hand is placed on the arm of the microscope, while the other supports the base. Always carry the microscope in the upright position so that it will not slip.

2 Use only lens paper to clean the lenses. Coarse paper will scratch the lens. Never allow your fingers to touch the lenses.

3 Never allow the lens to touch the cover slip of a slide or wet mount stains. Corrosive chemicals may destroy the lens.

4 Never attempt to repair your microscope. Always notify your teacher when problems arise. Microscopes require special repairs to ensure proper functioning.

5 Use stage clips or a mechanical stage to secure the slide.

6 Before returning your microscope to the storage area, remove any slides from the stage and rotate the nosepiece to the lowest-power objective. The low-power lens is farthest from the stage and, therefore, is less likely to be damaged. Slides left on the stage may damage lenses.

7 Remove the electrical cord from the socket by the plug. Do not pull on the cord.

Objective

To investigate the features and basic operations of the compound microscope.

Materials

compound microscope	cover slips
newspaper picture	medicine dropper
scissors	colored thread
lens paper	ruler
glass slides	small beaker (50 or 100 mL)

Procedure

Part 1: Parts of the Microscope

1 Locate the parts of the microscope shown in the diagram below.

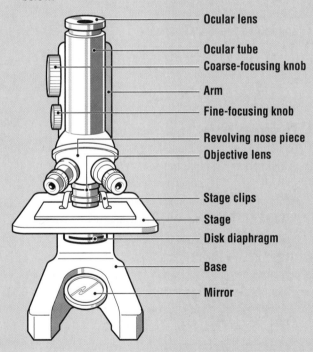

- Ocular lens
- Ocular tube
- Coarse-focusing knob
- Arm
- Fine-focusing knob
- Revolving nose piece
- Objective lens
- Stage clips
- Stage
- Disk diaphragm
- Base
- Mirror

Part 2: Resolving Power

2 Cut a 1 cm square from a newspaper picture.
a) Describe its appearance with the naked eye.
3 Obtain a glass slide, a cover slip, and a medicine dropper.

4 Make a temporary wet mount of the newspaper picture, as shown in the diagram. Add the cover slip.

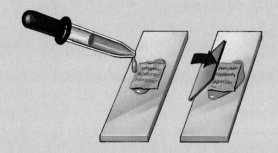

5 Clean the ocular and objective lenses with the lens paper. Adjust the nosepiece to the lowest-power objective lens. Make sure that the lens clicks into place.
6 Place the slide on the stage with the picture centered over the hole. Use the clips to hold the slide in position.
7 Rotate the coarse-adjustment knob forward to lower the low-power objective lens. If your microscope has an automatic stop, the lens will stop at a specific distance above the slide. If the microscope does not have a stop, do not let the objective lens get closer than 0.5 cm from the slide.
8 Using one eye, look through the ocular lens. (If you wear glasses, focusing with or without them on should not matter.) It is easier on your eyes if you keep both eyes open. You should be able to see a circular field of view.
9 Slowly rotate the coarse-adjustment focus backwards until the image is clear.
b) Describe the appearance of the picture under low-power magnification.

Part 3: Determining Focal Distance

10 Cut two different-colored threads and place them on a glass slide.
11 Make a temporary dry mount by placing one of the threads over the other in the form of an X. Cover the threads with a cover slip.

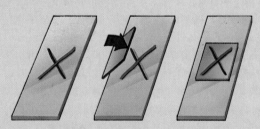

12 Place the slide on the stage so that the crossed threads are in the center over the hole. Use the clips to hold the slide in position.

13 Turn on the substage light. If your microscope has a mirror, make sure that it is adjusted to reflect the light through the objective lens.

14 Rotate the coarse-adjustment knob forward so that the low-power objective approaches the slide. Do not allow the objective to touch the slide.

15 View the crossed threads by slowly rotating the coarse-focus knob. At this point you may wish to adjust the diaphragm for better lighting.

c) Are both threads in focus?

16 Use the fine adjustment to focus on the top thread.

d) Measure and record the distance (in mm) between the bottom of the objective lens and the top of the cover slip.

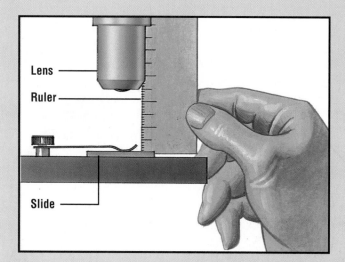

17 Rotate the revolving nosepiece to the medium-power objective. The objectives in most microscopes are *parfocal*, which means that once the low-power objective is in focus, the higher-power lenses are in focus. Usually some minor adjustment is required for sharp focusing. If your microscope is not parfocal, check with your teacher for special instructions.

18 Use the fine adjustment to focus on the upper thread. As a matter of procedure, always bring the image into focus with the low-power objective first and then proceed to use higher-power objectives. The coarse adjustment should only be used for low-power focusing.

e) Measure and record the distance between the cover slip and the objective lens.

19 Rotate the nosepiece to the high-power objective and fine focus.

f) Measure and record the distance between the cover slip and the objective lens.

g) As you move from the low- to higher-power objectives, describe the change in light intensity.

h) Which objective is the best for showing the detail of the threads?

i) Under which objective is the bottom thread clearest when the top thread is in focus? (You may wish to re-examine the threads with each of the magnifications.)

Part 4: The Field of View
The circle of light seen through the microscope is called the *field of view.* It represents the observed area.

20 Switch the nosepiece to the low-power objective and examine the length of thread seen.

21 Repeat the procedure for the medium- and high-power objectives.

j) Compare the length of thread seen under each objective.

22 Clean the slide and cover slip and return them to their appropriate location. Rotate the nosepiece to the low-power objective and return the microscope to the storage area.

Laboratory Application Questions

1 Explain why microscopes are stored with the low-power objective lens in position.

2 *Astigmatism* is a common disorder in which the lens of the eye has an asymmetrical shape. Most people have symmetrical lenses—the top half is identical in shape to the bottom half. Explain why individuals who have astigmatism may experience difficulties distinguishing fine detail with the naked eye.

3 Explain why resolving power decreases as the thickness of the objective lenses increases.

4 Why should the coarse-adjustment focus not be used with a high-power objective lens?

5 The microscope invented by van Leeuwenhoek consisted of a single lens. What advantages do compound microscopes have over single-lens microscopes? ■

ABIOGENESIS AND MICROBES

Even as the invention and refinement of the light microscope revolutionized the study of biology, scientists continued to make mistakes as they sought to interpret their observations. Such was the case of the English biologist John Needham (1713–1781) when he set out to re-examine the theory of abiogenesis. Needham observed that meat broth left unsealed soon changed color and gave off a putrid smell. Mold and bacteria were found growing in the rich nutrient, but it was unclear where these microbes came from. Unlike the early supporters of abiogenesis who used logical analysis, John Needham tested his hypothesis through experimentation. Experimentation had become an essential component of science.

Needham boiled flasks containing nutrient meat broth in loosely sealed flasks for a few minutes in order to kill the microbes. The solutions appeared clear after boiling. The flasks were then left for a few days and the murky contents were examined under a microscope. The broth was teeming with microorganisms. Could this mean that the broth had spontaneously created microbes? Needham rushed to retest the experiment, using different nutrient solutions. Despite the boiling, the microbes reappeared a few days later. Needham concluded that the microbes had come from nonliving things in the nutrient broth.

Needham's conclusions sent many scientists down the wrong pathway. Let us re-examine his experiment to understand why. One of the difficulties arose from the fact that the flasks were not sealed properly—the tiny microbes could have entered the flasks after boiling. Another difficulty resulted from the design of his experiment. The fact that the flasks appeared clear immediately after boiling did not mean that all the microorganisms were destroyed. If only a few of the tiny microbes had survived, they would be able to multiply to millions within a few days. Needham did not check the flasks for microbes immediately after boiling. Even if he had checked the flasks, it is unlikely that he would have found any of the remaining microbes. Each drop of the nutrient would have to be examined, and such an examination might even infect the flask.

Needham's conclusions were upheld for nearly 25 years. Lazzaro Spallanzani (1729–1799) repeated Needham's experiment, but boiled the flasks longer. Spallanzani also took special care to seal the flasks completely. No microorganisms were found; abiogenesis did not occur. Needham's supporters were cautious about Spallanzani's experiments. They suggested that because Spallanzani had completely sealed the jars, the active principle had been destroyed. You will recall that the active principle objection had been used to oppose the work of Francesco Redi about 100 years earlier. Others claimed that the boiling had destroyed the nutrients. Although Spallanzani did not believe that an active principle existed, he was unable to overcome the objections that centered on the active-principle hypothesis.

The final blow to the theory of abiogenesis was delivered by the great French scientist Louis Pasteur (1822–1895). In 1864, Pasteur had a glassworker develop a special swan-necked flask. Broth was placed in the flask and subsequently boiled to destroy the microbes. Air passed from the flask during boiling. Fresh air entered the flask as the flask cooled. However, the microbes were trapped in the curve of the flask and were not carried into the broth from the surrounding air. Because the broth remained clear, Pasteur predicted that microbes were not present.

Microscopic view after boiling

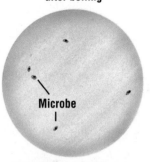

Microbe

Few microbes can be seen.

Microscopic view 2 days after boiling

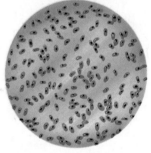

Many microbes can be seen.

Figure 1.5

Supplied with enough nutrients, microbes reproduce quickly.

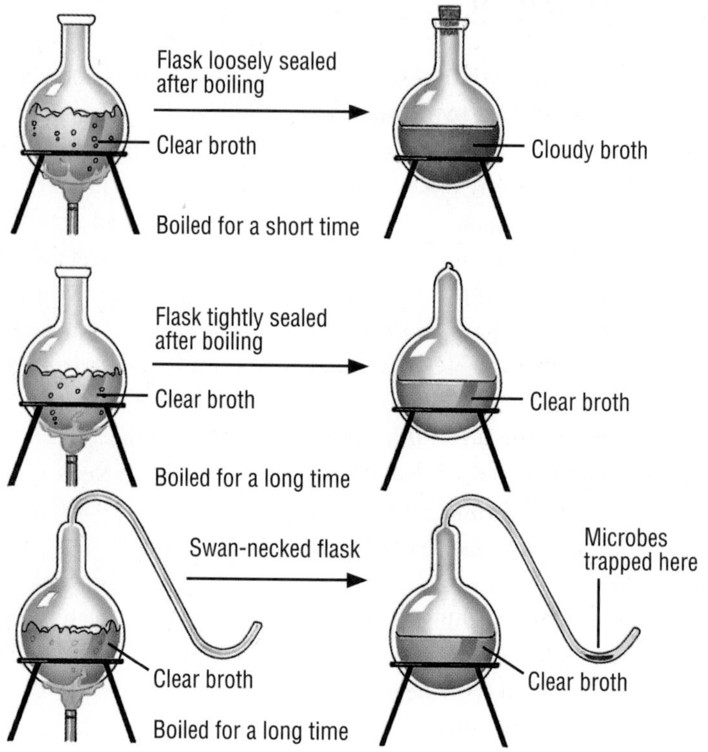

Needham

Flask loosely sealed after boiling

Clear broth

Boiled for a short time

Cloudy broth

Spallanzani

Flask tightly sealed after boiling

Clear broth

Boiled for a long time

Clear broth

Pasteur

Swan-necked flask

Clear broth

Boiled for a long time

Microbes trapped here

Clear broth

Figure 1.6

Pasteur's improvements to Needham's and Spallanzani's experiments dealt the final blow to the theory of abiogenesis.

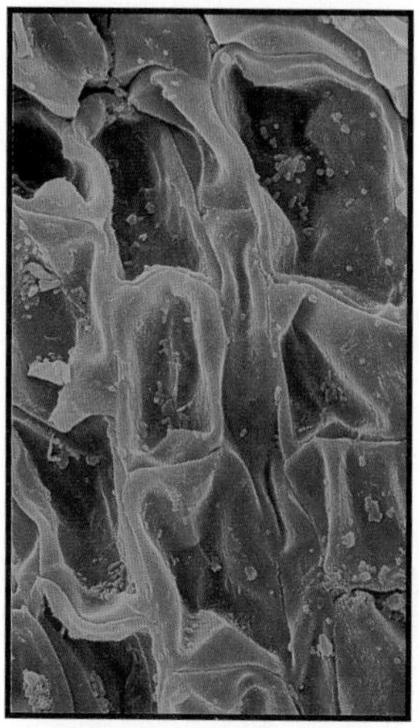

Figure 1.7

Cork cells seen through a microscope. The visible objects are the cell walls that surrounded the living cells.

A microscopic examination of the nutrient broth confirmed his prediction. Microbes could not be created from nonliving broth. As a finale, Pasteur tipped the broth in one of the flasks, allowing it to run into the curve of the swan-necked flask. As Pasteur had predicted, the broth became contaminated by the microorganisms trapped there. In a few days, the flask became cloudy.

EMERGENCE OF THE CELL THEORY

No one scientist developed the cell theory. Cells were probably first described in 1665, when the English scientist, Robert Hooke, noticed many repeating honeycomb-shaped structures while viewing a thin slice of cork under his primitive microscope. In his book, *Micrographia*, Hooke used the word "cell" to describe these structures. However, cork, the spongy tissue from cork trees, has few living cells. What Hooke observed were the rigid cell walls that surrounded the once-living plant cells.

A few years later, Anton van Leeuwenhoek observed living blood cells, bacteria, and single-cell organisms in a drop of water. As microscopes improved, more structures were described. Around 1820, Robert Brown described the appearance of a tiny sphere in plant cells. He called the structure the *nucleus* (plural: "nuclei"). Nuclei were soon discovered in animal cells. A zoologist, Theodor Schwann, and a botanist, Mathias Schleiden, concluded that plant and animal tissues are composed of cells. Schwann and Schleiden prepared the foundations for the modern *cell theory*. The modern cell theory states:

- All living things are composed of cells. The cell is the basic living unit of organization.
- All cells arise from pre-existing cells. Cells do not come from nonliving things.

Unit One:
Exchange of Energy and Matter in Cells

REVIEW QUESTIONS ?

1 What is a scientific hypothesis?
2 Why do scientists test hypotheses by experimentation?
3 What observations led to the theory of abiogenesis?
4 How did Francesco Redi's experiment challenge the theory of abiogenesis?
5 Using Redi's experiment, differentiate between experimental variables and controls.
6 Who was one of the first contributors to the development of the microscope?
7 How did the work of John Needham cause a resurgence of the theory of abiogenesis?
8 Explain how Pasteur refuted the theory of abiogenesis.
9 Why were Robert Hooke's discoveries important to the development of the cell theory?
10 What are the two components of the cell theory?

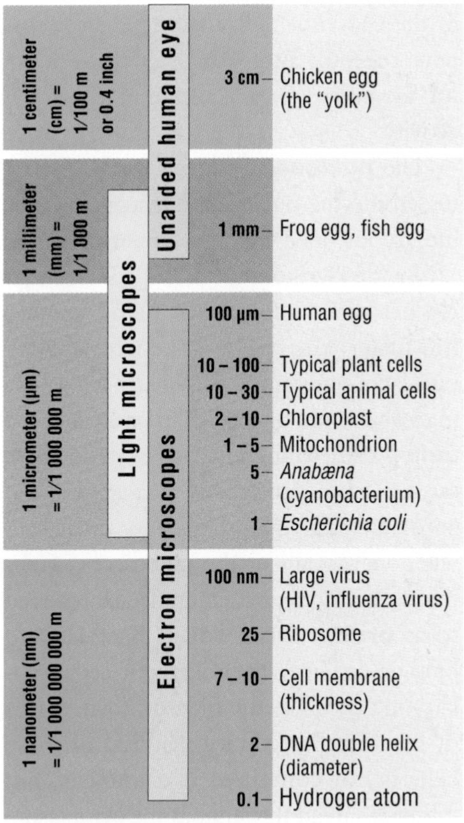

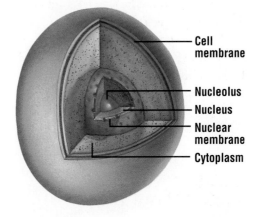

1 centimeter (cm) = 1/100 m or 0.4 inch	Unaided human eye	3 cm	Chicken egg (the "yolk")
1 millimeter (mm) = 1/1 000 m		1 mm	Frog egg, fish egg
1 micrometer (µm) = 1/1 000 000 m	Light microscopes	100 µm	Human egg
		10 – 100	Typical plant cells
		10 – 30	Typical animal cells
		2 – 10	Chloroplast
		1 – 5	Mitochondrion
		5	*Anabæna* (cyanobacterium)
		1	*Escherichia coli*
1 nanometer (nm) = 1/1 000 000 000 m	Electron microscopes	100 nm	Large virus (HIV, influenza virus)
		25	Ribosome
		7 – 10	Cell membrane (thickness)
		2	DNA double helix (diameter)
		0.1	Hydrogen atom

$1\ m = 10^2\ cm = 10^3\ mm = 10^6\ \mu m = 10^9\ nm$

Figure 1.8

Comparison of cell sizes.

OVERVIEW OF CELL STRUCTURE

Cells vary in size, shape, and function. Although there is no one common cell, all plant and animal cells are organized along a similar plan. Near the center of all cells is the nucleus, which is surrounded by fluid cytoplasm. The entire cell is covered by a membrane envelope, the cell membrane. Both the nucleus and the cytoplasm are collectively referred to as the **protoplasm.** The word protoplasm is a nonspecific term that refers to all substances within the cell.

The **cytoplasm** is the region of the protoplasm outside of the nucleus and the location where nutrients are absorbed, transported, and processed. During the processing of nutrients, waste products accumulate. The cytoplasm stores the

- Cell membrane
- Nucleolus
- Nucleus
- Nuclear membrane
- Cytoplasm

Figure 1.9

Basic cell structures of an animal cell.

wastes until proper disposal can be carried out.

The *cell membrane* is the outermost edge of the cell. Composed of a double layer of lipid or fat molecules and embedded proteins, the cell membrane provides the cell with a connection to the external environment. The membrane holds the contents of the cell in place and regulates the movement of molecules into and out

Protoplasm *refers to all the material within a cell. The protoplasm is composed of the nucleus and the cytoplasm.*

Cytoplasm *is the area of the protoplasm outside of the nucleus.*

of the cell. The cell membrane also contains receptor sites, which serve as docks for the entry of molecules that affect cell activity.

The **nucleus** is the cell's control center. Inside the nucleus the instructions for life are encoded within a molecule called DNA (deoxyribonucleic acid). DNA carries hereditary information. In most higher life forms, the nucleus is enclosed by a nuclear membrane. Cells that have a true nuclear membrane are called **eukaryotic cells.** Plant and animal cells are eukaryotic cells. The nucleus of a eukaryotic cell appears darker than the watery cytoplasm when viewed under the microscope. Cells that lack a nuclear membrane are referred to as **prokaryotic cells.** Bacteria and blue-green algae are prokaryotic cells. Prokaryotic cells are the oldest known forms of life. The hereditary material of these cells is not contained in a nucleus, but rather is spread throughout the cytoplasm.

The **nucleus** *is the control center for the cell and contains hereditary information.*

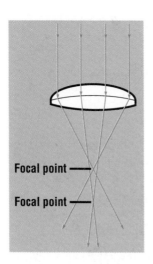

Figure 1.10

An image through a thick lens appears blurry.

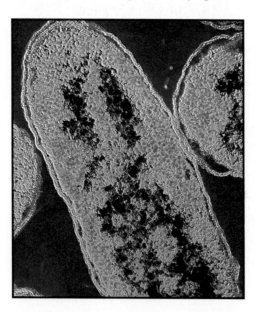

Figure 1.11

A prokaryotic cell, such as the bacterium *Escherichia coli,* lacks a nuclear membrane.

The hereditary or genetic material in the nucleus is organized into threadlike structures called **chromosomes.** Each chromosome contains a number of different characteristics, or genes. Genes are units of instruction that determine the specific traits of an individual. Some cells contain a **nucleolus,** a dark-stained, spherical structure within the nucleus. Although the entire function of the nucleolus is not known, scientists believe that it is involved with the synthesis of proteins in the cytoplasm.

THE SECOND TECHNOLOGICAL REVOLUTION

Oil-Immersion Lenses

As light rays move through the lenses of a microscope, the different wavelengths form various focal points. Because the image produced has many points of focus, it appears blurred. In order for things to appear larger, thicker lenses are required. But as lenses become thicker, light rays are bent even more, and the image becomes blurry. All microscopes sacrifice clarity, or resolution, in order to increase magnification.

The dilemma of increasing magnification while preserving resolution has plagued scientists for many years. One of the major breakthroughs in microscopy came when a drop of oil was placed on a slide, and the large, high-power objective was lowered into the oil. The denser oil limited the scattering of light and provided a clearer image. Oil-immersion lenses are still used today, especially for studying bacteria.

But oil-immersion lenses also have their limitations. Even the most sophisticated techniques limit the light microscope to $2000 \times$ magnification. For very tiny viruses or the details within a human cell to be seen, greater magnification is required. The electron microscope provides the answer.

Electron Microscope

A very crude electron microscope was invented in Germany in 1932. Although it provided 400× magnification, the image was grainy. In 1937, at the University of Toronto, James Hillier and Albert Prebus unveiled an electron microscope with 7000× magnification. Today, transmission electron microscopes are capable of 2 000 000× magnification.

The electron microscope uses electrons instead of light. Electrons are tiny subatomic particles that orbit around the nucleus of an atom. First described by J. J. Thompson in 1897, electrons move about in waves in a manner similar to visible light. However, the wavelength of the negatively charged electron is approximately 100 000 times shorter than that of visible light. Because this shorter wavelength is scattered less than longer wavelengths, a sharper image results.

The electron microscope looks much like a very large, upside-down light microscope. The object to be viewed is encased in thin plastic and placed on a stage. Electrons are beamed through the specimen and an image is projected on a screen. Because the electrons pass through the object, this microscope is often called a *transmission electron microscope*.

One of the greatest disadvantages of the electron microscope is that electrons are easily deflected or absorbed by other molecules. For this reason, the microscope's transmission chamber is equipped with a vacuum pump that removes air molecules. Because electrons pass more easily through single layers of cells, very thin sections of cells must be used. A thick specimen would absorb all the electrons and produce a blackened image. The cell sections are mounted in plastic, which means that only dead cells can be observed. Thus the transmission electron microscope is best suited for examining the structures within a nonfunctioning cell.

Scanning Electron Microscope

The scanning electron microscope provides even greater scope for scientific investigation. Made popular by physicist Victor Crewe in about 1970, this instrument produces a three-dimensional image. Electrons are passed through a series of magnetic lenses to a fine point. This fine point of electrons scans the surface of the specimen. The electrons are then reflected and magnified onto a TV screen where they produce an image. Specimens are often coated with a thin layer of gold to produce a sharper image.

Eukaryotic cells *are cells that have a true nucleus. The nuclear membrane surrounds a well-defined nucleus.*

Prokaryotic cells *are primitive cells that do not have a true nucleus or a nuclear membrane.*

Chromosomes *are threadlike structures of DNA that contain genes.*

The **nucleolus** *is a small spherical structure located inside the nucleus.*

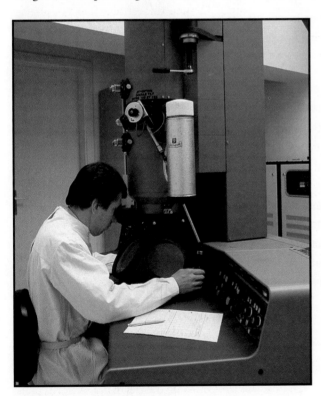

Figure 1.12

The transmission electron microscope permits scientists to view cell structures at extremely high magnification.

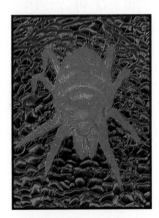

Figure 1.13

Scanning electron micrograph of the surface of a spider mite.

RESEARCH IN CANADA

James Hillier was born in Brantford, Ontario, in 1915 and trained as a physicist at the University of Toronto. Along with his colleague, Albert Prebus, Hillier has had a tremendous impact on the advancement of cell biology. Credited with designing the first truly functional transmission electron microscope, Hillier and Prebus permitted biologists to peer into a previously unseen world.

Dr. Thomas Chang, a scientist at McGill University, invented artificial cells in 1957 when he was still an undergraduate student. Artificial cells, which function in much the same way as natural cells, offer many opportunities for replacing the functions of biological cells.

Continuing research by Dr. Chang and other scientists around the world has resulted in artificial blood soon ready for human use. Artificial cells have also been tested for the treatment of diabetes and liver failure. Other types of artificial cells are being tested for the treatment of hereditary diseases and metabolic disorders. Researchers are also studying cell membranes of artificial cells to gather information about drug delivery systems. The possibilities of artificial cell technology seem almost limitless.

DR. JAMES HILLIER
Electron Microscope

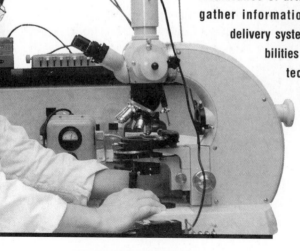

DR. THOMAS CHANG
Artificial Cells

THE CYTOPLASMIC ORGANELLES

A cell can be compared to a factory. Like factories, new cell structures are erected and damaged structures are repaired. Nutrient molecules are absorbed and fashioned into molecules essential for life. The cytoplasm is the area of the cell in which work is accomplished, while the nucleus provides the directions. The working area of the cell has specialized structures called organelles ("small organs"), which are visible when viewed under an electron microscope. The following sections discuss the organelles that play a major role in the proper functioning of the cell "factory."

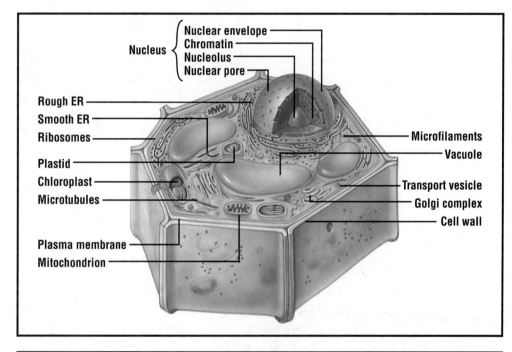

Figure 1.14

Plant cell and organelles.

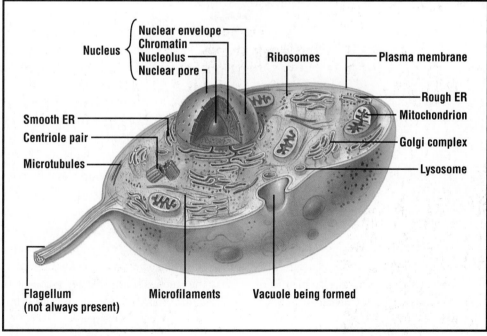

Figure 1.15

Animal cell and organelles.

Mitochondria

Tiny oval-shaped organelles called mitochondria (singular: "mitochondrion") are often referred to as the "power plant" of the cell. They provide the body with needed energy in a process called cell **respiration.** In this process, sugar molecules are combined with oxygen to form carbon dioxide and water. As the products are formed, energy is released.

BIOLOGY CLIP
Although most pictures show mitochondria as oval structures, scientists know that the shape can change quickly. Mitochondria swell and shrink in response to certain hormones or drugs.

$$C_6H_{12}O_6 + 6O_2 \longrightarrow 6CO_2 + 6H_2O + \text{energy (ATP)}$$
sugar   oxygen   carbon   water
                       dioxide

It is important to note that energy is not made in the mitochondria. Chemical bonds within the sugar molecule are broken and chemical energy is converted

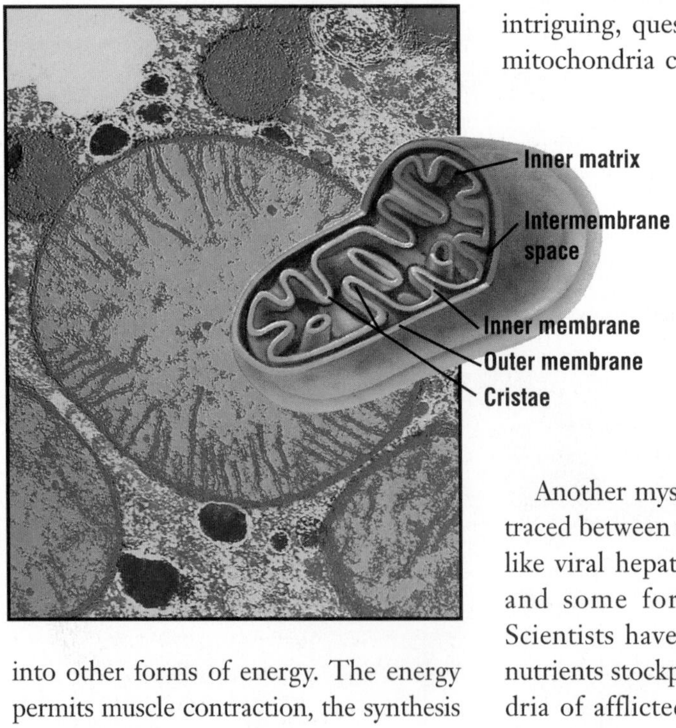

Figure 1.16

Electron micrograph of the mitochondria. Mitochondria are the largest of the cytoplasmic organelles.

Inner matrix
Intermembrane space
Inner membrane
Outer membrane
Cristae

into other forms of energy. The energy permits muscle contraction, the synthesis of new molecules, and the transport of certain molecules within the cell. Energy is essential to life.

The mitochondria have two separate membranes: a smooth outer membrane and a folding inner membrane. The inner membrane consists of fingerlike projections called *cristae.* Special proteins called **enzymes** are located on the cristae. These enzymes assist in the breakdown of the sugar molecules in the mitochondria. Approximately 36% of the energy from sugar molecules is converted into a molecule called *adenosine triphosphate (ATP),* a chemical storage compound. Most of the remaining 64% is converted into thermal energy. This means that animals that maintain a constant body temperature use a great number of sugar molecules to keep warm.

The origin of the mitochondria presents one of the most baffling, but intriguing, questions for biologists. The mitochondria contain their own hereditary material. However, the DNA of the mitochondria is not like that found in other eukaryotic cells. The DNA is much closer to bacterial DNA. Could this mean that the mitochondria were once separate organisms that invaded eukaryotic cells?

Another mysterious connection can be traced between mitochondria and diseases like viral hepatitis, obstructive jaundice, and some forms of muscle disease. Scientists have found large amounts of nutrients stockpiled within the mitochondria of afflicted cells. Scientists believe that the nutrients are waiting to be processed and that the enzymes required for this processing may be missing.

Ribosomes

Ribosomes are the organelles in which *proteins* are synthesized. Proteins are the molecules that make up cell structure. Cell growth and reproduction require the constant synthesis of protein. The nucleus provides the information for the type of protein needed. The chief building blocks of proteins, *amino acids*, are fused together by enzymes within the ribosome. There are 20 different amino acids. A change in position of a single amino acid can create a different protein.

Measuring just 20 nm (nanometers) in length, ribosomes are among the smallest organelles found in the cytoplasm. (There are one million nanometers in a single millimeter.) Yet, despite their minute size, ribosomes make up a great portion of the cytoplasm. It has been estimated that one-quarter of the cell mass of *Escherichia coli* are ribosomes. The large number of ribosomes permits the simultaneous construction of many proteins within a single cell.

Endoplasmic Reticulum

A series of canals carry materials throughout the cytoplasm. The canals, composed of parallel membranes, are referred to as *endoplasmic reticulum* or *ER*. The membranes can appear either rough or smooth when viewed under the electron microscope. The rough endoplasmic reticulum (RER) has many ribosomes attached to it. These ribosomes synthesize proteins. Rough endoplasmic reticulum is especially prevalent in cells that specialize in secreting proteins. For example, RER is highly developed in cells of the pancreas that secrete digestive enzymes. Smooth endoplasmic reticulum (SER) is free of ribosomes and is the area in which fats or lipids are synthesized. Smooth endoplasmic reticulum is prevalent in cells of developing seeds and animal cells that secrete steroid hormones.

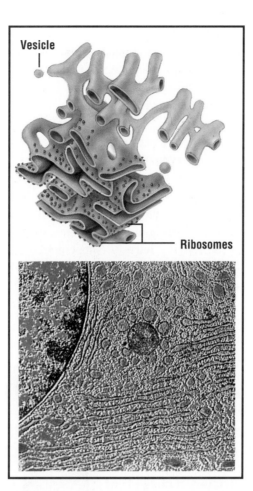

Vesicle

Ribosomes

Golgi Apparatus

The **Golgi apparatus** was first described by the Italian physician Camillo Golgi in 1898. Golgi had stained cells from a barn owl and found a new cytoplasmic structure. Half a century later, electron microscopy confirmed Golgi's observations. The structure appears like a stack of pancakes. The pancake-like structures are actually membranous sacs piled on top of each other. Each sac is called a Golgi body.

Protein molecules from the rough endoplasmic reticulum are stored within the Golgi apparatus. The packed protein membranes move toward the cell membrane. Once the Golgi apparatus fuses to the outer membrane, small packets, called *vesicles*, are released. This process, described as **exocytosis,** is the means by which large molecules such as hormones and enzymes are released from cells.

Figure 1.17

Diagram and electron micrograph of rough endoplasmic reticulum. The micrograph shows the endoplasmic reticulum as yellow and green parallel, linear structures.

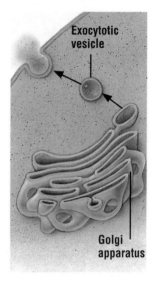

Exocytotic vesicle

Golgi apparatus

Figure 1.18

Large molecules are transported out of a cell by means of a vesicle.

The **Golgi apparatus** *is a protein-packaging organelle.*

Exocytosis *is the passage of large molecules through the cell membrane to the outside of the cell.*

Lysosomes

Lysosomes, formed by the Golgi apparatus, are saclike structures that contain digestive enzymes. Slightly smaller than the mitochondria, the lysosomes break down large molecules and cell parts within the cytoplasm. Food particles that are brought into the cell are broken down into the smaller molecules, which can then be used by the cell.

Lysosomes also play an important role in the human body's defense mechanism by destroying harmful substances that find their way into the cell. In the case of white blood cells that engulf invading bacteria, the lysosomes release their digestive enzymes, destroying both the bacterium and the white blood cell. The fluid and protein fragments that remain after the cells have been destroyed make up a substance called *pus*. The enzymes released from the lysosomes also destroy damaged or worn-out cells.

More than 30 different hereditary diseases have been linked to defective digestive enzymes in the lysosome. Tay-Sachs disease, for example, results from an enzyme deficiency that causes waste materials to accumulate. The build-up of waste products can cause brain damage.

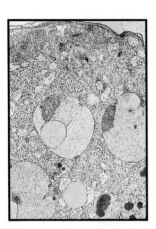

Figure 1.19

Acting as "suicide sacs," lysosomes release enzymes that destroy damaged or worn-out cells.

Plastids *are organelles that function as factories for the production of sugars or as storehouses for starch and some pigments.*

Chloroplasts *specialize in photosynthesis, and contain the green pigment chlorophyll found in plant cells.*

> **BIOLOGY CLIP**
>
> During the metamorphosis of insects, the membranes of the lysosomes become permeable, releasing digestive enzymes. The discovery of the reason behind this process may give scientists a better understanding of arthritis. The swelling and pain associated with arthritis have been linked to the seepage of lysosome enzymes into the cytoplasm. Drugs like cortisone, which reduce swelling, are known to strengthen lysosomal membranes.

Microfilaments and Microtubules

Microfilaments are pipelike structures that help provide shape and movement for the cells. Muscle cells have many microfilaments. Microtubules are tiny threadlike fibers that transport materials throughout the cytoplasm. Composed of proteins, the microtubules are found in structures called *cilia* and *flagella*. The cilia are tiny hairlike structures that aid in movement. The cells that line the windpipe, for example, use cilia to sweep foreign materials from the lungs. The flagella are longer whiplike tails that propel some cells forward. Sperm cells, for example, use flagella to move.

SPECIAL STRUCTURES OF PLANTS

The organelles discussed to this point are found in both plant and animal cells. However, plant cells differ from animal cells in that they can produce their own food. Specialized organelles, called **plastids,** are associated with the production and storage of food in plant cells. In addition, a large part of the cytoplasm of plant cells is composed of a fluid-filled space, called a *vacuole*. The vacuole serves as a storage space for sugars, minerals, and proteins. The vacuole also increases the size, and hence the surface area, of the cell, thereby increasing the rate of absorption of minerals necessary for plant nutrition.

Plastids

Plastids are chemical factories and storehouses for food and color pigments. **Chloroplasts** are plastids that contain the green pigment chlorophyll. They specialize in *photosynthesis*, a process by which plants combine carbon dioxide from the air with water from the roots in the presence of light, producing sugar and releasing oxygen.

Another type of plastid, called a **chromoplast,** stores the orange and yellow pigments found in numerous plants. Colorless plastids, called **amyloplasts,** are storehouses for starch. Potato tubers and seeds contain many amyloplasts.

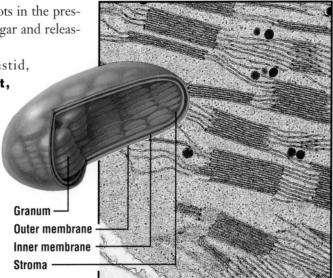

Granum —
Outer membrane —
Inner membrane —
Stroma —

Figure 1.20

Electron micrograph and diagram showing the general internal structure of a chloroplast.

Chromoplasts *store orange and yellow pigments.*

Amyloplasts *are colorless plastids that store starch.*

Cell Walls

Most plant cells are surrounded by a nonliving cell wall. Cell walls are composed of cellulose. Their main function is to protect and support plant cells. Some plants have a single cell wall, referred to as the *primary cell wall*, but others also have a *secondary cell wall*, which provides the cell with extra strength and support. The petals of a flower are composed of thin primary cell walls, as are cherries, strawberries, and lettuce. Particularly rigid secondary cell walls can be found in trees, even after the plant cells have died. In fact, most of a tree trunk is made up of hollow cell walls.

Plants cells are organized in regular patterns. The layer between the cell walls is referred to as the *middle lamella* (plural: "lamellae"). The middle lamella contains a sticky fluid, called *pectin*, that helps to hold the cells together. Have you ever noticed the sweet gooey material that forms on top of an apple pie after baking? This material, called pectin, is released during cooking, and forms a jellylike substance as it cools.

Table 1.1 Summary of Cell Components

Cell structure	Prokaryotic	Eukaryotic	
		Plant	Animal
Cell wall	yes	yes	no
Cell membrane	yes	yes	yes
Nucleus	no	yes	yes
DNA	yes	yes	yes
Mitochondria	no	yes	yes
Ribosomes	yes	yes	yes
Lysosomes	no	yes	yes
Golgi bodies	no	yes	yes
Endoplasmic reticulum	no	yes	yes
Plastids	some species	yes	no

■ REVIEW QUESTIONS ?

11 How do prokaryotic cells differ from eukaryotic cells?

12 Identify the two structures of protoplasm and give a generalized function of each.

13 What are chromosomes and genes?

14 Discuss the contributions of the Canadian scientist James Hillier to the study of cell biology.

15 List the cytoplasmic organelles found in animal cells and state the function of each.

16 What are chloroplasts?

17 Why would you expect to find more amyloplasts in a potato tuber (root) than in a potato leaf?

Primary cell wall —
Secondary cell wall

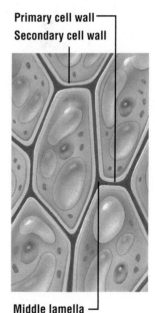

Middle lamella —

Figure 1.21

The middle lamella of plant cells contains pectin.

Objectives

To identify the components and compare the structures of plant and animal cells.

Materials

compound microscope	water
lens paper	cover slip
onion	iodine stain
scalpel	paper towels
forceps	prepared slide of human
medicine dropper	cheek epithelium
glass slide	50 mL beaker

Procedure

1 Use lens paper to clean the ocular and objective lenses of the microscope.

2 Obtain a 2 cm² section of a scale from an onion bulb.

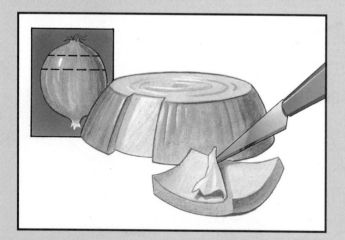

3 Using the scalpel, remove a single layer from the inner side of the scale leaf of the onion. If the extracted tissue is not transparent to light, try again.

4 Using forceps, place the tissue on a glass slide and add two drops of water. Holding the cover slip with your thumb and forefinger, touch it to the surface of the slide at a 45° angle. Gently lower the cover slip, allowing the air to escape. If air bubbles are present, gently tap the slide with the eraser end of a pencil to remove them.

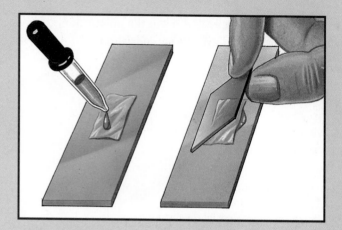

5 Focus the cells under low-power magnification. Identify a group of cells that you wish to study, and move the slide so that the cells are in the center of the field of view.

6 Rotate the nosepiece of the microscope to the medium-power objective and view the cells. Using your fine-adjustment focus, bring the cells into clear view.
   a) Draw and describe what you see.

7 Slowly decrease light intensity by adjusting the diaphragm of the microscope.
   b) Which light intensity reveals the greatest detail?

8 Remove the slide from the microscope and place a drop of iodine on one edge of the cover slip. Touch the opposite edge of the cover slip with the edge of a paper towel. This will draw the stain onto the slide.

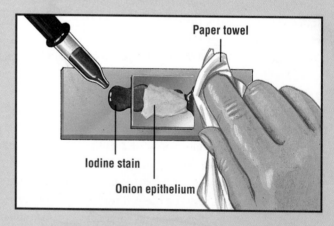

Paper towel

Iodine stain

Onion epithelium

9 Select a section of the tissue and view the cells under medium-power magnification.
   c) Why was iodine used?

10 Rotate the revolving nosepiece to high-power magnification. Using the fine adjustment, focus on a group of cells.
   d) Draw a four-cell grouping and label as many cell structures as you can see.
   e) Estimate the diameter of one cell.

11 Rotate the nosepiece back to low-power magnification and remove the slide containing the plant cells. Place the prepared slide of animal tissue epithelium on the stage of the microscope. Using the coarse-adjustment focus, locate a group of animal cells.

   f) How does the arrangement of plant and animal cells differ?

12 Locate a number of cells under medium power and focus using the fine-adjustment focus.

   g) Draw three different cells and label those cell structures that are visible.

   h) Estimate the diameter of the cells.

Laboratory Application Questions

1 What function is served by the cells of the epithelial tissue of plants and animals?

2 In what ways do the onion cells differ from those of the animal epithelium?

3 Explain why the cells of the onion bulb do not appear to have any chloroplasts.

4 Why are cells of the onion and animal epithelium classified as eukaryotic cells? ■

FRONTIERS OF TECHNOLOGY: THE THIRD REVOLUTION

The electron microscope opened a new window onto the cell, allowing scientists to observe things they had only been able to imagine before. As we discussed earlier, however, the transmission electron microscope has its limits. Because cell structures must be fixed in plastic to be viewed through this microscope, scientists are limited to observing nonliving things. A clearer understanding of a living cell requires different technologies.

Cell fractionation provides information about the thousands of chemical reactions that occur simultaneously in a cell. In this technique, cells are ground into fragments and placed in a test tube, which is then spun in a centrifuge. The centrifuge is a machine that rotates at high speeds to produce a force of gravity hundreds of thousands of times greater than normal. Heavy structures are driven to the bottom of the test tube, while lighter objects remain nearer the surface. Because each layer contains different cell

parts, the layers can be separated and the chemical reactions in each layer studied.

In another technique, living cells are treated with **radioisotopes.** The radioactivity emitted from the radioisotopes can be traced with special equipment. The radioactive chemicals can then be followed through chemical reactions. Radioisotopes permit scientists to follow specific chemical reactions in specific organelles.

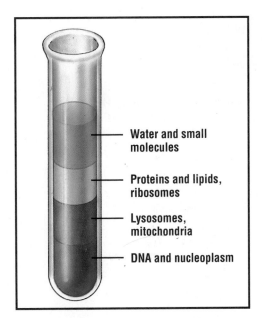

Water and small molecules

Proteins and lipids, ribosomes

Lysosomes, mitochondria

DNA and nucleoplasm

Figure 1.22

Cell fractionation.

Cell fractionation *is the process by which cell fragments are separated by centrifugation.*

Radioisotopes *are unstable chemicals that emit bursts of energy as they break down.*

THE LIVING CELL MEMBRANE

The cell membrane separates the protoplasm from the nonliving environment. However, the cell membrane is much more than a plastic envelope that holds the cytoplasmic organelles in place. It regulates what enters and leaves the cell. If the cell membrane is pierced with a pin, some cytoplasm will ooze out, but the puncture will soon be sealed. Cell membranes are living structures.

In order to survive, cells must extract nutrients from the nonliving environment. As nutrients are processed, wastes build up inside the cell. This time, the nonliving environment acts as a waste disposal site. Without it, the accumulation of poisonous wastes within a cell would soon cause cell death.

The fact that the cell membrane is involved in the transportation of materials into and out of the cell raises some questions: How does the cell know which materials must be absorbed and which must be excreted? How does the cell select molecules?

To understand the movement of molecular traffic across the cell membrane, the structure of the living membrane must be examined. The membrane appears as two layers of lipids. Each lipid molecule found in the cell membrane can be represented by a head and two tails. The head is hydrophilic, or "water loving," and is soluble in water. The tails are hydrophobic, or "water hating," and are not soluble in water. The water-soluble ends face the outer environment and the inner cell components. The lipid molecules are mobile, contributing to the movement within the cell membrane.

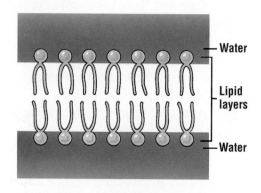

A variety of different protein molecules are embedded within the two layers of lipid. Most of these protein molecules carry a special sugar molecule and are thus referred to as **glycoproteins.** The sugar molecules provide the cell with a special signature. Like written signatures, the sugar molecules vary between different organisms, and even between individuals within the same species. These unique sugar proteins help distinguish a type A red blood cell from a type B red blood cell. Your immune system identifies foreign invaders by recognizing their unique structure on the cell membrane. This helps explain why transplanted organs are often rejected by recipients. Some of the lipid molecules also contain a specialized sugar called a **glycolipid.**

The protein molecules serve different functions. Some act as gatekeepers, opening and closing paths through the cell membrane. For example, specialized protein gatekeepers, located in nerve cells,

Glycoproteins *are compounds consisting of specialized sugar molecules attached to the proteins of the cell membrane. Many distinctive sugar molecules act as signatures that identify specialized cells.*

Glycolipids *are compounds consisting of specialized sugar molecules attached to lipids.*

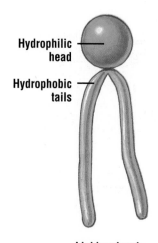

Hydrophilic head

Hydrophobic tails

Lipid molecule

BIOLOGY CLIP
Some scientists have speculated that many hormone disorders are due not to inadequate hormone production but to low numbers of receptor sites on the membranes of target cells. An excess number of receptor sites could also cause severe problems. Hypertension, or high blood pressure, may be related to an excess number of receptor sites for stress-related hormones.

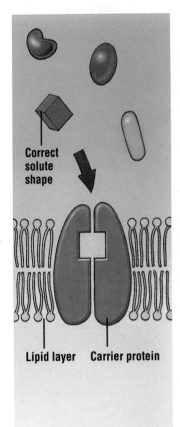
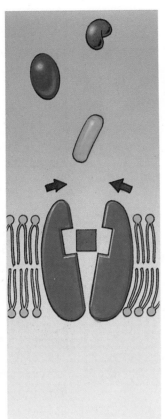
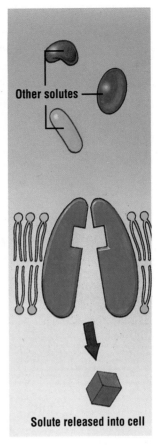

Correct solute shape

Lipid layer **Carrier protein**

Other solutes

Solute released into cell

Figure 1.28
Model of active transport.

Carrier protein molecules suspended within the cell membrane receive an energy boost that permits them to aid active transport. The energized proteins capture specific solute molecules and move them either into or out of the cell. Although the exact mechanism is not completely understood, scientists have devised models to help explain how transport molecules actively transport solutes against the concentration gradient. Figure 1.28 shows how the specific geometry of the protein molecule is well suited for trapping and then transporting solute molecules that have complementary shapes. Once the solute gains access to the binding site of the carrier protein, energy is released and the carrier molecule binds with another solute molecule. After the energy is consumed, the binding site loosens, and the solute molecule is released. Although the model presented shows the delivery of the solute into the cytoplasm, it is important to note that some carrier proteins actively pump materials out of cells.

Endocytosis and Exocytosis

Cells take in smaller solutes by means of transport carrier molecules, but how do cells absorb larger molecules? Some molecules, many of which are essential to life, do not fit between the pores of the cell membrane. The cell must expend energy to transport these larger substances.

Endocytosis is the process by which cells engulf large particles by extending their cytoplasm around the particle. As two cell membranes come together, the ingested particle is trapped within a pouch, or vacuole, inside the cytoplasm. Enzymes from the lysosomes are then often used to digest the large molecules absorbed by endocytosis.

Active transport *involves the use of cell energy to move materials across a cell membrane against the concentration gradient.*

Endocytosis *is the process by which particles too large to pass through cell membranes are transported within a cell.*

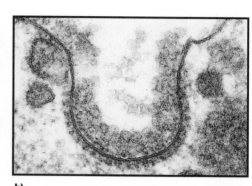

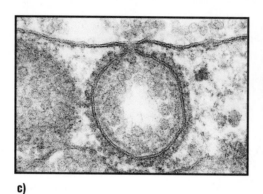
a)

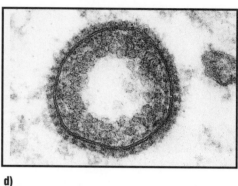
b)

c)

d)

Figure 1.29

Electron micrographs illustrate phagocytosis (a) Membrane begins to fold around molecules. (b) Membrane traps molecules. (c) Cell membranes come together and vacuole is formed. (d) Large molecules are digested within the vacuole.

Pinocytosis *is a form of endocytosis in which liquid droplets are engulfed by cells.*

Phagocytosis *is a form of endocytosis in which solid particles are absorbed by cells.*

There are two types of endocytosis. In **pinocytosis,** cells absorb liquid droplets by engulfing them. Cells of your small intestine engulf fat droplets by pinocytosis. **Phagocytosis** is the process by which cells engulf solid particles. Some white blood cells are often referred to as phagocytes (cell eaters) because they consume invading microbes by engulfing them. The microbe, trapped in the vacuole, is digested when the vacuole fuses with lysosomes.

Exocytosis is the process by which large molecules held within the cell are transported to the external environment. Waste materials are often released by exocytosis. Useful materials, like transmitter chemicals emitted from nerve cells, are also released by exocytosis. The Golgi apparatus holds the secretions inside the fluid-filled membranes. Small vesicles break off and move toward the cell membrane. The vesicles fuse with the cell membrane and the fluids are released.

■ REVIEW QUESTIONS ▐ ? ▌

18 How have cell fractionation techniques and the use of radioisotopes helped advance the knowledge of cell biology?

19 In what ways does a cell membrane differ from a cell wall?

20 Describe the structure of a cell membrane.

21 List three factors that alter diffusion rates.

22 Define isotonic, hypertonic, and hypotonic solutions.

23 How does facilitated diffusion differ from normal diffusion and osmosis?

24 Why does grass wilt if it is over-fertilized?

25 How does passive transport differ from active transport?

SOCIAL ISSUE:
Limits to Cell Technology

Dr. John Roder of Queen's University in Kingston has created a type of immortal human cell that churns out antibodies to fight cancer, leprosy, and tetanus.

Dr. Gerald Price, from McGill University, has fused conventional white blood cells with cancer cells. The resulting cells, called **hybridomas,** *produce huge amounts of antibodies for defense against disease.*

Other scientists are looking to lysosomes to unravel the mysteries of aging and disease. Lysosomes, the digestive packets held within the cells, release enzymes that destroy cells. Researchers believe that the inflammation and pain associated with arthritis may be caused by the leakage of enzymes from the lysosomes in white blood cells. This hypothesis is supported by the fact that a drug known to reduce swelling strengthens the lysosome membrane.

A Halifax hospital may soon begin transplanting brain tissue from aborted fetuses into patients with Parkinson's disease. This hereditary disease affects the nervous system, and is characterized by uncontrollable tremors. People with Parkinson's disease would be helped immensely by this new procedure.

Statement:

Cell research is progressing so quickly that many of the long-range effects of the research have not been considered. A private body, representing community values, should be established to oversee the technological applications of cell biology.

Point

- The general public has a right to know how scientific research will affect their lives. Research on lysosomes can forestall aging. Can you imagine the consequences of increasing life expectancy by 50%?
- Scientists have developed skills related to research. They are not trained to evaluate the moral, social, and economic consequences of their work.

Counterpoint

- Regulatory bodies have already been established by scientists who carefully monitor their own work. Greater interference by nonprofessionals would only slow the rate at which science progresses.
- There has been a long history of public monitoring of cell research. In 1978, the town council of Cambridge, Massachusetts, closed a cell biology lab at Harvard University. In the mid-1980s, a group in California worked to prevent the release of genetically engineered bacteria into the environment.

Research the issues.
Reflect on your findings.
Discuss the various viewpoints with others.
Prepare for the class debate.

CHAPTER HIGHLIGHTS

- A scientific hypothesis is an explanation based on the observation and interpretation of events in nature.
- A scientific hypothesis must be tested by experimentation.
- The theory of abiogenesis attempted to explain the origin of living things from a nonliving world.
- Francesco Redi used controlled experiments to refute the theory of abiogenesis.
- Pasteur and Spallanzani disproved the theory of spontaneous generation of microbes.
- The invention of the light microscope led to the discovery of cells and the development of the cell theory.
- The protoplasm of cells is composed of a nucleus and cytoplasm.
- Primitive prokaryotic cells do not contain a nuclear membrane or many of the organelles found within the more advanced eukaryotic cells.
- The transmission and scanning electron microscopes have made the study of cytoplasmic organelles possible.
- Cell fractionation techniques and the use of radioisotope tracers have enabled cell biologists to study the chemical reactions that occur within the cytoplasmic organelles.
- Plant cells contain cell walls and plastids.
- The cell membrane acts as a boundary between the cell and its external environment. The cell membrane regulates the movement of molecular traffic into and out of the cell.
- Materials are transported across cell membranes by either active or passive transport.
- Diffusion, osmosis, and facilitated diffusion are forms of passive transport in which molecules permeable through the cell membrane move from an area of higher concentration to an area of lower concentration.
- Materials can move against the concentration gradient if energy is used. The process of active transport requires the use of carrier molecules.
- Endocytosis and exocytosis are processes by which cells take in or release molecules too large to pass through the pores in cell membranes.

APPLYING THE CONCEPTS

1 Biology, like other sciences, progresses by observation. Unfortunately, the observations of nature are often flawed by interpretation. Provide two examples of faulty conclusions that supported the theory of abiogenesis and explain the source of the error.

2 Explain why John Needham came to an incorrect conclusion about abiogenesis. What was his error?

3 Those who supported the theory of abiogenesis used the critical factor of fresh air to refute Spallanzani's experiments. Explain why Spallanzani was unable to overcome the challenge from those who supported the theory of abiogenesis. How did Pasteur finally disprove the theory of abiogenesis?

4 A student suggests that flour sealed in a jar has been transformed into flour beetles after it has been sitting for six weeks. Provide a probable explanation for the flour beetles and then design an experiment to test your hypothesis.

5 A cell is viewed under low-power magnification. When the revolving nosepiece is turned to high-power magnification, the object appears to disappear, despite many attempts to refocus the slide. Provide a possible explanation for the disappearance of the cell.

6 By comparing a bee's body mass to its wing span, a physicist once calculated that bees should not be able to fly. Cell biologists have found that the muscles that control the wing of the bee have an incredible number of mitochondria. Indicate why this finding may help explain why bees can fly.

7 Explain why stomach cells have a large number of ribosomes and Golgi apparatus.

8 Hormones are the body's chemical messengers. Protein hormones, such as insulin, must attach themselves to a receptor site on the cell membrane; fat-soluble steroid hormones, such as sex hor-

mones, pass directly into the cell. Explain why steroid hormones pass directly into the cytoplasm of a cell.

9   Identify some limitations of the light microscope and explain why the transmission and scanning electron microscopes have had major impacts on the study of cell biology.

10 A marathon runner collapses after running on a hot day. Although the runner consumed water along the route, analysis shows that many of the runner's red blood cells had burst. Why did the red blood cells burst? (Hint: On hot days many runners consume drinks that contain sugar, salt, and water.)

CRITICAL-THINKING QUESTIONS

1   The statement "We are not alone" is often used in science-fiction movies. Explain how this statement applies to the cells within your body.

2   Early scientists often proposed hypotheses but did not test them by experimentation. Instead, they used logic to translate the hypotheses into an explanation or theory. Using the theory of abiogenesis, explain why this method of inquiry is susceptible to errors.

3   Why was Francesco Redi's experiment considered to be a significant turning point for the way in which scientific experiments were performed?

4   Many people argue that technology follows science. In many cases, this may be true; however, technology can assume a leading role. Using the development of the lens as an example, explain how technology changed the manner in which humans perceived themselves and their relationship to their environment.

5   Some unrelated diseases may have an interesting link: the mitochondria. Disorders of the liver (i.e., viral hepatitis, obstructive jaundice, and cirrhosis) and some muscle disorders (i.e., muscular dystrophy) are characterized by abnormally shaped mitochondria. Large amounts of unprocessed chemicals appear to accumulate in the mitochondria. Indicate why this link may be important. Does it provide any clues to the cause of the diseases? How might scientists use this information to develop a cure?

ENRICHMENT ACTIVITIES

1   Suggested reading:
    ● Chollar, Susan. "The Poison Eaters." *Discover*, April 1990, p. 76. A description of how microbial cells are used to solve the toxic-waste problem.
    ● Cook-Varsat, Alice, and Harvey F. Lodish. "How Receptors Bring Proteins and Particles into Cells." *Scientific American*, May 1984, p. 52. A technical presentation of transport systems.
    ● Kiester, Edward. "A Bug in the System." *Discover*, February 1991, p. 70. An investigation of how the DNA from mitochondria has been linked with hereditary disease.
    ● Murray, Mary. "Life on the Move." *Discover*, March 1991, p. 72. A description of microfilaments and movement.

2   Many industries and businesses use scientific research in their advertising to legitimize their products. Locate a magazine or TV advertisement that attempts to use scientific data to support a product. Does the ad present any data? Comment on the presentation of research.

3   Scientists at the Pasteur Institute in Paris were reported to be the first to grow HIV (the virus that causes AIDS) in tissue culture. Researchers from the United States indicated that they had also grown the virus, but insisted that they were first. Conduct research into the discovery of HIV.

Chemistry of Life

IMPORTANCE OF CELL CHEMISTRY

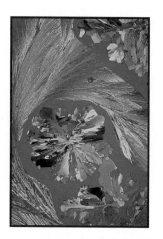

Defining the difference between living and nonliving things has presented an array of problems for scientists. During the 19th century, biologists such as Pasteur and Spallanzani performed experiments to dispel the belief that nonliving materials could be transformed into living organisms. Today there is little argument that living organisms only arise from other living organisms. Ironically, living things *are* composed of nonliving chemicals. Proteins, carbohydrates, lipids, and nucleic acids are often categorized as the chemicals of life despite the fact that none of them are capable of life by themselves.

Another group of scientists, called *vitalists*, believed that nonliving things were distinctly different from living things. Some even believed that life had a vital force that was neither chemical nor physical. They believed that living things had their own type of substance that was not found in nonliving things.

Scientific investigations have shown that the same principles of chemistry apply in both the physical world

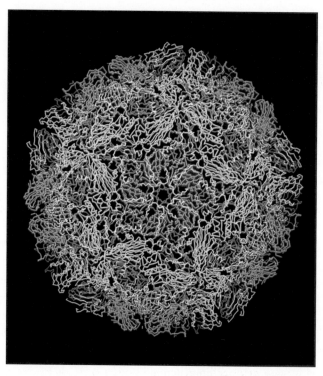

Figure 2.1

Computer graphics representation of the structure of a virus. Precise positions have been assigned to over 300 000 atoms that comprise the spherical coat that encloses the genetic material.

and the living world. An understanding of the chemistry of life comes in part from an understanding of how chemical reactions are regulated within cells. Chemicals from your surroundings are altered within your cells. Bonds are broken and new bonds are formed in a continuous cycling of matter. The absorption of nutrients, the synthesis of vital chemicals, and the excretion of wastes are vital to life. Consider the oxygen you breathe. Mitochondria use oxygen to break the bonds in nutrient sugar molecules. The breakdown of the sugar yields energy, carbon dioxide, and water. Plants and some bacteria utilize the water and the carbon dioxide released from animal cells to make sugars and to release oxygen. An oxygen molecule that you have absorbed may once have been part of a potato, a giant maple tree, or it may be the same oxygen molecule once used by Julius Caesar!

It has been estimated that over 200 000 chemical reactions occur in the cells of your body. The host of chemical reactions that take place within cells are referred to as **metabolism.** Metabolic reactions can be classified as either **catabolic**—reactions in which large chemical complexes are broken down into smaller components—or **anabolic**—reactions in which complex chemicals are built from smaller components.

ORGANIZATION OF MATTER

Matter can be described as the material of which things are made. Matter occupies space and has mass, and can exist in the form of a solid, liquid, or gas. Matter is composed of tiny particles called **molecules.** In solids, the molecules are packed tightly together and are held in a relatively fixed position. The molecules vibrate, creating small spaces between themselves. In a liquid, molecules move more freely, and the spaces between them are greater. Molecules in gases can have very large spaces between them. If energy is added to liquid water molecules by heating, the movement of the molecules intensifies, and molecules are liberated from the surface of the liquid. Liquid water is then converted into vapor.

The search for the composition of matter began with the ancient Greeks. Around 450 B.C., the Greek philosopher Empedocles speculated that matter was composed of four **elements:** air, water, fire, and earth. The four elements were in turn produced by combinations of four properties: coldness, wetness, hotness, and dryness. Water was thought to be a combination of coldness and wetness, while air was said to be a combination of wetness and hotness. By 400 B.C., another Greek, Democritus, provided a theory that was to serve as the basis for the development of the modern scientific theory of matter. He believed that matter was composed of many small subunits that were indivisible. These particles were called **atoms.** The word atom comes from the Greek word *atomos*, meaning "indivisible." Democritus proposed that all matter is composed of essentially the same materials, but that atoms come in different sizes and shapes. He reasoned that a cube of lead would have different particles than a sphere of lead.

Today we know that atoms are composed of protons, neutrons, and electrons. Protons carry a positive charge, electrons have a negative charge, and neutrons, as the name suggests, carry no charge at all. The larger protons and neutrons are located at the center of the atom, called the *nucleus.* The much smaller electrons orbit around the nucleus in an energy level, or *electron cloud.* Protons and neutrons make up about 99.9% of the total

Metabolism *is the sum of all chemical reactions that occur within the cells.*

Catabolism *refers to reactions in which complex chemical structures are broken down into simpler molecules.*

Anabolism *refers to chemical reactions in which simple chemical substances are combined to form complex chemical structures.*

Molecules *are units of matter. A molecule consists of two or more atoms of the same or different elements bonded together.*

Elements *are pure substances that cannot be broken down into simpler substances by chemical means. There are 109 different elements.*

Atoms *are small particles of matter and are composed of smaller subatomic particles: neutrons, protons, and electrons.*

mass of an atom, yet only a very small part of the total volume. Most of the volume of an atom is taken up by the space in which electrons orbit. To get an idea of the relative dimensions of an atom, consider a bee in the center of the baseball field at Toronto's SkyDome. The bee would represent the size of the nucleus of the atom, while the rest of the stadium would represent the part of the atom in which the electrons are found.

Isotopes have the same number of protons but different numbers of neutrons.

Figure 2.2

The nucleus of an atom can be compared to a bee in the center of the SkyDome.

Figure 2.3

Hydrogen, helium, and lithium atoms.

Figure 2.4

Three forms of hydrogen.

Table 2.1 The Subatomic Particles of Atoms

Particle	Relative mass	Relative charge
Electron	1/1836	− 1
Proton	1	+ 1
Neutron	1	0

The number of protons and neutrons in an atom determines its *atomic mass.* Hydrogen, with one proton and no neutrons, has an atomic mass of 1. Helium, with two protons and two neutrons, has an atomic mass of 4. Figure 2.3 shows three different atoms. The proton number, or *atomic number*, defines the type of element. Any element with two protons is helium and only helium atoms have two protons. Similarly, any atom that has three protons must be lithium and only lithium atoms have three protons. Each chemical element is made up of a different type of atom.

Each element is represented by a symbol, as shown in Figure 2.3. Lithium is represented by the symbol Li, hydrogen by H, and helium by He. Did you notice that the number of protons in an atom is always equal to the number

of electrons? The result of this is that atoms have no net charge. The number of neutrons, however, can vary even within the same element. Figure 2.4 shows three different forms of hydrogen that exist because of differences in atomic mass. Protium, deuterium, and tritium all have one proton and therefore are forms of hydrogen. Each form has a different number of neutrons. Deuterium, with an atomic mass of 2, has one proton and one neutron; tritium, with an atomic mass of 3, has one proton and two neutrons. Deuterium and tritium are often described as heavy hydrogen. Atoms that have the same atomic number, but different atomic mass, are called **isotopes.** Protium, the most common type of hydrogen, comprises 99.985% of naturally occurring hydrogen. A few atoms of the deuterium isotope can be expected to be found in nature. That is why atomic masses are not whole numbers. Atomic mass is calculated by averaging the most commonly occurring form of the element with all of its less common isotopes. The atomic mass of hydrogen is actually 1.0079.

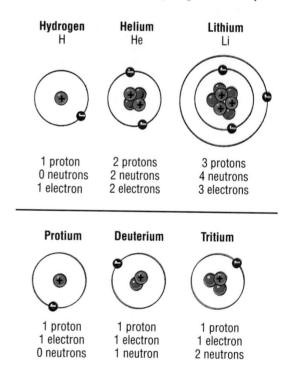

Hydrogen H	Helium He	Lithium Li
1 proton	2 protons	3 protons
0 neutrons	2 neutrons	4 neutrons
1 electron	2 electrons	3 electrons

Protium	Deuterium	Tritium
1 proton	1 proton	1 proton
1 electron	1 electron	1 electron
0 neutrons	1 neutron	2 neutrons

Most elements have at least one isotope. Tin has 11 different isotopes. Many isotopes are not stable and disintegrate spontaneously. These isotopes give off radiation as they decay. During decay, subatomic particles escape from the nucleus, and bursts of energy are emitted. For example, radioactive uranium is converted into lead after it decays. Not surprisingly, these isotopes are referred to as *radioactive isotopes*.

Devices like the *geiger counter* can detect the radiation released during radioactive decay. This permits scientists to follow the movement of radioactive molecules, making them ideal as tracers. Physicians, for example, can employ this technology by adding a radioactive isotope of chromium to a red blood cell. Then they can follow the movement of the blood cell through the organs of the body. The tracer even allows scientists to monitor the life span of the cells.

BIOLOGY CLIP
How fast do electrons move? One estimate places their speed at about 13 000 km/h. An electron moving at this incredible speed would orbit the nucleus billions of times every second.

Chemical Bonding

The electron holds the key to understanding why some chemicals combine with others. If you look again at the diagram of hydrogen, helium, and lithium in Figure 2.3, you will notice that a maximum of two electrons fill the first electron cloud, or energy level. The third electron of lithium is found in an outer energy level. Table 2.2 presents some of the common elements that you will study in this unit.

Why do some elements react quickly, while others tend not to take part in any chemical reactions? To find an answer you must look to the group of elements called the *noble gases*. Helium, neon, and argon are examples of noble gases. The noble gases do not react with other elements. The stability of this group of elements can be explained by the fact that they contain a full complement of electrons in their outermost energy

Table 2.2 Common Elements

Element	Symbol	Atomic number	Electron number	Number of electrons in each energy level		
				1	2	3
Hydrogen	H	1	1	1	0	0
Helium	He	2	2	2	0	0
Lithium	Li	3	3	2	1	0
Beryllium	Be	4	4	2	2	0
Boron	B	5	5	2	3	0
Carbon	C	6	6	2	4	0
Nitrogen	N	7	7	2	5	0
Oxygen	O	8	8	2	6	0
Fluorine	F	9	9	2	7	0
Neon	Ne	10	10	2	8	0
Argon	Ar	18	18	2	8	8

level. Helium has two electrons in the outermost energy level, while neon and argon both have eight electrons in their outermost energy levels. Scientists discovered that elements without a full complement of electrons in the outermost energy level lose or gain electrons until the outermost energy level is full. Consider the options for the atoms in Figure 2.5.

2 electrons in outer energy level   4 electrons in outer energy level   7 electrons in outer energy level

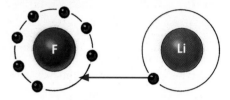

Figure 2.5

Elements lose or gain electrons from their outer energy level.

Beryllium
loses 2 electrons
gains 6 electrons

Carbon
loses 4 electrons
gains 4 electrons

Fluorine
loses 7 electrons
gains 1 electron

Beryllium can become stable by either losing two electrons or gaining six electrons. Which situation do you think is more likely? It is much easier to give up two electrons to another atom than it is to gain six electrons. Notice that once the beryllium gives up two electrons from its outermost energy level, the inner level acts as the outermost shell. This energy level carries a full complement of electrons. Beryllium is considered to be an *electron donor*. Carbon has two options: it can either lose four electrons or gain four electrons. Carbon is neither a strong electron donor nor a strong electron acceptor. Fluorine can either lose seven electrons or gain one electron. The fact that it is more likely to gain one electron makes fluorine a strong *electron acceptor*.

Atoms that have gained or lost electrons have an imbalance of charged particles. Atoms that donate electrons have an excess of positive charge. These atoms become positive **ions.** Atoms that accept

Ions *are atoms that have either lost electrons or gained electrons to become positively or negatively charged.*

Chemical compounds *are formed when two or more elements are joined by chemical bonds.*

Ionic bonds *are formed when electrons are transferred between two atoms.*

electrons have an excess of negative charge. These atoms become negative ions. Electron acceptors will accept electrons from strong electron donors. In the process, both atoms become stable.

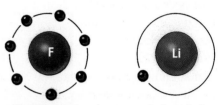

The fluorine atom has seven electrons in its outermost energy level, while the lithium atom has only one electron in its outermost level.

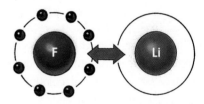

The lithium atom donates an electron to the fluorine atoms.

The negatively charged fluoride ion and positively charged lithium ion are drawn to each other by charge attraction.

Figure 2.6

Fluorine and lithium combine to form the compound lithium fluoride.

Chemical compounds are formed when two or more elements are joined. For example, lithium and fluorine combine to form the compound lithium fluoride, as shown in Figure 2.6. Lithium fluoride is classified as an ionic compound because it was formed from a metal (lithium) and a nonmetal (fluorine). The bond that joins the lithium and fluoride ions is referred to as an **ionic bond.** Ionic bonds are formed when electrons are transferred from strong electron donors to strong electron acceptors.

Another type of chemical bond is formed when electrons are shared rather than transferred. This type of bond is referred to as a **covalent bond.** Carbon, for example, which can either gain or lose four electrons, is capable of forming covalent bonds with other atoms.

Polar Molecules

Although electrons are shared between two atoms in a covalent bond, the sharing may not be equal. Water molecules are created by covalent bonds that join one oxygen and two hydrogen atoms. The electrons are drawn closer toward the oxygen atom, creating a region of negative charge near the oxygen end of the compound and a region of positive charge near the hydrogen end of the molecule. Despite the fact that the entire molecule is balanced in the number of positive and negative charges, the molecule has a positive pole and a negative pole. It is for this reason that water is referred to as a **polar molecule.**

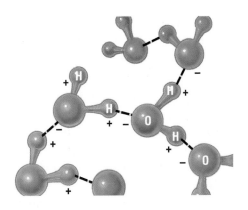

which has a single proton. Attraction between opposing charges of different polar molecules creates a special **hydrogen bond.** Hydrogen bonds pull polar molecules together.

Hydrogen bonding accounts for the strength of fibers in wood. It also helps explain some of the physical characteristics of water. Water boils at $100\,°C$ and freezes at $0\,°C$. By comparison, sulfur dioxide, a molecule of similar size, boils at $62\,°C$ and freezes at $-83\,°C$. The higher boiling point and melting point of water can be explained by the hydrogen bonds that help hold groups of water molecules together. These hydrogen bonds must be broken before water molecules can escape into the air. This requires energy in the form of heat. Molecules such as sulfur dioxide and carbon dioxide do not have hydrogen bonds and, consequently, require less energy to boil.

ACIDS AND BASES

Chemical reactions occur within the water of the cells and tissues. As mentioned before, water exists as two atoms of hydrogen attached to an atom of oxygen. However, a small number of water molecules dissociate into two separate

Figure 2.8

Hydrogen bonds are formed between the oxygen end of one water molecule and the hydrogen end of another water molecule.

Covalent bonds *are formed when electrons are shared between two or more atoms.*

Polar molecules *are molecules that have positive and negative ends.*

Hydrogen bonds *are formed between a hydrogen proton and the negative end of another molecule.*

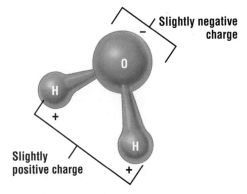

Slightly negative charge

Slightly positive charge

Figure 2.7

The electrons are pulled closer toward the oxygen end of the molecule. The single proton of the hydrogen causes a positively charged end near the hydrogen.

The negative end of a water molecule repels the negative end of another water molecule, but attracts its positive end

ions: a positive hydrogen ion and a negative hydroxide ion. The equation below shows the dissociation of water into ions:

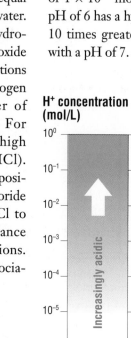

$$H_2O \longrightarrow H^+ + OH^-$$

water   hydrogen   hydroxide
ion   ion

The number of hydrogen ions is equal to the number of hydroxide ions in water. A solution in which the number of hydrogen ions equals the number of hydroxide ions is called a *neutral* solution. Solutions in which the concentration of hydrogen ions is greater than the number of hydroxide ions are called **acids.** For example, stomach juices contain high amounts of hydrochloric acid (HCl). Hydrochloric acid dissociates into positive hydrogen ions and negative chloride ions. Note that the addition of HCl to water would bring about an imbalance between hydrogen and hydroxide ions. The reaction below shows the dissociation of HCl in water:

$$HCl \longrightarrow H^+ + Cl^-$$

Bases are formed when the concentration of hydroxide ions is greater than the concentration of hydrogen ions. Sodium hydroxide, a base, dissociates into hydroxide and sodium ions. Should sodium hydroxide be added to water, the number of hydroxide ions would be greater than the number of hydrogen ions. The reaction below shows the dissociation of sodium hydroxide:

$$NaOH \longrightarrow Na^+ + OH^-$$

The **pH scale** *measures* how acidic or basic a solution is. The concentration of hydrogen ions in solution is used to measure the pH. The most useful part of the scale ranges from 1, indicating a very strong acid, to 14, indicating a very strong base. Water has a pH of 7, indicating a neutral solution— the concentration of

Acids are substances that release hydrogen ions in solution.

Bases are substances that release hydroxide ions in solution.

*The **pH scale** is used to measure how acidic or basic a solution is.*

***Neutralization** occurs when the pH is brought to 7, i.e., the [H⁺] = [OH⁻].*

hydrogen ions is equal to the concentration of hydroxide ions. The concentration of both ions would be expressed as 1×10^{-7} mol/L. The pH is calculated from the negative logarithm of the hydrogen ion concentration. A pH of 6, therefore, represents a hydrogen ion concentration of 1×10^{-6} mol/L. A solution that has a pH of 6 has a hydrogen ion concentration 10 times greater than that of a solution with a pH of 7.

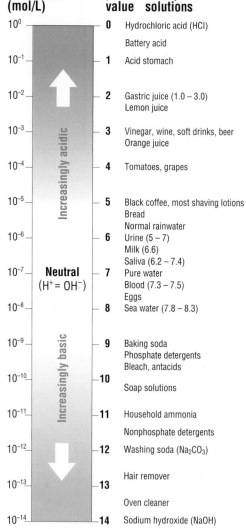

H⁺ concentration (mol/L)	pH value	Examples of solutions
10^0	0	Hydrochloric acid (HCl)
		Battery acid
10^{-1}	1	Acid stomach
10^{-2}	2	Gastric juice (1.0 – 3.0)
		Lemon juice
10^{-3}	3	Vinegar, wine, soft drinks, beer
		Orange juice
10^{-4}	4	Tomatoes, grapes
10^{-5}	5	Black coffee, most shaving lotions
		Bread
		Normal rainwater
10^{-6}	6	Urine (5 – 7)
		Milk (6.6)
		Saliva (6.2 – 7.4)
10^{-7}	7	Pure water
		Blood (7.3 – 7.5)
		Eggs
10^{-8}	8	Sea water (7.8 – 8.3)
10^{-9}	9	Baking soda
		Phosphate detergents
		Bleach, antacids
10^{-10}	10	Soap solutions
10^{-11}	11	Household ammonia
		Nonphosphate detergents
10^{-12}	12	Washing soda (Na_2CO_3)
10^{-13}	13	Hair remover
		Oven cleaner
10^{-14}	14	Sodium hydroxide (NaOH)

Increasingly acidic

Neutral ($H^+ = OH^-$)

Increasingly basic

Figure 2.9

Each increase of 1 in the pH scale actually represents a tenfold decrease in the concentration of hydrogen ions.

LABORATORY

IDENTIFICATION OF CARBOHYDRATES

Objective

To identify reducing sugars qualitatively and quantitatively.

Background Information

Benedict's solution identifies reducing sugars, and iodine solution identifies starches. Iodine turns blue-black in the presence of starches. The Cu^{2+} ions in the Benedict's solution are converted to Cu^+ ions if they react with a reducing sugar. Not all sugars are reducing sugars. All monosaccharides are reducing sugars, but some disaccharides will not react with Benedict's solution.

The chart summarizes the quantitative results obtained when a reducing sugar reacts with Benedict's solution:

Color of Benedict's solution	Approximate % of sugar
Blue	negative
Light green	0.5% – 1.0%
Green to yellow	1.0% – 1.5%
Orange	1.5% – 2.0%
Red to red brown	+2.0%

Materials

safety goggles	5% maltose solution
lab apron	5% starch solution
test-tube brushes	8 test tubes
detergent	test-tube rack
400 mL beaker	Benedict's solution
hot plate	wax pencil
thermometer	test-tube clamp
10 mL graduated cylinder	5 medicine droppers
distilled H_2O	depression plates
5% fructose solution	iodine solution
5% sucrose solution	solutions X, Y, and Z
5% glucose (dextrose) solution	

> **CAUTION:** The chemicals used are toxic and irritant. Avoid skin and eye contact. Wash all splashes off your skin and clothing thoroughly. If you get any chemical in your eyes, rinse for at least 15 min and inform your teacher.

Procedure

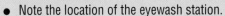

Before you begin:
- Make sure that all the glassware is clean and well rinsed.
- Note the location of the eyewash station.

Part I : Reducing Sugars

1 Prepare a water bath by heating 300 mL of tap water in a 400 mL beaker. Heat the water until it reaches approximately 80°C. (Use the thermometer to monitor the temperature.)

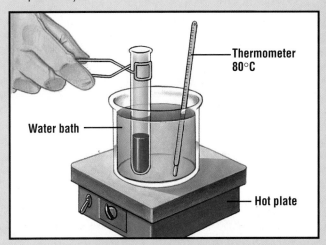

2 Using a 10 mL graduated cylinder, measure 3 mL each of distilled water, fructose, glucose, maltose, sucrose, and starch solutions. Pour each solution into a separate test tube. Clean and rinse the graduated cylinder after each solution. Add 1 mL of Benedict's solution to each of the test tubes and label using the wax pencil.

   a) Why should the graduated cylinder be cleaned and rinsed after the measurement of each solution?

3 Using a test-tube clamp, place each of the test tubes in the hot water bath. Observe for six minutes.

   b) Record any color changes in a chart like the one below.

Solution	Initial color	Final color	% sugar
Water	Blue		
Glucose			
Fructose			
Maltose			
Sucrose			
Starch			

Part II: Iodine Test

4 Using a medicine dropper, place a drop of water on a depression plate and add a drop of iodine.

   c) Record the color of the solution.

5 Repeat the procedure, this time using drops of starch, glucose, maltose, and sucrose instead of water.

   d) Record the color of the solutions. Which solutions indicate a positive test?

Part III: Checking Unknown Solutions

6 Test the unknown solutions for reducing sugars and starches. Design your own table, showing both qualitative and quantitative data.

   e) Record your data.

Laboratory Application Questions

1 Which test tube served as a control in the test for reducing sugars and starches?

2 What laboratory data suggest that not all sugars are reducing sugars?

3 A student decides to sabotage the laboratory results of his classmates and places a sugar cube into solution Z. Explain the effect of dissolving a sugar cube in the solution.

4 A drop of iodine accidently falls on a piece of paper. Predict the color change, if any, and provide an explanation for your prediction. ∎

LIPIDS

Triglycerides *are lipids composed of glycerol and three fatty acids.*

Fats *are lipids composed of glycerol and saturated fatty acids. They are solid at room temperature.*

Oils *are lipids composed of glycerol and unsaturated fatty acids. They are liquid at room temperature.*

Phospholipids *are the major components of cell membranes in plants and animals. Phospholipids have a phosphate molecule attached to the glycerol backbone, making the molecule polar.*

Lipids are nonpolar compounds that are insoluble in polar solvents like water. You may have noticed how fat floats on the surface of water while you are doing dishes. Many lipids are composed of two structural units: glycerol and fatty acids. Like complex carbohydrates, glycerol and fatty acids can be combined by dehydration synthesis. An important function of lipids is the storage of energy. Glycogen supplies are limited in most animals. Once glycogen stores have been built up, excess carbohydrates are converted into fat. This helps explain why the eating of carbohydrates can cause an increase in fat storage. Other lipids serve as key components in cell membranes, act as cushions for delicate organs of the body, serve as carriers for vitamins A, D, E, and K, and act as the raw materials for the synthesis of hormones and other important chemicals. A layer of lipids at the base of the skin insulates you against the cold. The thicker the layer of fat, the better the insulation. Taking their cue from whales, seals, and other marine mammals, marathon swim-

mers often coat their bodies with a layer of fat just before entering cold water.

Triglycerides are formed by the union of glycerol and three fatty acids. Triglycerides that are solid at room temperature are called **fats.** Most of the fatty acids in animal fats are said to be *saturated.* This means that only single covalent bonds exist between the carbon atoms. Because the single covalent bonds tend to be stable, animal fats are difficult to break down. Animal fats are usually solid or semi-solid at room temperature. Triglycerides that are liquid at room temperature are called **oils.** The fatty-acid components of most plants have some double bonds between carbon atoms. Plant oils are often described as *polyunsaturated.* Polyunsaturated fats have many double bonds between the carbon atoms that comprise the fatty acid portion. The unsaturated double bonds are somewhat reactive, and, therefore, plant oils are more easily broken down than animal fats.

A second group of lipids, called **phospholipids,** have a phosphate molecule attached to the glycerol backbone of the molecule. The negatively charged phosphate replaces one of the fatty acids, providing a polar end to the lipid. The polar end of phospholipids is soluble in water, while the nonpolar end is insoluble. These special properties make phospholipids well suited for cell membranes.

Waxes make up a third group of lipids. In waxes, long-chain fatty acids are joined to long-chain alcohols or to carbon rings. These long, stable molecules are insoluble in water, making them well suited as a waterproof coating for plant leaves or animal feathers and fur.

Fats and Diet

Almost everyone likes the taste of french fries. Unfortunately, when you eat french fries, you are also eating the fat they were cooked in. Although fats are a required part of your diet, problems arise when you consume too much fat. Doctors recommend that no more than 30% of total energy intake should be in the form of fats.

Fats are concentrated packages of energy containing more than twice as much energy as an equivalent mass of carbohydrate or protein. By eating 100 g of fat, you take in about 3780 kJ of energy. (The kilojoule, kJ, is a unit used to measure food energy.) By comparison, 100 g of carbohydrates or protein yield 1680 kJ of energy. When energy input or consumption exceeds energy output, the result is a weight gain. If you were to drink a 350 mL chocolate milkshake containing about 2200 kJ of energy, you would have to jog for about one hour to burn off the energy taken in.

Heart disease has been associated with

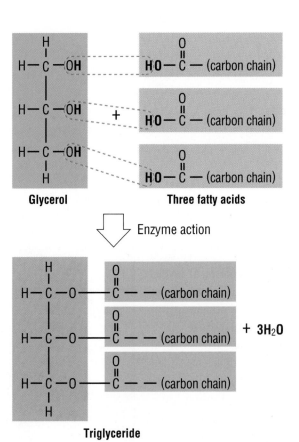

Glycerol   **Three fatty acids**

Enzyme action

Triglyceride

$+ \ 3H_2O$

Figure 2.15

Triglycerides are formed by the union of glycerol and three fatty acids. Note the removal of water in the synthesis. The terms monoglyceride or diglyceride are used to describe the joining of glycerol with one or two fatty acids respectively.

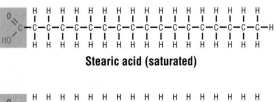

Stearic acid (saturated)

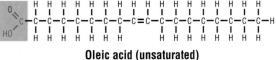

Oleic acid (unsaturated)

Figure 2.16

Saturated fats do not have double bonds between carbon atoms. Therefore, these compounds tend to be stable.

Waxes *are long-chain lipids that are insoluble in water.*

diets high in saturated fats. Recall that the single bonds between the carbon molecules make the fats stable. The stable fats tend to remain intact inside the cells of the body much longer than more reactive macromolecules. High-fat diets and obesity have also been linked to certain types of cancer, such as breast cancer, cancer of the colon, and prostate cancer. Obesity has also been linked to high blood pressure and adult diabetes. According to one report, over 80% of people with adult-onset diabetes are overweight.

The Cholesterol Controversy

Heart disease, the number-one killer of North Americans, can be caused by the accumulation of cholesterol in the blood vessels. Scientific research about cholesterol has changed direction in recent years. Lipid-rich foods, such as fish and olive oil, were once thought to raise blood cholesterol levels. Currently, most scientists believe that these foods may actually reduce blood cholesterol levels. Similarly, alcohol, in moderate consumption may contribute to a decrease in blood cholesterol levels. Added to this confusion is the fact that genes play a major role in determining cholesterol levels. Research indicates that people with a certain genetic makeup are predisposed to **atherosclerosis.** Individuals who are not susceptible to atherosclerosis can tolerate higher levels of cholesterol in their blood.

Not all cholesterol is bad. Cholesterol is found naturally in cell membranes and acts as the raw material for the synthesis of certain hormones. The difference between males and females would be less pronounced if it were not for cholesterol—sex hormones are made from cholesterol.

All of the cholesterol that the body needs can be obtained from the fats consumed. The cells of the body package cholesterol in a water-soluble protein in order to transport it in the blood. Scientists have identified two very important types of cholesterol packaging: low-density lipoproteins and high-density lipoproteins. Low-density lipoproteins, or LDLs, are considered to be "bad" cholesterol. About 70% of cholesterol intake is in the form of LDLs. High levels of LDLs have been associated with the clogging of arteries—LDL particles bind to receptor sites on cell membranes and are removed from the blood (principally by the liver). However, as the levels of LDLs increase and exceed the number of receptor sites, excess LDL-cholesterol begins to form deposits on the walls of arteries. The accumulation of cholesterol and other lipids on the artery walls is known as *plaque*. Unfortunately, plaque restricts blood flow to the heart and brain and can lead to heart attack or stroke.

The second type of cholesterol—high-density lipoproteins, or HDLs—is often called "good" cholesterol. The HDLs carry bad cholesterol back to the liver, which begins breaking it down. The HDLs lower blood cholesterol. Most researchers now believe that the balance between LDL and HDL is critical in assessing the risks of cardiovascular disease. A desirable level of HDL is 35 mg/100 mL of blood or higher. Some researchers believe that exercise increases the level of HDLs. Strong evidence also supports the theory that fiber or cellulose in the diet helps reduce cholesterol. It is believed that fiber binds to cholesterol in the gastrointestinal tract. However, it should be pointed out that fiber does not affect everyone the same way.

The controversy over the ideal diet continues. The most important thing to keep in mind is maintaining a balance between food intake and energy needs.

> **BIOLOGY CLIP**
> Have you ever gone on a diet only to discover that after days of reducing your food intake, you have actually gained weight? This can occur because as fat is used up by the body, it is replaced by water. Since water weighs more than fat, you may experience a temporary gain in weight. Eventually, the water will leave the tissues and your body weight will decrease.

Atherosclerosis *is a disorder of the blood vessels characterized by the accumulation of cholesterol and other fats along the inside lining.*

LABORATORY

IDENTIFICATION OF LIPIDS AND PROTEINS

Objective

To identify lipids and proteins.

Background Information

Proteins can be identified by means of the biuret test. The biuret reagent reacts with the peptide bonds that join amino acids together, producing color changes from blue, indicating no protein, to pink (+), to violet (++), and to purple (+++). The (+) sign indicates the relative amounts of peptide bonds. Lipids can be identified by a Sudan IV solution, which is soluble in nonpolar solvents. Lipids turn from a pink to a red color, but polar compounds will not assume the pink color of the Sudan IV solution. Lipids can also be identified by unglazed brown paper. Because lipids allow the transmission of light through the brown paper, the test is often called the *translucence test*.

Materials

safety goggles	8 test-tube stoppers
lab apron	5 medicine droppers
test-tube rack	5% glucose (dextrose)
test-tube brush	solution
8 test tubes	5% gelatin solution
wax pencil	liquid soap
10 mL graduated cylinder	egg albumin
distilled water	liquid detergent
vegetable oil	simulated blood plasma
Sudan IV	solutions X, Y, and Z
unglazed brown paper	biuret reagent

CAUTION: Avoid skin and eye contact. Wash all splashes off your skin and clothing thoroughly. If you get any chemical in your eyes, rinse for at least 15 min and inform your teacher.

Procedure

Part I: Sudan IV Lipid Test

1 Obtain two test tubes. Label one C, for control, and the other T, for lipid test. Add 3 mL of water to test tube C. Add 3 mL of vegetable oil to test tube T.

2 Add 6 drops of Sudan IV solution to each test tube. Place stoppers in the test tubes and shake them vigorously for two minutes.

   a) Record the color of the mixtures.

Part II: Translucence Lipid Test

3 Label a 10-cm square piece of unglazed paper C, for control. Place a drop of water on the paper. Label a second piece of paper T for lipid test. Place a drop of vegetable oil on the paper labelled T.

4 Wave papers C and T in the air until the water from paper C has evaporated. Hold both papers to the light and observe.

   b) Record whether or not the papers appear translucent.

Part III: Biuret Test for Proteins

5 Using a wax pencil, label two test tubes C and T. Add 2 ml of water to test tube C. Add 2 mL of gelatin to test tube T. Add 2 mL of biuret reagent to each of the test tubes, then tap the test tubes with your fingers to mix the contents.

   c) Record any color changes.

Part IV: Testing for the Presence of Protein and Lipid

6 Use the appropriate solution to test the following: 5% glucose(dextrose), 5% gelatin, liquid soap, egg albumin, liquid detergent, simulated blood serum, solution X, solution Y, solution Z.

   d) Record your results.

Laboratory Application Questions

1 Why were controls used for the experiments?

2 A student heats a test tube containing a large amount of protein and notices a color change in the test tube. Explain why heating causes a color change.

3 Explain the advantage of using two separate tests for fats.

4 Would you expect to find starches and sugars in the blood plasma? Explain your answer. ■

NUCLEIC ACIDS

Nucleotide

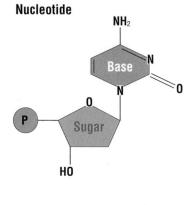

***Nucleotides** are the functional units of nucleic*

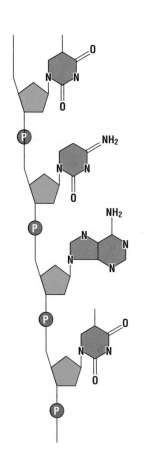

acids.

Figure 2.22

Nitrogen bases, shown in purple, are either single-ring molecules

The hereditary material found within the genes of chromosomes comprises yet another type of organic compound. Deoxyribonucleic acid, or DNA, as it is more commonly known, belongs to a group of compounds called nucleic acids. Nucleic acids are nitrogen-containing compounds and are composed of functional units called **nucleotides.** Each nucleotide is composed of a five-carbon sugar unit, a phosphate group, and a nitrogen base. DNA nucleotides contain the five-carbon sugar deoxyribose, while RNA nucleotides are made up of the five-carbon sugar ribose .

Nucleic acids are either double-stranded (DNA) or single-stranded (RNA) molecules whose nucleotides are linked by sugar and phosphate groups. DNA and RNA each have four of the five nitrogen bases attached to their sugar-phosphate backbones. (See Figure 2.22.) Adenine and guanine are purines: cytosine, thymine and uracil are pyrimidines. DNA contains thymine; RNA contains uracil. The sequencing of the nitrogen bases determines the genetic code.

◼ REVIEW QUESTIONS ◼ ?

9 Differentiate between organic and inorganic molecules. Provide an example of each.

10 Name three classes of carbohydrates.

11 What are the two structural components of triglycerides?

12 In what ways do plant and animal fats differ?

13 Differentiate between monoglycerides, diglycerides, and triglycerides.

14 What is cholesterol, and why has it become so important to health-conscious consumers?

15 State the function of HDLs and LDLs.

16 In what ways do proteins and nucleic acids differ from carbohydrates and lipids?

17 Why do so many different proteins exist?

18 Using examples of carbohydrates, lipids, and proteins, explain the process of dehydration synthesis.

19 Define denaturation and coagulation.

20 What are nucleotides?

ANABOLIC STEROIDS

Steroids are lipids produced by the cells of the body. Many steroids act as chemical messengers, regulating cell function. Anabolic steroids are tissue-building messengers. Most anabolic steroids are either synthetic or natural versions of the male sex hormone, testosterone. Male sex hormones in general are known as *androgens*.

At puberty, the level of testosterone increases dramatically in males, triggering muscle development and other secondary sex characteristics. Androgens from the blood attach themselves to a binding site on muscle cell membranes, causing the muscle cells to grow and develop. Women can increase muscle development by injecting male sex hormones. Although they can grow bigger and stronger, many experience side effects associated with male characteristics.

Males do not escape without side effects either. Males who rely on large doses of testosterone cannot utilize all of the testosterone. Consequently, the excess testosterone is gradually converted into estrogen, the female sex hormone. Although every male contains some estrogen, few contain the same levels as women. Breast enlargement, a shrinking of the testes, and liver and kidney dysfunctions have been associated with steroid use.

Unit One:
Exchange of Energy and Matter in Cells

SOCIAL ISSUE:
Athletes and Performance-Enhancing Drugs

In 1988, Canadian sprinter Ben Johnson was stripped of his gold medal at the summer Olympics after a drug test indicated anabolic steroids were in his system. At the subsequent Dubin Inquiry, held in Toronto, it was revealed that steroid use was commonplace among the world's top athletes.

Anabolic steroids, stimulants, and painkillers have improved performance levels of athletes. In many respects, today's athletes are a product of better chemistry. However, the use of performance-enhancing drugs raises some important issues. Current studies show that prolonged use of steroids can result in certain side effects, such as the accumulation of plaque inside the blood vessels. On a wider scale, it could be contended that the use of drugs has corrupted athletic competition and that only those countries with the best technology can win medals.

Statement:

Major athletic competitions such as the Olympics have become meaningless because of some athletes' use of performance-enhancing drugs.

Point

- The use of anabolic steroids, stimulants, and painkillers has improved performance levels of athletes. In many respects, these athletes are a product of better chemistry. How can other athletes afford to compete fairly?
- Improvements in drug testing will only lead to the use of better masking agents that disguise the drugs, or to the use of new synthetics that are more difficult to detect. A new peptide hormone, used in place of conventional anabolic steroids, has surfaced already. The peptide hormones are very difficult to detect.

Counterpoint

- The fact that some athletes have misused drugs should not take away from the nature of athletic competition. Improved drug testing at the Olympic games and other major sporting events will curtail the use of performance-enhancing drugs.
- Any chemical can be detected if enough care is exercised during the testing. However, testing must be combined with an education program. If performance-enhancing drugs are not accepted, their use will soon decrease.

Research the issue.
Reflect on your findings.
Discuss the various viewpoints with others.
Prepare for the class debate.

CHAPTER HIGHLIGHTS

- Metabolism is the sum of all chemical reactions that occur within cells. Metabolism can be divided into two main groups: catabolism, which involves the breakdown of macromolecules into smaller subunits, and anabolism, which involves the synthesis of large macromolecules from their component subunits.

- Molecules are the units of matter. Molecules are composed of two or more atoms of the same or different elements bonded together.

- Elements are pure substances that cannot be broken apart into simpler substances by chemical means. There are 109 elements, each with a specific number of protons.

- Chemical compounds are created by the joining of two or more elements.

- Ionic bonds are formed by the transfer of electrons from an electron donor to an electron acceptor. Ions of opposing charge attract.

- Covalent bonds form when electrons are shared by two or more atoms.

- Acids are solutions that release hydrogen ions in solution. Bases are substances that combine with hydrogen ions. How acidic or basic a solution is can be measured by the pH scale.

- Organic compounds contain carbon; inorganic compounds do not contain carbon. Living things are made up of organic compounds.

- Carbohydrates are complex molecules that contain hydrogen, carbon, and oxygen. Carbohydrates are the preferred source of energy for cells. Cells break down carbohydrates readily.

- Lipids are organic compounds formed from glycerol and fatty acids. Lipids are energy-storage compounds.

- Proteins are large molecules constructed of many amino acids. Proteins are the structural components of cells.

- Nucleic acids are composed of carbon, oxygen, hydrogen, nitrogen, and phosphorus. Nucleic acids are the molecules of heredity.

- Steroids are lipids formed from carbon rings. Cholesterol and sex hormones are steroids.

APPLYING THE CONCEPTS

1 Explain why lithium is more likely to react with fluorine than with beryllium or sodium. Use the information provided in the chart below.

Element	Symbol	Atomic number	Electron number	Energy level		
				1	2	3
Helium	He	2	2	2	0	0
Lithium	Li	3	3	2	1	0
Beryllium	Be	4	4	2	2	0
Fluorine	F	9	9	2	7	0
Sodium	Na	11	11	2	8	1

2 Use proteins to provide examples of catabolism and anabolism.

3 Explain why marathon runners consume large quantities of carbohydrates a few days prior to a big race.

4 Indicate some of the symptoms of individuals who are deficient in carbohydrates, proteins, and lipids, respectively.

5 Why is cellulose, or fiber, considered to be an important part of the diet?

6 Why are cows able to digest plant matter more effectively than humans?

7 Margarine is processed by attaching hydrogen atoms to unsaturated double bonds of plant oils. The oil becomes solid or semi-solid. Have you ever noticed that some margarines are stored in plastic tubs, while others are stored in wax paper? Compare the two types of margarine. Which would you recommend?

8 Why are phospholipids well suited for cell membranes?

9 Explain why many physicians suggest that the ratio of HDL-cholesterol to LDL-cholesterol is more significant than the cholesterol level.

10 List some of the side effects of prolonged anabolic steroid use.

CRITICAL-THINKING QUESTIONS

1 A student believes that the sugar inside a diet chocolate bar is actually sucrose because a test with Benedict's solution yields negative results. How would you go about testing whether or not the sugar present is a nonreducing sugar?

2 Three different digestive fluids are placed in test tubes. The fluid placed in test tube #1 was extracted from the mouth. The fluids in test tubes #2 and #3 were extracted from what was believed to be the stomach. Five milliliters of olive oil are placed in each of the test tubes, along with a pH indicator. The initial color of the solutions is red, indicating the presence of a slightly basic solution. The solution in test tube #3 turns clear after 10 min, but all of the other test tubes remain red. State the conclusions that you would draw from the experiment and support each of the conclusions with the data provided. (Hint: Consider which substance is digested. What are the structural components?)

3 High cholesterol levels and high risk of atherosclerosis are in part related to diet and in part determined by genetics. The LDL-cholesterol receptor sites located on cell membranes are controlled by genetics. Explain how the number of receptor sites may cause a predisposition to atherosclerosis.

4 Why do some athletes feel compelled to take anabolic steroids?

5 See the diagrams below. Three experimental designs are presented to determine the best pH for the hydrolysis of starch to smaller glucose units. Specified quantities of starch and 2 mL of an enzyme capable of initiating starch breakdown are added to each of the test tubes. Which experimental design would you choose? Justify your answer.

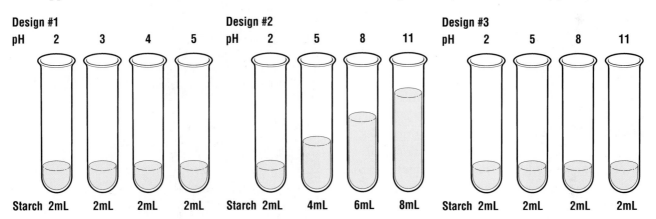

ENRICHMENT ACTIVITIES

Suggested reading:
- Beckett, Arnold. "Philosophy, Chemistry and the Athlete." *New Scientist*, August 1984, pp. 18–19.
- Blumenthal, Daniel. "Do You Know Your Cholesterol Level?" *FDA Consumer*, U.S. Food and Drug Administration, March 1989, pp. 24–27.
- Franklin, D. "Steroids Heft Heart Risk in Iron Pumpers." *Science News* 126 (21 July 1984): p. 38.
- Fritz, Sandy. "Drugs and Olympic Athletes." *Scholastic Science World* 40(16): pp. 7–14 (13 April, 1984).
- Harland, Susan. "Biotechnology: Emerging and Expanding Opportunities for the Food Industry." *Nutrition Today* (22)3: p. 21 (July/August).
- Harper, Alfred. "Killer French Fries." *Sciences*, Jan/Feb 1988, pp. 21–27.
- Monmaney, Terrence, and Karen Springen. "The Cholesterol Connection." *Newsweek*, 8 February 1988, pp. 56–58.
- Pertutz, Max. "The Birth of Protein Engineering." *New Scientist*, June 1985, pp. 12–16.
- Sperryn, Perry. "Drugged and Victorious: Doping and Sport." *New Scientist*, August 1984, pp. 16–18.

Energy within the Cell

IMPORTANCE OF CELL ENERGY

Energy can be defined as the ability to do work. For living things, the definition of work must be considered in a broad sense. Cell reproduction, the synthesis of cytoplasmic organelles, the repair of damaged cell membranes, movement, and active transport are just a few of the things that can be considered work.

Energy is found everywhere you look. Light, sound, and electricity are all forms of energy. The food that you ate for breakfast this morning provides the energy to move your limbs, maintain a constant body temperature, manufacture new chemicals, and maintain the cells of your body. This energy comes from the breakdown of macromolecules into their component parts.

There are two ways in which organisms can acquire food. *Autotrophic* organisms are organisms that are capable of making their own food. Plants use the energy from sunlight to make carbohydrates through a process called **photosynthesis.** In this process, energy from the sun converts low-energy compounds (carbon diox-

Photosynthesis *is the process by which plants and some bacteria use chlorophyll, a green pigment, to trap sunlight energy. The energy is used to synthesize carbohydrates.*

Figure 3.1

The sun is the source of life on earth. Sunlight provides the energy that powers the process of photosynthesis.

ide and water) into high-energy compounds (carbohydrates). *Heterotrophic* organisms are not capable of making their own food so they must take in food that is already made. All animals and fungi as well as many bacteria and protists are heterotrophs. Their food comes either directly from plants or from other animals that have eaten plants. Heterotrophs break chemical bonds in the large food molecules, thereby returning carbon dioxide and water to the environment and releasing energy. The energy-generating process is referred to as **cellular respiration.** (It should be noted that plants also carry out respiration to provide cell energy.)

Energy Systems

Energy is most often investigated within **energy systems.** Energy systems involve energy input, energy transformations, and energy output. Energy conversion or transformation occurs within cells.

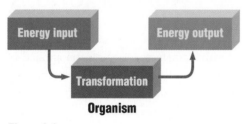

Figure 3.2

An energy system.

Unlike matter, energy does not cycle between autotrophs and heterotrophs. When living things die, their chemicals are returned to the soil. Decomposers ensure the cycling of matter. However, not all of an organism's energy is returned to the soil. The energy an organism uses for warmth and movement can never be recaptured and used by another organism. This energy is lost from the ecosystem. Fortunately, energy in the form of sunlight is constantly added, preventing the energy system from expiring.

LAWS OF THERMODYNAMICS

The study of energy relationships is called thermodynamics. The **first law of thermodynamics** states that energy cannot be created or destroyed, but energy can change forms. The amount of energy within a closed system remains constant. In other words, the same amount of energy was present at the beginning of the universe and will continue to be present until the end of time. The first law of thermodynamics has two essential parts. The first part indicates that energy cannot be destroyed. The energy held within a plant is not destroyed when it is eaten by a heterotroph; it merely changes form. The second part of the law indicates that energy may take a number of different forms. For example, plants use light energy to make sugar. In turn, the sugar is converted into starch for storage. During the process, light energy is converted into chemical energy.

The **second law of thermodynamics** states that all conversions of energy produce heat, which is not useful energy. Heat, which is a by-product of energy conversion, can be described as thermal exhaust, or waste energy. The fact that your muscles get warm during strenuous exercise attests to the fact that not all of the energy you generate is used for muscle contraction; some is wasted as heat.

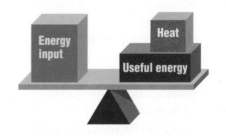

The amount of energy unavailable to do work is referred to as **entropy.** First described in 1850 by Rudolf Clausius, a

Figure 3.3

The second law of thermodynamics.

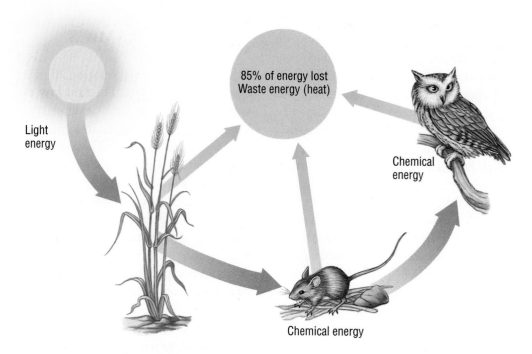

Figure 3.4

Much of the light energy reaching the plant is transformed into non-useful heat.

Light energy

85% of energy lost
Waste energy (heat)

Chemical energy

Chemical energy

Exergonic reactions
release energy.

German physicist, entropy is most often associated with non-useful heat. Clausius described entropy as the "running down of the universe" or the "heat death of the universe." Entropy reflects the disorder of a system. The greater the disorder, the greater the entropy.

As stated previously, plants use energy from sunlight to make sugars, another form of energy. However, not all the light energy that reaches the plant is transformed into chemical energy. Some of the light energy warms the soil and the air surrounding the plant. In turn, not all of the chemical energy held within the plant is transferred to the animal. The mouse in Figure 3.4 uses energy in moving toward the plant and in chewing. The mouse also uses a significant amount of food energy just to maintain a constant body temperature. The owl will eat the mouse, but again not all of the chemical energy can be converted into useful energy. The transfer of energy is incomplete. As heat escapes from the mouse and the owl, energy is lost from the system. Where does the energy go? The energy may be lost to the owl and the mouse, but it is not

destroyed. The heat from the owl warms the air surrounding the owl. Eventually, the energy is radiated into space.

METABOLIC REACTIONS

A comparison of input and output energy during transformations can be applied to the chemical reactions that occur in cells. Reactions that release energy are referred to as **exergonic reactions.** These reactions show a net loss of energy because the products of the reaction have less energy than the reactants. Some chemical energy may be lost as heat, as light, or as sound.

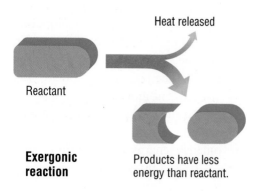

Heat released

Reactant

Exergonic reaction

Products have less energy than reactant.

LABORATORY

ENZYMES AND H₂O₂

Objective

To identify factors that affect the rate of enzyme-catalyzed reactions.

Background Information

Organisms that live in oxygen-rich environments need the catalase enzyme. The catalase enzyme breaks down hydrogen peroxide (H_2O_2), a toxin that forms readily from H_2O and O_2. The reaction below describes the effect of the catalase enzyme.

The formation of hydrogen peroxide:

$$2H_2O + O_2 \longrightarrow 2H_2O_2 \text{(hydrogen peroxide)}$$

The effect of catalase enzyme:

$$2H_2O_2 \xrightarrow[\text{enzyme}]{\text{catalase}} 2H_2O + O_2$$

Materials

safety goggles
lab apron
6 test tubes
wax marker
3% hydrogen peroxide
tweezers or forceps
10 mL graduated cylinder

fine sand
scalpel
potato
chicken liver (fresh)
stirring rod
mortar and pestle

CAUTION: Hydrogen peroxide is a strong irritant. Avoid skin and eye contact. Wash all splashes off your skin and clothing thoroughly. If you get any chemical in your eyes, rinse for at least 15 min and inform your teacher.

Procedure:

Part I: Identifying the Enzyme

1 Label three clean test tubes #1, #2, and #3.

2 Using a 10 mL graduated cylinder, measure 2 mL of hydrogen peroxide and add it to test tube #1. Add a sprinkle of sand to the test tube and observe.

3 Add 2 mL of H_2O_2 to test tubes #2 and #3. Using the scalpel, remove a slice of potato approximately the size of a raisin and add it to test tube #2. Observe the reaction. Repeat the procedure once again, but this time add a piece of liver to test tube #3. Observe the reaction.

4 Compare the reaction rates of the three test tubes. Use 0 to indicate little or no reaction, 1 to indicate slow, 2 to indicate moderate, 3 for fast, and 4 for very fast.

   a) Record your results in a table similar to the following.

Test-tube number	Reaction rate	Product of reaction
1		
2		
3		

Part II: Factors that Affect Reaction Rates

5 Divide the hydrogen peroxide used in test tube # 3 into two clean test tubes. Label one of the test tubes #4 and the other #5. Using tweezers or forceps, remove the liver from test tube #3 and divide it equally into test tubes #4 and #5. Add a second piece of liver to test tube #4 and observe. Add 1 mL of new hydrogen peroxide to test tube #5 and observe.

   b) Record your results.

6 Using a scalpel, cut another section of liver the size of a raisin. Add sand to a mortar and grind the liver into smaller pieces with the pestle. Remove the liver and place it in a clean test tube labelled #6. Add 2 mL of H_2O_2 and compare the reaction rate of the liver in test tube #6 with the uncrushed liver in test tube #3.

   c) Record your results.

Laboratory Application Questions

1 In part I , which test tube served as the control?

2 Account for the different reaction rates between the liver and potato.

3 Explain the different reaction rates in test tubes #4 and #5.

4 Why did the crushed liver in test tube #6 react differently from the uncrushed liver in test tube #3?

5 Predict what would happen if the liver in test tube #3 were boiled before adding the H_2O_2. Give the reasons for your prediction. ■

ENERGY STORAGE AND TRANSFORMATION

Energy storage and conversion processes are critical for sustaining life. Imagine the difficulty you would have storing solar energy in a box. If you opened the box one week later, would you be able to put the solar energy to use?

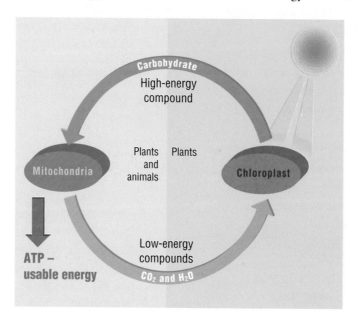

In this chapter you will study two important types of energy storage and transformation: photosynthesis and cellular respiration. During photosynthesis, solar energy is converted to chemical bond energy within carbohydrates. During cellular respiration, the energy released by the breakdown of carbohydrates becomes stored in **adenosine triphosphate, or ATP.** The energy stored in ATP is used for the synthesis of needed chemicals within cells, the active transport of materials across cell membranes, and the contraction of muscles.

Adenosine triphosphate is composed of a five-carbon sugar (ribose), a nitrogen base (adenine), and three phosphate groups. A diagram of the ATP molecule is shown in Figure 3.16.

ATP can be thought of as a gold coin that cells use to pay for work. The energy in ATP is found within the bonds between the phosphate groups. In much the same way as you draw money from your bank account, special enzymes can break one of these bonds and remove a phosphate group from ATP and, in doing so, release energy. The new product is called *adenosine diphosphate*, or *ADP*, because it has only two phosphate groups. ADP has less energy than the ATP molecule. The ADP molecule can be thought of as a silver coin. Although the silver coin is capable of paying for energy demands, it has less value than the ATP gold coin. Occasionally, a second phosphate bond can be broken and the energy used. In this step, ADP is converted to *adenosine monophosphate*, or *AMP*. The third phosphate group will not be removed. To return to the money analogy, once the silver coins have been spent, no additional energy can be purchased. The bank account is empty. The reaction in Figure 3.15 summarizes how ATP provides energy for a cell.

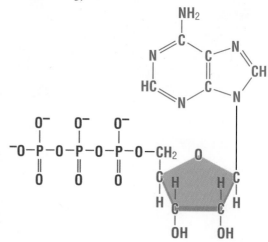

ATP *is a compound that stores chemical energy.*

Figure 3.14

Two important organelles for energy transformation. The chloroplasts are the sites of photosynthesis and the mitochondria are the sites of cellular respiration.

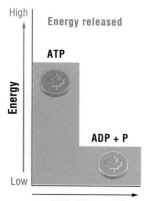

$$ATP \rightarrow ADP + P + energy$$

Figure 3.15

ATP can be thought of as a gold coin and ADP as a silver coin.

Figure 3.16

The structural formula for adenosine triphosphate. The ribose sugar is shown in green, the adenine molecule in red, and the three phosphate groups in purple.

In the same way as you must deposit more money into your bank account to continue withdrawing, ATP supplies must be constantly replenished. Energy must be added to refill ATP supplies, but where does the energy come from? Exergonic chemical reactions supply needed energy for the synthesis of ATP. A tremendous amount of potential energy is stored in chemical bonds. The conversion of high-energy molecules into lower-energy molecules releases energy, which, in turn, can be used to form high-energy phosphate bonds that convert ADP to ATP. The addition of a phosphate group is called **phosphorylation.**

Regulating Energy Input for ATP Production

An electrical spark can be used to combine hydrogen and oxygen. As shown in Figure 3.17, such a combination produces water. Because water contains much less energy than the reactants (hydrogen and oxygen), energy is released. Unfortunately, this rapid release of energy is not ideally suited for cells. In many situations, cells could be damaged by the uncontrolled release of heat. The heat may actually distort proteins and damage the cell. However, a step-by-step release of energy allows cells time to transform and store much of the energy.

Most energy releases in cells take place within the **electron transport system.** In many ways, the electron transport system resembles the set of stairs shown in Figure 3.17. During cellular respiration, the electrons move from high energy levels to lower energy levels, releasing energy at each step. The final electron acceptor, oxygen, combines with the proton of hydrogen and its electron to form a low-energy compound, water.

Phosphorylation *is the addition of one or more phosphate groups to a molecule.*

The **electron transport system** *is a series of progressively stronger electron acceptors. Each time an electron is transferred, energy is released.*

Figure 3.17

(a) The combination of oxygen and hydrogen releases energy as a burst of heat. (b) The step-by-step release of energy by electron transport systems enables cells to convert much of the energy into ATP.

a)

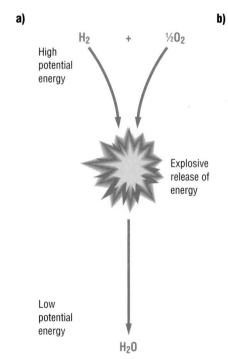

b)

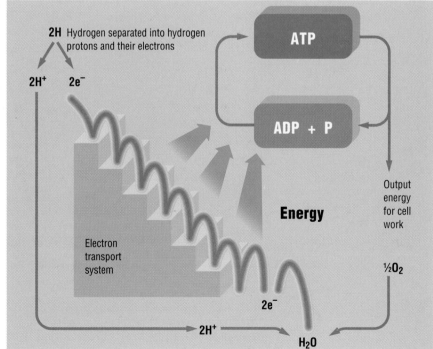

Oxidation *occurs when an atom or molecule loses electrons.*

Reduction *occurs when an atom or molecule gains electrons.*

NAD^+ *is a hydrogen acceptor important for electron transport systems in cellular respiration. NADH is the reduced form of NAD^+.*

$NADP^+$ *is a hydrogen acceptor important for electron transport systems in photosynthesis. NADPH is the reduced form of $NADP^+$.*

The stairway analogy may help you picture how a ball moves from one level to another, but why do electrons move? Elements such as lithium, calcium, and sodium are strong electron donors. This means that they tend to lose electrons to other elements in an attempt to become stable. Elements such as fluorine, chlorine, and oxygen are strong electron acceptors. By accepting electrons, these elements fill their outermost energy levels, thereby becoming stable. The process in which elements lose electrons in order to become stable is called **oxidation.** The process in which elements accept electrons is called **reduction.** When organic molecules are oxidized, the hydrogen nuclei and their electrons (H^+ and e^-) are lost. Organic molecules are reduced when they gain hydrogen nuclei and their electrons.

BIOLOGY CLIP

Some 40 to 60% of the world's photosynthesis occurs in the oceans. It is estimated that microscopic ocean plants produce one hundred billion tonnes of oxygen each year

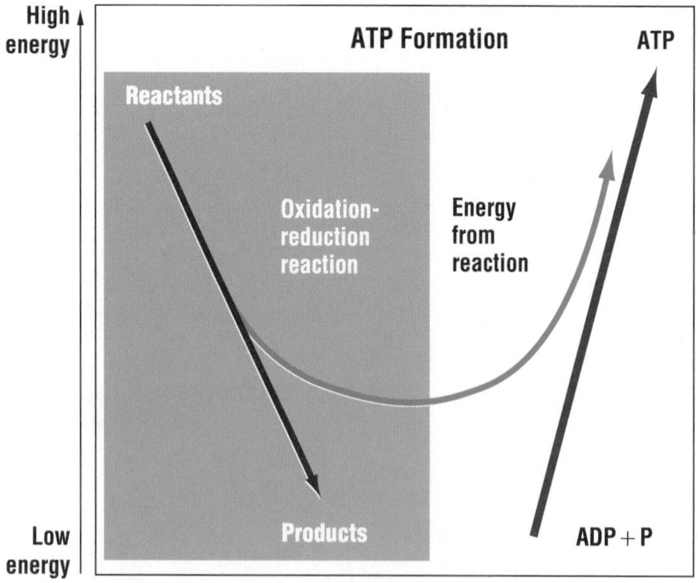

Figure 3.18

The energy released from the oxidation-reduction reaction is used to attach a free phosphate to ADP to make ATP. Note that ATP is a high-energy compound.

The transfer of electrons from one reactive atom to another produces more stable ions or compounds. The fact that the products have less energy than the reactants indicates that energy is released during the oxidation reaction. This energy can be used to make ATP. The diagram in Figure 3.18 shows how the energy from an oxidation-reduction reaction is used to attach phosphates to ADP. The product, ATP, is a high-energy compound.

Each time electrons are transferred in oxidation-reduction reactions, energy is made available for the cell to make ATP. Electron transport systems shuttle electrons from one molecule to another. Electron and proton carriers like **NAD^+** (nicotinamide adenine dinucleotide), an important coenzyme for cellular respiration, and **$NADP^+$** (nicotinamide adenine dinucleotide phosphate), an important coenzyme in photosynthesis, act as electron acceptors. NADH and NADPH are very good reducing agents. They readily lose electrons and protons. The electron acceptors strip 2 hydrogen protons and their electrons from a number of organic compounds. When the NADH or NADPH coenzymes are oxidized to NAD^+ or $NADP^+$, electrons and protons are released and ATP synthesis occurs. Eventually, the coenzyme loses electrons to an even stronger electron acceptor. Each time the electron is transferred in the electron transport system, energy is released.

LABORATORY
PHOTOSYNTHESIS

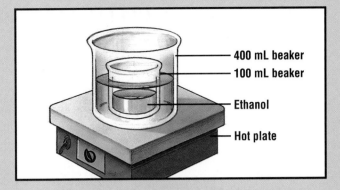

400 mL beaker
100 mL beaker
Ethanol
Hot plate

Objective

To demonstrate the importance of sunlight during photosynthesis.

Materials

safety goggles	400 mL beaker
lab apron	beaker tongs
10 mL graduated cylinder	hot plate
distilled water	denatured ethanol
5% starch solution	variegated coleus plant
2 test tubes	2 100-mL beakers
iodine solution	forceps
medicine dropper	petri dish

> **CAUTION:** The chemicals used are toxic and an irritant. Avoid skin and eye contact. Wash all splashes off your skin and clothing thoroughly. If you get any chemical in your eyes, rinse for at least 15 min and inform your teacher. Ethanol is flammable. Do not use near open flames.

Procedure

Part I: Testing for Starch

1 Using a 10 mL graduated cylinder, add 3 mL of distilled water and 3 mL starch solution to two separate test tubes. Add four drops of iodine to each test tube and observe.
  a) Record the color of each test tube.

Part II: Sunlight and Photosynthesis

2 Add 200 mL of water to a 400 mL beaker and, using beaker tongs, place the beaker on a hot plate.

3 Add 40 mL of ethanol to a 100 mL beaker. Using beaker tongs, place the 100 mL beaker inside the 400 mL beaker. Turn on the hot plate and bring the water to a boil.

> **CAUTION:** Always use a water bath.

4 Remove the leaf of a coleus plant that has been exposed to direct light for at least 72 h.
  b) Sketch the leaf, and label areas of red and green pigment.

5 Using forceps, place the leaf in boiling water for three seconds and then immerse it in the boiling alcohol for approximately five minutes, or until the pigment disappears completely from the leaf.

6 Once again, immerse the leaf in boiling water for approximately three seconds.
  c) Describe the appearance of the leaf.

7 Place the leaf in an open petri dish and carefully spread out the leaf. Add 20 to 30 drops of iodine, completely covering the leaf.
  d) Describe the appearance of the leaf.
  e) Sketch the leaf, and label the areas that contain starch.

8 Repeat the same procedure for a plant that has been kept in total darkness for 72 h.
  f) Describe the appearance of the leaf.
  g) Sketch the leaf, labelling the areas that contain starch.

9 Ask your teacher to dispose of the alcohol.

Laboratory Application Questions

1 Why was iodine used in the experiment?
2 From the data that you have collected, support the fact that light is required for photosynthesis.
3 Explain why coleus was used rather than a plant that contains only green pigments.
4 Predict how the results would have been affected if the plant exposed to light had been covered by a transparent green bag. State the reasons for your prediction.
5 Design an experiment to determine which visible colors provide the optimal energy for photosynthesis.
6 Why was a Bunsen burner not used as a heat source for the ethanol? ■

CELLULAR RESPIRATION IN PLANTS AND ANIMALS

Cellular respiration includes most of the chemical reactions that provide energy for life. Carbohydrates, most notably in the form of glucose, are the most usable source of energy. Only after glucose supplies have been depleted do cells turn to other fuels. For animals, glycogen, a storage carbohydrate composed of many glucose units, breaks down, releasing single glucose units into the blood in an attempt to maintain blood glucose levels. Once glycogen supplies from the liver and muscles are depleted, fat becomes the preferred energy source. Proteins, the organic compounds of cell structure, are used as a final resort once fat supplies have been exhausted. The utilization of proteins for energy means that the cell begins breaking down its own structures in order to obtain energy. Plants use starch as an energy storage compound in much the same manner as animals use glycogen. When needed, starch can be broken down to maltose, a disaccharide sugar. Unlike starch, maltose is soluble. Eventually, the disaccharide can be broken down into monosaccharide units.

Cellular respiration is the oxidation of glucose to produce ATP.

During respiration, chemical bonds are broken. Chemical energy is stored in these bonds. It has been estimated that, at best, only 36% of the energy is used to make ATP; the remaining 64% is released as heat. The thermal energy helps maintain a constant body temperature in mammals and birds. ATP, as mentioned before, has a form of stored energy that is readily accessed for active transport, the synthesis of needed chemicals, the contraction of muscle fibers, and other energy-requiring functions.

Oxidation and Phosphorylation

As the glucose molecule is oxidized, energy is released. Hydrogen and its electron move from a weak electron acceptor to a stronger electron acceptor. NAD⁺ (nicotinamide adenine dinucleotide) is a typical weak electron acceptor in cells, accepting hydrogen and its electron from activated carbohydrates. Electrons can move from one electron acceptor to progressively stronger acceptors. In much the same way as you descend a flight of stairs, the electron moves from one acceptor to another, losing energy at each step. Electrons found on the top step are able to provide the greatest amount of energy for ATP formation. Electrons found on the lower steps are less able to provide energy for ATP formation.

During oxidation, high-energy compounds like glucose are converted into low-energy compounds, such as carbon dioxide and water. Each time a hydrogen proton and its electron move from glucose, energy is released. Naturally, the greater the number of hydrogen protons extracted, the greater the amount of

Figure 3.24

NAD⁺ acts as electron acceptor. During the breakdown of glucose, NADH is generated by the reduction of NAD⁺. The final products— carbon dioxide and water—have less energy. The inset shows glucose burning. Oxygen removes hydrogen and its electron from glucose. A high-energy compound, glucose, is converted to low-energy compounds, carbon dioxide and water.

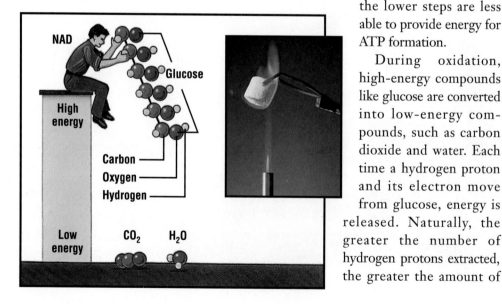

Unit One:
Exchange of Energy and Matter in Cells

energy released. The release of energy during an oxidation reaction might lead you to conclude that energy is created; however, as you read earlier, energy cannot be created—the total amount of energy remains constant within the system. The energy released during the chemical reactions comes from the chemical bonds of the high-energy molecules. Thus, the products of oxidation contain less energy than the reactants. The fact that the products have less energy can be easily proved. Place glucose in one test tube and light the glucose. The glucose burns. The substitution of glucose with carbon dioxide produces different results. Carbon dioxide does not burn. Oxygen from the atmosphere is capable of pulling electrons from glucose but is incapable of oxidizing carbon dioxide. Carbon dioxide has already been oxidized.

The energy released during the oxidation of glucose comes from the formation of new low-energy products. The chemical energy of glucose can be converted into heat or other forms of chemical energy storage such as ATP. However, glucose and other sugars will not release hydrogen and its electrons without first being activated. Consider the following scenario. The lid from a sugar bowl is removed exposing the sugar to oxygen, a strong electron acceptor. Does the sugar become oxidized? The answer is no. The prospect of a bowl of sugar bursting into flames every time the top is removed would make our morning cups of coffee extremely dangerous! Sugar molecules must be activated before they can react.

The match provides the activation energy

needed to initiate the burning of sugar. However, activating sugar in this manner is unacceptable for your cells.

ATP is used to activate glucose by a process known as *phosphorylation*. During phosphorylation, a high-energy phosphate bond is linked with a glucose molecule. As in other chemical reactions, activation energy must be reached before the oxidation of glucose begins. Energy must be used before greater energy can be released.

ANAEROBIC RESPIRATION

Anaerobic respiration takes place in the absence of oxygen. In animal cells, glucose can be partially broken down to lactic acid, which contains less energy than glucose; however, many hydrogen atoms remain attached to the carbon atoms. If oxygen is available, even more energy can be released as lactic acid is oxidized. Another type of anaerobic respiration, *fermentation*, occurs in yeast. The products of fermentation are carbon dioxide and alcohol. During both alcoholic fermentation and lactic acid anaerobic respiration a limited number of chemical bonds are broken. A total of two ATP are produced by each process.

Anaerobic respiration *takes place in the absence of oxygen.*

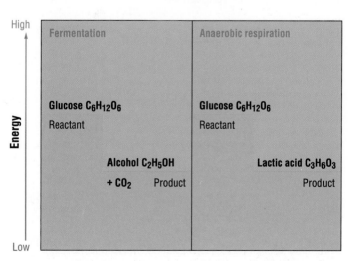

Figure 3.25

Higher-energy glucose is oxidized to lower-energy products. Alcohol and lactic acid have less potential energy than glucose.

Figure 3.26

Yeast fermentation produces alcohol and carbon dioxide.

Glycolysis is the process in which ATP is formed by the conversion of glucose into pyruvic acid.

Aerobic respiration refers to the complete oxidation of glucose in the presence of oxygen.

Glycolysis is the first step in cellular respiration and does not require oxygen. High-energy phosphate groups from the ATP are used to activate the six-carbon sugar, glucose. Figure 3.27 below summarizes the events of glycolysis in animals.

The final step shown in Figure 3.27 indicates that pyruvic acid is converted into lactic acid. However, lactic acid is not normally a product of fermentation. Yeast cells contain enzymes that extract carbon dioxide before lactic acid can be formed. The end-products of fermentation, therefore, are alcohol and carbon dioxide.

Lactic acid will accumulate in muscles during strenuous exercise if sufficient amounts of oxygen are not delivered to the tissues. Have you ever felt a sharp pain in your side while running? This pain may have been caused by a build-up of lactic acid. If you continue to run, the pain often intensifies. Anaerobic respiration does not supply enough energy to meet the demands that you are placing on your body, and the depletion of ATP reserves causes fatigue. The pain is merely providing a warning message.

Fermentation occurs in the cytoplasm of yeast cells. Like anaerobic lactic acid respiration, fermentation yields two ATP. However, special enzymes extract carbon dioxide from PGA, thereby creating different products. Bread dough rises when carbon dioxide gases are released during fermentation. The bubbles released in champagne are caused by the same chemical process. Both champagne and bread dough also contain alcohol as a product of fermentation, although most of the alcohol that would be found in bread evaporates during the baking.

AEROBIC RESPIRATION

Aerobic respiration requires oxygen and involves the complete oxidation of glucose. Carbon dioxide, water, and 36 ATP are the products of aerobic respiration. The reaction below summarizes aerobic respiration:

$$C_6H_{12}O_6 + 6O_2 \longrightarrow 6CO_2 + 6H_2O + 36\ ATP$$

Krebs Cycle

As in anaerobic respiration, glycolysis in aerobic respiration occurs within the cytoplasm. However, pyruvic acid does

Figure 3.27

Glycolysis is the first step in cellular respiration. During glycolysis, glucose is broken down to lower energy molecules. The energy released is used to synthesize ATP.

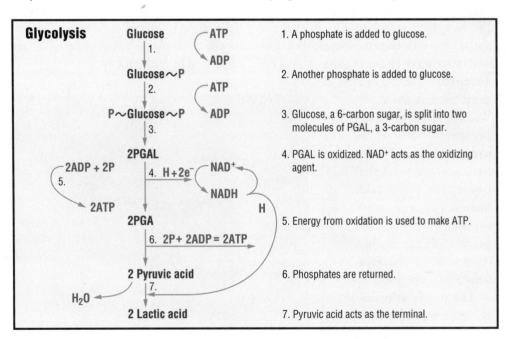

Glycolysis	Glucose	

1. A phosphate is added to glucose.

2. Another phosphate is added to glucose.

3. Glucose, a 6-carbon sugar, is split into two molecules of PGAL, a 3-carbon sugar.

4. PGAL is oxidized. NAD⁺ acts as the oxidizing agent.

5. Energy from oxidation is used to make ATP.

6. Phosphates are returned.

7. Pyruvic acid acts as the terminal.

$$C_6H_{12}O_6 + 6O_2 \longrightarrow 6CO_2 + 6H_2O$$

glucose   oxygen   carbon   water

dioxide

typical energy yield = 36 ATP

Figure 3.28

Vast energy supplies come from ATP, which is synthesized during aerobic respiration; however, when energy demands exceed the oxygen supply, some ATP is formed by anaerobic respiration. The product of anaerobic respiration is lactic acid. It is now known that some athletes can tolerate higher than average levels of lactic acid, enabling them to continue energy production longer.

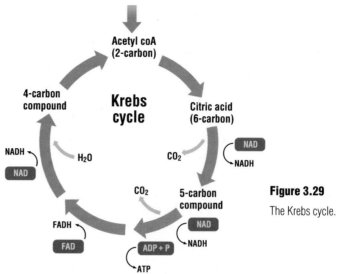

Figure 3.29

The Krebs cycle.

not serve as the final electron acceptor. Unstable pyruvic acid, in the presence of oxygen, breaks down to acetic acid. The acetic acid becomes attached to coenzyme A to form a complex called acetyl coenzyme A, or coA for short.

The acetyl coenzyme A enters a series of chemical reactions referred to as the *Krebs cycle*, during which additional hydrogen and carbon dioxide molecules are extracted. The hydrogen atoms move through the electron transport chain within the mitochondria and eventually combine with oxygen atoms to form water. The final products of the Krebs cycle are carbon dioxide and water. Figure 3.29 summarizes the chemical changes that take place during the Krebs cycle.

Electron Transport System

The key to understanding why aerobic respiration yields greater amounts of energy than anaerobic respiration is found within the electron transport system, which accepts electrons generated by the Krebs cycle. Located in the mitochondria, a series of enzymes pass hydrogen and its electron along a series of progressively stronger electron acceptors. Each time hydrogen and its electron are passed from one electron acceptor to another, energy is released. The energy released by these oxidation reactions is used to make ATP.

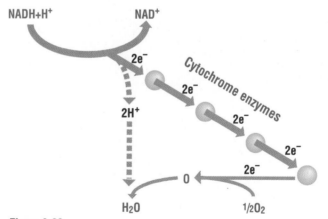

Figure 3.30

The electron transport system. As electrons are passed from one electron acceptor to the next, energy is released.

Oxygen acts as the terminal electron acceptor for aerobic respiration, combining with two hydrogen atoms and their electrons to form water. Without a constant supply of oxygen, the electrons could not be passed down the chain of oxidizing agents, and aerobic respiration would soon stop.

INTERMEDIARY PATHWAYS

Glucose is the main source of energy, but many other organic compounds are consumed. How is energy derived from fats or proteins? Figure 3.31 indicates how excess fat can enter either glycolysis or the Krebs cycle. Similarly, proteins can be converted into amino acids and enter aerobic metabolic pathways. The products of metabolism can also be used as raw materials to form fatty acids or amino acids. These biosynthesis pathways are indicated by green arrows, while energy-releasing pathways are indicated by red arrows.

Figure 3.31

Intermediary metabolic pathways.

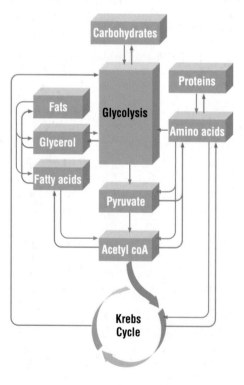

COMPARISON OF PHOTOSYNTHESIS AND RESPIRATION

The following table provides a comparison of photosynthesis and respiration.

Aerobic Respiration	Photosynthesis
Energy produced	Energy required
Oxidation	Reduction
High-energy reactants	Low-energy reactants
Low-energy products	High-energy products
Oxygen required	Oxygen released
Glucose required	CO_2 and H_2O required
CO_2 and H_2O produced	Glucose produced

■ REVIEW QUESTIONS ■ ?

23 Contrast ATP production in anaerobic and aerobic respiration.
24 Why is the phosphorylation of glucose necessary?
25 How does the Krebs cycle provide additional ATP?
26 How does the electron transport system provide the energy for the synthesis of ATP?
27 Under what conditions do plant cells undergo cellular respiration?

CAREER INVESTIGATION CAREER

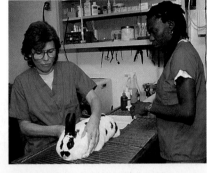

*Cells are essential to all forms of life on earth.
We must understand the mechanisms of cell function if we are to
properly utilize our life-form resources.*

Cosmetologist

The science of skin and hair care is big business. Designing products that protect and preserve hair and skin requires a knowledge of how the hair and skin cells work, as well as how they are affected by environmental factors.

Veterinarian

A veterinarian needs to be familiar with many different species. Often diseases will have different symptoms and treatments for each species. To provide effective care to animals, a veterinarian must thoroughly understand how nutrients are exchanged in the animal.

Athlete

In order to obtain the greatest benefit from exercise and training, an athlete must know how the human body reacts to exercise and the nutrients the body requires to perform at peak levels. This requires a knowledge of how cells in the body exchange both information and material.

- Identify a career that requires a knowledge of cells and energy.
- Investigate and list the features that appeal to you about this career. Make another list of features that you find less attractive about this career.
- Which high-school subjects are required for this career? Is a post-secondary degree required?
- Survey the newspapers in your area for job opportunities in this career.

Pharmacist

A pharmacist's background in chemistry and biology comes into use daily. Many of the drugs we use were discovered accidentally. However, research into drugs designed specifically to combat certain diseases is now an important field.

*Chapter Three:
Energy within the Cell*

97

SOCIAL ISSUE:
Interdependence of Cellular Respiration and Photosynthesis

The interaction of plants and animals has largely been ignored. The interdependence of respiration and photosynthesis illustrates the importance of plants. Any substantial changes in plant population must be considered in a wider sphere. Any reduction in forest areas will have an impact on human populations. Satellite sensing shows that tropical rain forests are disappearing at a rate of 171 000 km^2 each year.

Statement:

Legislation should be introduced to reduce fossil fuel emissions and the cutting of rain forests.

Point

- The constantly increasing human population has altered the balance between the carbon dioxide produced by combustion (oxidation) and the oxygen liberated by photosynthesis. The burning of fossil fuels is increasing levels of carbon dioxide. The climate is changing.

- Each year, the more than five billion people on our planet use about 40% of the organic matter fixed by plants. One estimate suggests that we consume two tonnes of coal and produce 150 kg of steel annually for every person on our planet. Carbon dioxide is produced by each of these reactions. Governments must institute a carbon tax to reduce carbon emissions.

- Wealthy countries should pay into a fund so that poor nations are not tempted to cut their trees.

> *Research the issue.*
> *Reflect on your findings.*
> *Discuss various viewpoints with others.*
> *Prepare for the class debate.*

Counterpoint

- The climate has changed in the past. Suggestions that increasing carbon dioxide levels have created an imbalance between the products of respiration and photosynthesis are not cause for alarm.

- The fact that some governments have considered taxing polluters does not mean that they understand the problem. The most important supply of oxygen is the ocean. Algae use carbon dioxide to synthesize sugars and oxygen. Should extreme measures be instituted, a country's economy would be seriously hindered.

- Education is the answer, not increasing taxes or bringing in legislation.

CHAPTER HIGHLIGHTS

- Chemical energy is stored in chemical bonds.
- Activation energy is required to begin many chemical reactions in cells.
- Chemical reactions within cells are regulated by enzymes. Enzymes are protein catalysts that lower activation energy.

- Enzymes permit chemical reactions within your body to proceed at low temperatures.
- Cofactors are inorganic molecules that help enzymes combine with substrate molecules.
- Coenzymes are organic molecules that help enzymes combine with substrate molecules.

98

Unit One:
Exchange of Energy and Matter in Cells

- A competitive inhibitor has a shape complementary to a specific enzyme, thereby permitting it access to the active site of the enzyme. Inhibitors block chemical reactions.
- Feedback inhibition is the inhibition of the first enzyme in a metabolic pathway by the final product of that pathway.
- Precursor activity is the activation of the last enzyme in a metabolic pathway by the initial reactant.
- Photosynthesis is the process by which plants and some bacteria use chlorophyll, a green pigment, to trap sunlight energy. The energy is then used to synthesize carbohydrates.
- Photosynthetic reactions can be classified according to two main phases; light-dependent reactions and light-independent, or carbon-fixation, reactions.
- Cellular respiration is the process by which organisms obtain energy from the breakdown of high-energy compounds.
- The first law of thermodynamics states that energy can be neither created nor destroyed.

- The second law of thermodynamics states that all conversions of energy produce some heat, which is not useful energy.
- Electron transport systems are a series of progressively stronger electron acceptors. Each time an electron is transferred, energy is released.
- Oxidation occurs when an atom loses electrons. Reduction occurs when an element gains electrons.
- NAD^+ is a strong electron acceptor important for electron transport systems in cellular respiration. NADH is the reduced form of NAD^+.
- $NADP^+$ is a strong electron acceptor important for electron transport systems in photosynthesis. NADPH is the reduced form of $NADP^+$.
- Chlorophyll is the green pigment found in plants. Chlorophyll traps sunlight energy for photosynthesis.
- Cellular respiration is the oxidation of glucose to produce ATP.
- Phosphorylation is the process by which high-energy phosphate groups are added to molecules.
- Anaerobic respiration takes place in the absence of oxygen. Aerobic respiration refers to the oxidation of glucose in the presence of oxygen.

APPLYING THE CONCEPTS

1 Explain how enzymes work in the lock-and-key model. How has the "induced-fit" model changed the way in which biochemists describe enzyme activities?

2 Using the information that you have gained about enzymes, explain why high fevers can be dangerous.

3 Cyanide attaches to the active site of a cytochrome enzyme. How does cyanide cause death?

4 Use the metabolic pathway shown below to explain feedback inhibition.

5 What is the source of the oxygen released during photosynthesis? What is the source of carbon dioxide fixed during photosynthesis?

6 How does light intensity affect the rate of photosynthesis?

7 A photosynthesizing plant is exposed to radioactively labelled carbon dioxide. In which compound would the radioactive carbon first appear? Explain your answer.

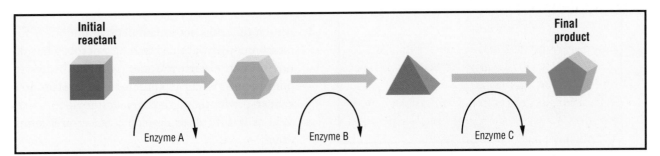

8 The removal of carbon dioxide from pyruvic acid distinguishes fermentation from lactic acid anaerobic respiration. What would occur if an enzyme in your body removed the carbon dioxide from pyruvic acid before lactic acid is formed?

9 Copy the chart below and place the correct answer in the appropriate column.

Characteristic	Anaerobic	Aerobic
1 Amount of ATP produced		
2 Terminal electron acceptor		
3 Site of activity in cell		
4 Final products		

10 Compare respiration and photosynthesis by completing a chart like the one below. Place a ✔ in the correct column.

Characteristic	Photosynthesis	Respiration
1 ATP used to initiate reaction		
2 ATP produced during reaction		
3 Oxidation reaction		
4 Reduction reaction		
5 Oxygen is a product		
6 Occurs in plant cells		
7 Occurs in animal cells		
8 Carbohydrate is reactant		

CRITICAL-THINKING QUESTIONS

1 To study the splitting of water in the process of photosynthesis, a student uses a heavy isotope of oxygen in the form of water ($H_2^{18}O$). If the following apparatus were used, explain where you would expect to find the heavy isotope of oxygen.

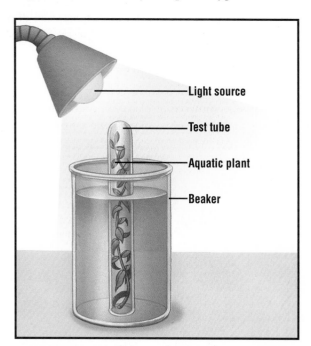

2 The following graph shows the rate at which products are formed for an enzyme-catalyzed reaction. By completing the graph, predict how a competitive inhibitor added at time X would affect the reaction.

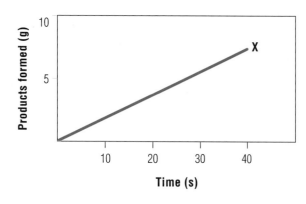

3 Antibiotic drugs act as competitive inhibitors for metabolic pathways in invading microbes but do not interfere with the metabolic pathways in human cells. Using this information, explain how cancer-preventing drugs are designed to work. Why is it difficult to develop a cancer-preventing drug?

4 The earth's early atmosphere was rich in carbon dioxide and low in oxygen. With the evolution of plankton, the composition of the atmosphere changed. Through photosynthesis, carbon dioxide in the atmosphere was fixed by the plankton, and oxygen was released. As plants became more abundant, carbon dioxide levels began to fall and oxygen levels began to rise. A number of scientists have suggested that increasing algae growth in the oceans may reverse increasing carbon dioxide levels and reduce atmospheric warming. One suggestion is to seed the oceans with iron and phosphates—nutrients that will increase plankton blooms. Comment on this strategy.

5 The following experiment was designed to demonstrate the relationship between plants and animals. Bromthymol blue indicator was placed in each of the test tubes. High levels of carbon dioxide will combine with water to form carbonic acid. Acids will cause the bromthymol blue indicator to turn yellow. The initial color of the test tubes is blue.

Predict the color change, if any, in each of the test tubes. Explain your predictions.

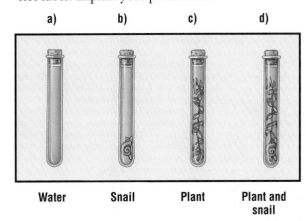

a)	b)	c)	d)
Water	Snail	Plant	Plant and snail

ENRICHMENT ACTIVITIES

Suggested reading:
- Doolittle, G. "Proteins." *Scientific American*, 253(4) (October 1988): pp. 88–99.
- Hinkle, P., and R. McCarthy. "How Cells Make ATP." *Scientific American* 238(23) (March 1978): p. 104.
- Kunzig, Robert. "Invisible Garden." *Discover* 11(4) (April 1990): p. 66.
- Lovelock, James. *The Ages of Gaia: A Biography of Our Living Earth*. New York: W.W. Norton Company, 1988.

- Miller, K. "Three-Dimensional Structure of Photosynthetic Membrane." *Nature* 300 (1982): p. 5887.
- Moore, P. "The Varied Ways Plants Trap the Sun." *New Scientist*, February 1981, p. 394.
- Steinberg, S. "Genes Shed Light on Photosynthesis." *Science News*, August 1983, p. 102.

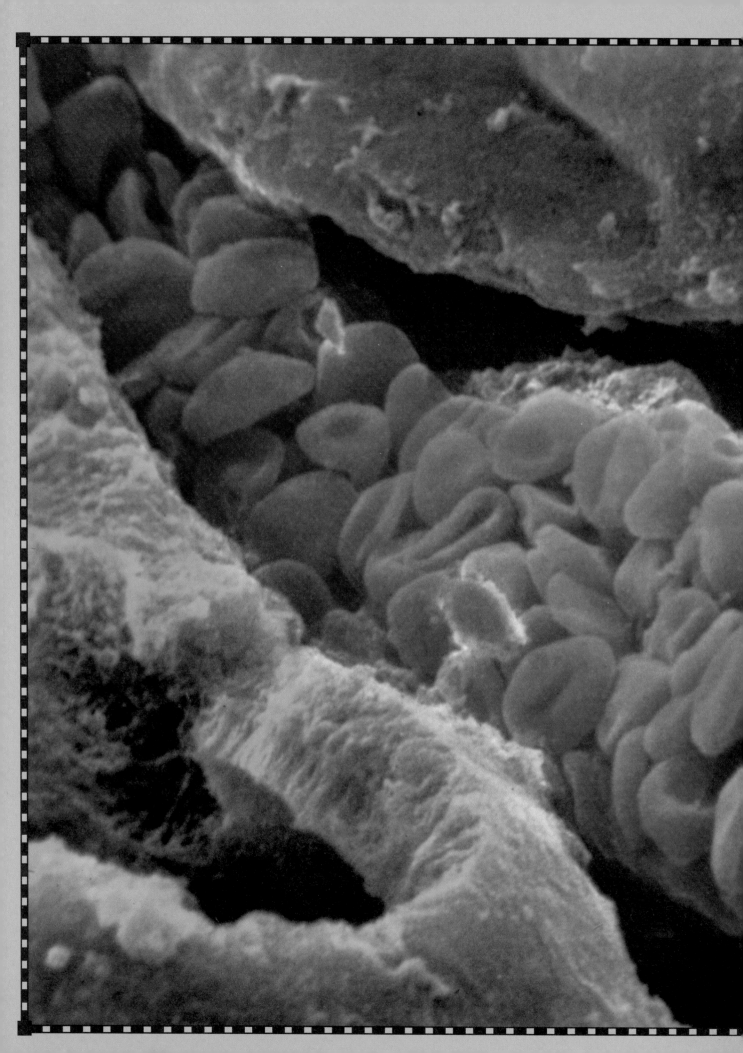

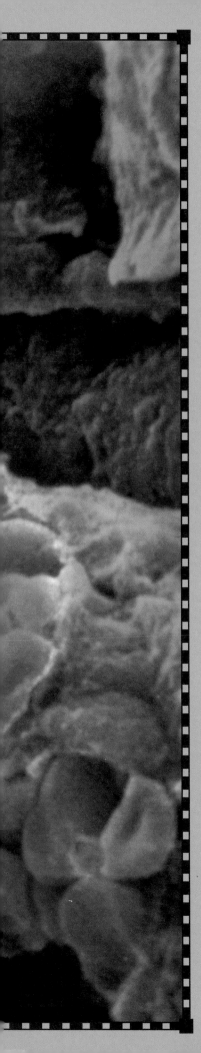

Exchange of Matter and Energy in Humans

.

UNIT 2

Organs and Organ Systems

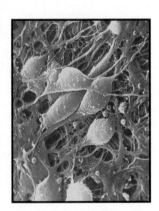

Nerve cells from the cerebral cortex. Nerve cells exist in varying sizes and shapes, but all have the same basic structure.

IMPORTANCE OF ORGANS AND ORGAN SYSTEMS

Your body is made up of trillions of cells. Unlike single-celled organisms, which appear somewhat alike, the cells of your body display remarkable variation. Your cells are organized into groups that have distinctive shapes and functions. Muscle cells, for example, are relatively long and contain special protein filaments that slide over each other when stimulated. These specialized filaments cause the muscle cell to shorten during muscle contractions. Fat cells contain unusually large vacuoles that store the high-energy fat compounds. Long, thin nerve cells, some of them nearly 1 m in length but less than 0.001 mm in diameter, are covered with a glistening sheath that helps provide insulation and speeds the movement of impulses. Specialized ends of the motor nerve cells either receive nerve messages or produce transmitter chemicals that permit the passing of the impulse to other nerves.

Like multicellular organisms, single-cell organisms must move, coordinate information from their surroundings, and store energy, but unlike multicellular

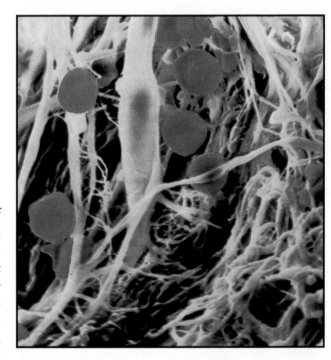

Figure 4.1

Human connective tissue, showing collagen fibers. Red blood cells are interspersed between the fibres. Connective tissue provides metabolic and structural support for other tissues and organs.

organisms they must accomplish these tasks without the benefit of specialization.

Imagine how difficult your life would be if specialists did not exist. Could you repair a car's transmission by yourself? Would you be able to diagnose your own illnesses? Would you be able to produce the food required for a well-balanced diet? The division of labor permits greater efficiency in function. By specializing, cells become very good at performing particular tasks. For example, muscle cells are well suited for contracting, but are poorly suited for other tasks.

LEVELS OF CELL ORGANIZATION

Tissues

Cells similar in shape and function work together to ensure the survival of the organism. These cells make up the body's **tissues.** Cells within a tissue are capable of recognizing similar cells and sticking to them. This point is well illustrated by a simple experiment that was performed on two types of sponges. Although sponges do not have organs or organ systems, their cells are arranged in two layers of similar tissue. The cells of red and yellow sponges were separated from neighboring cells by passing them through a fine-mesh sieve. Researchers noted that some cells adhered to each other, but not to other cells. With further investigation, the researchers were able to determine that cells from the red sponge only adhered to other cells from the red sponge. Similarly, cells from the yellow sponge formed clusters with other cells from the yellow sponge. As the clusters became larger, researchers noted that the cells began to arrange themselves into patterns characteristic of an intact sponge.

In the human body there are four primary kinds of tissue: epithelial, connective, muscle, and nervous. **Epithelial tissue** is a covering tissue that protects not only the outer surface of your body, but your internal organs as well. It also lines the inside of your digestive tract and blood vessels. A cell's size, shape, and arrangement in the tissues is consistent with its designated function. For example, a single layer of cells is well suited for transporting nutrients or wastes within the body. Multiple layers form a barrier to protect organs from invading microbes and disease-causing agents. Layers of thin, flat cells, referred to as *squamous epithelium*, provide maximum coverage and protection for the outer layers of your skin. Similarly shaped cells also line blood vessels and comprise the air sacs of the lung. Cube-shaped epithelial tissue provides the framework for the kidney tubules, while pillar-shaped cells, referred to as *columnar epithelium*, line the stomach and the respiratory tract. The cube-shaped cells are well suited for providing strong structural support, while the much longer cells within columnar epithelium often produce secretions.

Connective tissue is made up of fibrous proteins and a material called the **matrix,** which is found between the cells. Cells called *fibroblasts* produce the solid, semi-solid, or fluid matrix and fibrous proteins. The components of the connective tissue are arranged according to function. For example, fibrous protein structures are commonly found in muscles, which move limbs; tendons, which anchor muscle to bone; and ligaments, which bind bones at the joints. Connective tissues are also found in blood, adipose (fat), cartilage, bone, and lymph tissue. Blood and lymph tissue contain a fluid matrix that is well-suited for transporting materials throughout the body. A more solid matrix is found in bone and cartilage, where support is necessary.

Tissues *are groups of similarly shaped cells that work together to carry out a similar function.*

Epithelial tissue *is a covering tissue that protects organs, lines body cavities, and covers the surface of the body.*

Connective tissue *provides support and holds various parts of the body together.*

The **matrix** *is the noncellular material secreted by the cells of connective tissue.*

Figure 4.2

Adipose connective tissue contains many cells with fat-storage vacuoles.

Muscle tissue is composed of cells containing special contractile proteins and accounts for approximately 40 to 50% of your total body mass. Muscle tissues shorten when these proteins contract. This contraction enables you to move.

Nerve tissue may be the most diverse and complex of all tissues. The brain, the spinal cord, and the sense organs are primarily composed of nerve tissue. Charged with the task of communication, nerve tissue is essential for the growth and development of all other tissues. Other tissues rely on nerve tissue for information about the environment in order to coordinate responses to environmental changes.

Organs

Organs are groups of different tissues specialized to carry out particular functions. The heart is an excellent example of a complex organ. The heart's outer structure is covered by epithelial tissue, which also lines the inside of the heart's chambers and blood vessels. Nerve tissue initiates and synchronizes cardiac muscle

contractions and relays information to the brain about the strength of these contractions and the relative condition of the muscle cells. The blood that pulses through the heart is connective tissue. Your hands, stomach, and kidneys are further examples of complex organs. Although each organ is composed of a variety of different tissues, the tissues act together to accomplish a common goal.

Organ Systems

Organ systems are groups of organs that have related functions. Although your heart can be considered the focal point of blood transportation, it is but one of the many important organs that make up your circulatory system. The circulatory system consists of the heart, the arteries (which carry blood rich in oxygen and other nutrients to the tissues), the capillaries (the site of nutrient and waste exchange), and the veins (which carry wastes away from the cell). Table 4.1 outlines the levels of cell organization for the body's important organ systems.

Organs *are structures composed of different tissues specialized to carry out a specific function.*

Organ systems *are groups of organs that have related functions. Organ systems often interact.*

Figure 4.3

Human organ systems.

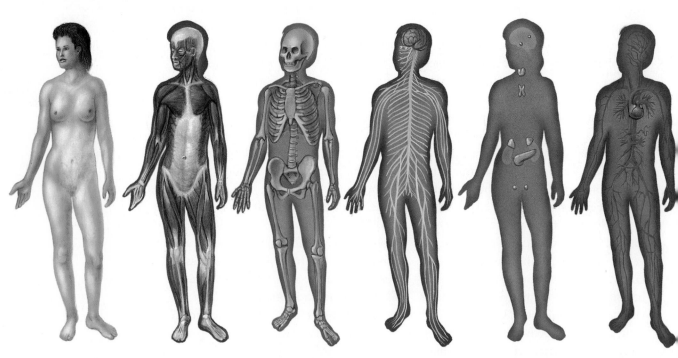

| Integumentary system | Muscular system | Skeletal system | Nervous system | Endocrine system | Circulatory system |

Table 4.1 Levels of Cell Organization

Organ System	Organs	Tissues
Nervous	brain, spinal cord, eye, ear, peripheral nerves	nerve, connective, epithelial
Excretory	kidney, bladder, ureter, urethra	epithelial, nerve, connective, muscle
Circulatory	heart, blood vessels	epithelial, nerve, connective, muscle
Digestive	esophagus, stomach, intestines	epithelial, nerve, connective, muscle
Reproductive	testes, vas deferens, ovary, uterus, Fallopian tubes, glands	epithelial, nerve, connective, muscle
Respiratory	lungs, windpipe, blood vessels	epithelial, nerve, connective, muscle
Endocrine	pancreas, adrenal glands, pituitary	epithelial, nerve, connective

Organ systems can be classified in many different ways. Part of the difficulty in establishing a classification system stems from the way in which body systems interact with each other. For example, the body's circulatory system would not be able to function properly if the respiratory system did not provide adequate gas exchange. The functions of the circulatory and respiratory systems are so intertwined that some scientists choose to refer to a single cardiopulmonary system.

A second problem in classifying organ systems arises because some organs are classified according to anatomy rather than function. Consider the kidneys and large intestine, which both remove wastes from the body. The kidneys remove wastes from the blood, while the large intestine concentrates and stores undigested matter. Although the kidney is classified as part of the excretory system, the large intestine is often described as the last segment of the digestive system, even though it does not digest food. Figure 4.3 shows an overview of human organ systems.

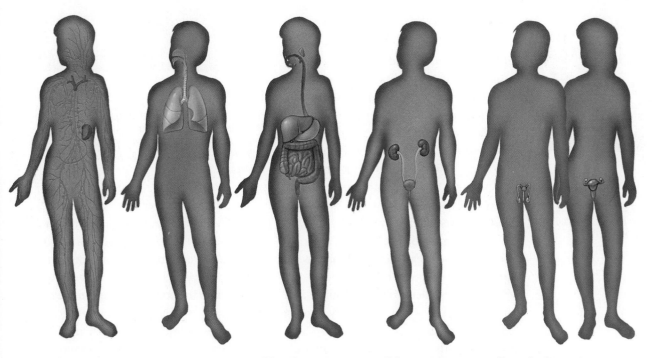

Lymphatic system　　**Respiratory system**　　**Digestive system**　　**Urinary system**　　**Reproductive system**

MONITORING ORGANS

The first X-ray photograph was taken in 1895 when the German physicist Wilhelm Roentgen used high-energy electromagnetic waves to take a picture of his wife's hand. X rays pass through soft tissue like muscle, but are absorbed by denser bone.

Walter Cannon was the first person to harness X rays to diagnose soft tissue. Cannon developed a stain that contained bismuth, a nontoxic mineral that is opaque to X rays. Because the X ray could not penetrate the stained organ, the organ became visible as a white image on a black background. For the first time, organ structures could be observed without surgery.

Today, scientists have combined X-ray technology with computer technology to view body organs in ways that Roentgen and Cannon never dreamed possible. Computer-assisted tomography, or CAT scans, allow doctors to view many different X rays from numerous angles. The computer interprets the X-ray images and reconstructs them to provide three-dimensional representations of any body organ.

In the CAT scan procedure, the X-ray machine rotates around the patient, taking hundreds of individual pictures. The images are stored in a computer along with their location and angle. The computer can reassemble the pictures to provide thin cross-sectional views, which the physician can rotate 180°. The computer organizes the pictures automatically, permitting three-dimensional imaging; the organ can then be viewed section by section.

The imaging of the CAT scan is so accurate that it can detect abrasions as small as one millimeter. The scanner can also distinguish between gases, fluids, and solid tissues, and is able to identify tumors imbedded in the brain or liver. CAT scans are particularly useful as a diagnostic tool

to assess head injuries, which can often be life threatening because blood masses, called *hematomas*, impair the flow of blood through the blood vessels of the brain.

FRONTIERS OF TECHNOLOGY: NUCLEAR MEDICINE

Nuclear imaging is a valuable diagnostic tool that allows doctors to view a beating heart or detect bone cancer without resorting to surgery. Unlike the CAT scan, which uses external radiation to produce an image, nuclear imaging measures the radiation emitted from within the body. Nuclear imaging also provides information about the function of the organ as opposed to its structure.

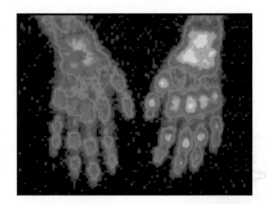

Figure 4.6

Hands of a person with extensive rheumatoid arthritis. The arthritic joints appear as brighter areas. The image records the distribution and intensity of gamma rays emitted by a tiny amount of radioisotope injected into the patient.

Nuclear imaging employs radioisotopes much like X rays use opaque dyes to identify organs. Radioisotopes, now called *radionuclides*, are unstable atoms that emit rays of energy. The radionuclides are injected into the body and collect in the target organs. A scanner, called a gamma camera, records the release of the energy from the radionuclides and produces a picture.

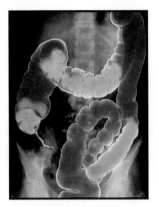

Figure 4.4

The penetrating properties of X rays, combined with stains, are used to monitor the state of internal organs. This X ray shows the large intestine.

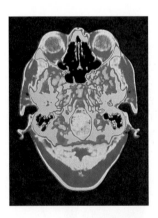

Figure 4.5

CAT scan of a cross section through the base of the skull, showing the eyeballs at the top.

10 Locate the liver near the anterior of the abdominal cavity. Using a probe, lift the lobes and locate the saclike gall bladder. Follow the thin duct from the gall bladder to the coiled small intestine. Bile salts, produced in the liver, are stored in the gall bladder. The bile duct conducts the fat-emulsifying bile salt to the small intestine.

e) How many lobes does the liver have?

f) Describe the location of the gall bladder.

11 Locate the J-shaped stomach beneath the liver. Using forceps and a probe, lift the stomach and locate the esophagus near the anterior junction. Locate the small intestine at the posterior junction of the stomach. The coiled small intestine is held in place by *mesentery*, a thin, somewhat transparent connective tissue. Note the blood vessels that transport digested nutrients from the intestine to the liver.

12 Using a dissecting needle and forceps, lift the junction between the stomach and small intestine, removing supporting tissue. Uncoil the junction and locate the creamy-white gland called the pancreas. The pancreas produces a number of digestive enzymes and a hormone called insulin, which helps regulate blood sugar.

g) Describe the appearance of the pancreas.

13 Locate the spleen, the elongated organ found around the outer curvature of the stomach. The spleen stores red and white blood cells. The spleen also removes damaged red blood cells from the circulatory system.

14 Using a scalpel, remove the stomach from the pig by making transverse cuts near the junction of the stomach and the esophagus, and near the junction of the stomach and small intestine. Make a cut along the mid-line of the stomach, open the cavity, and rinse. View the stomach under a dissecting microscope.

h) Describe the appearance of the inner lining of the stomach.

Laboratory Application Questions

1 State the function of the following organs.

Organ	Function
Stomach	
Liver	
Small intestine	
Gall bladder	
Pancreas	
Large intestine	
Spleen	

2 What is the function of the mesentery?

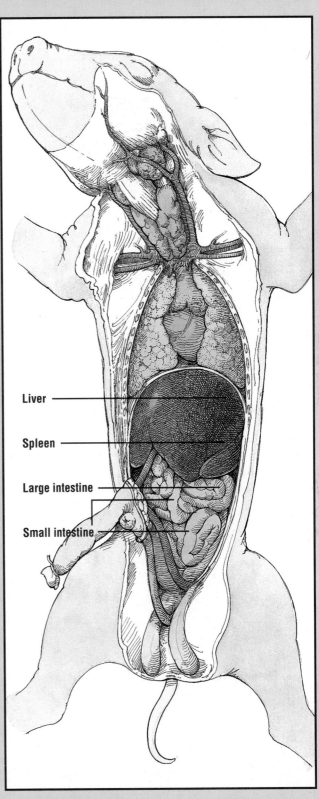

Abdominal cavity and thoracic cavity of the fetal pig. Organs of the digestive system and circulatory system are highlighted in the diagram.

LABORATORY

LAB #2: FETAL PIG DISSECTION

Objective

To study the thoracic cavity and urogenital system of the fetal pig.

Materials

preserved pig	dissecting tray
scissors	dissecting pins
scalpel	probe
forceps	ruler
dissecting gloves	dissecting microscope or hand lens
lab apron	safety goggles

Safety Notes

- **Wear safety goggles and an apron at all times.**
- **Wear plastic gloves when performing a dissection to prevent any chemicals from coming in contact with your skin.**
- **Fetal pigs are preserved in a formalin-based or other preservative solution. Wash all splashes off your skin and clothing immediately. If you get any chemical in your eyes, rinse for at least 15 min.**
- **Work in a well-ventilated area.**
- **When you are finished the activities, clean your work area, wash your hands thoroughly, and dispose of all specimens, chemicals, and materials as instructed by your teacher.**

Procedure

Part I: Thoracic Cavity

1 Carefully fold back the flaps of skin that cover the thoracic cavity. You may use dissecting pins to attach the ribs to the dissecting tray.

   a) What organs are found in the thoracic cavity?

2 Locate the heart. Using forceps and a dissection probe, remove the *pericardium* from the outer surface of the heart. The large blood vessel that carries blood from the liver to the right side of the heart is called the *inferior vena cava*. (The right side refers to the pig's right side.)

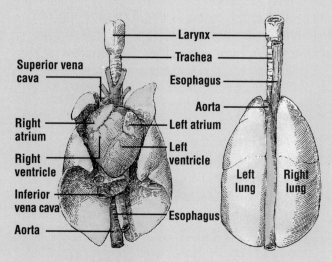

(a) Ventral view of heart and lungs.   (b) Dorsal view of heart and lungs.

3 Blood from the head enters the right side of the heart through the *superior vena cava*. Both the superior and inferior venae cavae are considered to be veins because they bring blood to the heart.

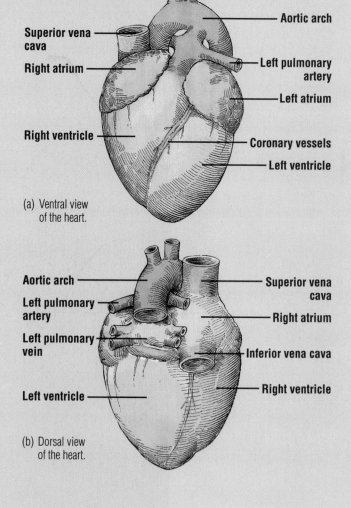

(a) Ventral view of the heart.

(b) Dorsal view of the heart.

4 Trace blood flow through the heart. Blood entering the right side of the heart collects in the *right atrium*. Blood from the right atrium is pumped into the *right ventricle*. Upon contraction of the right ventricle, blood flows to the lungs by way of the *pulmonary artery*. Arteries carry blood away from the heart. Blood, rich in oxygen, returns from the lungs by way of the pulmonary veins and enters the left atrium. Blood is pumped from the left atrium to the left ventricle and out the *aorta*.

5 Make a lateral incision across the heart and expose the heart chamber.

   b) Compare the size of the wall of a ventricle with that of an atrium.

   c) Why does the left ventricle contain more muscle than the right ventricle?

6 Locate the spongy lungs on either side of the heart and the *trachea* leading into the lungs.

   d) Why do the lungs feel spongy?

7 Place your index finger on the trachea and push downward.

   e) Describe what happens.

   f) What function do the cartilaginous rings of the trachea serve?

Part II: Urogenital System

8 Using scissors, remove the intestines and what remains of the stomach.

9 Refer to the diagram of the thoracic cavity and urogenital system and locate the kidney. Using forceps and a scalpel, carefully remove the fat deposits that surround the kidney.

   g) Describe the shape and color of the kidneys.

10 Locate the thin tube leading from the kidneys. The *ureter* carries urine from the kidney to the bladder. Cut into the kidney and note the large number of tubules.

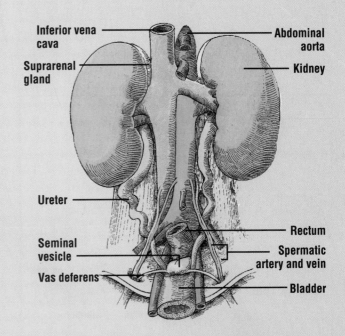

Urinary system of a fetal pig.

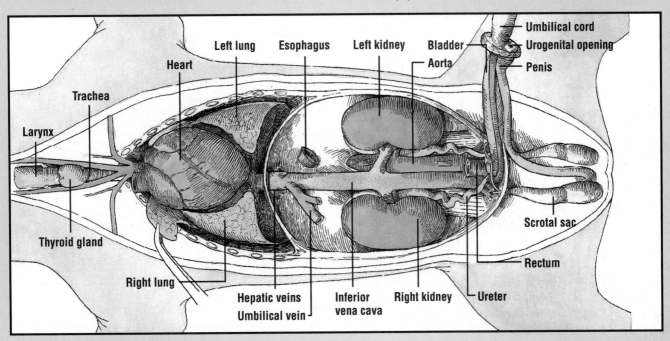

Thoracic cavity and urogenital system.

If your pig is a male, refer to steps 11 and 12. If your pig is a female, refer to steps 13 and 14. Make sure that you also view the organs of a pig of the opposite sex from your specimen.

11 Use the diagram of the male urogenital system to locate the testes, which produce the sperm cells. If your fetal pig is advanced in development, the testes may have descended into the scrotum; however, they will probably be found in the *inguinal canal*. Like the ovaries, the testes develop inside the body cavity. The lower temperatures of the scrotum promote the proper development of sperm cells.

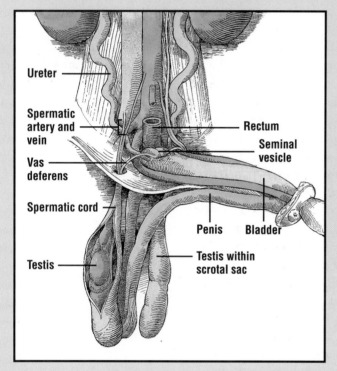

Urogenital system of a male pig.

12 Use the diagram to locate the vas deferens, which conducts sperm cells from the testes to the urethra.

13 Use the diagram of the female reproductive system to locate the ovaries, which produce the egg cells. The ovaries can be found immediately posterior to the kidneys.

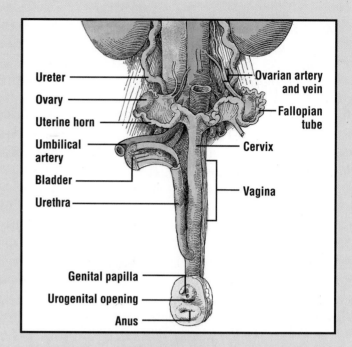

Ventral view of the female reproductive system.

14 Locate the Fallopian tubes leading from the ovaries. The Fallopian tubes are supported by the broad ligaments. Follow the Fallopian tubes, which meet to form the uterus, or womb. The Fallopian tube is the site of fertilization. Once fertilized, the egg travels to the uterus, the site of embryo and fetal development. Locate the vagina and follow the canal into the uterus. The constriction that marks the division between the vagina and uterus is known as the *cervix*.

Laboratory Application Questions

1 State the function of the following organs.

Organ	Function
Heart	
Kidney	
Ureter	
Urethra	
Testes	
Ovaries	
Uterus	

2 How is the abdominal cavity separated from the thoracic cavity?

3 Indicate why the male reproductive system is often referred to as the *urogenital system* while the female's is not. ■

HOMEOSTASIS AND CONTROL SYSTEMS

Your body works best at 37°C, with a 0.1% blood sugar level, and at a blood pH level of 7.35. However, the external environment does not always provide the ideal conditions for life. Air temperatures in Canada can fluctuate between −40°C and +40°C. Foods are rarely 0.1% glucose and rarely have a pH of 7.35. You also place different demands on your body when you take part in various activities, such as playing racketball, swimming, or digesting a large meal. Your body systems must adjust to these variations to maintain a reasonably constant internal environment. The system of active balance requires constant monitoring or feedback about body conditions. Information about blood sugar, body temperature, blood pressure, and oxygen levels, to name a few, are relayed to a coordinating center once they move outside the normal limits. From the coordinating center, regulators bring about the needed adjustments. An increase in the heart rate during exercise or the release of glucose from the liver to restore blood sugar levels are but a few of the adjustments made by regulators.

> **BIOLOGY CLIP**
> Sweating is an important homeostatic mechanism. Because evaporation requires heat, your body cools when you perspire. Following extreme exercise, sweat is produced all over the surface of your body. Fear or nervousness produces sweat mainly on the palms and soles.

The term **homeostasis** is most often used to describe the body's attempt to adjust to a fluctuating external environment. The word is derived from the Greek words *homoios*, meaning "similar" or "like," and *stasis*, which means "standing still." The term is appropriate because the body maintains a constant balance, or steady state, through a series of monitored adjustments.

Special receptors located in the organs of the body signal a coordinating center once an organ begins to operate outside its normal limits. The coordinating center relays the information to the appropriate regulator, which helps restore the normal balance. For example, pressure receptors located in the arteries of your neck become distended when blood pressure exceeds normal limits. A nerve is excited and a message is sent to the brain, which relays the information, by way of another nerve, to the heart. The pace and strength of heart contractions are reduced, thereby lowering blood pressure. A system of turning down or turning up the force and rate of heart contractions is used to maintain homeostasis.

Homeostasis *is a process by which a constant internal environment is maintained despite changes in the external environment.*

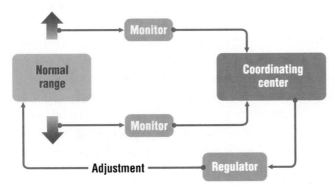

Figure 4.10

A control system.

FRONTIERS OF TECHNOLOGY: ARTIFICIAL BODY PARTS

In the 1987 movie *RoboCop*, a police officer is killed in the line of duty. Scientists use his remains to construct a cyborg—part man, part machine— that is virtually indestructible. Although the film may have been ahead of its time in showing how science can rebuild human body parts, it did reflect developments in a rapidly growing field.

Artificial hips became practical during the late 1960s, elbows during the mid-1970s, and wrists during the late 1970s. People with rheumatoid arthritis and degenerative bone disorders have been given new hope. Microelectronics has provided a new generation of artificial limbs for amputees. Special electrodes set in artificial limbs can be activated either subconsciously or consciously by the user.

Skin substitutes have provided a temporary solution for burn victims. Unlike skin transplants from pigs or cadavers, new artificial grafts do not present rejection problems. A polymer film, applied to burn victims as a gauze, provides a good seal, protecting the living cells from infections and offering a moist surface to promote new cell growth.

The University of Western Ontario has established itself as a world leader in the development of artificial eyes. Researchers placed a wafer containing 65 electrodes on the visual cortex of a blind man. By connecting wires from the electrodes to a computer, these researchers were able to produce visual images for the man, the first flashes of light he had seen since an accident blinded him. By altering the firing pattern, researchers were then able to produce letters and objects of various shapes. This experiment may well be the first step toward providing artificial sight.

Artificial organs have a long history. Dialysis machines have enabled individuals with severe kidney damage to filter wastes from their blood. Artificial hearts have also been used, but with much less success. Recently, drugs such as cyclosporin have increased success. Artificial voice boxes have been developed for people whose larynx has been removed because of cancer. In this operation, a 3 cm rubber tube is fitted into the person's windpipe. A small connection is made between the windpipe and the esophagus. A one-way valve permits air to enter the windpipe and prevents it from leaking back. To speak, the person closes a small hole in the neck and exhales. Air passes into the esophagus and makes a sound. Although the sound is somewhat different from that made by the larynx, it does allow the person to communicate.

RESEARCH IN CANADA

Investigating
Wheelchair Safety

An estimated 84,000 Canadians (about one in three hundred people) go about their daily routines from wheelchairs, but in many respects the wheelchair can be both friend and foe. The wheelchair provides opportunities for enhanced mobility, but it also presents opportunities for accidents. According to statistics analyzed by a team of Halifax investigators, an estimated 50 Americans die from wheelchair-related accidents and an estimated 30,000 Americans are treated in emergency wards for associated injuries each year. Although no records of wheelchair accidents have been recorded in Canada, the Halifax group estimates that about 1/10th the United States number would be reasonable.

The research team from Dalhousie University and the Technical University of Nova Scotia is investigating wheelchairs to improve the safety for an increasingly active group of citizens. Lee Kirby, a specialist in Physical Medicine and Rehabilitation, heads the team, which includes epidemiologists Murray Brown and Susan Kirkland, engineers Peter Gregson, Robert Baird, Adam Bell, and Biman Das, consumer advocate Alice Loomer, and a number of enthusiastic students.

The research team is in the early stages of developing and refining a registry of wheelchair users in Nova Scotia in order to establish a data base that will record the types of wheelchair-related accidents and the causes of the accidents over a number of years.

DR. LEE KIRBY

SOCIAL ISSUE:
Artificial Organs

In 1966, a plastic artificial heart was implanted in a human at a hospital in Houston, Texas. An artificial pancreas was developed by researchers at the University of Toronto. Gordon Murray, a Toronto researcher, developed the first artificial kidney to be used successfully in North America. Artificial limbs, eyes, and skin are further examples of synthetic body organs that have been implanted, with varying degrees of success, in humans.

Although the prospects of successful implantation of organs may have sounded like science fiction two decades ago, many people with disabilities today are able to lead longer and more comfortable lives as a result of artificial organ technology. However, the union of science and technology raises many ethical issues that must be confronted by society. Are there boundaries to artificial organ research? What are the implications if artificial organs one day exceed the capabilities of nature's organs?

Statement:

Research into artificial organs should be carefully controlled by an impartial committee representing community standards.

Point

- If artificial organs that exceed the capacity of nature's organs ever become available, millions of people who just want to enhance their physical or mental abilities will demand to have them. The boundaries of research must be defined by members of society outside the medical and health-care establishment to ensure that developments are acceptable to the community.

- The cost of producing artificial organs and giving them to patients could bankrupt the health-care system. Only the rich will be able to afford them. Can our society as a whole bear the costs of research from which only a few people will benefit?

- The body was not designed to receive artificial organs. The side effects for patients could be serious. It can take years to discover that a procedure that was thought to be safe is actually dangerous.

Counterpoint

- Laws can always be passed to ensure that artificial organs are only available to those who genuinely need them. But if research is too strictly controlled, the good and useful work will never be done.

- The argument of cost ignores the long-term benefits of artificial organ research in improving the quality and length of life for people who might otherwise remain disabled or even die.

- Caution is always necessary, but the benefits outweigh the dangers. For example, not too many years ago artificial kidneys and heart pacemakers seemed impractical and even dangerous. Now they are an accepted part of medical treatment. We must explore the opportunities that technology offers us.

Research the issue.
Reflect on your findings.
Discuss the various viewpoints with others.
Prepare for the class debate.

CHAPTER HIGHLIGHTS

- Tissues are groups of similarly shaped cells that work together to carry out a similar function.
- Epithelial tissue is a covering tissue that protects organs and lines body cavities.
- Connective tissue is a group of cells that provides support and holds various parts of the body together.
- Organ systems contain organs that have related functions. Organ systems often interact.

- Nuclear imaging techniques use radioisotopes to view organs and tissues of the body.
- Nuclear magnetic resonance techniques employ magnetic fields and radio waves to determine the behavior of molecules in soft tissue.
- Homeostasis is the system by which a constant internal environment is maintained despite changes in the external environment.

APPLYING THE CONCEPTS

1 Describe the advantages associated with cell specialization.
2 Using epidermal tissue as an example, explain the relationship between cell shape and tissue function.
3 List seven organ systems and provide at least two examples of organs that belong to those systems.

4 Explain the advantages of the CAT scan over conventional X rays.
5 Explain the concept of homeostasis by describing how your body adjusts to cold environmental temperatures.

CRITICAL-THINKING QUESTIONS

1 The cost of nuclear medicine, CAT scans, and artificial body parts is extremely high and places a heavy financial burden on the health care system. Can we continue to support such expensive research projects?

2 Liposuction is a fat-reducing technique in which fat cells are mechanically sucked out of the lower layer of the skin. Fees for this kind of surgery can range from $750 to over $4000, depending on the type of procedure. Discuss the moral and economic aspects of medical procedures that are designed to improve appearance.

ENRICHMENT ACTIVITIES

1 Research career opportunities in the field of medical engineering.
2 Locate photographs of X rays, NMR, and nuclear imaging. Which organs can you identify?

3 Survey dissection guides for chordates other than a pig. What anatomical similarities and differences can you observe?
4 Research Leonardo da Vinci's pioneering work in the field of human anatomy.

CHAPTER 5

Digestion

IMPORTANCE OF DIGESTION

Unlike plants, which make their own food, heterotrophs must consume organic compounds to survive. These organic compounds, called **nutrients,** are digested in the gastrointestinal tract, absorbed, and transported by the circulatory system to the cells of the body. Once inside the cells, the nutrients supply the body with energy or the raw materials for the synthesis of essential chemical compounds. The chemical compounds are used for growth, maintenance, and tissue repair.

The digestive system is responsible for the breakdown of large, complex organic materials into smaller components that are utilized by the tissues of the body. Every organ system depends on the digestive system for nutrients, but the digestive system also depends on other organ systems. Muscles and bones permit the ingestion of foods. The circulatory system transports oxygen and other needed materials to the digestive organs. The circulatory system complements the digestion process by transporting the absorbed foods to the tissues of the body. The nervous and endocrine systems coordinate and regulate the actions of the digestive

Nutrients *are chemicals that provide nourishment. Nutrients provide energy or are assimilated to form protoplasmic structures.*

Figure 5.1

The sight and aroma of food can activate the salivary glands, setting the digestive process into action.

organs. In many respects, the study of the digestive organs is a study of the interacting body systems.

ORGANS OF DIGESTION

In this chapter you will study four components of digestion. *Ingestion* involves the taking in of nutrients. *Digestion* involves the breakdown of organic molecules into smaller complexes. *Absorption* involves the transport of digested nutrients to the tissues of the body. *Egestion* involves the removal of materials from the food that the body cannot digest. The assimilation of nutrients within the cells of the body is dealt with in other chapters.

The digestive tract, or *alimentary canal*, is an open-ended muscular tube. Measuring between 6.5 and 9 m in adults, the digestive tract stores and breaks down organic molecules into simpler components. Physical digestion begins in the mouth, where food is chewed into a bolus (the Greek word for "ball"). Salivary gland secretions activate the taste buds and lubricate the passage of food. **Amylase enzymes** contained in the saliva break down starches to smaller-chain carbohydrates called *dextrins*. The watery fluids produced by the salivary glands also dissolve food particles. Flavor can only be detected if food particles are in solution. Food particles dissolved in solution penetrate the cells of the taste buds located along the tongue. Different types of

nerve cells respond to specific flavors. For example, sweet flavors are detected by taste buds near the tip of the tongue, while bitter flavors are detected by taste buds near the middle of the tongue. You can see the importance of dissolving foods by drying your tongue and then placing a few grains of sugar or salt on it. You will not detect any flavor until the crystals dissolve.

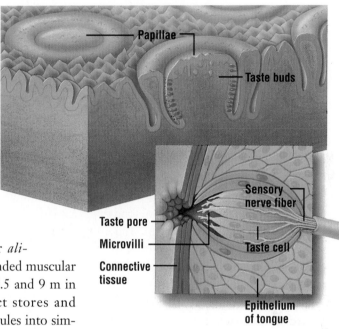

Once swallowed, food travels from the mouth to the stomach by way of the **esophagus.** The bolus of food stretches the walls of the esophagus, activating smooth muscles, which set up waves of rhythmic contractions called **peristalsis.** Involuntary peristaltic contractions move food along the entire gastrointestinal tract. Voluntary control of food movement is only exercised during swallowing and during the last phase, egestion. Peristaltic action will move food or fluids from the esophagus to the stomach even if you stand on your head.

> **BIOLOGY CLIP**
> Salivary glands become swollen in someone infected with the mumps virus.

Amylase enzymes *hydrolyze (break down) complex carbohydrates.*

The **esophagus** *is a tube that carries food from the mouth to the stomach.*

Figure 5.2

Taste buds are located along the tongue.

Contraction of smooth muscle

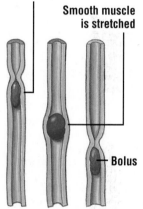

Phase 1   Phase 2   Phase 3

Figure 5.3

Rhythmic contractions of the smooth muscle move food along the digestive tract.

Peristalsis *is the rhythmic, wavelike contraction of smooth muscle that moves food along the gastrointestinal tract.*

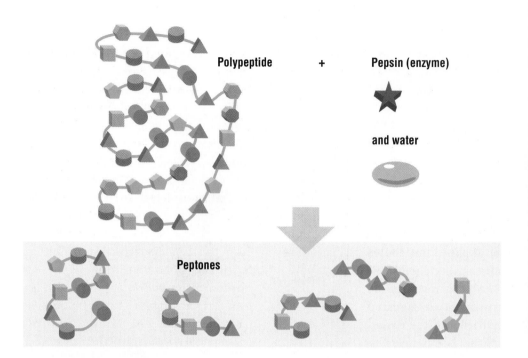

Polypeptide + **Pepsin (enzyme)**

and water

Peptones

Figure 5.4

Proteins are composed of many amino acids. Digestion reactions are referred to as *hydrolysis reactions*. The water molecule is used to break bonds within large organic molecules.

STOMACH

The stomach is the site of food storage and initial protein digestion. **Sphincter** muscles regulate the movement of food to and from the stomach. These circular muscles act like the drawstrings of a purse. The contraction of the *cardiac sphincter* closes the opening to the stomach; its relaxation allows food to enter. A second sphincter, the *pyloric sphincter*, regulates the movement of foods and stomach acids to the small intestine.

The J-shaped stomach can store about 1.5 L of food. Millions of secretory cells line the inner wall of the stomach. Approximately 500 mL of gastric fluids are produced following a large meal. Mucous cells secrete a protective coating; parietal cells secrete hydrochloric acid. Peptic cells secrete a protein-digesting enzyme called *pepsinogen*. The active form of the enzyme, called **pepsin,** breaks proteins into peptones, shorter chains of amino acids.

The storage of a highly corrosive acid (hydrochloric acid) and a protein-digesting enzyme in a stomach that is composed of cells made of protein poses a major engineering problem. The pH inside the stomach normally ranges between 2.0 and 3.0, but may approach pH 1.0. Acids with a pH of 2.0 can dissolve fibers in a rug! Although the mechanism of HCl formation is not fully understood, a model has been proposed to explain its role in the digestive process. As shown in Figure 5.5, HCl forms in the *lumen*, or gut cavity. The protective mucous lining prevents the HCl from dissolving cells. In turn, the HCl destroys invading microbes.

Sphincters are constrictor muscles that surround a tube-like structure.

Pepsin is a protein-digesting enzyme produced by the cells of the stomach.

Figure 5.5

Proposed mechanism to explain how HCl is synthesized.

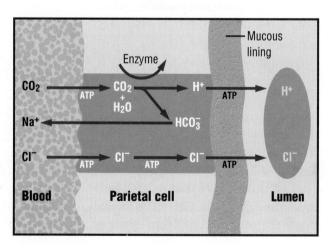

other functions. Similarly, wastes produced in all cells must be transported to cells specialized for waste removal. Intercellular transport systems allow cells to specialize and work together.

The importance of a circulatory system can be emphasized by examining higher organisms, which have three distinct cell layers. Because the middle layer of cells cannot receive nutrients directly from seawater, an efficient circulatory system is required. Your circulatory system is composed of 96 000 km of blood vessels that supply the estimated 60 trillion cells of your body with nutrients. No cell is further than two cells away from a blood vessel. Every minute, five liters of blood cycle from the heart to the lungs, where oxygen is received, and then back to the heart. The heart pumps oxygen-rich blood to the tissues of the body. Here, oxygen and nutrients are given up, and wastes return to the heart. Glucose, the primary fuel of cells, is broken down by oxygen into carbon dioxide and water. The conversion of high-energy glucose into low-energy compounds releases energy, which is used by cells to build new materials, repair pre-existing structures, or for a variety of other energy-consuming reactions. It is apparent that the conversion of cell energy depends on an adequate supply of glucose and oxygen. It is here that the value of the circulatory system becomes evident. The more effective the delivery system, the greater the energy available, as shown in Table 6.1.

Your circulatory system carries nutrients to cells, wastes away from cells, and chemical messengers from cells in one part of the body to distant target tissues. It distributes heat throughout the body and, along with the kidneys, maintains acceptable levels of body fluid. Defense against invading organisms is also associated with the circulatory system.

Table 6.1 Various Types of Transport Systems

Organism	Function	Oxygen used (μg/g body mass/h)
Jellyfish	Two cell layers, bathed by seawater	5
Earthworm	Three cell layers, 5 primitive hearts	60
Cockroach	Open circulatory system	450
Goldfish	Closed circulatory system, two-chambered heart	420
Mouse (resting)	Closed circulatory system, four-chambered heart	2000

Whether simple or complex, a circulatory system is vital to an organism's survival. In this chapter, you will investigate human circulation. Then, in the next chapter, you will study the circulating fluid, the blood, in greater detail.

EARLY THEORIES OF CIRCULATION

The ancient Greeks believed that the heart was the center of human intelligence, an "innate heat" that generated four humors: black and yellow bile, phlegm, and blood. Galen, the personal physician of Roman emperor Marcus Aurelius in the second century A.D., influenced early physiology. Although he provided many enlightening theories, Galen is best known for steering scientists in the wrong direction. Galen believed that blood did not circulate. Although he believed that blood might ebb like the tides, he never thought of the heart as a pump. Galen's theory was generally accepted until the 17th century.

Some science historians have suggested that his failure to consider the pumping action of the heart could be attributed to a lack of a technical model: the water pump had not been invented when Galen applied his theory.

William Harvey (1578–1657), the great English physiologist, questioned Galen's hypothesis. Harvey, like many Europeans during that period, was influenced by the astronomer Galileo. Galileo's new principles of dynamics became the foundation of Harvey's work. By applying Galileo's theories of fluid movement to that of blood, Harvey reasoned that blood must circulate.

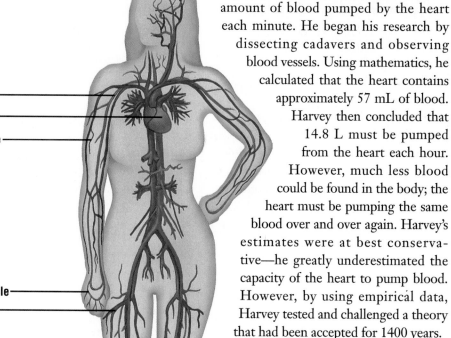

Artery
Heart
Venule
Arteriole
Vein

Figure 6.2
William Harvey (1578–1657).

Figure 6.3
Circulatory system.

> **B I O L O G Y   C L I P**
> Evidence from an ancient Egyptian papyrus discovered in the 19th century suggests that the Egyptians correctly mapped the flow of blood from the heart 3300 years before William Harvey.

Harvey attempted to quantify the amount of blood pumped by the heart each minute. He began his research by dissecting cadavers and observing blood vessels. Using mathematics, he calculated that the heart contains approximately 57 mL of blood. Harvey then concluded that 14.8 L must be pumped from the heart each hour. However, much less blood could be found in the body; the heart must be pumping the same blood over and over again. Harvey's estimates were at best conservative—he greatly underestimated the capacity of the heart to pump blood. However, by using empirical data, Harvey tested and challenged a theory that had been accepted for 1400 years.

BLOOD VESSELS

Although William Harvey was convinced that blood must pass from the arteries to the veins, there was no visible evidence of how this was accomplished. The solution came four years after his death, when an Italian physiologist, Marcello Malpighi, used a microscope to observe the tiniest blood vessels, the capillaries (from the Latin, meaning "hairlike"). Harvey had speculated that blood vessels too small to be seen by the human eye might explain how blood circulates. Now, with Malpighi's observations, Harvey's theory of circulation was confirmed.

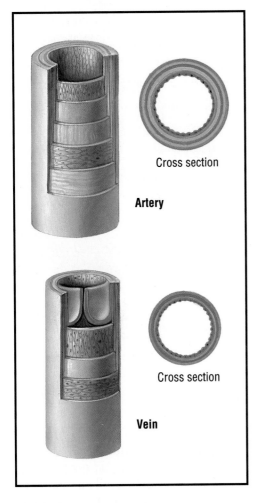

Figure 6.4

Simplified diagrams of an artery and a vein. Arteries have strong walls capable of withstanding great pressure. The middle layer of arteries contains both muscle tissue and elastic connective tissue. The low-pressure veins have a thinner middle layer.

Arteries

Arteries are the blood vessels that carry blood away from the heart. They have thick walls composed of three distinct layers. The outer and inner layers are primarily rigid connective tissue. The middle layer is made up of muscle fibers and elastic connective tissue. Every time the heart contracts, blood surges from the heart and enters the arteries. The arteries stretch to accommodate the inrush of blood. The **pulse** you can feel near your wrist and on either side of your neck is created by the changes in the diameter of the artery near the surface of your body following heart contractions. Heart contraction is followed by a relaxation phase. During this phase, pressure drops and elastic fibers in the walls of the artery recoil. It is interesting to note that the many cells of the artery are themselves supplied with blood vessels that provide nourishment.

A birth defect or injury can cause the inner wall of the artery to bulge. Known as an **aneurysm,** this condition is infrequent in young people, but can lead to serious problems. In much the same way as the weakened wall of an inner tube begins to bulge, the weakened segment of the artery protrudes as blood pulses through. The problem escalates as the thinner wall offers less support and eventually ruptures. A weakened artery in the brain is one of the conditions that can lead to a stroke. Cells die because less oxygen and nutrients are delivered to the tissues.

Blood from the arteries passes into smaller arteries, called **arterioles.** The middle layer of arterioles is composed of elastic fibers and smooth muscle. The diameter of the arterioles is regulated by nerves from the autonomic nervous system. A sympathetic nerve impulse causes smooth muscle in the arterioles to contract, reducing the diameter of the blood vessel. This process is called **vasoconstriction.** Vasoconstriction decreases blood flow to tissues. Relaxation of the smooth muscle causes dilation of the arterioles, and blood flow increases. This process is called **vasodilation.** Vasodilation increases the delivery of nutrients to tis-

Arteries are high-pressure blood vessels that carry blood away from the heart.

A pulse is caused by blood being pumped through an artery.

An aneurysm is a fluid-filled bulge found in the weakened wall of an artery.

Arterioles are fine branches from arteries.

Vasoconstriction is the narrowing of a blood vessel. Less blood goes to the tissues when the arterioles constrict.

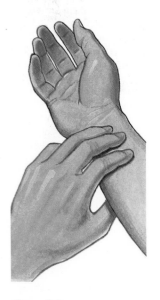

Figure 6.5

Arteries near the surface of the body provide a pulse. Galen believed that 27 different types of pulses could be produced; however, he never related pulse to the heart or to the movement of blood.

sues. This, in turn, increases the capacity of the cells in that localized area to perform energy-consuming tasks.

Precapillary sphincter muscles regulate the movement of blood from the arterioles into capillaries. Blushing is caused by vasodilation of the arterioles leading to skin capillaries. Red blood cells close to the surface of the skin produce the pink color. Have you ever noticed someone's face turn a paler shade when they are frightened? The constriction of the arteriolar muscle diverts blood away from the outer capillaries of the skin toward the muscles. The increased blood flow to the muscles provides more oxygen and glucose for energy to meet the demands of the "flight-or-fight" response.

Arterioles leading to capillaries open only when cells in that area require blood. It has been estimated that 200 L of blood would be required if the arterioles could not selectively open and close gateways to the capillaries. Although the majority of brain capillaries remain open, as few as 1/50 of the capillaries in resting muscle remain open.

Fat in the Arteries: Atherosclerosis

Anyone who has ever washed dishes is aware of how fat floats on water. You may have noticed that when one fat droplet meets another they stick together and form a larger droplet. Unfortunately, the same thing can happen in your arteries. As fat droplets grow into larger and larger blockages, they slowly close off the opening of the blood vessel. Calcium and other minerals deposit on top of the lipid, forming a fibrous net of plaque. This condition, known as **atherosclerosis,** can narrow the artery to one-quarter of its original diameter and lead to high blood pressure. To make matters worse, blood clots, a natural *life-saving* property of blood, form around the fat deposits. As fat droplets accumulate, adequate amounts of blood and oxygen cannot be delivered to the heart muscle, resulting in chest pains.

Every year heart disease kills more Canadians than any other disease. Lifestyle changes must accompany any medical treatment. A low-fat diet, plus regular, controlled exercise are keys to prevention.

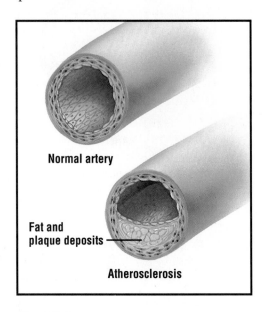

Normal artery

Fat and plaque deposits

Atherosclerosis

Figure 6.6

Fat deposits have narrowed the passageway.

Capillaries

The **capillary,** composed of a single layer of cells, is the site of fluid and gas exchange between blood and body cells. No cell is further than two cells away from a capillary, and many active cells, such as muscle cells, may be supplied by more than one capillary. Most capillaries extend between 0.4 and 1.0 mm, but have a diameter of less than 0.005 mm. The diameter is so small that red blood cells must travel through capillaries in single file.

The single cell layer, although ideal for diffusion, creates problems. Capillary beds are easily destroyed. High blood pressure or any impact, such as that caused by a punch, can rupture

Vasodilation *is the widening of the diameter of the blood vessel. More blood moves to tissues when arterioles dilate.*

Atherosclerosis *is a degeneration of the blood vessel caused by the accumulation of fat deposits along the inner wall.*

Capillaries, *tiny blood vessels that connect arteries and veins, are the site of fluid and gas exchange.*

the thin-layered capillary. Bruising occurs when blood rushes into the **interstitial spaces.**

Oxygen diffuses from the blood into the surrounding tissues through the thin walls of the capillaries. Oxygenated blood, which appears red in color, becomes a purple-blue color as it leaves the capillary. The deoxygenated blood collects in small veins called venules and is carried back to the heart. Some protein is also exchanged, but the process is believed to involve endocytosis and exocytosis rather than diffusion. Water-soluble ions and vitamins are believed to pass through the spaces between capillary cells. The fact that some spaces are wider than others may explain why some capillaries seem to be more permeable than others.

Veins

Capillaries merge and become progressively larger vessels, called **venules.** Unlike capillaries, the walls of venules are lined with smooth muscle. Venules merge into **veins,** which have greater diameter. Gradually the diameter of the veins increases as blood is returned to the heart. However, the very return of blood to the heart poses a problem. Blood flow through the arterioles and capillaries is greatly reduced. The passage of blood through incrementally narrower vessels reduces fluid pressure. By the time blood enters the venules, the pressure is between 15 mm Hg and 20 mm Hg. These pressures, however, are not enough to drive the blood back to the heart, especially from the lower limbs.

How, then, does blood get back to the heart? Let us return to William Harvey's experiments to answer that question. In one of his experiments, Harvey tied a band around the arm of one of his subjects, restricting venous blood flow. The veins soon became engorged with blood, and swelled. Harvey then placed his fin-

ger on the vein and pushed blood toward the heart. The vein collapsed. Harvey repeated the procedure, but this time he pushed the blood back toward the hand. Bulges appeared in the vein at regular intervals. What caused the bulges? Dissection of the veins confirmed the existence of valves.

The valves open in one direction, steering blood toward the heart. By attempting to push blood toward the hand, Harvey closed the valves, causing blood to pool in front of the valve. The pooling of blood caused the vein to become distended. However, by directing blood toward the heart, Harvey opened the valves, and blood flowed from one compartment into the next.

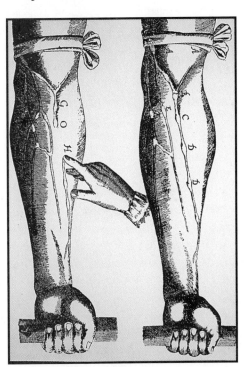

Skeletal muscles also aid venous blood flow. Venous pressure increases when skeletal muscles contract and push against the vein. The muscles bulge when they contract, thereby reducing the vein's diameter. Pressure inside the vein increases and the valves open, allowing blood to flow toward the heart. Sequen-

Figure 6.7

The one-way valves direct blood flow back to the heart. William Harvey's teacher, Hieronymus Fabricius, had already discovered the valves, but he did not fully investigate their function. He, like many others, still believed in Galen's ebb-and-flow theory.

Interstitial spaces *are the spaces between the cells.*

Venules *are small veins.*

Veins *carry blood back to the heart.*

tial contractions of skeletal muscle create a massaging action that moves blood back to the heart. This may explain why you feel like stretching first thing in the morning. It also provides clues as to why some soldiers faint after standing at attention for long periods of time. Blood begins to pool in the lower limbs. The movement of the leg muscles is required to move blood back to the heart.

The veins serve as more than just low-pressure transport canals—they are also important blood reservoirs. As much as 50% of your total blood volume can be found in the veins. During times of stress, venous blood flow can be increased to help you meet increased energy demands. Nerve impulses cause smooth muscle in the walls of the veins to contract, increasing fluid pressure. Increased pressure drives more blood to the heart, increasing heart filling.

Unfortunately, veins, like other blood vessels, are subject to problems. Large volumes of blood can distend the veins. In most cases, veins return to normal diameter, but if the pooling of blood occurs over a long period of time, the one-way valves are damaged. Without proper functioning of the valves, gravity carries blood toward the feet and greater pooling occurs. Surface veins gradually become larger and begin to bulge. The disorder is known as **varicose veins.** Although a genetic link to a weakness in the vein walls exists, lifestyle can accelerate the damage. Prolonged standing, especially with restricted movement, increases pooling of blood. Prolonged compression of the superficial veins in the leg can contribute to varicose veins.

Figure 6.8

Venous valves and skeletal muscle work together in a low-pressure system to move blood back to the heart.

Varicose veins *are distended veins.*

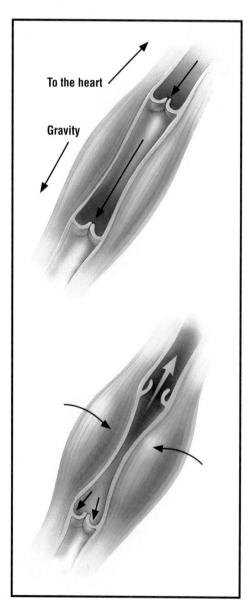

To the heart

Gravity

■ REVIEW QUESTIONS ■ ?

1 Why do multicellular animals need a circulatory system?
2 Explain the importance of William Harvey's theory that blood circulated?
3 How do arteries differ from veins?
4 What causes a pulse?
5 Why are aneurysms dangerous?
6 Define vasodilation and vasoconstriction.
7 Why are fat deposits in arteries dangerous?
8 What is the function of capillaries?
9 Fluid pressure is very low in the veins. Explain how blood gets back to the heart.

Blood samples can also be taken with the catheter to determine how much oxygen is in the blood in the different chambers. This tells the physician how well the blood is being oxygenated in the lungs. Low levels of oxygen in the left side of the heart can provide information about the teamwork of the circulatory and respiratory systems. The catheter can even be used to monitor pressures in each of the heart chambers.

SETTING THE HEART'S TEMPO

Heart or cardiac muscle differs from other types of muscle. Like skeletal muscle, cardiac muscle appears striated when viewed under a microscope. But, unlike skeletal muscle, cardiac muscle displays a branching pattern. The greatest difference stems from the ability of this muscle to contract without external nerve stimulation. Muscle with this ability is called **myogenic muscle.** This latter ability explains why the heart will continue to beat, at least for a short time, when removed from the body.

The remarkable capacity of the heart to beat can be illustrated by a simple experiment. A frog's heart is removed and placed in a salt solution that simulates the minerals found within the body. The heart is then sliced into small pieces. Incredible as it may seem, each of the pieces continues to beat, although not at the same speed. Muscle tissue from the ventricles follows a slower rhythm than muscle tissue from the atria. Muscle tissue closest to the entry port of the venae cavae has the faster tempo. The unique

> **BIOLOGY CLIP**
> Many old-time medical treatments are now being supported by science. Foxglove, a popular garden plant in England, has long been used in tea as a tonic. Scientists have found that the active ingredient in the plant, digitalis, initiates strong regular heart contractions and is now used to treat congestive heart failure.

nature of the heart becomes evident when two separated pieces are brought together. The united fragments assume a single beat. The slower muscle tissue assumes the tempo set by the muscle tissue that beats more rapidly.

The heart's tempo or beat rate is set by the **sinoatrial,** or **SA node.** This bundle of specialized nerves and muscle is located where the venae cavae enter the right atrium. The sinoatrial node acts as a pacemaker, setting a rhythm of about 70 beats per minute for the heart. Nerve impulses are carried from the pacemaker to other muscle cells by modified muscle tissue. Originating in the atria, the contractions travel to a second node, the *atrioventicular*, or *AV node*. The AV node serves as a conductor, passing nerve impulses along special tracts through the dividing septum toward the ventricles. Both right and left atria contract prior to the contraction of the right and left ventricles.

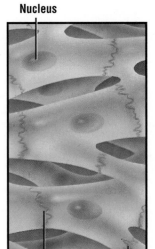

Nucleus

Cell

Figure 6.14

The heart is composed of cardiac muscle. The branching pattern is unique to cardiac muscle.

Myogenic muscle *tissue contracts without external nerve stimulation. Cardiac muscle sets a beat.*

The sinoatrial node *is the heart's pacemaker.*

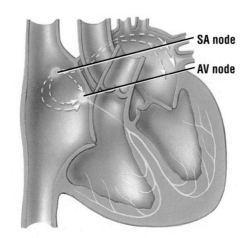

SA node

AV node

Figure 6.15

The pacemaker initiates heart contractions. Modified muscle tissue passes a nerve impulse from the pacemaker down the dividing septum toward the ventricles.

One of the greatest challenges for surgeons performing open-heart surgery is to make incisions at the appropriate location. A scalpel placed in the wrong spot could cut conducting fibers.

Electrical fields within the heart can be mapped by a device called the **electrocardiograph.** Electrodes placed on the body surface are connected to a recording device. The electrical impulses are displayed on a graph called an electrocardiogram. Changes in electrical current reveal normal or abnormal events of the cardiac cycle. The first wave, referred to as the P wave, monitors atrial contraction. The larger spike, referred to as the QRS wave, records ventricular contraction. A final T wave signals that the ventricles have recovered.

Doctors use electrocardiograph tracings to diagnose certain heart problems. A patch of dead heart tissue, for example, will not conduct impulses, and produces abnormal line tracings. By comparing the tracings, doctors are able to locate the area of the heart that is damaged.

The electrocardiograph is especially useful for monitoring the body's response to exercise. Stress tests are performed by monitoring a subject who is riding a sta-tionary bike or running on a treadmill. Some heart malfunctions remain hidden during rest, but can be detected during vigorous exercise.

Heart rate is influenced by autonomic or automatic nerves. Two regulatory nerves—the sympathetic and parasympathetic nerves—conduct impulses from the brain to the pacemaker. Stimulated during times of stress, the sympathetic nerve increases heart rate. This increases blood flow to tissues, enabling the body to meet increased energy demands. Conditions in which the heart rate exceeds 100 beats per minute are referred to as *tachycardia*. Tachycardia can result during exercise or from the consumption of such drugs as caffeine or nicotine. During times of relaxation, the parasympathetic nerve is stimulated. The parasympathetic nerve slows heart rate. The condition in which the heart beats very slowly is referred to as *bradycardia*.

*An **electrocardiograph** is an instrument that monitors the electrical activity of the heart.*

Figure 6.16

Electrocardiograph tracing of a single normal heartbeat.

Figure 6.17

An abnormal electrocardiograph tracing. Can you determine what has gone wrong?

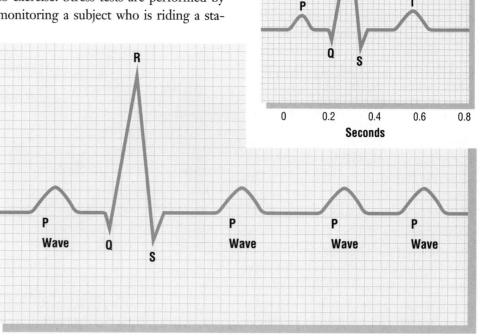

Unit Two:
Exchange of Matter and Energy in Humans

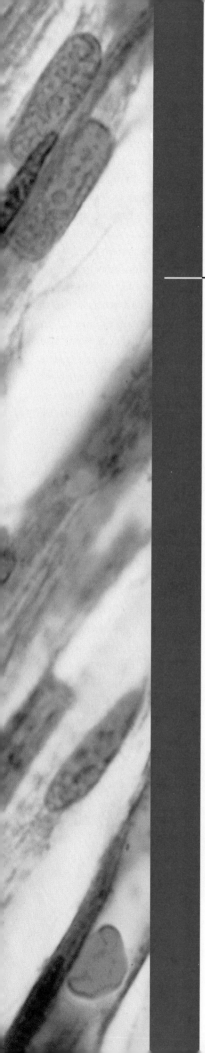

RESEARCH IN CANADA

For those who have never had heart problems, the regular, rhythmic beating of the heart is usually taken for granted. Like the regulatory functions of other organ systems, the importance of the electrical activity of the heart muscle only becomes noticeable when a malfunction occurs. At Halifax's Dalhousie University, Dr. Gerald Stronik heads a research team that is studying electrical abnormalities of the heart. The team is hoping to capitalize on two techniques, the electrocardiograph and the magnetocardiograph, that will provide them with a map of the heart's activities. Dr. Stronik and his associates are attempting to find answers to such questions as: How do the electrical activity and magnetic activity of diseased and normal heart tissue differ? How do heart attacks affect normal electrical activity? How do heart attacks disrupt the normal electrical activities of a functioning heart?

A group of researchers from McMaster University in Hamilton, Ontario, is investigating the effectiveness of an implantable device capable of restoring normal cardiac rhythm following cardiac arrest. Drs. Stuart Connolly, Michael Gent, and Robin Roberts head the research study, which is being conducted at 16 different centers and will involve more than 400 patients. The technological device, known as a defibrillator, is implanted into patients who have had a heart attack and are at risk of another attack. The defibrillator monitors heartbeat and delivers an electrical stimulus to restore normal heart rhythm if another attack occurs. The device will be compared with drugs to determine if it is more effective than drug therapy.

Cardiac Rhythm

DR. STUART CONNOLLY

DR. GERALD STRONIK

HEART SOUNDS

The familiar "lubb-dubb" heart sounds are caused by the closing of the heart valves. Contraction of the muscular walls of the atria increases pressure and forces blood through AV valves into the ventricles. As stated before, the cardiac muscle contraction proceeds from the atria to the ventricles. Atria begin to fill with blood as they relax. The term **diastole** is used to describe this relaxation. As the ventricles begin to contract, blood is forced up the sides of the ventricles and the AV valves close, producing a lubb sound. Ventricular contraction increases pressure in the chambers, forcing blood through the semilunar valves and out of the arteries. The term **systole** is used to describe this contraction. As the ventricles begin to relax, the volume of the chambers increases. With increased volume, pressure in the ventri-

Diastole refers to heart relaxation.

Systole refers to heart contraction.

Murmurs are caused by faulty heart valves, which permit the backflow of blood into one of the heart chambers.

> ### BIOLOGY CLIP
> In 1816 Rene Laennec, a young physician, was examining a patient for heart distress. The common practice at the time was for the doctor to place his ear on the patient's chest and listen for the lubb-dubb sounds. However, Laennec found that because of the patient's heavy bulk, the heart sounds were muffled. As the examination became more and more awkward, Laennec decided to try another avenue. He rolled up a paper and placed it to the patient's chest. Much to his relief, the heart sounds became clearer. Later, wooden cylinders were used, eventually to be replaced by the modern Y-shaped stethoscope, which literally means "chest-viewer."

cles begins to decrease and blood is drawn from the arteries toward the area of lower pressure. However, the blood is prevented from re-entering the ventricles by the semilunar valves. The blood causes the semilunar valves to close, creating the dubb sound.

Occasionally, the valves do not close completely. This condition, referred to as a heart **murmur,** occurs when blood leaks past the closed heart valve because of an improper seal. The AV valves, especially the left AV valve, or *mitral valve*, must withstand great pressure and are especially susceptible to defects. The rush of blood from the ventricle back into the atrium produces a gurgling sound that can be detected by a stethoscope. Blood flowing back toward the atrium is not directed to the systemic or pulmonary systems, but the hearts of individuals who experience murmurs do compensate for

Figure 6.18

Right and left atria contract in unison, pushing blood into the right and left ventricles. Ventricular contractions close the AV valves and open the semilunar valves. The relaxation of the ventricles lowers pressure and draws blood back to the chamber. The closing of the semilunar valves prevents blood from re-entering the ventricles.

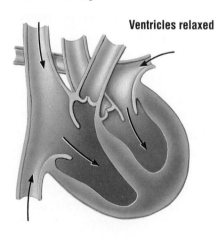
Ventricles relaxed

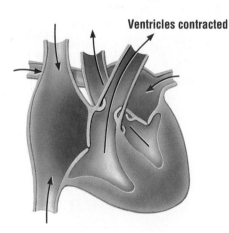
Ventricles contracted

decreased oxygen delivery by beating faster. You will learn more about this compensating mechanism in the next section.

A second compensatory mechanism helps increase blood flow. Like an elastic band, the more cardiac muscle is stretched, the stronger is the force of contraction. When blood flows from the ventricle back into the atrium, blood volume in the atrium increases. The atrium accepts the normal filling volumes but, in addition, accepts blood from the ventricle. This causes the atrium to stretch and the blood to be driven to the ventricle with greater force. Subsequently, increased blood volume in the ventricle causes the ventricle to contract with greater force, driving more blood to the tissues.

CARDIAC OUTPUT

Cardiac output is defined as the amount of blood that flows from each side of the heart per minute. Unless some dysfunction occurs, the amount of blood pumped from the right side of the heart is equal to the amount of blood pumped from the left side of the heart. Two factors affect cardiac output: stroke volume and heart rate.

Stroke volume is the quantity of blood pumped with each beat of the heart. The stronger the heart contraction, the greater the stroke volume. Approximately 70 mL of blood per beat leave each ventricle while you are resting. *Heart rate* is the number of times the heart beats per minute. The equation below shows how cardiac output is determined by stroke volume and heart rate.

$$\text{Cardiac output} = \text{heart rate} \times \text{stroke volume}$$
$$= 70 \text{ beats/min} \times 70 \text{ mL/beat}$$
$$= 4900 \text{ mL/min}$$

Individuals who have a mass of 70 kg must pump approximately 5 L of blood per minute. Smaller individuals require less blood and therefore have lower cardiac outputs. Naturally, cardiac output must be adjusted to meet energy needs. During exercise, heart rate may increase to 150 or 180 beats per minute to meet increased energy demands.

The cardiac output equation provides a basis for comparing individual fitness. Why do two people with the same body mass have different heart rates? If you assume that both people are at rest, both should require the same quantity of oxygen each minute. For example, Tom, who has a heart rate of 100 beats per minute, has a lower stroke volume. Lee, who has a heart rate of 50 beats per minute, has a higher stroke volume.

Cardiac output = stroke volume $\times$ heart rate

Tom	Lee
5 L = 50 mL/beat $\times$ 100 beats/min	5 L = 100 mL/beat $\times$ 50 beats/min

Lee's lower heart rate indicates a higher stroke volume. People who have well-developed hearts can pump greater volumes of blood with each beat. This is why athletes often have low heart rates. Those with weaker hearts are unable to pump as much blood per beat, but compensate by increasing heart rate to meet the body's energy demands. It is impor-

Cardiac output *is the amount of blood pumped from the heart each minute.*

Stroke volume *measures the quantity of blood pumped with each beat of the heart.*

tant to recognize that heart rate is only one factor that determines physical fitness. You may also find that your pulse rate will fluctuate throughout the day. Various kinds of food, stress, or a host of other factors can affect your heart rate.

BLOOD PRESSURE

Blood surges through the arteries with every beat of the heart. Elastic connective tissue and smooth muscle in the walls of the arteries stretch to accommodate the increase in fluid pressure. The arterial walls recoil much like an elastic band as the heart begins the relaxation phase characterized by lower pressure. Even the recoil forces help push blood through arterioles toward the tissues.

Blood pressure can be measured indirectly with an instrument called a **sphygmomanometer.** A cuff with an air bladder is wrapped around the arm. A small pump is used to inflate the air bladder, thereby closing off blood flow through the brachial artery, one of the major arteries of the arm. A stethoscope is placed below the cuff and air is slowly released from the bladder until a low-pitched sound can be detected. The sound is caused by blood entering the previously closed artery.

Each time the heart contracts, the sound is heard. A gauge on the sphygmomanometer measures the pressure that blood exerts during ventricular contraction. This pressure is called *systolic blood pressure*. Normal systolic blood pressure for young adults is about 120 mm Hg. The cuff is then deflated even more, until the sound disappears. At this point, blood flows into the artery during ventricular relaxation or filling. This pressure is called *diastolic blood pressure*. Normal diastolic blood pressure for young adults is about 80 mm Hg. Reduced filling, such as

that caused by an internal hemorrhage, will cause diastolic blood pressure to fall.

Figure 6.20 shows that fluid pressure decreases with distance from the ventricles. The aorta records the greatest pressure readings, despite the fact that they have the largest diameter. To help you understand why, imagine two hoses connected to a tap; the first hose is 1 m in length, the second is 100 m in length. The same amount of water is released from the tap into each hose. As water passes through the hoses, friction is created, slowing down its movement. The longer the hose, the greater the amount of friction and the slower the stream of water. This explains why the pulse in the carotid artery is stronger than the pulse detected near your wrist and why blood-pressure readings are not the same in all arteries.

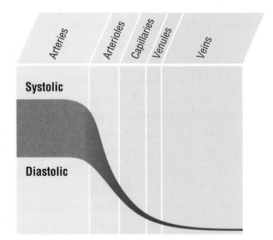

The artery acts as a reservoir for blood. Two factors regulate the amount of blood in the reservoir. The first is cardiac output. Any factor that increases cardiac output will increase blood pressure. The second factor is arteriolar resistance. You may recall that the diameter of the arterioles is regulated by coiling, smooth muscles. Constriction of the smooth muscles surrounding the arterioles closes the opening and reduces blood flow through the arteriole. With this reduced blood

Figure 6.19

This sphygmomanometer is calibrated in the nonmetric units of millimeters of mercury.

Figure 6.20

Fluid pressure decreases the further blood moves from the heart.

A **sphygmomanometer** *is a device used to measure blood pressure.*

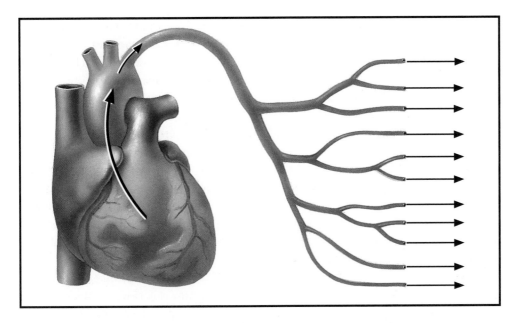

Figure 6.21

Blood pressure is measured in arteries. Two factors, cardiac output and arteriolar resistance, affect blood pressure.

flow, more blood is left in the artery. The increased blood volume in the artery produces higher blood pressure. Conversely, factors that cause arteriolar dilation increase blood flow from the arterioles, thereby reducing blood pressure.

The smooth muscles in the walls of the arterioles respond to nerve and endocrine controls that regulate blood pressure. However, the diameter of the arterioles also adjusts to metabolic products. For example, the cellular breakdown of sugar with oxygen produces carbon dioxide and water. The cellular breakdown of sugar in the absence of oxygen yields lactic acid. Carbon dioxide and lactic acid are two of the metabolic products that cause relaxation of smooth muscles in the walls of the arterioles, causing the arterioles to open. The dilation of the arterioles increases blood flow to local tissues. Arteriolar dilation, in response to increased metabolic products, provides a good example of homeostasis (see Figure 6.22). Because these products accumulate in the most active tissues, the increased blood flow helps provide greater nutrient supply, while carrying the potentially toxic materials away. Tissues that are less active produce fewer metabolic products.

These arterioles remain closed until the products accumulate.

Table 6.2 Factors that Affect Arteriolar Resistance

Factor	Effect
Epinephrine	arteriolar constriction, except to the heart, muscles, and skin
Sympathetic nerve stimulation	arteriolar constriction, except in skeletal and cardiac muscle
Acid accumulation	arteriolar dilation
CO_2 accumulation	arteriolar dilation
Lactic acid accumulation	arteriolar dilation

Homeostatic Adjustment

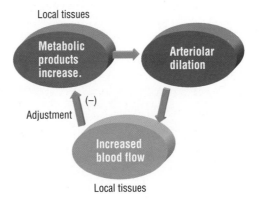

Figure 6.22

Dilation of the arterioles increases blood flow to local tissues.

10 Draw and label the major blood vessels and chambers of the heart. Trace the flow of deoxygenated and oxygenated blood through the heart.

11 Differentiate between the systemic circulatory system and the pulmonary circulatory system.

12 What causes the characteristic heart sounds?

13 What are coronary bypass operations and why are they performed?

14 Explain the function of the sinoatrial node.

15 What is an electrocardiogram?

16 Differentiate between systolic and diastolic blood pressure.

17 Define cardiac output, stroke volume, and heart rate.

18 How do metabolic products affect blood flow through arterioles?

Blood pressure receptors *are specialized nerve cells that are activated by high blood pressure.*

REGULATION OF BLOOD PRESSURE

Regulation of blood pressure is essential. Low blood pressure reduces your capacity to transport blood. The problem is particularly acute for tissue in the head where blood pressure works against the force of gravity. High blood pressure creates equally serious problems. High fluid pressure can weaken an artery and eventually lead to the rupturing of the vessel.

Special **blood pressure receptors** are located in the walls of the aorta and the carotid arteries, which are major arteries found on either side of the neck. These receptors, known as *baroreceptors*, are sensitive to high pressures. When blood pressure exceeds acceptable levels, the baroreceptors respond to the increased pressure on the wall of the artery. A nerve message travels to the medulla oblongata, the blood pressure regulator located at the stem of the brain. The sympathetic nerve—the "stress nerve"—is turned down and the parasympathetic nerve—the "slow down nerve"—is stimulated. By decreasing sympathetic nerve stimulation, arterioles dilate, increasing outflow of blood from the artery. Stimulation of the parasympathetic nerve causes heart rate and stroke volume to decrease. The decreased cardiac output slows the movement of blood into the arteries and consequently lowers blood pressure.

Low blood pressure is adjusted by the sympathetic nerve. Without nerve information from the pressure receptors of the carotid artery and aorta, the sympathetic nerve will not be "turned off." Under the influence of the sympathetic nerve, cardiac output increases and arterioles constrict. The increased flow of blood into the artery accompanied by decreased outflow raises blood pressure to acceptable levels.

Figure 6.23

Decreased cardiac output and arteriolar dilation decrease blood volume in arteries. The lower volume of blood in the artery decreases blood pressure.

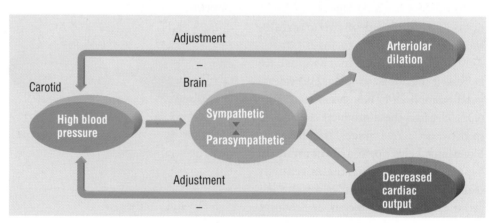

LABORATORY
EFFECTS OF POSTURE ON BLOOD PRESSURE

Objective

To determine how posture affects blood pressure.

Materials

sphygmomanometer   stethoscope
watch with second hand   alcohol

Procedure

1 Ask your partner to sit quietly for one minute. Clean the earpieces of the stethoscope with alcohol. Expose the arm of your partner and place the sphygmomanometer just above the elbow.
2 Close the valve on the rubber bulb on the end of the sphygmomanometer, and inflate it by squeezing the rubber ball until a pressure of 180 mm Hg registers.

> **CAUTION:** Do not leave the pressure on for longer than 1 min. If you are unsuccessful, release the pressure and try again.

3 Place the stethoscope bell, or diaphragm, on the inside of the arm immediately below the cuff.
4 Slowly release the pressure by opening the valve on the rubber ball of the sphygmomanometer. Listen for a low-pitched sound.
   a) Record the reading on the sphygmomanometer. This is the systolic blood pressure.

5 Continue releasing the pressure until the sound can no longer be heard.
   b) Record the reading on the dial of the sphygmomanometer when the sound disappears. This is the diastolic pressure.
6 Completely deflate the sphygmomanometer and take your partner's pulse. Place your index and middle fingers on the arm near the wrist. Count the number of pulses in one minute.
   c) Record the pulse rate while seated.
7 Repeat the procedure while your partner is in a standing position and then in a lying position.
   d) Record your results in a table like the one below.

Position	Systolic B P (mm Hg)	Diastolic B P (mm Hg)	Pulse rate (beats/min)
Standing			
Sitting			
Lying			

Laboratory Application Questions

1 Would you expect blood pressure readings in all the major arteries to be the same? Explain your answer.
2 Why should the lowest systolic pressure be recorded while you are lying down?
3 Atherosclerosis, or hardening of the arteries, is a disorder that causes high blood pressure. Provide an explanation for this condition.
4 Predict how exercise would affect systolic blood pressure. Provide your reasons.
5 Why might diastolic blood pressure decrease as heart rate increases?
6 Design a procedure to investigate the role of exercise in influencing blood pressure. ■

ADJUSTMENT OF THE CIRCULATORY SYSTEM TO EXERCISE

Your body's response to exercise is an excellent example of a homeostatic mechanism. The demands placed on the circulatory system by tissues during exercise are considerable. The circulatory system does not act alone in monitoring the needs of tissues or in ensuring that adequate levels of oxygen and other nutrients are delivered to the active cells. The nervous and endocrine systems also play important roles in adjustment mechanisms.

The sympathetic nerve stimulates the adrenal glands. During times of stress, the hormone epinephrine is released from the adrenal medulla and travels in the blood to other organs of the body. Epinephrine stimulates the release of red blood cells from the spleen, a storage site. Although the significance of the response is not yet understood, it is clear that increased numbers of red blood cells aid oxygen delivery. Epinephrine and direct stimulation from the sympathetic nerve increase heart rate and breathing rate. The increased heart rate provides for faster oxygen transport, while the increased breathing rate ensures that the blood contains higher levels of oxygen. Both systems work together to improve oxygen delivery to the active tissues. A secondary, but equally important, function is associated with more effective waste removal from the active tissues.

Blood cannot flow to all capillaries of the body simultaneously. The effect of dilating all arterioles would be disastrous—blood pressure would plunge. Epinephrine causes vasodilation or widening of the arterioles leading to the heart, brain, and muscles. At the same time, epinephrine causes the constriction of blood vessels leading to the kidney, stomach, and intestines. The most active tissues receive priority in times of stress. Blood flow is diverted to the muscles and heart, enabling the organism to perform responses associated with flight-or-fight reactions. Organs from the digestive system and kidney are deprived of much of their required nutrients until the stress situation has been overcome.

CAPILLARY FLUID EXCHANGE

It has been estimated that nearly every tissue of the body is within 0.1 mm of a capillary. Earlier in the chapter you learned how capillaries provide cells with oxygen, glucose, and amino acids. Capillaries are also associated with fluid exchange between the blood and surrounding **extracellular fluid (ECF).** Most fluids simply diffuse through capillaries. The capillary cell membranes are permeable to oxygen and carbon dioxide. Water and certain ions are thought to pass through the clefts between the cells of the capillary. Larger molecules and a very small number of proteins are believed to be exchanged by endocytosis or exocytosis. This section will focus on the movement of water molecules.

Two forces regulate the movement of water between the blood and ECF: fluid pressure and osmotic pressure. The force that blood exerts on the wall of a capillary is about 35 mm Hg pressure at the arteriole end of the capillary and approximately 15 mm Hg pressure at the venous end. The reservoir of blood in the arteries creates pressure on the inner wall of the capillary. Much lower pressure is found in the ECF. Although fluids bathe cells, no force drives the extracellular fluids. Water moves from an area of high pressure—the capillary—into an area of low pressure—the ECF. The outward flow of water and small minerals ions is known as **filtration.** Because capillaries are selectively permeable, large materials such as proteins, red blood cells, and white blood cells remain in the capillary.

The movement of fluids from the capillary must be balanced with a force that moves fluid into the capillary. The fact that large proteins are found in the blood but not in the ECF may provide a hint as to the nature of the second force. Osmotic pressure draws water back into the capillary. The large protein molecules of the blood and dissolved minerals are primarily responsible for the movement of fluid into capillaries. The movement of

Extracellular fluids (ECF) *occupy the spaces between cells and tissues.*

Filtration *is the selective movement of materials through capillary walls by a pressure gradient.*

fluid into capillaries is called **absorption.**
Osmotic pressure in the capillaries is usually about 25 mm Hg, but it is important to note that the concentration of solutes can change with fluid intake or excess fluid loss caused by perspiration, vomiting, or diarrhea.

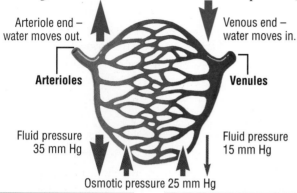

Arteriole end –
water moves out.

Arterioles

Venous end –
water moves in.

Venules

Fluid pressure
35 mm Hg

Fluid pressure
15 mm Hg

Osmotic pressure 25 mm Hg

Arteriole end		Venous end	
Fluid pressure	35 mm Hg	Osmotic pressure	25 mm Hg
Osmotic pressure	25 mm Hg	Fluid pressure	15 mm Hg
Net filtration	**10 mm Hg**	**Net filtration**	**10 mm Hg**

Application of the capillary exchange model provides a foundation for understanding homeostatic adjustments to a variety of problems. The balance between osmotic pressure and fluid pressure is upset during hemorrhage. The decrease in blood volume resulting from the hemorrhage affects blood pressure. The force that drives fluid from the capillaries diminishes, but the osmotic pressure, which draws water into the capillaries, is not altered. Although proteins are lost with the hemorrhage, so are fluids. Fewer proteins are present, but the concentration has not been changed. The force drawing water from the tissues and ECF is greater than the force pushing water from the capillary. The net movement of water into the capillaries provides a homeostatic adjustment. As water moves into the capillaries, fluid volumes are restored.

Individuals who are suffering from starvation often display tissue swelling, or **edema.** Plasma proteins are often mobilized as one of the last sources of energy. The decrease in concentration of plasma proteins has a dramatic effect on osmotic pressure, which draws fluids from the tissues and ECF into the capillaries. The decreased number of proteins lowers osmotic pressure, thereby decreasing absorption. More water enters the tissue spaces than is pulled back into the capillaries, causing swelling.

Why do tissues swell during inflammation or allergic reactions? When you eat a food to which you are allergic, endangered cells—or cells that "believe" they are endangered—release a chemical messenger, called *bradykinin*, which stimulates the release of another chemical stimulator, *histamine*. Histamine changes the cells of the capillaries, thereby increasing permeability. The enlarged capillary causes the area to redden. Proteins and white blood cells leave the capillary in search of the foreign invader, but, in doing so, they alter the osmotic pressure. The proteins in the ECF create another osmotic force that opposes the osmotic force in the capillaries. Less water is absorbed into the capillary, and tissues swell.

Figure 6.24

Fluid movement into and out of the capillaries.

Absorption *is the movement of fluids in the direction of a diffusion, or osmotic, gradient.*

Osmotic pressure *is the pressure exerted on the wall of a semipermeable membrane resulting from differences in solute concentration. (In this case, the more concentrated the plasma proteins, the greater is the osmotic pressure.)*

Edema *is tissue swelling caused by decreased osmotic pressure in the capillaries.*

Figure 6.25

The balance between osmotic pressure and fluid pressure is upset during hemorrhage, starvation, or inflammation.

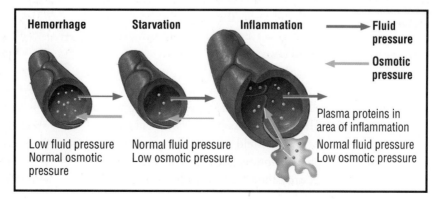

Hemorrhage

Starvation

Inflammation

Fluid
pressure

Osmotic
pressure

Low fluid pressure
Normal osmotic
pressure

Normal fluid pressure
Low osmotic pressure

Plasma proteins in
area of inflammation

Normal fluid pressure
Low osmotic pressure

THE LYMPHATIC SYSTEM

Normally, a small amount of protein leaks from capillaries to the tissue spaces. Despite the fact that the leak is very slow, the accumulation of proteins in the ECF would create a major problem: osmotic pressure would decrease and tissues would swell.

Figure 6.26

Lymph vessels (shown in green) are open-ended vessels.

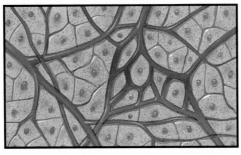

Figure 6.27

Lymph, or ECF, vessels carry lymph back to the circulatory system. Lymph nodes filter debris from the lymph. Insert shows lymph flow through a lymph node.

Lymph nodes *contain white blood cells that filter lymph.*

Lymph *is the fluid found outside capillaries. Most often, the lymph contains some small proteins that have leaked through capillary walls.*

The proteins are drained from the ECF and returned to the circulatory system by way of another network of vessels: the lymphatic system. **Lymph,** a fluid similar to blood plasma, is transported in open-ended lymph vessels in much the same way as veins. The low-pressure return system operates by slow muscle contractions against the vessels, which are supplied with flaplike valves that prevent the backflow of fluids. Eventually, lymph is returned to the venous system. In the previous chapter you read about specialized lymph vessels that carry fats from the small intestine. Called *lacteals*, these vessels provide the products of fat digestion access to your circulatory system.

Enlargements called **lymph nodes** are located at intervals along the lymph vessel. These house phagocytotic white blood cells that filter bacteria from lymph. Should bacteria be present, the phagocytotic white blood cells engulf and destroy them. The lymph nodes also fil-

ter damaged cells and debris from the lymph. Have you ever experienced swelling of the lymph nodes? The lymph nodes sometimes swell when you have a sore throat.

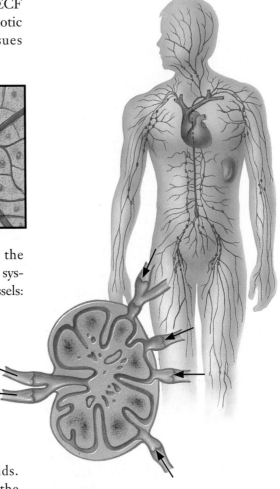

REVIEW QUESTIONS ?

19 How do blood pressure regulators detect high blood pressure?

20 Outline homeostatic adjustment to high blood pressure.

21 What two factors regulate the exchange of fluids between capillaries and ECF?

22 Why does a low concentration of plasma protein cause edema?

23 What are lymph vessels and how are they related to the circulatory system?

SOCIAL ISSUE:
Heart Care

Heart disease is the number-one killer of North Americans. Although congenital heart defects account for some of the problems, the incidence of heart disease can often be traced to lifestyle. The relationship between heart problems and smoking, excessive alcohol consumption, stress, high-cholesterol diets, and high blood pressure has long been established.

Statement:

People who refuse to alter lifestyles that are dangerous to their health should not be permitted equal access to health care.

Point

- People who deliberately ignore a doctor's advice and continue to place themselves at risk should not be permitted equal access to the health system. High-cost health care should be reserved for those who deserve it most.
- People who have a high-risk lifestyle should pay higher medical insurance and be placed lower on waiting lists for heart transplants. How can we be assured that these people will not abuse their transplanted heart?

Counterpoint

- People cannot always be held responsible for their behavior. Some people have a genetic link to obesity and a predisposition to alcoholism, indicating that behavior could be determined by factors other than environment.
- Health care should always be given on the basis of need. Medical authorities should decide who needs a transplant, regardless of the person's previous lifestyle or reliability.

Research the issue.
Reflect on your findings.
Discuss the various viewpoints with others.
Prepare for the class debate.

CHAPTER HIGHLIGHTS

- Multicellular organisms need a circulatory system.
- The heart pumps blood into the arteries of the pulmonary and systemic circulatory systems. Blood travels from arteries to arterioles, capillaries, venules, and finally veins, which carry blood back to the heart.
- Arteries are high-pressure blood vessels that carry blood away from the heart. Arteries stretch to accommodate high pressure, producing a pulse.
- The capillaries are vessels composed of a single layer of cells. The diffusion of materials between the blood and extracellular fluid occurs in the capillaries.
- Veins are low-pressure blood vessels that carry blood back to the heart. Valves located in the veins prevent the backflow of blood.
- Cardiac output is a function of stroke volume and heart rate.
- Sympathetic nerve stimulation and the hormone epinephrine increase heart rate and cardiac output. Parasympathetic nerve stimulation decreases heart rate and cardiac output.

- Blood pressure is measured in arteries. Systolic blood pressure occurs when blood surges into arteries following ventricular contraction. Diastolic blood pressure occurs while the heart is relaxing.
- Blood pressure is regulated by cardiac output and arteriolar resistance.

- Arterioles dilate in response to metabolic products.
- Lymph vessels complement the circulatory system by restoring osmotic pressure and transporting protein and other solutes back into the blood.

APPLYING THE CONCEPTS

1 Agree or disagree with the following statement and give reasons for your views: Oxygenated blood is found in all arteries of the body.

2 Why does the left ventricle contain more muscle than the right ventricle?

3 Why does blood pressure fluctuate in an artery?

4 Which area of the graph represents blood in a capillary? Explain your answer.

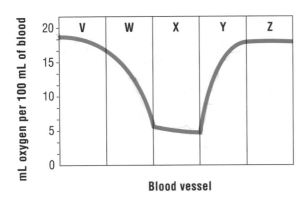

Blood vessel

5 Using a capillary exchange model, explain why the intake of salt is regulated for patients who suffer from high blood pressure. (Hint: The salt is absorbed from the digestive system into the blood.)

6 Why do some soldiers faint after standing at attention for a long time?

7 Explain why someone who suffers a severe cut might develop a rapid and weak pulse. Why might body temperature begin to fall?

8 Why does the blockage of a lymph vessel in the left leg cause swelling in that area?

9 A fetus has no need for pulmonary circulation. Oxygen diffuses from the mother's circulatory system into that of the fetus through the placenta. Therefore, the movement of blood through the heart is highly modified: blood flows from the right atrium through an opening in the septum to the left atrium and then to the left ventricle. The opening between the right and left atria becomes sealed at birth. Explain why any failure to seal the opening results in what has been termed a "blue baby."

10 A person's blood pressure is taken in a sitting position before and after exercise. Compare blood pressure readings before and after exercise as shown in the following chart.

Condition	Systolic B P (mm Hg)	Diastolic B P (mm Hg)	Pulse rate (beats/min)
Resting	120	80	70
After exercise	180	45	160

a) Why does systolic blood pressure increase after exercise?

b) Why does diastolic blood pressure decrease after exercise?

CRITICAL-THINKING QUESTIONS

1 **a)** Nicotine causes the constriction of arterioles. Using the information that you gained about fetal circulation from question 9 in Applying the Concepts, explain why pregnant women are advised not to smoke.

   b) Mothers who smoke give birth to babies who are, on average, 1 kg smaller than normal. Speculate on the relationship between the effects of nicotine on the mother's circulatory system and the lower body mass of babies.

2 Coronary heart disease is often related to lifestyle. Stress, smoking, alcohol consumption, and poor diet are considered to be contributing factors to heart problems. Should all people have equal access to heart transplants? Should people born with genetic heart defects be treated differently from people who have abused their hearts?

3 Heart disease is currently the number-one killer of middle-aged males, accounting for billions of dollars every year in medical expenses and productivity loss. Should males be required, by law, to undergo heart examinations?

4 Caffeine causes heart rate to accelerate; however, a scientist who works for a coffee company has suggested that blood pressure will not increase due to coffee consumption. This scientist states that homeostatic adjustment mechanisms ensure that blood pressure readings will remain within an acceptable range. Design an experiment that will test the scientist's hypothesis. For what other reasons do you think the scientist might have suggested that caffeine does not increase blood pressure?

5 It has been estimated that for every extra kilogram of fat a person carries, an additional kilometer of circulatory vessels is required to supply the tissues with nutrients. Indicate why obesity has often been associated with high blood pressure. Do only overweight people suffer from high blood pressure? Explain your answer.

ENRICHMENT ACTIVITIES

1 Suggested reading:
   - Brand, David. "Searching for Life's Elixir." *Time*, December 12, 1988, p. 60.
   - Eisenberg, M.S., et al. "Sudden Cardiac Death." *Scientific American* 25(5) (1986): p. 33.
   - Franklin, D. "Steroids Heft Heart Risk In Iron Pumpers." *Science News* 126(July 21, 1984): p. 38.
   - Fritz, Sand. "Drugs and Olympic Athletes." *Scholastic Science World*, 40(16) (April 13, 1984): pp. 7–14.
   - Harper, Alfred. "Killer French Fries." *Sciences*, Jan/Feb 1988, pp. 21–27.
   - Hastings, Paul. *Medicine: An International History*. London: Ernest Benn, 1974.
   - Kusinitz, Beryl. "The Artificial Heart." *Science World* 39(16) (1983): p. 8.
   - Miller, Jonathan. *The Body in Question*. London: Macmillan, 1978.
   - Monmaney, Terrence, and Karen Springen. "The Cholesterol Connection." *Newsweek*, February 8, 1988, pp. 56–58.
   - Robinson, T.E. et al. "The Heart as a Suction Pump." *Scientific American* 254(6) (1986): p. 84.
   - Sperryn, Perry. "Drugged and Victorious: Doping and Sport." *New Scientist*, August 1984, pp. 16–18.
   - ———. "New Medicine." *National Geographic*, January 1987.

2 View the following films:
   - TVOntario. *Medical*. BPN 179110.
   - National Geographic Society. *Circulatory and Respiratory System*. 51305. This excellent video examines the human transport system.

*B*lood and Immunity

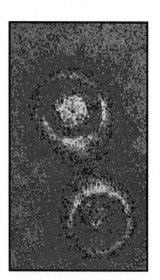

IMPORTANCE OF BLOOD AND IMMUNITY

In the last chapter you discovered that blood can be associated with many different functions. In addition to its transport functions, blood helps maintain the water balance of organ systems, body temperature, and pH balance. Blood is also an important part of the human immune system, protecting the body against a host of invaders.

Figure 7.1

David, the "boy in the plastic bubble," had severe combined immunodeficiency syndrome.

To appreciate the importance of the immune system, consider the story of David, "the boy in the plastic bubble." David was born without an immune system, which meant that his body was unable to produce the cells necessary to protect him from disease. As a result, David had to live in a virtually germ-free environment. People who came in contact with him had to wear plastic gloves. Eventually, David received a bone marrow transplant from his sister. Unfortunately, a virus was hidden in the bone marrow. The sister, who had a functioning immune system, was able to protect herself from the virus, but David was not.

Because blood cells are suspended in a watery fluid, blood has been described as a fluid tissue. Like other tissues, the individual cells in the blood work together for a common purpose. The watery nature of blood provides an interesting reminder of the origins of human life. Blood has the same ions in approximately the same relative concentration as the ancient seas from which humans evolved.

COMPONENTS OF BLOOD

The average 70 kg individual is nourished and protected by about 5 L of blood. Approximately 55% of the blood is fluid; the remaining 45% is composed of blood cells. The fluid portion of the blood is referred to as the **plasma.** Although it is approximately 90% water, the plasma also contains blood proteins, glucose, vitamins, minerals, dissolved gases, and waste products of cell metabolism.

The large plasma proteins play special roles in maintaining homeostasis. One group of proteins, the *albumins*, along with inorganic minerals, establishes an osmotic pressure that draws water back into capillaries and helps maintain body fluid levels. A second group of proteins, the *globulins*, produces antibodies that provide protection against invading microbes. You will learn more about antibodies later in the chapter. *Fibrinogens*, the third group of plasma proteins, are important in blood clotting. Table 7.1 summarizes the types of plasma proteins and their functions.

Erythrocytes

The primary function of red blood cells is the transport of oxygen. Referred to as erythrocytes (from the Greek *erythros*, meaning "red"), the red blood cells are packed with a respiratory pigment that absorbs oxygen. Oxygen diffuses from the air into the plasma, but the amount of oxygen that can be carried by the plasma is limited. At body temperature, 1 L of blood would carry about 3 mL of oxygen. An iron-containing respiratory pigment called **hemoglobin** greatly increases the capacity of the blood to carry oxygen. When hemoglobin is present, 1 L of blood is capable of carrying 200 mL of oxygen, a 70-fold increase. Without hemoglobin, your red blood cells would supply only enough oxygen to maintain life for approximately 4.5 s. With hemoglobin, life can continue for five minutes. Five minutes is not very long, but remember that the blood returns to the heart and is pumped to the lungs, where oxygen supplies are replenished. Anyone deprived of oxygen for longer than five minutes starts to experience cell death. This might indicate why people survive even when the heart stops for short periods of time. Children who have been immersed in cold water for longer than five minutes have survived with comparatively minor cell damage. Colder temperatures slow body metabolism, thereby decreasing oxygen demand.

Hematocrit

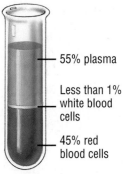

55% plasma

Less than 1% white blood cells

45% red blood cells

Figure 7.2

The percentage of red blood cells in the blood is called the hematocrit.

Plasma *is the fluid portion of the blood.*

Hemoglobin *is the pigment found in red blood cells.*

Table 7.1 Plasma Proteins

Type	Function
Albumins	osmotic balance
Globulins	antibodies, immunity
Fibrinogen	blood clotting

An estimated 280 million hemoglobin molecules are found in a single red blood cell. The hemoglobin is composed of *heme*, the iron-containing pigment, and *globin*, the protein structure. Four iron molecules attach to the folded protein structure and bind with oxygen molecules. The oxyhemoglobin complex gives blood its red color. (Actually, a single red blood cell appears pale orange—the composite of many red blood cells produces the red color.) Once oxygen is given up to cells of the body, the shape of the hemoglobin molecule changes, causing the reflection of blue light. This explains why blood appears blue in the veins of your arms and hands.

Red blood cells appear as biconcave (meaning concave on both sides) disks approximately 7.0 μm in diameter. The folded disk shape provides a greater surface area for gas exchange—between 20 and 30% more surface area than a similar sphere. Red blood cells do not contain a nucleus when mature; they are said to be **enucleated.** The absence of a nucleus provides more room for the cell to carry hemoglobin. This enucleated condition raises two important questions. Since cells by definition contain a nucleus, are red blood cells actually cells? The second question addresses cell reproduction. How do cells without a nucleus and chromosomes reproduce? Since red blood cells live only about 120 days, cell reproduction is essential. One estimate suggests that at least five million red blood cells are produced every minute of the day.

The answer to both of the above questions can be found in the bone marrow, the site of red blood cell reproduction, or **erythropoiesis** (the suffix *poiesis* means "to make"). Red blood cells begin as stem cells, which do contain a nucleus. The cells divide and shrink as they take up hemoglobin. Eventually, the nucleus dis-

appears and the cells are discharged into the blood. The mature red blood cell cannot undergo mitosis, but the immature form may divide many times.

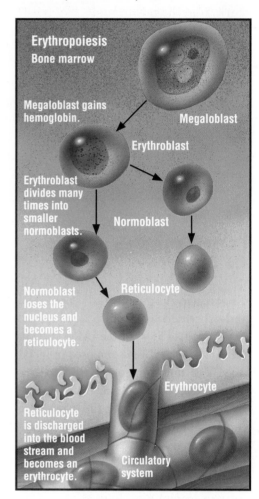

The average male contains about 5.5 million red blood cells per milliliter of blood, while the average female has about 4.5 million red blood cells per milliliter of blood. Individuals living at high altitudes can have as many as 8 million red blood cells per milliliter of blood. How does the body count red blood cells to ensure that adequate numbers are maintained? The problem is an immense one, just from a bookkeeping point of view. The outer membranes of red blood cells become brittle with age, causing them to rupture as they file through the narrow capillaries. Specialized white blood cells,

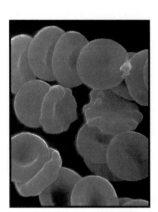

Figure 7.3

Red blood cells move through the capillaries in single file.

Figure 7.4

Immature red blood cells are found in the bone marrow. Stem cells in the bone marrow are continually dividing and forming megaloblasts, cells destined to develop into red blood cells.

Enucleated *cells do not contain a nucleus.*

Erythropoiesis *is the process by which red blood cells are made.*

located primarily in the spleen and liver, also monitor the age of red blood cells and remove debris from the circulatory system. Following the breakdown of red blood cells, the hemoglobin is released. Iron is recovered and stored in the bone marrow for later use. The heme portion of the hemoglobin is transformed into bile pigments.

The short lifespan of red blood cells and the body's changing needs mean that oxygen delivery must continually be monitored. However, red blood cell numbers are not monitored directly. Red blood cell reproduction is directed by oxygen levels. Any condition that lowers blood oxygen levels causes an increase in the rate of erythropoiesis. The kidneys respond to low levels of oxygen by releasing REF, or renal erythropoietic factor. The REF combines with liver globulins and forms erythropoietin, a chemical messenger that stimulates red blood cell production in the bone marrow. Individuals who live at high altitudes compensate for lower oxygen levels in the air by increasing red blood cell production. In a similar manner, red blood cell production is stimulated following blood transfusions or hemorrhaging.

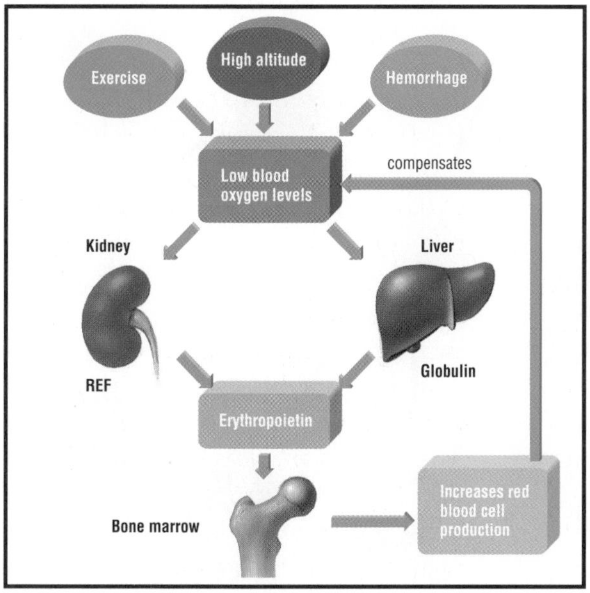

A deficiency in hemoglobin or red blood cells decreases oxygen delivery to the tissues. This condition, known as **anemia,** is characterized by low energy levels. The most common cause of a low red blood cell count is hemorrhage. Physical injury or internal bleeding caused by ulcers or hemorrhage in the lungs associated with tuberculosis can cause anemia. If more than 40% of the blood is lost, the body is incapable of coping. Anemia may also be associated with a dietary deficiency of iron, which, you will recall, is an important component of hemoglobin. The red blood cells must be packed with sufficient numbers of hemoglobin molecules to ensure adequate oxygen delivery. Raisins and liver are two foods rich in iron.

Leukocytes

White blood cells, or leukocytes, are much less numerous than red blood cells. It has been estimated that red blood cells outnumber white blood cells by a ratio of 700 to 1. White blood cells have a nucleus, making them easily distinguishable from red blood cells. In fact, the shape and size of the nucleus, along with the granules in the cytoplasm, have been used to identify different types of leukocytes. Figure 7.6 shows the different types of leukocytes. The *granulocytes* are classified according to small cytoplasmic granules that become visible when stained. The *agranulocytes* are white blood cells that do not have a granular cytoplasm. Granulocytes are produced in the bone marrow. Agranulocytes are also produced in the bone marrow, but are modified in the lymph nodes. Some leukocytes destroy invading microbes by phagocytosis,

Anemia *refers to the reduction in blood oxygen due to low levels of hemoglobin or poor red blood cell production.*

Figure 7.5

Low levels of blood oxygen stimulate the production of erythropoietin, which, in turn, triggers the production of red blood cells. With an increase in red blood cell production, oxygen delivery is improved.

squeezing out of capillaries and moving toward the microbe like an amoeba. This process is known as **diapedesis.** Once the microbe has been engulfed, the leukocyte releases enzymes that digest the microbe and the leukocyte itself. Fragments of remaining protein from the white blood cell and invader are called **pus.** Other white blood cells form special proteins, called *antibodies*, which interfere with foreign invading microbes and toxins. You will learn more about antibodies later in this chapter.

Platelets

Platelets, like red blood cells, do not contain a nucleus, and are produced from large nucleated cells in the bone marrow. Small fragments of cytoplasm break from the large megakaryocyte to form platelets. The irregularly shaped platelets move through the smooth blood vessels of the body but rupture if they strike a sharp edge, such as that produced by a torn

Diapedesis *is the process by which white blood cells squeeze through clefts between capillary cells.*

Pus *is formed when white blood cells engulf and destroy invading microbes. The white blood cell is also destroyed in the process. The remaining protein fragments are known as pus.*

blood vessel. The fragile platelets initiate blood-clotting reactions. You will learn more about blood-clotting reactions later in the chapter.

REVIEW QUESTIONS ?

1 Why is blood considered to be a tissue?
2 Name the two major components of blood.
3 List three plasma proteins and indicate the function of each.
4 What is the function of hemoglobin?
5 What is erythropoiesis?
6 List factors that initiate red blood cell production.
7 What is anemia?
8 How do white blood cells differ from red blood cells?
9 State two major functions associated with leukocytes.
10 What is the function of platelets?

Figure 7.6

Stem cells of the bone marrow give rise to blood cells. Two classes of white blood cells are shown. The agranulocytes include the monocytes and lymphocytes. The granulocytes include the eosinophils, basophils, and neutrophils.

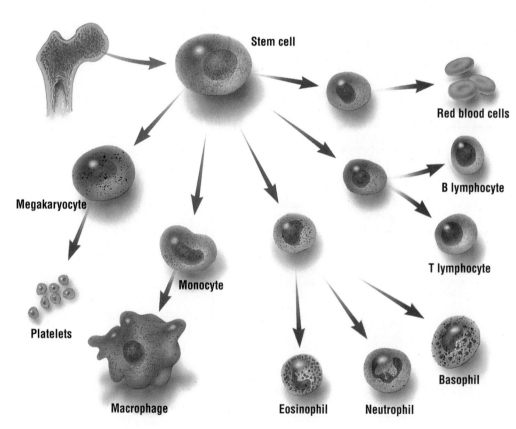

Stem cell

Red blood cells

B lymphocyte

T lymphocyte

Megakaryocyte

Monocyte

Platelets

Macrophage

Basophil

Eosinophil

Neutrophil

Unit Two:
Exchange of Matter and Energy in Humans

LABORATORY
MICROSCOPIC EXAMINATION OF BLOOD

Objective

To examine red and white blood cells.

Materials

prepared slide of fish blood   light microscope
prepared slide of human blood

Procedure

1 Before beginning the investigation, clean all microscope lenses with lens paper and rotate the nosepiece to the low-power objective. Place the slide of fish blood on the stage and focus under low power. Locate an area in which individual blood cells can be seen.

2 Rotate the revolving nosepiece to the medium-power objective, and focus. Red blood cells greatly outnumber white blood cells. Locate a single red blood cell in the center of the field of view and rotate the nosepiece to the high-power objective. Note the nucleus found in the red blood cells of the fish.
   a) Diagram the red blood cell of the fish.
   b) Estimate the size of the red blood cell.

3 Repeat the same procedure with the human blood slide.
   c) Diagram a single human red blood cell.
   d) Estimate the size of the human red blood cell.

4 Scan the field of view for different white blood cells. Using the classification of leukocytes provided in the chart below, classify the leukocytes and record your results in a table similar to the one below.

5 Repeat the procedure by scanning 10 different visual fields. Record your data in your chart.

Laboratory Application Questions

1 The red blood cells of fish contain a nucleus, while the human red blood cells do not. Indicate the advantage of the mammalian type of red blood cell over that of the fish.

    Blood tests are used to help diagnose different diseases. The chart below shows a few representative diseases. Use the chart to answer questions 2 to 4.

Leukocyte change	Associated conditions
Increased eosinophils	Allergic condition, chorea, scarlet fever, granulocyte leukemia
Increased neutrophils	Toxic chemical, newborn acidosis, hemorrhage, rheumatic fever, severe burns, acidosis
Decreased neutrophils	Pernicious anemia, protozoan infection, malnutrition, aplastic anemia
Increased monocytes	Tuberculosis (active), monocyte leukemia, protozoan infection, mononucleosis
Increased lymphocytes	Tuberculosis (healing), lymphocyte leukemia, mumps

2 Why would a physician not diagnose leukemia on the basis of a single blood test?

3 What information might a blood test provide a physician about a patient being treated for the lung disease tuberculosis? Why would blood tests be taken even after the disease has been diagnosed?

4 Leukemia can be caused by the uncontrolled division of cells from two different sites: the bone marrow or lymph nodes. Indicate how blood tests could be used to determine which of the sites harbors the cancerous tumor. ■

Classification of Leukocytes

Type	Description	Number	%
Granulocyte	Granular cytoplasm		
Neutrophil	Three-lobed nucleus, 10 nm (Wright's stain: purple nucleus, pink granules)		
Eosinophil	Two-lobed nucleus, 13 nm (Wright's stain: blue nucleus, red granules)		
Basophil	Two-lobed nucleus, 14 nm (Wright's stain: blue-black nucleus, blue-black granules)		
Agranulocyte	Nongranular cytoplasm		
Monocyte	U-shaped nucleus, 15 nm (Wright's stain: light bluish-purple nucleus, no granules)		
Lymphocyte (small)	Large nucleus, 7 nm (Wright's stain: dark bluish-purple nucleus, no granules)		
Lymphocyte (large)	Large nucleus, 10 nm (Wright's stain: dark bluish-purple nucleus, no granules)		

CASE STUDY

DIAGNOSIS USING HEMATOCRITS

Objective

To use hematocrits to diagnose various disorders.

Materials

metric ruler

Hematocrits

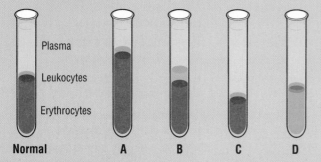

Plasma

Leukocytes

Erythrocytes

Normal   **A**   **B**   **C**   **D**

Procedure

1 Determine the normal hematocrit by using the following formula:

$$\text{Hematocrit} = \frac{\text{red blood cell volume}}{\text{total blood volume}} \times 100$$

a) Calculate the hematocrit of the normal subject.

2 A device called a hemacytometer is used to measure the amount of hemoglobin present. Red blood cells have the ability to concentrate hemoglobin to about 34 g/100 mL of blood. Readings below 15 g/100 mL of blood indicate anemia. Blood appears pale if hemoglobin levels are low.
b) Which subject do you believe has a low level of hemoglobin: A, B, C, or D?

Case-Study Application Questions

1 Cancer of the white blood cells is called leukemia. Like other cancers, leukemia is associated with rapid and uncontrolled cell production. Using the data in the case study, predict which subject might be suffering from leukemia. Give your reasons.

2 Although hematocrits provide some information about blood disorders, most physicians would not diagnose leukemia on the basis of one test. What other conditions might explain the hematocrit reading you chose in question 1? Give your reasons.

3 Lead poisoning can cause bone marrow destruction. Which of the subjects in the case study might have lead poisoning? Give your reasons.

4 Which subject lives at a high altitude? Give your reasons.

5 Recently, athletes have begun to take advantage of the benefits of extra red blood cells. Two weeks prior to a competition, a blood sample is taken and centrifuged, and the red blood cell component is stored. A few days before the event, the red blood cells are injected into the athlete. Why would athletes remove red blood cells only to return them to their body later? What problems could be created should the blood contain too many red blood cells? Give your reasons. ■

BLOOD CLOTTING

Blood clotting maintains homeostasis by preventing the loss of blood from torn or ruptured blood vessels. Blood clots also forestall the rupture of weakened blood vessels by providing additional support.

Trillions of fragile platelets move through the blood vessels. Should the platelet strike a rough surface, such as that created by a torn blood vessel in a cut or abrasion, the platelet breaks apart and releases a protein called *thromboplastin*. The thromboplastin, along with calcium ions present in the blood, activates a plasma protein called *prothrombin*. Prothrombin, along with another plasma protein, called fibrinogen, is produced by the liver. Under the influence of thromboplastin, prothrombin is transformed into thrombin. In turn, thrombin acts as an enzyme by splicing two amino acids from the fibrinogen molecule. Fibrinogen is converted into fibrin threads, which wrap around the damaged area, sealing the cut in the skin with a clot. Invading microbes cannot gain access. Although the threads prevent red blood cells from passing into the damaged area, they provide a framework that white blood cells crawl over.

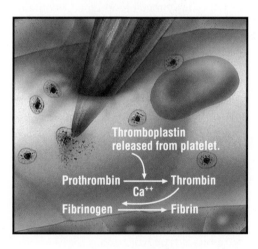

Although blood clotting preserves life, it can also result in life-threatening situations. A **thrombus** is a blood clot that seals a blood vessel. Because blood will not pass through the area, local tissues are not supplied with oxygen and nutrients. If a clot forms in the brain, cerebral thrombosis can cause a stroke. Coronary thrombosis—a clot in the coronary artery of the heart—can be equally dangerous.

Should a blood clot dislodge, it becomes an **embolus.** The embolus may travel through the body to lodge in a vital organ. Cerebral embolisms, coronary embolisms, and pulmonary embolisms can be life-threatening. What causes an embolus or thrombus is not completely understood, but scientists believe that genetic factors may be involved. It is known, however, that the incidence of thrombosis and embolisms becomes greater as people get older.

BLOOD GROUPS

In the 17th century, Jean-Baptiste Denis performed the first blood transfusion by injecting lamb's blood into a young boy. The youth survived, but a repeat of the experiment, on an older man, proved disastrous—the man died almost immediately. Denis attempted to explain what went wrong, but he lacked crucial information. (It was eventually revealed that the older man was being poisoned by his wife; Denis was exonerated.)

The idea that young people are healthier than older people became linked with the notion that older people have older blood. Occasionally, the transfusion of blood from a younger person to an older person worked: the greater oxygen-carrying capacity provided more energy. However, the recipient often died. Why do some transfusions help, while others kill?

At the turn of the 20th century, Karl Landsteiner discovered that different blood types exist. Therefore, the secret to successful transfusions was the correct matching of blood types. Special markers, called **glycoproteins,** are located on the membrane of some of the red

Figure 7.7

Platelets burst when they strike a sharp surface. The thromboplastin released from the platelet initiates a series of reactions that produce a blood clot.

*A **thrombus** is a blood clot that forms within a blood vessel.*

*An **embolus** is a blood clot that dislodges and is carried by the circulatory system to vital organs.*

Glycoproteins *are large chemical complexes composed of carbohydrates and protein. Glycoproteins can be found on cell membranes.*

blood cells. Individuals with blood type A have a special glycoprotein, the A marker, attached to their cell membrane. Individuals with blood type B have a special glycoprotein, the B marker, attached to their cell membrane. Individuals with blood type AB have both A and B markers attached to their cell membrane. Blood type O has no special marker.

Should an individual with blood type O receive blood from an individual with blood type A, the individual with blood type O would recognize the A marker as a foreign invader. The A marker acts as an **antigen** in the body of the individual with blood type O. Special proteins, called **antibodies,** are produced in response to a foreign invader. The antibodies attach to the antigen markers and cause the blood to clump. It is important to note that antigen A would not cause the same immune response if transfused into the body of an individual with blood type A. The marker associated with blood type A is not a foreign invader because it is part of the genetic makeup of that individual. A-type antigens are found on that individual's red blood cells. Table 7.2 summarizes the antigens and antibodies for the various blood groups.

The antibodies produced by the recipient act on the invading antigens. As shown in Figure 7.9, the antibodies cause the blood to **agglutinate,** or clump. The importance of the correct transfusion is emphasized by the fact that agglutinated blood can no longer pass through the tiny capillaries. The agglutinated blood therefore clogs local tissues and prevents the delivery of oxygen and nutrients. Individuals with type AB blood, in fact, possess both antigens and, therefore, are able to receive blood from any donor. Blood type AB is the universal acceptor. Blood type O is referred to as the universal donor because it can be received by individuals of all blood types. Since blood type O contains no antigen, it contains no special features not found in any of the other blood types. Although antibodies will not be produced against type O, individuals with blood type O can recognize antigens on other blood cells. In some ways, blood types O and AB provide a paradox. Blood type O, despite being the universal donor, may only accept blood from individuals with blood type O. Blood type AB, despite being the universal acceptor, may only donate blood to individuals with blood type AB.

> **BIOLOGY CLIP**
> Because red blood cells live only 120 days, they are continually breaking down and being replenished. The misconception that "young blood" is better than "old blood" persists even today. The blood of elderly people is virtually the same as the blood of young people.

*An **antigen** is a substance, usually protein in nature, that stimulates the formation of antibodies.*

***Antibodies** are proteins formed within the blood that react with antigens.*

***Agglutination** refers to the clumping of blood cells caused by antigens and antibodies.*

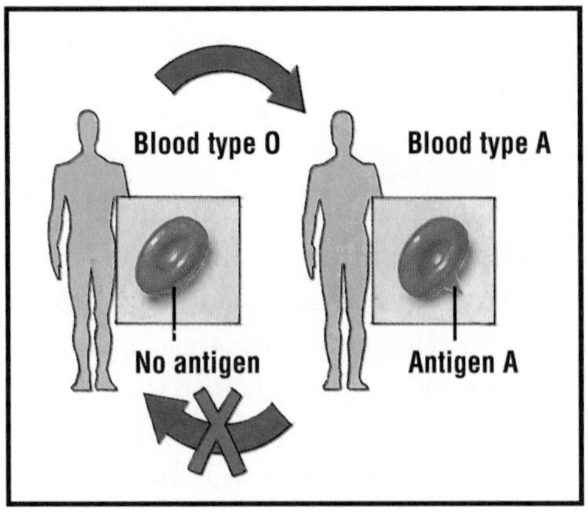

Figure 7.8

Individuals with blood type A can receive blood type O during a transfusion. However, individuals with blood type O cannot receive blood type A during a transfusion.

Table 7.2 Antigens and Antibodies Found in Blood Groups

Blood group	Antigen on red blood cell	Antibody in serum
O	none	A and B
A	A	B
B	B	A
AB	A and B	none

Blood type O — No antigen

Blood type A — Antigen A

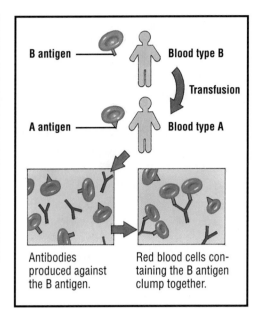

	Blood type of donor			
Blood type of recipient	**O**	**A**	**B**	**AB**
O				
A				
B				
AB				

B antigen → Blood type B

Transfusion

A antigen → Blood type A

Antibodies produced against the B antigen.

Red blood cells containing the B antigen clump together.

Figure 7.9

Agglutination response of blood type A (recipient) to blood type B (donor).

RHESUS FACTOR

During the 1940s scientists discovered another antigen on the red blood cell—the rhesus factor. Like the ABO blood groups, the rhesus factor is inherited. Individuals who have this special antigen are said to be Rh+. Approximately 85% of Canadians have the antigen. The remaining 15% of individuals who do not have the antigen are said to be Rh–. Individuals who are Rh– may donate blood to Rh+ individuals, but should not receive their blood. The human body has no natural antibodies against Rh factors, but antibodies can be produced following a transfusion. Although antibodies are produced in response to antigens, it should be pointed out that the immune reaction is subdued compared with that of the ABO group.

Rhesus-factor incompatibilities become important for Rh+ babies of Rh– mothers. If the baby inherits the Rh+ factor from the father, a condition called **erythroblastosis fetalis** can occur with the second and subsequent pregnancies. The first child is spared because the blood of the mother and blood of the baby are separated by the **placenta,** a thin membrane found between the developing embryo and mother. Located within the uterus, the placenta permits the movement of materials between mother and baby. Nutrients and oxygen move from the mother's blood into the baby, while wastes diffuse from the baby's blood into the mother for disposal. Although capillary beds from the mother and baby intertwine in the placenta, blood flows do not mix until birth. During birth, the placenta is removed from the uterus. Capillary beds rupture, and, for the first time, the blood of the baby comes in contact with the blood of the mother. The mother's immune system recognizes the Rh+ antigens and triggers the production of antibodies. By the time the antibodies are produced, the first baby, no longer connected to the placenta, has escaped the potentially dangerous environment.

A second pregnancy presents problems if the child is Rh+. The mother retains many of the antibodies from her first encounter with Rh+ blood. Should some of the antibodies move across the placenta, they attach to the antigen on the baby's red blood cells, causing them to clump. In this condition, red blood cells

Erythroblastosis fetalis, *or "blue baby," occurs when the mother's antibodies against Rh+ blood enter the Rh+ blood of her fetus.*

The **placenta** *is an organ made from the cells of the baby and the cells of the mother. It is the site of nutrient and waste exchange between mother and baby.*

are unable to pass through the narrow capillaries. Red blood cells jam the capillary entrances, and oxygen delivery is severely reduced. The body attempts to compensate by increasing red blood cell production, but even these new red blood cells contain antigens. Once again, the antibodies that have seeped into the baby's circulatory system agglutinate the new red blood cells. Low oxygen levels often cause the baby to turn blue, hence the name "blue baby."

Treatment for erythroblastosis fetalis involves a transfusion of Rh– blood. The antibodies from the mother will not attack the Rh– blood because it carries no antigens. Eventually, the baby begins producing Rh+ blood, but by then some of the mother's antibodies will have broken down—no protein lives forever. The effects of erythroblastosis fetalis may be decreased, if not eliminated altogether, by injecting the mother with a drug that inhibits the formation of antibodies against Rh+ antigens. The injection is given immediately after the woman's first birth. The mother's blood can then be monitored during the second pregnancy to determine whether or not antibodies are being developed.

FRONTIERS OF TECHNOLOGY: ARTIFICIAL BLOOD

On March 1, 1982, a precedent-setting legal case brought attention to an emerging medical technology. A man and woman, trying to push their car, were critically injured when they were hit by another car. Because of their religious beliefs, the couple refused a blood transfusion. During the legal dispute that ensued, the wife died, and the courts ruled that action must be taken to save the husband's life. Five liters of Fluosol—artificial blood—were transfused into the man over a period of five days. Doctors believed that the artificial blood could maintain adequate oxygen levels until the man's bone marrow began replenishing red blood cells.

Fluosol, a nontoxic liquid that contains fluorine, was developed in Japan. Fluosol carries both oxygen and carbon dioxide. It requires no blood matching and, when frozen, can be stored for long periods of time. Artificial blood, unlike human blood, does not have to undergo expensive screening procedures before being used in transfusions. Artificial blood will not carry the HIV or hepatitis viruses. However, despite its advantages, artificial blood is not as good as the real thing. Although it carries oxygen, it is ill-suited for many of the other functions associated with blood, such as blood clotting and immunity. The real value of artificial blood is that it provides time until natural human blood can be administered. It could also serve as a supplement for patients with diseases like thalassemia (Cooley's anemia) or aplastic anemia, which require multiple transfusions. Artificial blood might also help prevent an overload of iron.

IMMUNE RESPONSE

The body's first line of defense against foreign invaders is largely physical. The skin provides a protective barrier; only a few bacteria and parasites are specialized enough to break through the skin's barrier. In the respiratory passage, invading microbes and foreign debris become trapped in a layer of mucus and are swept away from the lungs by tiny hairlike structures called *cilia*.

Chemical protection is provided by the stomach, which secretes a strong acid that destroys many of the marauding cells.

SOCIAL ISSUE:
Smoking

The correlation between cigarette smoking and lung cancer has been established. Smoking has also been linked to high blood pressure, heart disease, and a host of other diseases. Individuals who smoke two or more packs of cigarettes a day have about a 25 times greater chance of contracting lung cancer than a nonsmoker. With increasing evidence of the dangers of second-hand smoke, nonsmokers' rights groups and health officials have successfully lobbied for a ban on smoking in the workplace and other public areas.

In 1986, the Canadian tobacco industry employed 6000 full-time and 51 000 seasonal workers in addition to the thousands of wholesale and retail workers who derive profits from the sale of tobacco products. Both federal and provincial taxes on tobacco products account for 67% of the retail price of a pack of cigarettes. In 1986 the federal government received over $1.8 billion in taxes from the sale of tobacco products, with the provinces earning a similar amount.

In 1992, workers in the tobacco industry protested that with its policy of increased taxes on tobacco products, the government was costing workers their jobs. The government responded that the cost of health care for people with tobacco-related illnesses far surpasses the revenues it generates from tobacco taxes. The government claims that by increasing the price of cigarettes, more people will stop smoking and the general health of the populace will improve.

Statement:

The government should ban the sales of tobacco products.

Point

- Smoking leads some people to addiction. Such people have great difficulty stopping smoking. Making tobacco unavailable would help them to quit. It would also save them a lot of money.
- According to one estimate, the costs of treating patients with respiratory and circulatory diseases brought on by smoking exceed one billion dollars every year. Smokers also miss more days of work and tend to be less productive on the job.
- The tobacco industry provides industry high profit returns for investment. This wealthy industry has a formidable lobby group.

Counterpoint

- Individuals in our society value freedom. If governments force people to give up smoking, what other actions might they take to restrict our civil liberties?
- The major portion of the cost of cigarettes is in the form of government taxes. These taxes help finance health and other government programs. If governments are concerned about our health, perhaps they should outlaw alcohol or overeating as well.
- Tobacco companies have been disclosing the dangers of smoking for many years. People who still choose to smoke should be permitted to do so. The legitimate profits made by the industry keep farmers on the farm and workers on the job.

> **Research the issue.**
> **Reflect on your findings.**
> **Discuss the various viewpoints with others.**
> **Prepare for the class debate.**

CHAPTER HIGHLIGHTS

- Respiration involves the intake, transportation, and utilization of oxygen by cells to produce energy.
- Air passes from the atmosphere to the pharynx, trachea, bronchi, bronchioles, and finally to the alveoli.
- Cilia are tiny hairlike protein structures found in eukaryotic cells. Cilia sweep foreign debris from the respiratory tract.
- The larynx, also called the voice box, is responsible for producing sound.
- Rib and diaphragm movements change pleural pressure. Inhalation and exhalation are regulated by changes in pleural pressure. The diaphragm is a sheet of muscle that separates the organs of the chest cavity from those of the abdominal cavity.
- Breathing movements are regulated by CO_2 and O_2 chemoreceptors. CO_2 receptors are more sensitive.
- Bronchial asthma is characterized by a narrowing of the bronchial passage.
- Emphysema involves an overinflation of the alveoli. Continued overinflation can lead to the rupture of the alveoli.

APPLYING THE CONCEPTS

1   Why do breathing rates increase in crowded rooms?

2   A patient is given a sedative that inhibits nerves leading to the pharynx, including those that control the epiglottis. What precautions would you take with this patient? Give your reasons.

3   A man has a chest wound. The attending physician notices that the man is breathing rapidly and gasping for air.

   a) Why does the man's breathing rate increase? Why does he gasp?

   b) What could be done to restore normal breathing?

4   During mouth-to-mouth resuscitation, exhaled air is forced into the victim's trachea. As you know, exhaled air contains higher levels of CO_2 than atmospheric air. Would the higher levels of CO_2 create problems or would they be beneficial? Provide your reasons.

5   Prior to swimming underwater, a diver breathes deeply and rapidly for a few seconds. How has hyperventilating helped the diver hold her breath longer?

   Use the following data table to answer questions 6, 7, and 8.

Individual	Breathing rate (breaths/min)	Hemoglobin (g/100 mL blood)	O₂ Content (mL/100 mL)
A (normal)	15	15.1	19.5
B	21	8	13.7
C	12	17.9	22.1
D	22	16.0	14.1

6   Which individual has recently moved from Calgary to Halifax? (Note: Halifax is at or near sea level, while Calgary is at a much higher altitude.) Give your reasons.

7   Which individual is suffering from dietary iron deficiency? Give your reasons.

8   Which individual has been exposed to low levels of CO? Give your reasons.

   Use the graph to answer question 9. The graph compares fetal and adult hemoglobin.

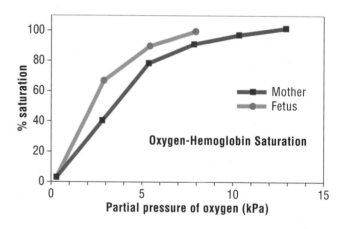

9   Which hemoglobin is more effective at absorbing oxygen? What adaptive advantage is provided by a hemoglobin that readily combines with oxygen?

Unit Two:
Exchange of Matter and Energy in Humans

CRITICAL-THINKING QUESTIONS

1 A scientist sets up the following experimental design.

Sodium hydroxide absorbs carbon dioxide. Limewater turns cloudy when it absorbs carbon dioxide.

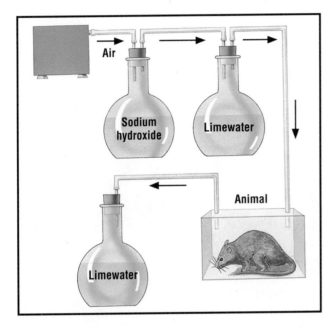

a) Indicate the purpose of the experiment.

b) Which flask acts as a control?

c) Why is sodium hydroxide used?

2 The following chart shows a comparison of inhaled and exhaled gases. On the basis of the experimental data, a scientist concludes that nitrogen from the air is used by the cells of the body. Critique the conclusion.

	Percentage by volume	
Component	Inhaled air	Exhaled air
Nitrogen	78.62%	74.90%
Oxygen	20.85%	15.30%
Carbon dioxide	0.03%	3.60%
Water (vapor)	0.50%	6.20%

3 Nicotine causes blood vessels, including those in the placenta, to constrict. Babies born to women who smoke are, on average, about 1 kg smaller than normal. This may be related to decreased oxygen delivery. Speculate about other problems that may face developing embryos due to the constriction of blood vessels in the placenta.

ENRICHMENT ACTIVITIES

1 Allan Becker, of the University of Manitoba, is studying dogs to learn more about how asthma works in people. Asthma is frequently associated with allergies, especially in children. Dr. Becker is especially interested in studying how allergic reactions affect airway function. Research how allergies have been linked with asthma. Indicate the benefits of using modelling experiments on dogs. Are there any disadvantages?

2 Suggested reading:
- Arehart-Treichel, Joan. "Lubricating Distressed Lungs." *Science News*, March 5, 1983, p. 150.
- Kanigel, Robert. "Nicotine Becomes Addictive." *Science Illustrated*, Oct/Nov, 1988, p. 32.
- Fletcher, William. "Asbestos." *Medical Self Care*, Nov/Dec, 1988, p. 32.

CHAPTER 9

Kidneys and Excretion

IMPORTANCE OF THE KIDNEYS

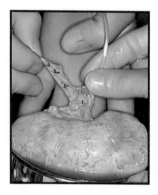

The cells of your body obtain energy by converting complex organic compounds into simpler compounds. Unfortunately, many of these simpler compounds can be harmful. To maintain life processes, the body must eliminate waste products. The lungs eliminate carbon dioxide, one of the products of cellular respiration. The large intestine removes toxic wastes from the digestive system. The liver transforms ingested toxins such as alcohol and heavy metals into soluble compounds that can be eliminated by the kidneys. The liver also transforms the hazardous products of protein metabolism into metabolites, which are then eliminated by the kidneys. In this chapter you will concentrate on the role that the kidneys play in removing waste, balancing blood pH, and maintaining water balance.

The average Canadian consumes more protein than is required to maintain tissues and promote cell growth. Excess protein is often converted into carbohydrates.

Figure 9.1

Kidney transplants are performed on patients with chronic renal failure. Transplants extend the life expectancy for these patients. However, survival is limited by the rejection of the transplanted organ by the immune system.

Proteins, unlike carbohydrates, contain a nitrogen molecule. The nitrogen and two attached hydrogen molecules, characteristic of amino acids, must be removed. (Recall that amino acids are the building blocks of proteins.) This process, referred to as **deamination**, occurs in the liver. The by-product of deamination is ammonia. Ammonia, like carbon dioxide, is a water-soluble gas. However, ammonia is extremely toxic—as little as 0.005 mg can kill you. Fish are able to avoid ammonia build-up by continually releasing it through their gills. Land animals, however, do not have the ability to release small quantities of ammonia throughout the day—wastes must be stored. Once again, the liver is called into action. In the liver, two molecules of ammonia combine with another waste product, carbon dioxide, to form **urea.** Urea is 1000 times less toxic than ammonia. The blood can dissolve 33 mg of urea per 100 mL of blood. A second waste product, **uric acid,** is formed by the breakdown of nucleic acids.

The kidneys also help maintain water balance. Although it is possible to survive for weeks without food, humans cannot survive more than a few days without water. Humans deplete their water reserves faster than their food reserves. The average adult loses about 2 L of water every day through urine, perspiration, and exhaled air. Greater volumes are lost when physical activity increases. For the body to maintain water balance, humans must consume 2 L of fluids daily. A drop in fluid intake by as little as 1% of your body mass will cause thirst, a decrease of 5% will bring about extreme pain and collapse, while a decrease of 10% will cause death.

> **BIOLOGY CLIP**
>
> Uric acid is only found in the urine of humans, higher apes, and Dalmatian dogs. The uric acid molecule has a structure similar to that of caffeine.

URINARY SYSTEM

Renal arteries branch from the aorta and carry blood to the paired kidneys. With a mass of about 0.5 kg, the fist-shaped kidneys may hold as much as one-quarter of the body's blood at any given time. Wastes are filtered from the blood by the kidneys and conducted to the urinary bladder by **ureters.** A sphincter muscle located at the base of the urinary bladder acts as a valve, permitting the storage of urine. When approximately 200 mL of urine has been collected, the bladder stretches slightly, and nerves signal the brain about the condition of the bladder. Should the bladder fill to 400 mL, more stretch receptors are activated and the message becomes more urgent. If you continue to ignore the messages, the bladder continues to fill, and after about 600 mL of urine accumulates, voluntary control is lost. The sphincter relaxes, urine enters the **urethra** and is voided.

Differences between males and females become evident in this last structure of the urinary system. In males, the urethra exits by way of the penis, providing a common pathway for sperm and urine from the body. In females, the reproductive and excretory functions remain distinct. The urethra of females lies within the vulva, the external genital organ, but has no connection to reproductive organs. The urethra of males is longer than that found in women. This may account for the fact that females are more prone to bladder infections than are males.

A cross section of a kidney reveals three different structures. An outer layer of connective tissue, the **cortex,** encircles the kidney. An inner layer, the

Deamination *is the removal of an amino group from an organic compound.*

Urea *is a nitrogen waste formed from two molecules of ammonia and one molecule of carbon dioxide.*

Uric acid *is a waste product formed from the breakdown of nucleic acids.*

Ureters *are tubes that conduct urine from the kidneys to the bladder.*

The **urethra** *is a tube that carries urine from the bladder to the exterior of the body.*

The **cortex** *is the outer layer of the kidney.*

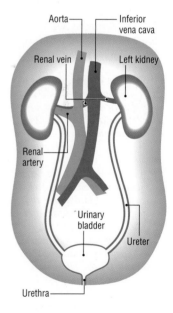

Figure 9.2

Simplified diagram of the human urinary system.

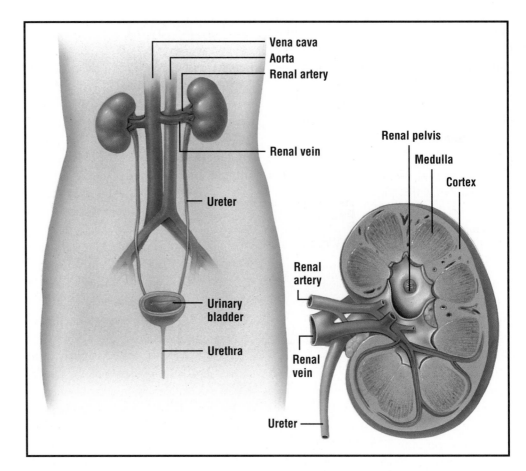

Vena cava
Aorta
Renal artery
Renal vein
Ureter
Urinary bladder
Urethra

Renal pelvis
Medulla
Cortex
Renal artery
Renal vein
Ureter

Figure 9.3

Structure of the human excretory system (left). Microscopic view of the kidney.

*The **medulla** is the area inside of the cortex.*

*The **renal pelvis** is the area in which the kidney joins the ureters.*

***Nephrons** are the functional units of the kidneys.*

*The **afferent arteriole** carries blood to the glomerulus.*

*The **glomerulus** is a high-pressure capillary bed that is surrounded by Bowman's capsule. The glomerulus is the site of filtration.*

*The **efferent arteriole** carries blood away from the glomerulus to a capillary net.*

***Bowman's capsule** is a cuplike structure that surrounds the glomerulus. The capsule receives filtered fluids from the glomerulus.*

medulla, is found beneath the cortex. A hollow chamber, the **renal pelvis,** joins the kidney with the ureter.

NEPHRONS

Approximately one million slender tubules, called the **nephrons,** are the functional units of the kidneys. Small branches from the renal artery, the **afferent arterioles,** supply the nephrons with blood. The afferent arterioles branch into a capillary bed, called the **glomerulus.** Unlike other capillaries, the glomerulus does not transfer blood to a venule. Blood leaves the glomerulus by way of another arteriole,

the **efferent arteriole.** Blood is carried from the efferent arteriole to a capillary net that wraps around the kidney tubule.

The glomerulus is surrounded by a funnel-like part of the nephron, called **Bowman's capsule.** Bowman's capsule, the afferent arteriole, and the efferent arteriole are located in the cortex of the kidney. Fluids to be processed into urine enter Bowman's capsule from the blood. The capsule tapers to a thin tubule, called the **proximal tubule.** Urine is carried from the proximal tubule to the *loop of Henle,* which descends into the medulla of the kidney. Urine moves through the **distal tubule,** the last seg-

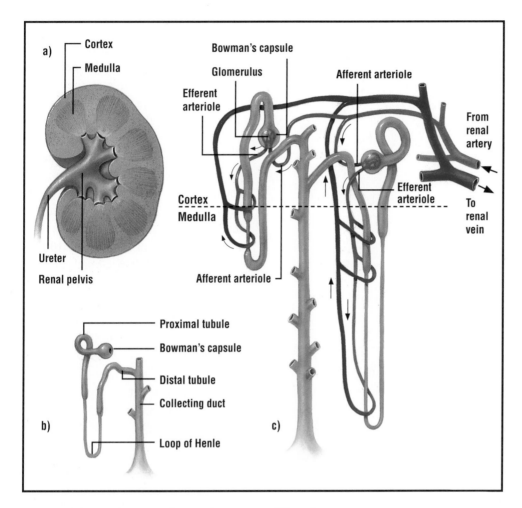

a)
Cortex
Medulla
Ureter
Renal pelvis

Bowman's capsule
Glomerulus
Efferent arteriole
Cortex
Medulla
Afferent arteriole

Afferent arteriole
From renal artery
Efferent arteriole
To renal vein

b)
Proximal tubule
Bowman's capsule
Distal tubule
Collecting duct
Loop of Henle

c)

Figure 9.4

(a) Anatomy of the kidney showing cortex and medulla.
(b) Diagram of the nephron.
(c) Blood vessels associated with the kidney.

*The **proximal tubule** is a section of the nephron joining Bowman's capsule with the loop of Henle. The proximal tubule is found within the cortex of the kidney.*

*The **distal tubule** conducts urine from the loop of Henle to the collecting duct.*

*The **collecting duct** receives urine from a number of nephrons and carries urine to the pelvis.*

***Filtration** occurs during the movement of fluids from the glomerulus into Bowman's capsule.*

*A **micropipette** is a thin glass tube that can be used to extract urine from the nephron.*

ment of the nephron, and into the **collecting ducts.** As the name suggests, the collecting ducts collect urine from many nephrons, which, in turn, merge in the pelvis of the kidney.

FORMATION OF URINE

Urine formation depends on three functions: filtration, reabsorption, and secretion. Filtration is accomplished by the movement of fluids from the blood into Bowman's capsule. Reabsorption involves the transfer of essential solutes and water from the nephron back into the blood. Secretion involves the movement of materials from the blood back into the nephron.

Filtration

Each nephron has an independent blood supply. Blood moves through the afferent arteriole into the glomerulus, a high-pressure filter. Normally, pressure in a capillary bed is about 2 kPa. The pressure in the glomerulus is 8 kPa. Dissolved solutes pass through the walls of the glomerulus into Bowman's capsule. While materials move from areas of high pressure to areas of low pressure, not all materials enter the capsule. Scientists have extracted fluids from the glomerulus and Bowman's capsule using a thin glass tube called a **micropipette.** Table 9.1 compares sample solutes extracted from the glomerulus and Bowman's capsule. Which solutes are filtered? Provide a hypothesis that explains selective permeability.

Table 9.1 Comparison of Solutes

Solute	Glomerulus	Bowman's capsule
Water	yes	yes
NaCl	yes	yes
Glucose	yes	yes
Amino acids	yes	yes
Hydrogen ions	yes	yes
Plasma proteins	yes	no
Erythrocytes	yes	no
Platelets	yes	no

Plasma protein, blood cells, and platelets are too large to move through the walls of the glomerulus. Smaller molecules pass through the cell membranes and enter the nephron.

Reabsorption

The importance of reabsorption can be emphasized by examining changes in the concentrations of fluids as they move through the kidneys. On average, about 600 mL of fluid flows through the kidneys every minute. Approximately 20% of the fluid, or about 120 mL, is filtered into the nephron. Imagine what would happen if none of the filtrate was reabsorbed. You would form 120 mL of urine each minute. You would also have to consume at least 1 L of fluids every 10 min to maintain homeostasis. Much of your day would be concerned with regulating water balance. Fortunately, only 1 mL of urine is formed for every 120 mL of fluids filtered into the nephron. The remaining 119 mL of fluids and solutes are reabsorbed.

Selective reabsorption occurs by both active and passive transport. Carrier molecules move Na^+ ions across the cell membranes of the cells that line the

Figure 9.5

Carrier molecules transport Na^+ ions from the nephron to the blood.

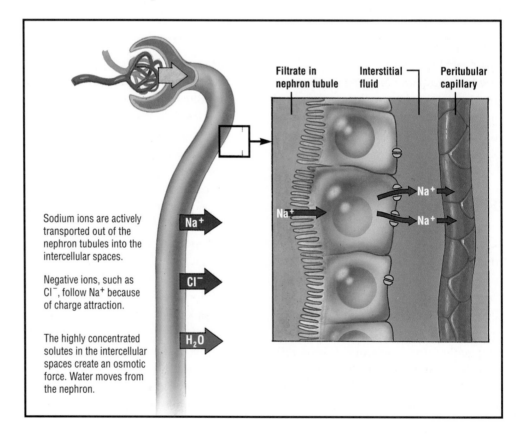

Sodium ions are actively transported out of the nephron tubules into the intercellular spaces.

Negative ions, such as Cl^-, follow Na^+ because of charge attraction.

The highly concentrated solutes in the intercellular spaces create an osmotic force. Water moves from the nephron.

Filtrate in nephron tubule **Interstitial fluid** **Peritubular capillary**

Na⁺   Cl⁻   H₂O

nephron. Negative ions, such as Cl^- and HCO_3^-, follow the positive Na^+ ions by charge attraction. Numerous mitochondria supply the energy necessary for active transport. However, the energy supply is limited. Reabsorption occurs until the **threshold level** of a substance is reached. Excess NaCl remains in the nephron and is excreted with the urine.

Other molecules are actively transported from the proximal tubule. Glucose and amino acids attach to specific carrier molecules, which shuttle them out of the nephron and into the blood. However, the amount of solute that can be reabsorbed is limited. For example, excess glucose will not be shuttled out of the nephron by the carrier molecules. This means that individuals who have diabetes mellitus, a disease characterized by high blood glucose, will lose excess glucose in their urine. Individuals who consume large amounts of simple sugars also excrete some of the excess glucose.

The solutes that are actively transported out of the nephron create an osmotic gradient that draws water from the nephron. A second osmotic force, created by the proteins not filtered into the nephron, also helps reabsorption. The proteins remain in the bloodstream and draw water from the intercellular spaces into the blood. As water is reabsorbed from the nephron, the remaining solutes become more concentrated. Molecules such as urea and uric acid will diffuse from the nephron back into the blood, although less is reabsorbed than was originally filtered.

Secretion

Secretion is the movement of wastes from the blood into the nephron. Nitrogen-containing wastes, histamine, excess H^+ ions, and other minerals are balanced by secretion. Even drugs such as penicillin can be secreted. Cells loaded with mito-

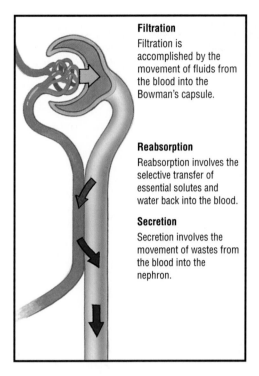

Filtration
Filtration is accomplished by the movement of fluids from the blood into the Bowman's capsule.

Reabsorption
Reabsorption involves the selective transfer of essential solutes and water back into the blood.

Secretion
Secretion involves the movement of wastes from the blood into the nephron.

Threshold level, in terms of reabsorption, is the maximum amount of material that can be moved across the nephron.

Figure 9.6
Overview of the steps in urine formation: filtration, reabsorption, and secretion.

chondria line the distal tubule. Like reabsorption, tubular secretion occurs by active transport, but, unlike reabsorption, molecules are shuttled from the blood into the nephron.

■ **REVIEW QUESTIONS** ■ ?

1 List the three main functions of the kidneys.
2 What is deamination and why is it an important process?
3 How does the formation of urea prevent poisoning?
4 Diagram and label the following parts of the excretory system: kidney, renal artery, renal vein, ureter, bladder, and urethra. State the function of each organ.
5 Diagram and label the following parts of the nephron: Bowman's capsule, proximal tubule, loop of Henle, distal tubule, and collecting duct. State the function of each of the parts.
6 List and describe the three processes involved in urine formation.

CASE STUDY

COMPARING SOLUTES IN THE PLASMA, NEPHRON, AND URINE

Objective

To compare solutes along the nephron.

Procedure

Micropipettes were used to draw fluids from Bowman's capsule, the glomerulus, the loop of Henle, and the collecting duct. The data are displayed in the table below. Unfortunately, some of the data were not recorded. The absence of data is indicated on the table.

Solute	Glomerulus	Bowman's capsule	Loop of Henle	Collecting duct
Protein	8.0	0	0	0
Urea	0.05	0.05	1.50	2.00
Glucose	0.10	no data	0	0
Chloride	0.37	no data	no data	0.6
Ammonia	0.0001	0.0001	0.0001	0.04
Substance X	9.15	0	0	0

Note: Quantities are recorded in grams per 100 mL.

Case-Study Application Questions

1 Which of the solutes was not filtered into the nephron? Explain your answer.
2 Unfortunately, the test for glucose was not completed for the sample taken from the glomerulus. Predict whether or not glucose would be found in the glomerulus. Provide reasons for your prediction.
3 Why do urea and ammonia levels increase after filtration occurs?
4 Chloride ions (Cl^-) follow actively transported Na^+ ions from the nephron into the blood. Would you not expect the Cl^- concentration to decrease as fluids are extracted along the nephron? What causes the discrepancy noted? ■

WATER BALANCE

You adjust for increased water intake by increasing urine output. Conversely, you adjust for increased exercise or decreased water intake by decreasing urine output. The kidneys are involved in regulating body fluid levels. The adjustments involve the interaction of the body's two communication systems: the nervous system and the endocrine system.

Regulating ADH

A hormone called the **antidiuretic hormone (ADH)** helps regulate the osmotic pressure of body fluids by acting on the kidneys to increase water reabsorption. When ADH is released, a more concentrated urine is produced, thereby conserving body water. ADH is produced by specialized nerve cells in the brain in an area called the hypothalamus. ADH moves along specialized fibers from the hypothalamus to the pituitary gland. The pituitary stores and releases ADH into the blood.

Specialized nerve receptors, called *osmoreceptors*, located in the hypothalamus detect changes in osmotic pressure. When you decrease water intake or

*The **antidiuretic hormone (ADH)** acts on the kidneys to increase water reabsorption.*

increase water loss, by sweating for example, the solutes of the blood become more concentrated. This increases the blood's osmotic pressure. Consequently, water moves into the bloodstream, causing the cells of the hypothalamus to shrink. When this happens, a nerve message is sent to the pituitary, signalling the release of ADH, which is carried by the bloodstream to the kidneys. By reabsorbing more water, the kidneys produce a more concentrated urine, preventing the osmotic pressure of the body fluids from increasing any further.

The shrinking of the hypothalamus also initiates a behavioral response: the sensation of thirst. If more water is taken in, it is absorbed by the blood and the concentration of solutes in the blood decreases. The greater the volume of water consumed, the lower the osmotic pressure of the blood. As the blood becomes more dilute, fluids move from the blood into the hypothalamus. The hypothalamus swells, and nerve messages to the pituitary stop. Less ADH is released, and thus less water is reabsorbed from the nephrons.

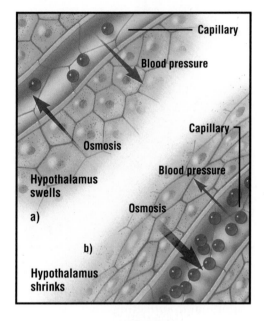

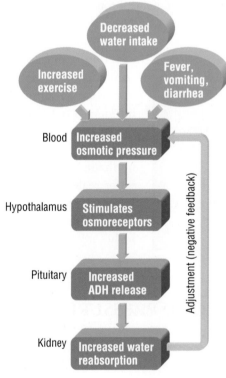

Figure 9.8

Physiological response to increased osmotic pressure of body fluids. High concentrations of solutes will cause the release of ADH. By increasing water reabsorption in the kidney, ADH helps conserve body water.

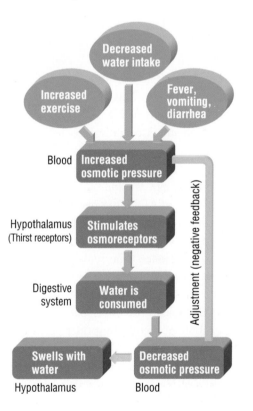

Figure 9.7

(a) Solutes are dilute. Osmotic pressure is balanced by blood pressure, and the hypothalamus will not lose water. (b) If the solute concentration increases, osmotic pressure of the blood increases, more water is drawn into the blood, and the cells of the hypothalamus shrink. ADH is released when the hypothalamus shrinks.

Figure 9.9

The special receptors within the hypothalamus cause the thirst response. If water is consumed, the concentration of blood solutes decreases.

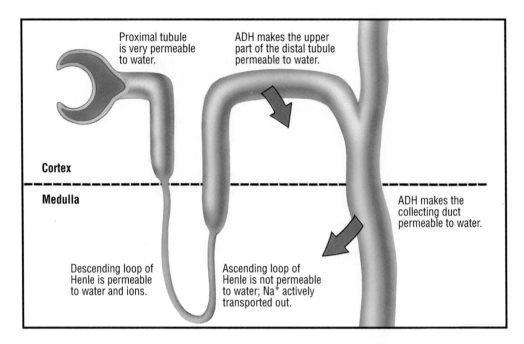

Proximal tubule is very permeable to water.

ADH makes the upper part of the distal tubule permeable to water.

Cortex

Medulla

ADH makes the collecting duct permeable to water.

Descending loop of Henle is permeable to water and ions.

Ascending loop of Henle is not permeable to water; Na$^+$ actively transported out.

Figure 9.10

ADH acts on the upper section of the distal tubule and collecting duct by increasing permeability to water. Regulation of water reabsorption in these two areas allows the kidney to balance the osmotic pressure of body fluids.

Aldosterone *is a hormone that increases Na$^+$ reabsorption from the distal tubule and collecting duct.*

ADH and the Nephron

Approximately 85% of the water filtered into the nephron is reabsorbed in the proximal tubule. Although the proximal tubule is very permeable to water, this permeability does not extend to other segments of the nephron. The descending loop of Henle is permeable to water and ions, but the ascending tubule is only permeable to NaCl. Active transport of Na$^+$ ions from the ascending section of the loop concentrates solutes within the medulla of the kidney. Without ADH, the rest of the tubule remains impermeable to water, but continues to actively transport Na$^+$ ions from the tubules. The remaining 15% of the water filtered into the nephron will be lost if no ADH is present.

ADH makes the upper part of the distal tubule and collecting duct permeable to water. When ADH makes the cell membranes permeable, the high concentration of NaCl in the intercellular spaces

BIOLOGY CLIP

Alcohol consumption decreases the release of ADH. Some of the symptoms experienced the day following excessive alchohol consumption can be attributed to increased water loss by urine and decreased body fluid levels.

creates an osmotic pressure that draws water from the upper section of the distal tubule and collecting duct. As water passes from the nephron to the intercellular spaces and the blood, the urine remaining in the nephron becomes more concentrated. It is important to note that the kidneys only control the last 15% of the water found in the nephron. By varying water reabsorption, the kidneys regulate the fluid volumes of the body.

KIDNEYS AND BLOOD PRESSURE

The kidneys also play a role in the regulation of blood pressure by adjusting for blood volumes. A hormone called **aldosterone** acts on the nephrons to increase Na$^+$ reabsorption. The hormone is produced in the *adrenal cortex*, which lies in the outer core of the adrenal gland, above the kidneys. Not surprisingly, as NaCl reabsorption increases, the osmotic gradient increases and more water moves out of the nephron by osmosis.

Conditions that lead to increased fluid loss can decrease blood pressure. Blood pressure receptors in the *juxtaglomerular apparatus* detect low blood pressure. The juxtaglomerular apparatus, as the name suggests, is found near the glomerulus. Specialized cells within the structure release *renin*, an enzyme that converts angiotensinogen into angiotensin. *Angiotensinogen* is a plasma protein produced by the liver. *Angiotensin*, the activated form, has two important functions. First, the activated enzyme causes constriction of blood vessels. Blood pressure increases when the diameter of blood vessels is reduced. Second, angiotensin initiates the release of aldosterone from the adrenal gland. The aldosterone is then carried in the blood to the kidneys, where it acts on the cells of the distal tubule and collecting duct to increase Na^+ transport. This causes the fluid level of the body to increase.

KIDNEY DISEASE

Proper functioning of the kidneys is essential if the body is to maintain homeostasis. The multifunctional kidneys are affected when other systems break down. Similarly, kidney dysfunctions have an impact on other systems. A variety of kidney disorders can be detected by urinalysis.

Diabetes Mellitus

Diabetes mellitus is caused by inadequate secretion of insulin from islet cells in the pancreas. Without insulin, blood sugar levels tend to rise. The cells of the proximal tubule are supplied with enough ATP to reabsorb 0.1% blood sugar, but in diabetes mellitus much higher blood sugar concentrations are found. The excess sugar remains in the nephron. This excess sugar provides an osmotic pressure that opposes the osmotic pressure created by other solutes that have been actively transported out of the nephron. Water remains in the nephron and is lost with the urine. Individuals with diabetes mellitus void large volumes of sweet urine. This explains why individuals with diabetes mellitus are often thirsty. The water lost with the excreted sugar must be replenished.

Diabetes Insipidus

The destruction of the ADH-producing cells of the hypothalamus or the destruction of the nerve tracts leading from the hypothalamus to the pituitary gland can cause diabetes insipidus. Without ADH to regulate water reabsorption, urine output increases dramatically. In

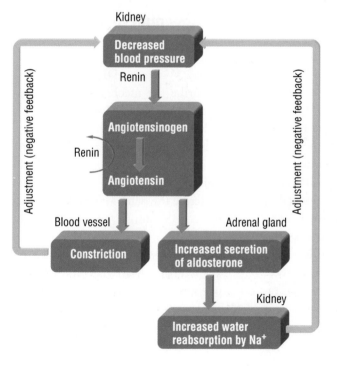

Figure 9.11

The hormone aldosterone maintains homeostasis by increasing water reabsorption.

extreme cases, as much as 20 L of urine can be produced each day, creating a strong thirst response. A person with diabetes insipidus must drink large quantities of water to replace what he or she has not been able to reabsorb.

Bright's Disease
Named after Richard Bright, a 19th-century English physician, this disease is also called *nephritis*. Nephritis is not a single disease, but a broad description of many diseases characterized by inflammation of the nephrons. One type of nephritis affects the tiny blood vessels of the glomerulus. It is believed that toxins produced by invading microbes destroy the tiny blood vessels, altering the permeability of the nephron. Proteins and other large molecules are able to pass into the nephron. Because no mechanism is designed to reabsorb protein, the proteins remain in the nephron and create an osmotic pressure that draws water into the nephron. The movement of water into the nephron increases the output of urine.

Kidney Stones
Kidney stones are caused by the precipitation of mineral solutes from the blood. Kidney stones can be categorized into two groups: alkaline and acid stones. The sharp-sided stones may become lodged in the renal pelvis or move into the narrow ureter. Delicate tissues are torn as the stone moves toward the bladder. The stone may move farther down the excretory passage and lodge in the urethra, causing excruciating pain as it moves.

> **BIOLOGY CLIP**
> Operations to remove kidney stones were performed in the time of Hippocrates, the Greek physician considered to be the father of medicine (ca 460–ca 377 B.C.).

FRONTIERS OF TECHNOLOGY: BLASTING KIDNEY STONES

The traditional treatment for kidney stones has been surgical removal, followed by a period of convalescence. A technique developed by Dr. Christian Chaussy in Germany has greatly improved prospects for kidney stone patients. The technique utilizes high-energy sound waves to smash the kidney stones. No surgery is required, and recovery time is greatly reduced. Sound waves bounce off soft tissue but penetrate the stone. The collision of waves that bounce off soft tissue with those that penetrate the stone cause the stone to explode. After a few days, tiny granules from the stone can be voided through the excretory system.

■ REVIEW QUESTIONS ■ ?

7 What is ADH and where is it produced? Where is ADH stored?

8 Describe the mechanism that regulates the release of ADH.

9 Where is the thirst center located?

10 Describe the physiological adjustment to increased osmotic pressure in body fluids.

11 Describe the behavioral adjustment to increased osmotic pressure in body fluids.

12 Discuss the mechanism by which aldosterone helps maintain blood pressure.

13 What are kidney stones?

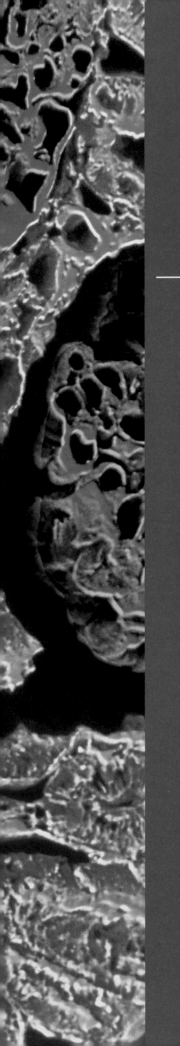

RESEARCH IN CANADA

Dr. Linda Peterson, a research scientist at the University of Ottawa, is studying the way in which the kidneys regulate water balance. The aim of Dr. Peterson's research is to investigate the interactions of hormones, such as ADH and aldosterone, and electrolytes, such as sodium, calcium, and chloride ions, in the conservation of salt by the ascending limb of the nephron.

Kidneys and Water Balance

Dr. Peterson's earlier studies have shown that ADH stimulates salt conservation by the ascending limb. One of the major questions yet to be answered is how calcium affects the ascending limb. Although a number of other studies have indicated that high blood calcium levels impair the conservation of water, the exact mechanism is still unknown. Dr. Peterson has demonstrated that high blood calcium levels increase the production of a hormone-like substance called prostaglandin E2. Since prostaglandin inhibits the effects of ADH in other segments of the nephron, high levels of the hormone may be responsible for impairing the activity of cells in the ascending limb.

A better understanding of water conservation will offer scientists a more complete picture of how humans maintain a constant internal environment.

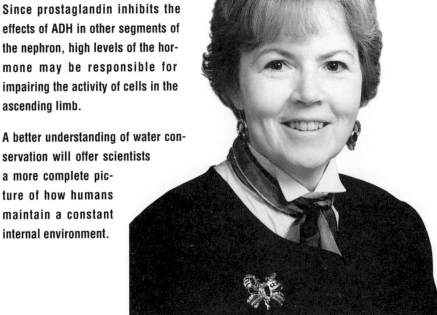
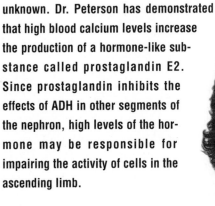

DR. LINDA PETERSON

LABORATORY

DIAGNOSIS OF KIDNEY DISORDERS

Objective

To identify kidney disorders by urinalysis.

Materials

Combustrip
10 mL graduated cylinder
distilled water in wash bottle
4 small test tubes
4 urine samples (simulated) W, X, Y, and Z

Procedure

1 Using a graduated cylinder, measure 2 mL of urine sample W and place it in a test tube. Label the test tube W. Rinse the graduated cylinder with distilled water and repeat the procedure for samples X, Y, and Z.
   a) Why was the graduated cylinder washed?

2 Place separate Combustrips in each of the test tubes and leave for one min.

3 Compare the color bars for glucose, protein, and pH with the chart provided in the package. (Note: Different companies use slightly different charts.)

b) Record the values for each of the samples in a table similar to the one below.

Sample	Glucose	Protein	pH
W			
X			
Y			
Z			

Laboratory Application Questions

1 Which subject do you suspect has diabetes mellitus? Provide your reasons.

2 Which subject do you suspect has diabetes insipidus? Provide your reasons.

3 Which subject do you suspect has Bright's disease? Provide your reasons.

4 Which subject do you suspect has lost a tremendous amount of body water while exercising? Provide your reasons. ■

INVESTIGATION
CAREER

Our aging population will require a larger and more complete health care system than is in place today. The emphasis in the field is expected to shift from the treatment of diseases to the prevention of disease and the improvement in the quality of life for the elderly. In the long run, it will be more efficient and less costly to prevent problems than to cure them with drugs and expensive surgery.

Geriatric specialist

Geriatrics is the branch of medicine that deals with health problems related to old age. The population of Canada is expected to age dramatically over the next few decades. A greater percentage of the population will be retired, and will require specialized health care.

Dietician/Nutritionist

Our society has never been more aware that a healthy diet is a major part of a healthy lifestyle. Dieticians and nutritionists are required by hospitals, retirement homes, and food manufacturers.

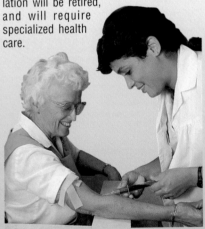

Home health aide

More people are choosing to recover from injuries and illnesses in their own home rather than in a hospital. Not only does this save money, but the familiar and comfortable surroundings of a person's home also speed recovery. This practice should lead to a high demand for home health-care aides. A home health aide requires a general understanding of the human body and a knowledge of emergency procedures such as cardiopulmonary resuscitation (CPR).

Physiotherapist

Physiotherapists help people who have been in accidents or who have degenerative diseases regain their health through treatments such as ultrasound, laser therapy, and massage. Physiotherapy allows people to lead more active lives during their recovery period and decreases the amount of time needed for rehabilitation. A physiotherapist must have a thorough understanding of how the body works.

- Identify a career that requires a knowledge of the exchange of energy and matter in humans.
- Investigate and list the features that appeal to you about this career. Make another list of features that you find less attractive about this career.
- Which high-school subjects are required for this career? Is a post-secondary degree required?
- Survey the newspapers in your area for job opportunities in this career.

SOCIAL ISSUE:
Fillings and Kidney Disease

*Experiments by Drs. Murray Vimy and Fritz Lorscheider at the University of Calgary Medical School point out the potential risks of using mercury fillings for teeth. Studies carried out on sheep indicated that half of their kidneys became dysfunctional within 30 days after 12 mercury amalgam fillings were placed in their mouths. The results of this study were published in the prestigious journal, **The Physiologist,** in August 1990, and presented to the American Physiology Society in October of that year. The two researchers believe that mercury in the amalgam filling can leak into the bloodstream, where it can do harm to the kidneys.*

Statement:

Alloy fillings that contain mercury should be banned, pending further research.

Point

- Mercury has long been perceived as dangerous when used in dental fillings. Dentists should discontinue its use until conclusive evidence proves otherwise.

- Dentists can use alternative substances for fillings, such as gold, ceramic, and different kinds of resins.

Counterpoint

- Not all researchers agree with the interpretations of Drs. Vimy and Lorscheider. Their studies were not carried out on humans. Amalgam fillings have been used for many years on humans, the vast majority of whom have well-functioning kidneys.
- How do we know that the replacements are less harmful? Dr. Bill Long, past president of the Calgary and District Dental Society, has stated that the amalgam fillings are safe and that replacement materials are not as effective.

Research the issue.
Reflect on your findings.
Discuss the various viewpoints with others.
Prepare for the class debate.

CHAPTER HIGHLIGHTS

- The liver processes wastes, making them soluble.
- The excretory system rids the body of wastes carried in the blood.
- The kidneys filter wastes from the blood, help regulate blood pH, and regulate water balance.
- The nephron is the functional unit of a kidney.
- Urine formation involves filtration, reabsorption, and secretion.

- ADH regulates water balance by controlling absorption of the remaining 15% of the water filtered into the nephron. ADH acts on the cells of the distal tubule and collecting duct.
- Aldosterone regulates the reabsorption of sodium in the distal tubule and collecting duct.
- A number of diseases affect proper kidney function including diabetes mellitus and Bright's disease.

APPLYING THE CONCEPTS

1 Why is the formation of urea by the liver especially important for land animals?

2 Predict how a drop in blood pressure would affect urine output. Give reasons for your prediction.

3 A drug causes dilation of the afferent arteriole and constriction of the efferent arteriole. Indicate how the drug will affect urine production.

4 Why do the walls of the proximal tubule contain so many mitochondria?

5 In an experiment, the pituitary gland of a dog is removed. Predict how the removal of the pituitary gland will affect the dog's regulation of water balance.

6 How does excessive salt intake affect the release of ADH from the pituitary gland?

7 A drug that inhibits the formation of ATP by the cells of the proximal tubule is introduced into the nephron. How will the drug affect urine formation? Provide a complete physiological explanation.

8 A blood clot lodges in the renal artery and restricts blood flow to the kidney. Explain why this condition leads to high blood pressure.

9 For every 100 mL of salt water consumed, 150 mL of body water is lost. The solute concentration found in seawater is greater than that found in the blood. Provide a physiological explanation to account for the loss of body water. (Hint: Consider the threshold level for salt reabsorption by the cells of the nephron.)

10 Explain why the presence of proteins in the urine can lead to tissue swelling, or *edema*.

CRITICAL-THINKING QUESTIONS

1 Alcohol is a diuretic, a substance that increases the production of urine. It also suppresses the production and release of ADH. Should individuals who are prone to developing kidney stones consume alcohol? Explain.

2 In an experiment, four subjects each consumed four cups of black coffee per hour for a two-hour duration. A control group each consumed four cups of water per hour over the same two hours. It was noted that the group consuming coffee had a greater urine output than the control group. Provide a hypothesis that accounts for the data provided. Note: Formulate your hypothesis with an "If ... then ..." statement.

3 Design an experiment that would test the hypothesis developed in question 2.

4 Diseases such as syphilis, a venereal disease, can cause glomerular nephritis. (Other factors can also give rise to this disorder.) Glomerular nephritis is associated with the destruction of the high-pressure capillaries that regulate filtration. This occurs when proteins and other large molecules enter the filtrate. Once the filter is destroyed, it cannot be repaired. When a significant number of nephrons are destroyed, kidney function fails. Patients who suffer kidney failure can be treated by artificial dialysis or kidney transplants, two extremely expensive techniques. Should everyone who needs a kidney transplant be given the same priority?

ENRICHMENT ACTIVITIES

1 Research how kidney machines remove toxins from the blood.

2 The following article will provide you with more information on fillings:
- Wilkerfield, Irene, and Hawley Truax. "The Fear of Fillings." *Environmental Action*, Nov/Dec 1987, p. 10.

3 Using dialysis tubing, construct a working model of an artificial kidney machine. Does your kidney machine use force filtration?

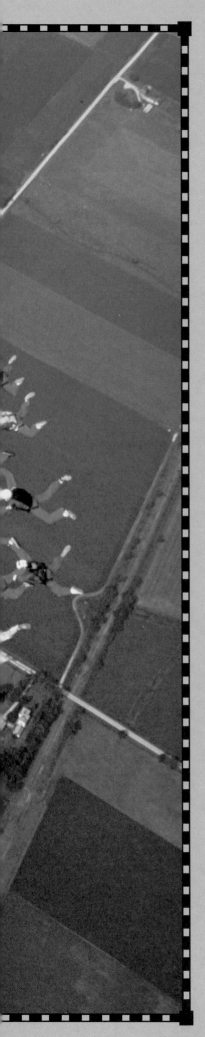

Coordination and Regulation in Humans

■

UNIT **3**

*T*he Endocrine System and Homeostasis

Hormones *are chemicals released by cells that affect cells in other parts of the body. Only a small amount of a hormone is required to alter cell metabolism.*

Endocrine hormones *are chemicals secreted by glands directly into the blood.*

Target tissues *have specific receptor sites that bind with the hormones.*

IMPORTANCE OF THE ENDOCRINE SYSTEM

The trillions of cells of the body all interact with each other—no cell operates in isolation. The integration of body functions depends on chemical controls. **Hormones** are chemical regulators produced by cells in one part of the body that affect cells in another part of the body. Chemicals produced in glands and secreted directly into the blood are referred to as **endocrine hormones.** The circulatory system carries these hormones to the various organs of the body.

Hormones can be classified according to their activation site. Hormones such as growth hormone, which regulates the development of the long bones; insulin, which regulates blood sugar; and epinephrine (adrenaline), which is produced in times of stress, affect many cells throughout the body. These hormones are referred to as *nontarget hormones*. Hormones such as parathyroid hormone, which regulates calcium levels in the body, and gastrin, which stimulates cells of the stomach to produce digestive enzymes, affect specific cells, or **target tissues.**

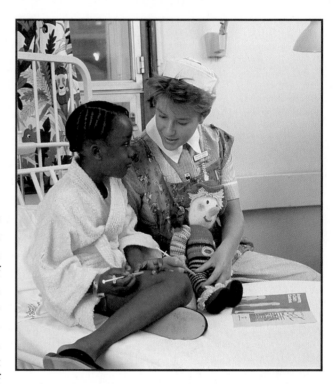

Figure 10.1

A nurse teaching a child with diabetes mellitus to inject herself with insulin.

CHEMICAL CONTROL SYSTEMS

Along with the nervous system, the endocrine system provides integration and control of the various organs and tissues. The malfunction of one organ affects other organs. However, an animal can continue to function because of compensations made by the two control systems. The nervous system enables the body to adjust quickly to changes in the environment. The endocrine system is designed to maintain control over a longer duration. Growth hormone and the various sex hormones, for example, regulate and sustain development for many years.

The division between the nervous system and endocrine system is most obscure in an area of the brain called the *hypothalamus*. The hypothalamus regulates the pituitary gland, referred to as the master gland of the body, through nerve stimulation. However, the endocrine glands, stimulated by the pituitary, secrete chemicals that affect the nerve activity of the hypothalamus.

The name hormone comes from the Greek *hormon*, meaning to excite or set into motion. Hormones do not serve the body as chemical products, but rather as regulators, speeding up or slowing down certain bodily processes. Long before scientists began to study the clinical effects of hormones, farmers knew that castrated bulls (steers) produced better meat. Today we know that chemicals within the testes of the bull are associated with the animal's aggressive nature and with tougher meat. Steers have greater value because of their docility and increased body mass.

This relationship between chemical messengers and the activity of organ systems within the body was established by experiment. In 1899, Joseph von Merring and Oscar Minkowski showed that a chemical messenger produced in the pancreas was responsible for the regulation of blood sugar. After removing the pancreases from a number of dogs, the two scientists noticed that the animals began to lose weight very quickly. Within a few hours the dogs became fatigued and displayed some of the symptoms that are now associated with human sugar diabetes. By chance, the two scientists also noted that ants began gathering in the kennels where the sick dogs were kept. No ants, however, were found in the kennels of healthy dogs. What caused the ants to gather? The scientists analyzed the urine of the sick dogs and found that it contained glucose, a sugar, while the urine of the healthy dogs did not. The ants had been attracted to the sugar. The experiment provided evidence that a chemical messenger, produced by the pancreas, was responsible for the regulation of blood sugar. This chemical is the hormone called **insulin.**

The von Merring and Minkowski experiment typified classical approaches to uncovering the effect of specific hormones. In many cases a gland or organ was removed and the effects on the organism were monitored. Once the changes in behavior were noted, chemical extracts from the organ were often injected into the animal, and the animal's activities monitored. By varying dosages of the identified chemical messenger, scientists hoped to determine how it worked. Although effective to a certain degree, classical techniques were limited. No hormonal response works independently. The concentrations of other hormones often increase to help compensate for a disorder. Some glands produce many different hormones. Therefore, the effect cannot be attributed to a single hormone. For example, early experimenters who attempted to uncover the function of the thyroid gland unwittingly

Insulin *is a hormone produced by the islets of Langerhans in the pancreas. Insulin is secreted when blood sugar levels are high.*

removed the parathyroid gland along with the thyroid. It might be expected that these tiny glands are part of the thyroid and are related to its function. However, although the parathyroid is embedded in the tissue of the thyroid, it is a separate organ. What the scientists failed to realize is that many illnesses they associated with low thyroid secretions were actually created by the parathyroid.

The main problem for early researchers was obtaining and isolating the actual messenger among the other chemicals found within the removed organ. Most hormones are found in very small amounts. Furthermore, the concentration of hormone varies throughout the day. The prediction of site and time was often a matter of mere luck.

In the past few years, technological improvements in chemical analysis techniques and microscopy have vastly increased our knowledge of the endocrine system. Radioactive tracers enable scientists to follow messenger chemicals from the organ in which they are produced to the target cells. The radioactive tracers also allow researchers to discern how the chemical messenger is broken down into other compounds and removed as waste. With new chemical analysis equipment, scientists can determine and measure the concentration of even the smallest amounts of a hormone as the body responds to changes in the external environment. In addition, high-power microscopes provide a clearer picture of the structure of cell membranes and along with it a better understanding of how chemical messengers attach themselves to target sites.

Steroid hormones *are made from cholesterol. This group includes male and female sex hormones and cortisol.*

Protein hormones *are composed of chains of amino acids. This group includes insulin, growth hormone, and epinephrine.*

Cyclic AMP *is a secondary chemical messenger that directs the synthesis of protein hormones by ribosomes.*

BIOLOGY CLIP
Protein hormones that do not attach to receptor molecules on target cells are removed from the body by the liver or kidney. The presence of these hormones can be monitored by urinalysis.

CHEMICAL SIGNALS

How do hormones signal cells? First, it is important to note that hormones do not affect all cells. Cells may have receptors for one hormone but not another. The number of receptors found on individual cells may also vary. For example, liver cells and muscle cells have many receptor sites for the hormone insulin. Fewer receptor sites are found in less active cells such as bone cells and cartilage cells.

Second, there are two different types of hormones that differ in chemical structure and action. **Steroid hormones,** which include both male and female sex hormones and cortisol, are made from cholesterol, a lipid compound. Steroid molecules are composed of complex rings of carbon, hydrogen, and oxygen molecules. Steroid molecules are not soluble in water, but are soluble in fat. The second group, **protein,** or **protein-related hormones,** includes insulin, growth hormone, and epinephrine. These hormones contain chains of amino acids of varying length and are soluble in water.

Steroid hormones diffuse from the capillaries into the interstitial fluid, and then into the target cells, where they combine with receptor molecules located in the cytoplasm. The hormone-receptor complex then moves into the nucleus and attaches to a segment of chromatin that has a complementary shape. The hormone activates a gene that sends a message to the ribosomes in the cytoplasm to begin producing a specific protein.

Protein hormones exhibit a different action. Unlike steroid hormones, which diffuse into the cell, protein hormones combine with receptors on the cell mem-

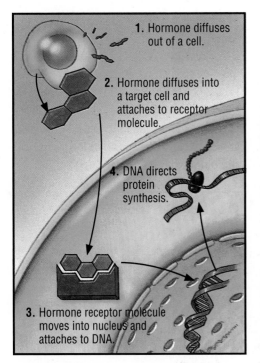

1. Hormone diffuses out of a cell.

2. Hormone diffuses into a target cell and attaches to receptor molecule.

4. DNA directs protein synthesis.

3. Hormone receptor molecule moves into nucleus and attaches to DNA.

Figure 10.2

The steroid hormone molecule passes into the cell, combines with a receptor molecule, and then activates a gene in the nucleus. The gene directs the production of a specific protein.

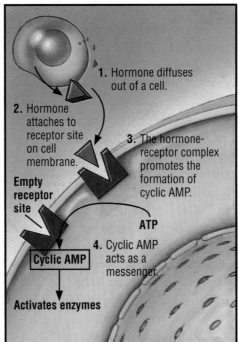

1. Hormone diffuses out of a cell.

2. Hormone attaches to receptor site on cell membrane.

3. The hormone-receptor complex promotes the formation of cyclic AMP.

Empty receptor site

ATP

4. Cyclic AMP acts as a messenger.

Cyclic AMP

Activates enzymes

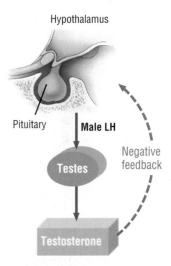

Figure 10.3

The protein hormone combines with specific receptor sites and triggers the formation of cyclic AMP from ATP. Cyclic AMP acts as a secondary messenger, activating enzymes within the cell.

Hypothalamus

Pituitary

Male LH

Testes

Negative feedback

Testosterone

Figure 10.4

High levels of testosterone inhibit the release of LH from the pituitary.

*A **negative-feedback** system is a control system designed to prevent chemical imbalances in the body. The body responds to changes in the external or internal environment. Once the effect is detected, receptors are activated and the response is inhibited, thereby maintaining homeostasis.*

brane. Specific hormones combine at specific receptor sites. The hormone-receptor complex activates the production of an enzyme called *adenyl catalase*. The adenyl catalase causes the cell to convert ATP (adenosine triphosphate), the primary source of cell energy, into **cyclic AMP** (adenosine monophosphate). The cyclic AMP functions as a messenger, activating enzymes in the cytoplasm to carry out their normal functions. For example, when thyroid-stimulating hormone (TSH) attaches to the receptor sites in the thyroid gland, cyclic AMP is produced in thyroid cells. Cells of the kidneys and muscles are not affected because they have no receptors for thyroid-stimulating hormone. The cyclic AMP in the thyroid cell activates enzymes, which begin producing thyroxine, a hormone that regulates metabolism. You will learn more about thyroxine later in the chapter.

NEGATIVE FEEDBACK

Hormone production must be regulated. Once the hormone produces the desired effect, hormone production must be decreased. Consider, for example, testosterone, the male sex hormone, which is responsible for the development of secondary male characteristics. The development of facial hair, sex drive, and lowering of the voice are all associated with the production of the hormone. The hormone itself is regulated by a hormone from the pituitary gland, called the male luteinizing hormone (LH), which activates the testosterone-producing cells of the male testes. Once LH (the male hormone) is produced, testosterone secretions begin. However, once testosterone reaches acceptable levels, secretions must be turned off. Testosterone exerts a negative effect on male LH: high levels of testosterone inhibit the release of male LH. This feedback control system is referred to as **negative feedback.**

The regulation of other hormones such as growth hormone and epinephrine (adrenaline) are equally important. Gigantism results when the production of growth hormone fails to turn off after adequate levels have been reached. Unregulated epinephrine production is equally dangerous. Epinephrine enables the body to respond to stress situations. The hormone causes heart and breathing rates to accelerate and blood sugar levels to rise, among other responses. All actions are designed to allow the body to respond to stress in what has come to be known as the "flight-or-fight" response. However, once the stressful situation is gone, the body returns to normal resting levels. Once again, negative-feedback action is required to restore homeostasis. Throughout this chapter you will study other specific examples of negative feedback.

Figure 10.5

Hormones secreted by nerve cells of the hypothalamus are stored in the posterior pituitary.

produced by the hypothalamus. Antidiuretic hormone acts on the kidneys and helps regulate body water. Oxytocin initiates strong uterine contractions during labor. The hormones travel by way of specialized nerve cells from the hypothalamus to the pituitary. The pituitary gland stores the hormones, releasing them into the blood when necessary.

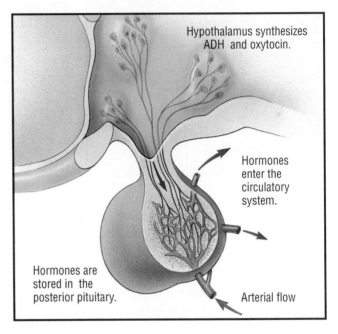

Hypothalamus synthesizes ADH and oxytocin.

Hormones enter the circulatory system.

Hormones are stored in the posterior pituitary.

Arterial flow

THE PITUITARY: THE MASTER GLAND

The pituitary gland is often referred to as the "master gland" because it exercises control over other endocrine glands. This small saclike structure is connected by a stalk to the hypothalamus, the area of the brain associated with homeostasis. The interaction between the nervous system and endocrine system is evident in this hypothalamus-pituitary complex.

The pituitary gland is actually composed of two separate lobes: the posterior lobe and the anterior lobe. The posterior lobe of the pituitary stores and releases hormones such as antidiuretic hormone (ADH) and oxytocin, which have been

The anterior lobe of the pituitary, unlike the posterior lobe, produces its own hormones. However, like the posterior lobe, the anterior lobe of the pituitary is richly supplied with nerves from the hypothalamus. The hypothalamus regulates the release of hormones from the anterior pituitary. Hormones are secreted from the nerve ends of the cells of the hypothalamus and transported in the blood to the pituitary gland. Most of the hormones target specific cells in the pituitary, causing the release of pituitary hormones, which are then carried by the blood to target tissues. Two hypothalamus-releasing factors inhibit the release of hormones from the anterior lobe of the pituitary. The releasing factor dopamine inhibits the secretion of prolactin, a pituitary hormone that stimulates milk production in pregnant women. The hormone somatostatin inhibits the secretion of somatotropin, a pituitary hormone associated with growth of the long bones.

A number of different regulator hormones are stored in the anterior lobe of the pituitary gland. Thyroid-stimulating hormones (TSH), as the name implies, stimulate the thyroid gland to produce its hormone thyroxine. The anterior pituitary also releases reproductive-stimulating hormones, growth-stimulating hormones, prolactin, as well as the hormone that stimulates the adrenal cortex. Table 10.1 summarizes the hormones released from the pituitary gland.

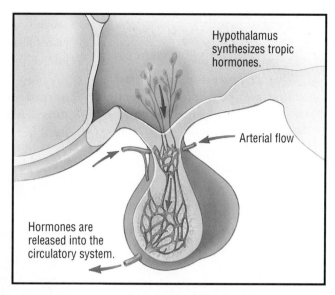

Figure 10.6

Releasing hormones secreted by nerve cells of the hypothalamus regulate hormones secreted by the anterior pituitary.

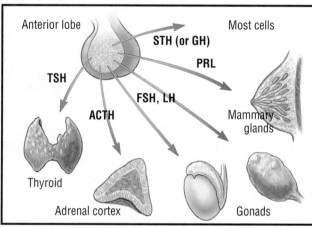

Figure 10.7

Regulator hormones of the anterior lobe of the pituitary and their target organs.

Table 10.1 Pituitary Hormones

Hormone	Target	Primary function
Anterior lobe		
Thyroid-stimulating hormone (TSH)	Thyroid gland	Stimulates release of thyroxine from thyroid. Thyroxine regulates cell metabolism.
Corticotropin adrenal steroid (ACTH)	Adrenal cortex	Stimulates the release of hormones involved in stress responses.
Somatotropin (STH), or growth hormone (GH)	Most cells	Promotes growth.
Follicle-stimulating hormone (FSH)	Ovaries, testes	In females, stimulates follicle development in ovaries. In males, promotes the development of sperm cells in testes.
Luteinizing hormone (LH)	Ovaries, testes	In females, stimulates ovulation and formation of the corpus luteum. In males, stimulates the production of the male sex hormone, testosterone.
Prolactin (PRL)	Mammary glands	Stimulates and maintains milk production in females.
Posterior lobe		
Oxytocin	Uterus	Initiates strong contractions.
	Mammary glands	Triggers milk production.
Antidiuretic hormone (ADH)	Kidney	Increases water reabsorption by kidney.

REVIEW QUESTIONS ?

1 Define the term hormone.
2 What are target tissues or organs?
3 How are the nervous system and endocrine system specialized to maintain homeostasis?
4 Why is the pituitary called the "master gland"?
5 Describe the signalling action of steroid hormones and protein hormones.
6 What is negative feedback?

GROWTH HORMONE

The effects of **growth hormone,** or somatotropin, are most evident when the body produces too much or too little of it. Low secretion of growth hormone during childhood can result in dwarfism; high secretions during childhood can result in gigantism. Although growth hormone affects most of the cells of the body, the effect is most pronounced on cartilage cells and bone cells. If the production of growth hormone continues after the cartilaginous growth plates have been fused, other bones respond. Once the growth plates have fused, the long bones can no longer increase in length, but bones of the jaw, forehead, fingers, and toes increase in width. The disorder, referred to as *acromegaly*, causes a broadening of facial features.

Figure 10.8

The effects of low and high secretions of growth hormone are apparent in people who have the conditions called gigantism and dwarfism.

Figure 10.9

Acromegaly is caused by high secretion of growth hormone during adulthood. Note the progressive widening of bones in the face.

ADRENAL GLANDS

The adrenal glands are located above each kidney. (The word adrenal comes from the Latin *ad*, meaning "to" or "at," and *renes*, meaning "kidneys.") Each adrenal gland is made up of two glands encased in one shell. The inner gland, the adrenal medulla, is surrounded by an outer casing called the adrenal cortex. The medulla is regulated by the nervous system, while the adrenal cortex is regulated by hormones.

The **adrenal medulla** produces two hormones: epinephrine (adrenaline) and norepinephrine (noradrenaline). The link between the nervous system and the adrenal

Unit Three:
Coordination and Regulation in Humans

medulla lies in the fact that both produce epinephrine. The hormone-producing cells within the adrenal gland are stimulated by sympathetic nerves in times of stress.

In a stress situation, **epinephrine** and **norepinephrine** are released from the adrenal medulla into the blood. Under their influence, the blood sugar level rises. Glycogen, a carbohydrate storage compound in the liver and muscles, is converted into glucose, a readily usable form of energy. The increased blood sugar level ensures that a greater energy reserve will be available for the tissues of the body. The hormones also increase heart rate, breathing rate, and cell metabolism. Blood vessels dilate, allowing more oxygen and nutrients to reach the tissues. Even the iris of the eye dilates, allowing more light to enter the retina—in a stress situation, the body attempts to get as much visual information as possible.

BIOLOGY CLIP

The adrenal medulla can be surgically removed without any apparent ill effects. If the adrenal cortex were removed, however, death would result.

The **adrenal cortex** produces three different types of hormones: the glucocorticoids, the mineralocorticoids, and small amounts of sex hormones. The glucocorticoids are associated with blood glucose levels. One of the most important of the glucocorticoids, **cortisol,** increases the level of amino acids in the blood in an attempt to help the body recover from stress. The amino acids are converted into glucose by the liver, thereby raising the level of blood sugar. Increased glucose levels provide a greater energy source, which helps cell recovery. Any of the amino acids not converted into glucose are available for protein synthesis. The proteins can be used to repair damaged cells.

Negative-feedback control of cortisol is shown in Figure 10.10. Stressful situations are identified by the brain. The hypothalamus sends a releasing hormone to the anterior lobe of the pituitary, stimulating the pituitary to secrete corticotropin, or, as it is also called, adrenocorticotropin hormone (ACTH). The ACTH is carried by the blood to the target cells in the adrenal cortex. Under the influence of ACTH, the cells of the adrenal cortex secrete cortisol, which is carried to target cells in the liver and muscles. As cortisol levels rise, cells within the hypothalamus and pituitary decrease the production of regulatory hormones, and eventually the levels of cortisol begin to fall.

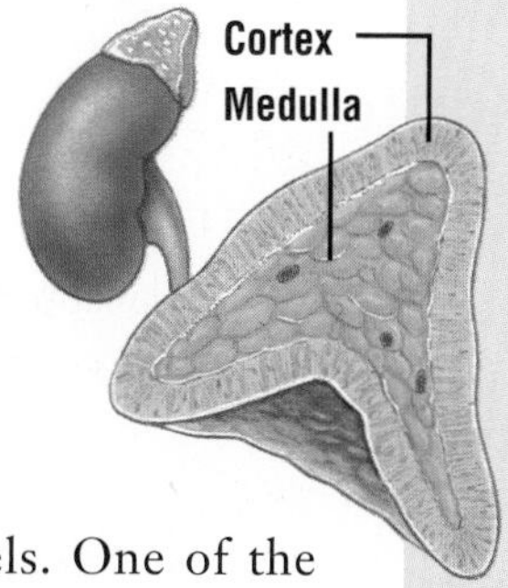

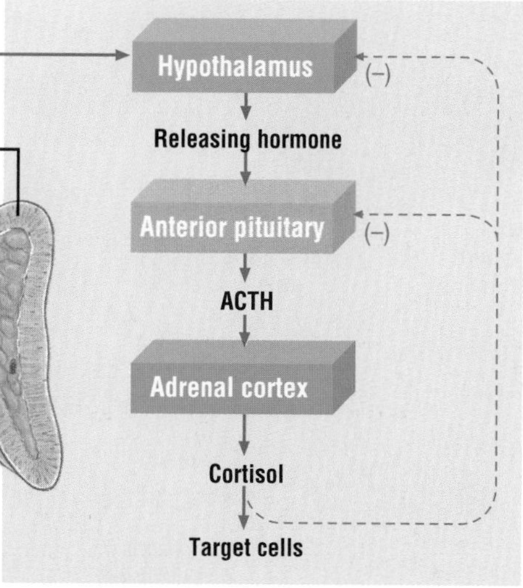

Figure 10.10

Negative feedback of cortisol.

Aldosterone is the most important of the mineralocorticoids, the second major group of hormones produced by the adrenal cortex. Secretions of aldosterone increase sodium retention and water reabsorption by the kidney and thereby help maintain body fluid levels.

Epinephrine *is a hormone produced by the adrenal medulla that initiates the flight-or-fight response.*

Norepinephrine *is a hormone produced by the adrenal medulla that initiates the flight-or-fight response.*

The **adrenal cortex** *is the outer region of the adrenal gland. It produces glucocorticoids and mineralocorticoids.*

Cortisol *is a hormone that stimulates the conversion of amino acids to glucose by the liver.*

Aldosterone *is a hormone produced by the adrenal cortex. It helps regulate water balance in the body by increasing sodium and water reabsorption by the kidneys.*

INSULIN AND THE REGULATION OF BLOOD SUGAR

The pancreas contains two types of cells: one type produces digestive enzymes, the other type produces hormones. The hormone-producing cells are located in structures called the *islets of Langerhans*, named after their discoverer, Paul Langerhans. Over 2000 tiny islets scattered throughout the pancreas are responsible for the production of two hormones: insulin and glucagon.

Insulin is produced in the beta cells of the islets of Langerhans and is released when the blood sugar level is high. Insulin increases glucose utilization by making many cells of the body permeable to glucose. After a meal, when the blood sugar level rises, insulin is released. This insulin causes cells of the liver and muscles, among other organs, to become permeable to the glucose. In the liver, the glucose is converted into glycogen, the primary storage form for glucose. By storing excess glucose in the form of glycogen, the insulin enables the blood sugar level to return to normal. Insulin helps maintain homeostasis.

Glucagon and insulin work in complementary fashion. Insulin causes a decrease in blood sugar level, and glucagon causes an increase in blood sugar level. Produced by the alpha cells of the islets of Langerhans, glucagon is released when blood sugar levels are low. After periods of fasting, when blood sugar dips below normal levels, glucagon is released. Glucagon promotes the conversion of glycogen to glucose, which is absorbed by the blood. As glycogen is converted to glucose in the liver, the blood sugar level returns to normal.

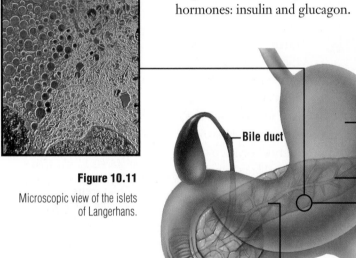

Figure 10.11

Microscopic view of the islets of Langerhans.

Bile duct

Stomach

Pancreas

Islets of Langerhans

Small intestine

Pancreatic duct

Figure 10.12

Insulin is released when blood sugar levels are high. Insulin increases the permeability of cells to glucose. Glucose is converted into glycogen within the liver, thereby restoring blood sugar levels. Glucagon is released when blood sugar levels are low. Glucagon promotes the conversion of liver glycogen into glucose, thereby restoring blood sugar levels.

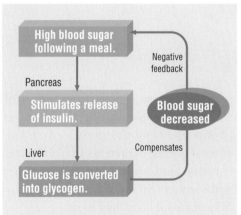

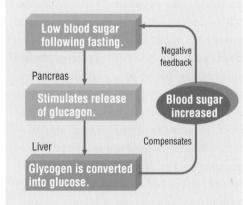

SUGAR DIABETES

Diabetes mellitus is a genetic disorder associated with inadequate production of insulin. Although the exact cause of diabetes is not known, it is generally believed to occur when insulin-producing cells within the islets of Langerhans deteriorate. Without adequate levels of insulin, blood sugar levels tend to rise very sharply following meals. This condition is known as *hyperglycemia*, or high blood sugar (from *hyper*, meaning "too much," *glyco*, meaning "sugar," and *emia* referring to a condition of the blood).

A variety of symptoms are associated with high blood sugar. Because the kidney is unable to reabsorb all of the blood glucose that is filtered through it, glucose appears in the urine. The appearance of sugar in the urine accounts for the name *diabetes mellitus*, which literally means "going through honey-sweet." The loss of glucose in the urine creates yet another symptom. Since the excretion of glucose draws water from the body by osmosis, diabetics excrete unusually large volumes of urine and are often thirsty.

Diabetics also experience low energy levels. Remember, insulin is required for the cells of the body to become permeable to glucose. Despite the abundance of glucose in the blood, little is able to move into the cells of the body. Cells of diabetics soon become starved for glucose and must turn to other sources of energy. Fats and proteins can be metabolized for energy, but, unlike carbohydrates, they are not an easily accessible energy source. The switch to these other sources of energy creates a host of problems for diabetics. Acetone, an interme-

diary product of fat metabolism, can change blood pH. In severe cases, doctors are able to smell acetone on the breath of people with diabetes.

There are at least two different types of diabetes mellitus: juvenile, or early-onset diabetes, and adult, or maturity-onset, diabetes. Juvenile diabetes is caused by the early degeneration of the beta cells in the islets of Langerhans and can only be treated by insulin injections. Maturity-onset diabetes is associated with decreased insulin production. Insulin-producing cells do not disappear as they do in juvenile diabetes, but become less effective. People with maturity-onset diabetes can be helped by oral drugs known as sulfonamides. It is believed that these drugs, which are not effective against juvenile diabetes, stimulate the residual function of the islets of Langerhans in older people.

> ### BIOLOGY CLIP
> Not all of the body's cells depend on insulin. Nerve cells and blood cells are able to absorb glucose without insulin. Muscle cells, which make up nearly 50% of your body mass, depend on insulin.

Glucagon *is a hormone produced by the pancreas. When blood sugar levels are low, glucagon promotes the conversion of glycogen to glucose.*

Diabetes mellitus *is a genetic disorder characterized by high blood sugar levels.*

■ REVIEW QUESTIONS ■ ?

7 How would decreased secretions of growth hormone affect an individual?

8 How would an increased secretion of growth hormone affect an individual after puberty?

9 Name two regions of the adrenal gland, and list two hormones produced in each area.

10 How would high levels of ACTH affect secretions of cortisol from the adrenal glands? How would high levels of cortisol affect ACTH?

11 Where is insulin produced?

12 How does insulin regulate blood sugar levels?

13 How does glucagon regulate blood sugar levels?

RESEARCH IN CANADA

Frederick Banting (1891–1941) served in World War I as a doctor. On returning from the war, he became interested in diabetes research. At the time, diabetes was thought to be caused by a deficiency of a hormone located in specialized cells of the pancreas. However, extracting the hormone from the pancreas presented a problem since the pancreas also stores digestion enzymes capable of breaking down the protein hormone.

The Discovery of Insulin

In 1921, Banting approached John J.R. MacLeod, a professor at the University of Toronto, with his idea for isolating the hormone. MacLeod assigned him a makeshift laboratory as well as an assistant, Charles Best, who was a graduate student in biochemistry. Banting and Best tied the pancreatic duct of experimental dogs, and waited seven weeks for the pancreas to shrivel. Although the cells producing digestive enzymes had deteriorated, cells from the islets of Langerhans remained. The hormone was then extracted from the pancreas. When the hormone was injected into dogs who had had their pancreases removed, symptoms of diabetes ceased. Banting and Best wanted to call the hormone "isletin," but MacLeod insisted that it be called "insulin." In 1923 Banting and MacLeod were awarded the Nobel Prize for physiology and medicine. Banting was furious. Charles Best, his co-worker had not been included and MacLeod, the professor who had contributed laboratory space, had been included.

DR. FREDERICK BANTING
DR. CHARLES BEST

LABORATORY

IDENTIFICATION OF HYPERGLYCEMIA

Objective

To use urinalysis techniques to identify diabetes mellitus.

Materials

4 samples of simulated urine (labelled A, B, C, D)

hot plate	wax pen
400 mL beaker	beaker tongs
Clinitest tablets	forceps
Benedict's solution	medicine dropper
pipettes	distilled water
10 mL graduated cylinder	test-tube clamp
test-tube rack	goggles
lab apron	

CAUTION: Benedict's solution is toxic and an irritant. Avoid skin and eye contact. Wash all splashes off your skin and clothing thoroughly. If you get any chemical in your eyes, rinse for at least 15 min and inform your teacher.

Procedure

Part I: Benedict's Test

Benedict's solution identifies reducing sugars. Cupric ions in the solution combine with sugars to form cuprous oxides, which produce color changes. See the table below for color changes.

Benedict Test

Color of solution	Glucose concentration
Blue	0.0%
Light green	0.15% – 0.5%
Olive green	0.5% – 01.0%
Yellow-green to yellow	1.0% – 1.5%
Orange	1.5% – 2.0%
Red to red-brown	2.0%

1 Label the four test tubes A, B, C, and D. Using a 10 mL graduated cylinder, measure 5 mL of Benedict's solution into each test tube.

2 Using a medicine dropper, add 10 drops of urine from sample A to test tube A. Rinse the medicine dropper and repeat for samples B, C, and D.
   a) Why must the medicine dropper be rinsed?

3 Fill a 400 mL beaker with approximately 300 mL of tap water. Using beaker tongs, position the beaker on a hot plate. The beaker will be used as a hot-water bath. Using the test-tube clamp, place the test tubes in the hot-water bath for 5 min.

4 Using the test tube clamp, remove the samples and record the final colors of the solutions.
   b) Provide your data.

Part II: Clinitest Tablet Method

The reducing sugar in the urine will react with copper sulfate to reduce cupric ions to cupric oxide. The chemical reaction is indicated by a color change. The table below provides quantitative results.

Color of solution	Glucose concentration
Blue	0.0%
Green	0.25% – 0.5%
Green to green-brown	0.5% – 1.0%
Orange	2.0% – greater

5 Clean the four test tubes and place 10 drops of distilled water into each of the test tubes.

6 Add 5 drops of urine to each of the appropriately labelled test tubes. Place each of the test tubes in a test-tube rack.

7 Using forceps, place one Clinitest tablet in each of the test tubes. Observe the color.
   c) Provide your data.

Laboratory Application Questions

1 Why is insulin not taken orally?

2 Explain why diabetics experience the following symptoms: low energy levels, large volumes of urine, the presence of acetone on the breath, and acidosis (blood pH becomes acidic).

3 Why might the injection of too much insulin be harmful?

4 How would you help someone who had taken too much insulin? Explain your answer. ∎

FRONTIERS OF TECHNOLOGY: ISLET TRANSPLANTS

Juvenile diabetes is the second leading cause of blindness in Canada. Other side effects of the disease, such as kidney and heart failure, stroke, and peripheral nerve damage, affect over 50 000 Canadians.

Why do doctors not just replace defective cells from the islets of Langerhans with ones that are working properly? In theory, islet transplants sound rather straightforward, but nothing could be further from the truth. A number of important questions must be addressed. First, how can cells be removed from the donor? Organs can be fixed into place, but what about cells? Can the islets develop a new blood supply? As in all living cells, nutrients are required and wastes must be eliminated. Will the transplanted islet cells be rejected by the recipient? Will the transplanted islet cells actually produce insulin?

Researchers around the world have been busily searching for answers to these questions. Unlike insulin therapy, islet transplants have the potential to reverse the effects of diabetes. Although insulin injection provides some regulation of blood sugar, it will not necessarily prevent many of the serious complications, such as blindness and stroke. Insulin therapy requires monitoring of blood sugar level and balancing injections of insulin with carbohydrate intake and exercise. Transplanted islet cells, however, could replace the body's natural mechanism for monitoring and producing insulin.

On February 24, 1989, Jim Connor became Canada's first patient to receive transplanted islets of Langerhans cells. Dr. Garth Warnock, a surgeon at the University of Alberta Hospital, transplanted millions of isolated cells from the pancreas of a donor. A month later, the procedure was once again successfully completed. Although other transplants had taken place before, none had reported the same degree of success. Why had the team at the University of Alberta Hospital triumphed where others failed? First, Dr. Ray Rajotte, the team leader, had developed frozen preservation methods used for tissue banks. Second, Dr. Garth Warnock devised more successful ways to harvest and purify islets. Third, Dr. Norman Kneteman improved immunosuppression therapy, thereby preventing the rejection of the transplanted cells.

It is important to note that cell transplants, although a significant step above insulin therapy, do not provide a cure for juvenile diabetes. A cure would prevent diabetes from occurring. Transplant therapy, like insulin therapy, controls the effects. Currently, the transplants are only being done on people who must undergo kidney transplants and therefore must already take immunosuppressive drugs. The researchers are guardedly optimistic about the future of cell transplants, but also warn that transplant therapy is not suitable for all patients.

THYROID GLAND

The thyroid gland is located at the base of the neck, immediately in front of the trachea or windpipe. Two important thyroid hormones, thyroxine and triiodothyronine, regulate body metabolism and the growth and differentiation of tissues. Approximately 65% of thyroid secretions are thyroxine; however, both hormones appear to have the same function. Most of this discussion will focus on the principal hormone, **thyroxine.**

Have you ever wondered why some people seem to be able to consume fantastic amounts of food without any weight change, while others appear to gain

Thyroxine *is a hormone secreted by the thyroid gland that regulates the rate of body metabolism.*

weight at the mere sight of food? Part of this anomaly can be explained by thyroxine and the regulation of metabolic rate. Individuals who secrete higher levels of thyroxine oxidize sugars and other nutrients at a faster rate. Approximately 50% of the glucose oxidized in the body is released as heat (which explains why these individuals usually feel warm). The remaining 50% is converted to ATP, the storage form for cell energy. This added energy reserve is often consumed during activity. Therefore, these individuals tend not to gain weight.

Individuals who have lower levels of thyroxine do not oxidize nutrients as quickly, and therefore tend not to break down sugars as quickly. Excess blood sugar is eventually converted into liver and muscle glycogen. However, once the glycogen stores are filled, excess sugar is converted into fat. It follows that the slower the blood sugar is used, the faster the fat stores are built up. People who secrete low amounts of thyroxine tend to be less active, intolerant of cold, and have skin that tends to dry out quickly. It is important to note that not all types of weight gain are due to *hypothyroidism* (low thyroid secretions). In many cases, weight gain reflects a poor diet.

Control of thyroid hormones, like many other hormones, is accomplished by negative feedback. Should the metabolic rate decrease, receptors in the hypothalamus are activated. Nerve cells release thyroid-releasing factor (TRF), which stimulates the pituitary to release thyroid-stimulating hormone (TSH). Thyroid-stimulating hormone is carried by way of the blood to the thyroid gland, which, in turn, releases thyroxine. Thyroxine raises metabolism by stimulating increased sugar utilization by the cells of the body. Higher levels of thyroxine cause the pathway to be "turned off." Thyroxine inhibits the release of the thyroid-releasing factor from the hypothalamus, thereby turning off the production of TSH from the pituitary.

THYROID DISORDERS

Iodine is an important component of both thyroid hormones. A normal component of your diet, iodine is actively transported from the blood into the follicle cells of the thyroid. The concentration of iodine in the cells can be 25 times greater than that of the blood. Problems are created when iodine levels begin to fall. When adequate amounts of iodine are not obtained from the diet, the thyroid enlarges, producing a **goiter.**

The presence of a goiter emphasizes the importance of a negative-feedback control system. Without iodine, thyroid production and secretion of thyroxine drops. This causes more and more TSH to be produced, and, consequently, the thyroid is stimulated more and more. Under the relentless influence of TSH, cells of the thyroid continue to develop, and the thyroid enlarges. Goiters were once prevalent in areas where the soil lacked iodine salts. This is why table salt has been iodized.

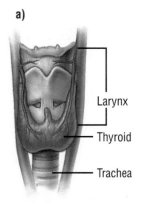

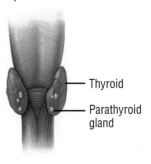

Figure 10.13

(a) Anterior view of thyroid gland. (b) Posterior view of thyroid gland.

Figure 10.14

Control of thyroid hormones.

*A **goiter** is an enlargement of the thyroid gland.*

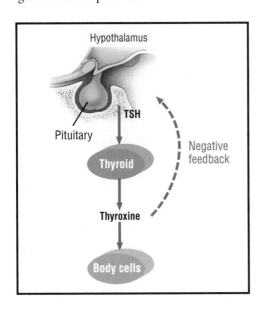

CASE STUDY

THE EFFECTS OF HORMONES ON BLOOD SUGAR

Objective

To use experimental data to investigate the effects of various hormones on blood sugar levels.

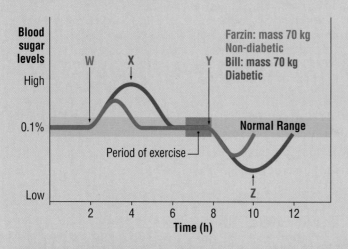

Procedure

Blood sugar levels of a diabetic and a nondiabetic patient were monitored over a period of 12 h. Both ate an identical meal and performed 1h of similar exercise. Use the data in the graph to answer the Case-Study Application Questions.

Case-Study Application Questions

1 Which hormone injection did Bill receive at the time labelled X? Provide your reasons.
2 What might have happened to Bill's blood sugar level if hormone X had not been injected? Provide your reasons.
3 Explain what happened at time W for Bill and Farzin?
4 Explain why blood sugar levels begin to fall after time Y?
5 What hormone might Bill have received at time Z? Explain your answer.
6 Why is it important to note that both Farzin and Bill had the same body mass?
7 What differences in blood sugar levels are illustrated by the data collected from Bill and Farzin.
8 Why do Bill and Farzin respond differently to varying levels of blood sugar? ■

PROSTAGLANDINS

Prostaglandins are hormones that have a pronounced effect in a small localized area.

Local responses to changes in the immediate environment of cells are detected by mediator cells, which produce **prostaglandins.** More than 16 different types of prostaglandins alter cell activity in a manner that counteracts or adjusts for the change. Generally, prostaglandins are secreted in low concentrations by mediator cells, but secretions increase when changes take place. Most of the molecules released during secretions, even in time of change, tend to be absorbed rapidly by surrounding tissues. Few of the prostaglandin molecules are absorbed by capillaries and carried in the blood.

Two different prostaglandins can adjust blood flow in times of stress. Stimulated by

the release of epinephrine, the hormones increase blood flow to local tissues. Other prostaglandins respond to stress by triggering the relaxation of smooth muscle in the passages leading to the lung. Prostaglandins are also released during allergic reactions.

■ REVIEW QUESTIONS   ?

14 How does thyroxine affect blood sugar?
15 List the symptoms associated with hypothyroidism and hyperthyroidism.
16 How do the pituitary and hypothalamus interact to regulate thyroxine levels?
17 What is a goiter?
18 What are prostaglandins and what is their function?

SOCIAL ISSUE:
Growth Hormone–The Anti-aging Drug

*In July 1990, Dr. Daniel Rudman published a study in the prestigious **New England Journal of Medicine** proposing that injections of growth hormone could slow the aging process. According to Dr. Rudman's findings, injections of growth hormone in older people actually increase the development of muscle tissue and the disappearance of fat. In many ways, growth hormone appears to be able to reverse the effects of decades of aging.*

Although researchers warn that the drug's long-term effects have not been documented and that the drug may not be suitable for everyone, speculation about the potentials of an anti-aging drug abounds in both scientific and nonscientific communities.

Statement:

Widespread use of anti-aging drugs could cause serious problems for society.

Point

- At 1990 prices, injections of growth hormone for a 70 kg man would be about $14 000 per year. Only the richest people in society would be able to pay for such treatments.

- It is very possible that growth hormone, like other steroids, might have negative long-term effects on individuals.

- People are already living longer and thus putting a strain on our health services and pension plans. If more people live even longer, the social system could collapse.

Counterpoint

- Although growth hormone is expensive today, the price will likely come down once the procedure becomes established. Plastic surgery was once available only to the rich, and it has become commonplace.

- With proper studies and control of the use of growth hormone, potential users would be protected and well informed. Ultimately, it is up to the individual to choose whether or not to risk any possible negative effects.

- Far from causing the social system to collapse, anti-aging therapy would produce economic benefits. People would work longer and generally experience a better quality of life.

Research the issue.
Reflect on your findings.
Discuss the various viewpoints with others.
Prepare for the class debate.

CHAPTER HIGHLIGHTS

- Hormones are chemicals released by cells that affect cells in other parts of the body. Only small amounts of hormones are required to alter cell metabolism.
- Endocrine hormones are secreted from glands directly into the blood.
- Integration of organs is accomplished by the nervous system and endocrine system. Chemical transmitters act on specific tissues. Target tissues have specific receptor sites that bind with the hormones.
- A negative-feedback system is a control system designed to prevent chemical imbalances in the body. The body responds to changes in the external or internal environment. Once the effect is detected, receptors are activated and the response is inhibited.
- Growth hormone is produced by the cells of the anterior lobe of the pituitary. Prior to puberty, the hormone promotes growth of the long bones.
- The adrenal gland secretes hormones. The adrenal medulla, found at the core of the adrenal gland, produces epinephrine and norepinephrine. The adrenal cortex, the outer region of the adrenal gland, produces glucocorticoids and mineralocorticoids.
- Insulin is a hormone produced by the pancreas. Secreted when blood sugar levels are high, insulin promotes carbohydrate storage.
- Glucagon is a hormone produced by the pancreas. Secreted when blood sugar levels are low, glucagon promotes the conversion of glycogen to glucose.
- Diabetes mellitus is a genetic disorder characterized by high blood sugar levels.
- Thyroxine is a hormone secreted by the thyroid gland that regulates the rate of body metabolism.
- Prostaglandins are hormones that have a pronounced effect in a small localized area.

APPLYING THE CONCEPTS

1 Referring to the interaction between the hypothalamus and pituitary, indicate how the nervous system and endocrine system complement each other.

2 With reference to the adrenal glands, explain how the nervous system and endocrine system interact in times of stress.

3 A disorder called testicular feminization syndrome occurs when the receptor molecules to which testosterone binds are defective. Predict the effect of testicular feminization syndrome and explain how normal steroid hormone action is altered.

4 A rare virus destroys cells of the anterior lobe of the pituitary. Predict how the destruction of the pituitary cells would affect blood sugar. Explain.

5 Why do insulin levels increase during times of stress?

6 Provide an explanation for the following symptoms associated with diabetes mellitus: lack of energy, increased urine output, and thirst.

7 A tumor on a gland can increase the gland's secretions. Explain how increases in the following hormones affect blood sugar levels: insulin, epinephrine, and thyroxine.

8 A physician notes that her patient is very active and remains warm on a cold day even when wearing a light coat. Further discussion reveals that even though the patient's daily food intake exceeds that of most people, the patient remains thin. Why might the doctor suspect a hormone imbalance? Which hormone might the doctor suspect?

9 With reference to negative feedback, provide an example of why low levels of iodine in your diet can cause goiters.

10 Three classical methods have been used to study hormone function:
- The organ that secretes the hormone is removed. The effects are studied.
- Grafts from the removed organ are placed within a gland. The effects are studied.
- Chemicals from the extracted gland are isolated and injected back into the body. The effects are studied.

Explain how each of these procedures could be used to investigate how insulin affects blood sugar.

Many axons are covered with a glistening white coat of a fatty protein called the **myelin sheath,** which acts as insulation for the neurons. Formed by the Schwann cells, the myelin sheath acts as an insulator by preventing the loss of charged ions from the nerve cell. The areas between the sections of myelin sheath are known as the **nodes of Ranvier.** Axons that have a myelin covering are said to be *myelinated.* Nerve impulses jump from one node to another, thereby speeding the movement of nerve impulses. Not surprisingly, nerve impulses move much faster along myelinated nerve fibers than unmyelinated ones. The speed of the impulse along the nerve fiber is also affected by the diameter of the axon. Generally, the smaller the diameter of the axon, the faster the speed of the nerve impulse.

All nerve fibers found within the peripheral nervous system contain a thin membrane, called the **neurilemma,** that surrounds the axon. The neurilemma promotes the regeneration of damaged axons. This explains why feeling gradually returns to your finger following a paper cut—severed neurons can be rejoined. However, not all nerve cells contain a neurilemma and a myelin sheath. Nerves within the brain that contain myelinated fibers and a neurilemma are called *white matter.* Other nerve cells within the brain and spinal cord, referred to as the *gray matter*, lack a myelin sheath, and will not be regenerated after injury. Damage to the gray matter is permanent.

> **BIOLOGY CLIP**
>
> Multiple sclerosis is caused by the destruction of the myelin sheath that surrounds the nerve axons. The myelinated nerves in the brain and spinal cord are gradually destroyed as the myelin sheath hardens and forms scars, or plaques. The destruction of the sheath results in short circuits. Often referred to as MS, multiple sclerosis can produce symptoms of double vision, speech difficulty, jerky limb movements, and partial paralysis of the voluntary muscles.

NEURAL CIRCUITS

If you have ever touched a hot stove, you probably didn't think about how your nervous system told you it was hot! The sensation of heat is detected by specialized temperature receptors in your skin, and a nerve impulse is carried to the spinal cord. The sensory nerve passes the impulse on to an interneuron, which, in turn, relays the impulse to a motor neuron. The motor nerve causes the muscles in the hand to contract and the hand to pull away. All this happens in a split second, long before the information travels to the brain. A few seconds later, the sensation of pain becomes noticeable and you may let out a scream. Reflexes are involuntary and often unconscious. Imagine how badly you would have burned yourself if you had to wait for the sensation of pain before removing your hand. The damage would have been much worse if you had attempted to gauge the intensity of pain and to contemplate the appropriate action. Time is required for nerve impulses to move through the many circuits of the brain.

The simplest nerve pathway is the *reflex arc.* Reflexes occur without brain coordination. They contain five essential components: the receptor, the sensory neuron, the interneuron in the spinal cord, the motor neuron, and the effector. Figure 11.5 on the following page shows a reflex arc. The touch receptor in the finger detects the tack. Sensory information is transmitted to the CNS.

*The **myelin sheath** is a fatty covering over the axon of a nerve cell.*

*The **nodes of Ranvier** are the regularly occurring gaps between sections of myelin sheath along the axon.*

*The **neurilemma** is the delicate membrane that surrounds the axon of nerve cells.*

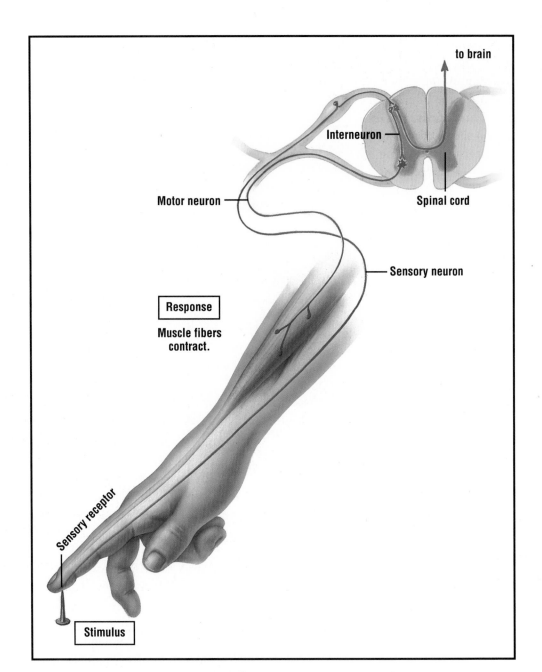

to brain

Interneuron

Motor neuron

Spinal cord

Sensory neuron

Response

Muscle fibers contract.

Sensory receptor

Stimulus

Figure 11.5

Sensory information is relayed from the sensory neuron (in purple) to the spinal cord. Interneurons in the spinal cord receive the information from the sensory neuron and relay it to the motor neuron shown in red. The motor neuron activates the muscle cell, causing it to contract.

▧ REVIEW QUESTIONS ▧ ?

1 Differentiate between the PNS and CNS.

2 What are neurons?

3 Differentiate between sensory and motor nerves.

4 Briefly describe the function of the following parts of a neuron: dendrites, myelin sheath, Schwann cells, cell body, and axon.

5 What is the neurilemma?

6 Name the essential components of a reflex arc, and state the function of each of the components.

LABORATORY
REFLEX ARCS AND REACTION RATE

Objective

To investigate different reflex arcs and reaction rate.

Materials

rubber reflex hammer   penlight or microscope light
30 cm ruler

Procedure

Knee-jerk

1 Have your subject sit with his or her legs crossed on a chair. The subject's upper leg should remain relaxed.
2 Locate the position of the kneecap on the upper leg and feel the large tendon below the midline of the knee cap.
3 Using a reflex hammer, gently strike the tendon below the kneecap.
   a) Describe the movement of the leg.
4 Ask the subject to clench a book with both hands, then strike the tendon of the upper leg once again.
   b) Compare the movement of the leg while the subject is clenching the book with the movement in the previous procedure.

Achilles Reflex

5 Remove the subject's shoe. Ask your subject to kneel on a chair so that the feet hang over the edge of the chair. Push the toes toward the legs of the chair and then lightly tap the Achilles tendon with the reflex hammer.
   c) Describe the movement of the foot.

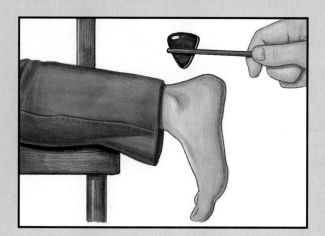

Babinski Reflex

6 Ask the subject to remove a shoe and sock. Have the subject sit in a chair, placing the shoeless foot on another chair for support. Quickly slide the reflex hammer across the sole of the subject's foot, beginning at the heel and moving toward the toes.
   d) Describe the movement of the toes.

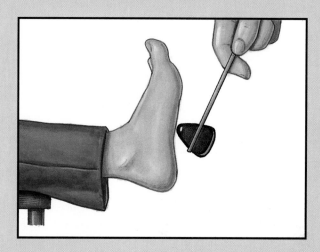

Pupillary Reflex

7 Have the subject close one eye for approximately one minute. Ask him or her to open the closed eye; now compare the size of the pupils.
   e) Which pupil is larger?
8 Ask the subject to close both eyes for one minute. Have the subject open both eyes; now shine a penlight in one of the eyes.
   f) Describe the changes you observe in the pupil.

Reaction Rate

9 Ask your subject to place his or her forearm flat on the surface of a desk. The subject's entire hand should be extended over the edge of the desk.
10 Ask the subject to place the index finger and thumb approximately 2 cm apart. Place a 30 cm ruler between the thumb and forefinger of the subject. The lower end of the ruler should be even with the top of the thumb and forefinger.

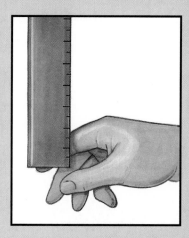

11 Indicate when ready, and release the ruler within the next 30 s. Measure the distance the ruler falls before being caught with the subject's thumb and forefinger. Repeat the procedure for the left hand. Record your data in a table similar to the one below.

g)

Trial	Distance– right hand (cm)	Distance– left hand (cm)
1		
2		
3		
Average		

Laboratory Application Questions

1 A student touches a stove, withdraws his or her hand, and then yells. Why does the yelling occur after the hand is withdrawn? Does the student become aware of the pain before the hand is withdrawn?

2 The neuron is severed at point X. Explain how the reflex arc would be affected.

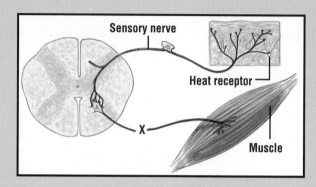

3 Explain how the knee-jerk and Achilles reflexes are important in walking.

4 While examining the victim of a serious car accident, a physician lightly pokes the patient's leg with a needle. The light pokes begin near the ankle and gradually progress toward the knee. Why is the physician poking the patient? Why begin near the foot?

5 Why is the knee-jerk reflex exaggerated when the subject is clenching the book? ∎

ELECTROCHEMICAL IMPULSE

In 1791, the Italian scientist Luigi Galvani discovered that the calf muscle of a dead frog could be made to twitch under electrical stimulation. Galvani concluded that the "animal electricity" was produced by the muscle. Although Galvani's conclusion was incorrect, it spawned a flood of research that led to the development of theories about how electrical current is generated in the body. In 1840, Emil DuBois-Reymond set about refining instruments that would enable him to detect the passage of currents in nerves and muscles. By 1906, the Dutch physiologist Willem Einthoven began recording the movement of electrical impulses in heart muscle. The electrocardiogram, or ECG, has been refined many times since 1906 and is still used today to diagnose heart problems. In 1929, Hans Berger placed electrodes on the skull and measured electrical changes that accompany brain activity. The electroencephalograph, or EEG, is used to measure brain-wave activity.

Figure 11.6

Mapping electrical current has diagnostic value.

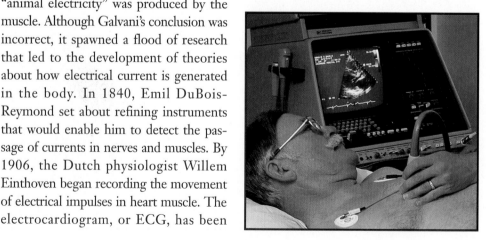

As research continued, the difference between electricity and nerves soon became evident. Current travels along a wire much faster than the impulse travels across a nerve. In addition, the cytoplasmic core of a nerve cell offers great resistance to the movement of electrical current. Unlike electrical currents, which diminish as they move through a wire, nerve impulses remain as strong at the end of a nerve as they were at the beginning. One of the greatest differences between nerve impulses and electricity is that nerves use cellular energy to generate current. By comparison, the electrical wire relies on some external energy source to push electrons along its length. As early as 1900, Julius Berstein suggested that nerve impulses were an electrochemical message created by the movement of ions through the nerve cell membrane. Evidence supporting Berstein's theory was provided in 1939 when K.S. Cole and J.J. Curtis placed a tiny electrode inside the large nerve cell of a squid. A rapid change in charge across the membrane was detected every time the nerve became excited. The **resting membrane** normally had a negative charge somewhere near −70 mV; however, when the nerve became excited, the charge on the inside of the membrane registered +40 mV. This reversal of charges is described as an **action potential.** Cole and Curtis noticed that the +40 mV did not last more than a few milliseconds before the charge on the inside of the nerve cell returned to −70 mV.

Nerve cells are different from other cells in that no other cell is charged. How do nerve cell membranes become charged? To find the answer, we must examine the nerve cell on a molecular level. Unlike most cells, neurons have a rich supply of positive and negative ions both inside and outside the neuron. Although it might seem surprising, the negative ions do little to create a charged membrane. The electrochemical event is caused by an unequal concentration of positive ions across the nerve membrane. A potassium pump, located in the cell membrane, pulls potassium ions into the nerve

> **BIOLOGY CLIP**
> Nerve impulses move much slower than electricity. It has been estimated that most nerve impulses travel at speeds between 0.5 and 1.0 m/s. Electrical current moves at the speed of 3×10^5 km/s.

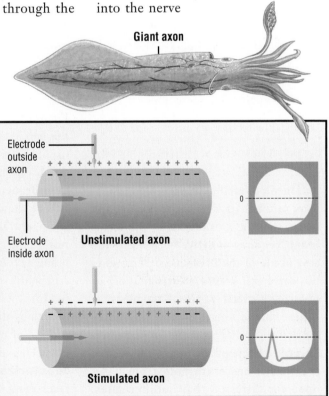

Giant axon

Electrode outside axon

Electrode inside axon

Unstimulated axon

Stimulated axon

Figure 11.7

The squid has a large axon. A miniature electrode is placed inside the giant axon of a squid. The inside of the resting membrane is negative with respect to the outside of the membrane. When stimulated, the charges across the nerve membrane temporarily reverse.

cell, while a sodium pump relays sodium ions outside the nerve cell. It has been estimated that for every 150 potassium ions on the inside of the cell membrane there are 5 potassium ions on the outside of the cell membrane. The sodium pump is nearly as effective: for every 150 sodium ions on the outside of the nerve cell membrane, 15 can be found on the inside of the nerve cell membrane. The highly concentrated potassium ions on the inside of the nerve cell have a tendency to diffuse outside the nerve cell. Similarly, the highly concentrated sodium ions on the outside of the nerve cell have a tendency to diffuse into the nerve cell. As potassium diffuses outside of the neuron, sodium diffuses into the neuron. Therefore, positively charged ions move both into and out of the cell. However, the diffusion of sodium ions and potassium ions is not equal. In a resting membrane, more channels are open to the potassium ions. Consequently, more potassium ions diffuse out of the nerve cell than sodium ions diffuse into the nerve cell.

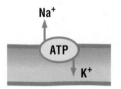 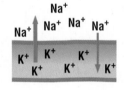

Sodium is pumped outside of the nerve cell, while potassium is pumped into the nerve cell.

Potassium diffuses out of the nerve cell faster than sodium diffuses into the nerve cell.

The more rapid diffusion of potassium ions outside of the nerve membrane means that the area outside of the membrane becomes positive relative to the area inside of the membrane. Therefore, the nerve cell loses a greater number of positive ions than it gains. Biologists now believe that ion gates control the movement of ions across the membrane. According to popular theory, the resting membrane has more potassium gates open for diffusion than sodium gates.

Excess positive ions accumulate along the outside of the nerve membrane, while excess negative ions accumulate along the inside of the membrane. The resting membrane is said to be charged, or **polarized**. The separation of electrical charges by a membrane has the potential to do work, which is expressed in millivolts (mV). A charge of -70 mV indicates the difference between the number of positive charges found on the inside of the nerve membrane relative to the outside. (A charge of -90 mV on the inside of the nerve membrane would indicate even fewer positive ions inside the membrane relative to the outside.)

Upon excitation, the nerve cell membrane becomes more permeable to sodium than potassium. Scientists believe that sodium gates are opened in the nerve membrane, while potassium gates close. The highly concentrated sodium ions rush into the nerve cell by diffusion and charge attraction. The rapid inflow of sodium causes a charge reversal, or **depolarization**. Once the voltage inside of the nerve cell becomes positive, the sodium gates slam closed and the inflow of sodium is halted. The potassium gates now open and potassium ions once again begin to diffuse out of the nerve cell. Eventually, the flow of potassium out of the nerve cell restores the original polarity of the membrane. However, the Na^+ and K^+ are now on the side of the resting membrane opposite to their position before depolarization occurred. Once again, the sodium-potassium pump will restore the condition of the resting membrane by transporting Na^+ out of the neuron while moving K^+ ions inside the neuron. The energy supply from ATP maintains the polarized membrane. The process of restoring the original polarity of the nerve membrane is referred to as **repolarization**. The process usually takes about 0.001 s.

Polarized membranes *are charged membranes. Polarization is caused by the unequal distribution of positively charged ions.*

Depolarization *is caused by the diffusion of sodium ions into the nerve cell. Excess positive ions are found inside the nerve cell.*

Repolarization *is a process in which the original polarity of the nerve membrane is restored. Excess positive ions are found outside the nerve membrane.*

Figure 11.8

The potassium and sodium pumps of the nerve cell are highly effective.

Unit Three:
Coordination and Regulation in Humans

Nerves conducting an impulse cannot be activated until the condition of the resting membrane is restored. The period of depolarization must be completed and the nerve must repolarize before a second action potential can be conducted. The period of time required for the nerve cell to become repolarized is called the **refractory period.** The refractory period usually lasts 1 to 10 ms.

*The **refractory period** is the recovery time required before a neuron can produce another action potential.*

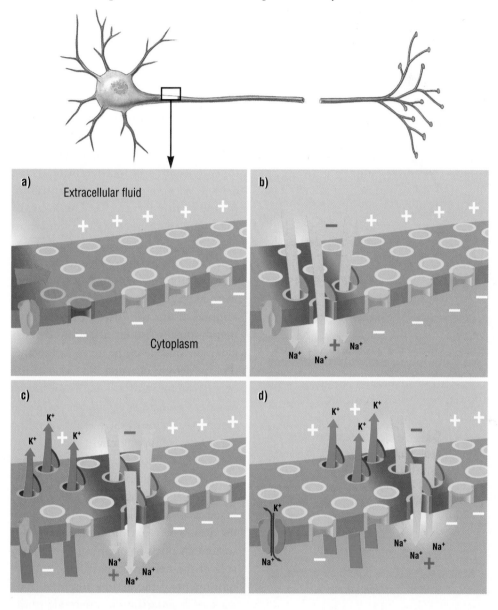

Figure 11.9

(a) The resting membrane is more permeable to potassium than sodium. Potassium ions diffuse out of the nerve faster than sodium ions diffuse into the nerve. The outside of the nerve cell becomes positive relative to the inside. (b) A strong electrical disturbance, shown by the darker coloration of the cell membrane, moves across the cell membrane. The disturbance opens sodium ion gates, and sodium ions rush into the nerve cell. The membrane becomes depolarized. (c) Depolarization causes the sodium gates to close, while the potassium gates are opened once again. Potassium follows the concentration gradient and moves out of the nerve cell by diffusion. Adjoining areas of the nerve membrane become permeable to sodium ions, and the action potential moves away from the site of origin. (d) The electrical disturbance moves along the nerve membrane in a wave of depolarization. The membrane is restored as successive areas once again become more permeable to potassium. The sodium and potassium pumps restore and maintain the polarization of the membrane.

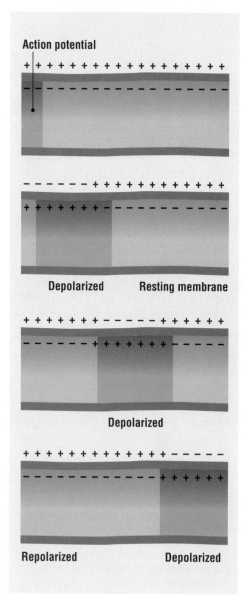

Refractory area   **Action potential**   **Resting membrane**

Extracellular fluid

Cytoplasm of nerve cell

| Repolarized area has recovered. | Depolarized area of nerve cell membrane | Adjacent area to be depolarized |

Direction of nerve impulse →

Figure 11.10

The movement of a nerve impulse.

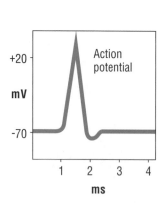

Figure 11.11

The action potential moves along the nerve cell membrane, creating a wave of depolarization and repolarization.

Action potential

Depolarized   **Resting membrane**

Depolarized

Repolarized   **Depolarized**

Movement of the Action Potential

The movement of sodium ions into the nerve cell causes a depolarization of the membrane and signals an action potential in that area. However, for the impulse to be conducted along the axon, the impulse must move from the zone of depolarization to adjacent regions.

It is important to understand the action potential. The action potential is characterized by the opening of the sodium channels in the nerve membrane. Sodium ions rush into the cytoplasm of the nerve cell, diffusing from an area of high concentration (outside of the nerve cell) to an area of lower concentration (inside of the nerve cell). The influx of the positively charged sodium ion causes a charge reversal, or depolarization, in that area. The positively charged ions that rush into the nerve cell are then attracted to the adjacent negative ions, which are aligned along the inside of the nerve membrane. A similar attraction occurs along the outside of the nerve membrane. The positively charged Na^+ ions of the resting membrane are attracted to the negative charge that has accumulated along the outside of the membrane in the area of the action potential.

The flow of positively charged Na^+ ions from the area of the action potential toward the adjacent regions of the resting membrane causes a depolarization in the adjoining area. The electrical disturbance causes sodium channels to open in the adjoining area of the nerve cell membrane and the movement of the action potential. As a wave of depolarization moves along the nerve membrane, the initiation point of the action potential enters a refractory period as the membrane once again becomes more permeable to K^+ ions. Depolarization of the membrane causes the sodium channels to close and the potassium channels to reopen. The wave of depolarization is followed by a wave of repolarization.

Threshold Levels and the All-or-None Response

A great deal of information about nerve cells has been acquired through laboratory experiments. Nerve cells respond to changes in pH, changes in pressure, and to specific chemicals. However, mild electrical shock is most often used in experimentation because it is easily controlled and its intensity can be regulated.

In a classic experiment, a single neuron leading to a muscle is isolated and a mild electrical shock is applied to the neuron. A special recorder measures the strength of muscle contraction. For this example, stimuli less than 2 mV will not produce any muscle contraction. A potential stimulus must be above a critical value in order to produce a response. The critical intensity of the stimulus is known as the **threshold level**. Stimuli below threshold levels do not initiate a response. A

threshold level of 2 mV is required to produce a response in the data shown in Figure 11.12; however, threshold levels are different for each neuron.

A second, but equally important, conclusion can be drawn from the experimental data. Increasing the intensity of the stimuli above the critical threshold value will not produce an increased response—the intensity of the nerve impulse and speed of transmission remain the same. In what is referred to as the **all-or-none response**, neurons either fire maximally or not at all.

How do animals detect the intensity of stimuli if nerve fibers either fire completely or not at all? Experience tells you that you are capable of differentiating between a warm object and one that is hot. To explain the apparent anomaly, we must examine the manner in which the brain interprets nerve impulses. Although stimuli above threshold levels produce nerve impulses of identical speed and intensity, variation with respect to frequency does occur. The more intense the stimulus, the greater the frequency of impulses. Therefore, when a warm glass rod is placed on your hand, sensory impulses may be sent to the brain at a slow rate. A hot glass rod placed on the same tissue will also cause the nerve to fire, but the frequency of impulses is greatly increased. The brain interprets the frequency of impulses.

The different threshold levels of neurons provide a second way in which the intensity of stimuli can be detected. Each nerve is composed of many individual nerve cells or neurons. A glass rod at 40°C may cause a single neuron to reach threshold level, but the same glass rod at 50°C will cause two or more neurons to fire. The second neuron has a higher threshold level. The greater the number of impulses reaching the brain, the greater the intensity of the response.

Threshold level *is the minimum level of a stimulus required to produce a response.*

The **all-or-none response** *of a nerve or muscle fiber means that the nerve or muscle responds completely or not at all to a stimulus.*

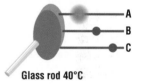

Glass rod 40°C

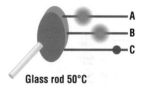

Glass rod 50°C

Figure 11.13

A glass rod at 40°C will elicit a response from neuron A. A glass rod heated to 50°C will elicit a response from neuron A and neuron B. Neuron B has a higher threshold level than neuron A, and will not fire until the glass rod is heated above 40°C. The brain interprets both the number of neurons excited and the frequency of impulses.

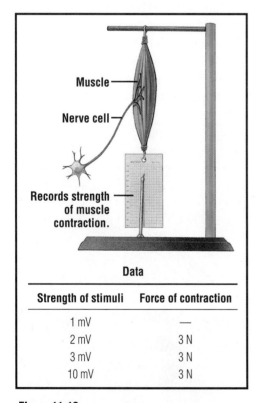

Muscle

Nerve cell

Records strength of muscle contraction.

Data

Strength of stimuli	Force of contraction
1 mV	—
2 mV	3 N
3 mV	3 N
10 mV	3 N

Figure 11.12

The threshold level for this neuron is 2 mV. Different neurons have different threshold levels.

THE SYNAPSE

Synapses *are the regions between neurons or between neurons and effectors.*

Presynaptic neurons *are the neurons that carry impulses to the synapse.*

Postsynaptic neurons *are the neurons that carry impulses away from the synapse.*

Acetylcholine *is a transmitter chemical released from vesicles in the end plates of neurons. Acetylcholine makes the postsynaptic membranes more permeable to Na⁺ ions.*

Cholinesterase *is an enzyme released from vesicles in the end plates of neurons shortly after acetylcholine. Cholinesterase breaks down acetylcholine.*

Small spaces between neurons or between neurons and effectors are known as **synapses.** A single neuron may branch many times at its end plate and join with many different neurons. Synapses rarely involve just two nerves. Small vesicles containing transmitter chemicals are located in the end plates of axons. The impulse moves along the axon and releases transmitter chemicals from the end plates. The transmitter chemicals are released from the **presynaptic neuron** and diffuse across the synapse, creating a depolarization of the dendrites of the **postsynaptic neuron.** Although the space between neurons is very small—approximately 20 nm (nanometers)—the nerve transmission slows across the synapse. Diffusion is a slow process. Not surprisingly, the greater the number of synapses, the slower the speed of transmission over a specified distance. This may explain why you react so quickly to a stimulus in a reflex arc, which has few synapses, while solving biology problems requires greater time.

Acetylcholine is a typical transmitter chemical found in the end plates of many nerve cells. Acetylcholine can act as an excitatory transmitter chemical on many postsynaptic neurons by opening the sodium ion channels. Once the channels are opened, the sodium ions rush into the postsynaptic neuron, causing depolarization. The reversal of charge causes the action potential. However, acetylcholine also presents a problem. By opening the sodium channels, the postsynaptic neuron would remain in a constant state of depolarization. How would the nerve ever respond to a second impulse if it did not recover? The release of the enzyme **cholinesterase** follows the acetylcholine and destroys it. Once acetylcholine is destroyed, the sodium channels are closed, and the neuron begins its recovery phase. Many insecticides take advantage of the synapse by blocking cholinesterase. The heart of an insect, unlike the human heart, is totally under nerve control. The next time you use an insecticide, consider that the insect's heart responds to the nerve message by contracting, but it never relaxes.

Figure 11.14

(a) Branching end plates synapse with the dendrites from many different neurons. (b) Synaptic vesicles in the end plates of the presynaptic neuron release transmitter chemicals by exocytosis. (c) The transmitter chemical attaches itself to the postsynaptic membrane, causing it to depolarize. The action potential continues along the postsynaptic neuron.

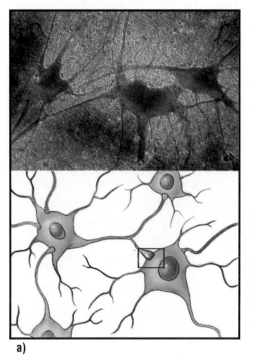

a)

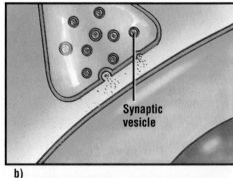

Synaptic vesicle

b)

Receptor on postsynaptic membrane

Transmitter molecules in synaptic vesicle

Synaptic cleft

c)

Unit Three:
Coordination and Regulation in Humans

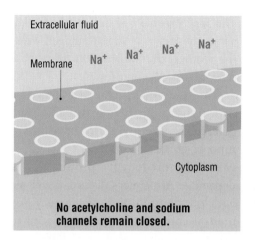

Extracellular fluid

Membrane

Na⁺ Na⁺ Na⁺ Na⁺

Cytoplasm

No acetylcholine and sodium channels remain closed.

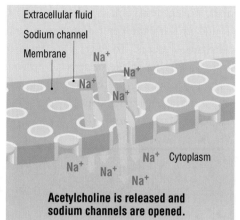

Extracellular fluid

Sodium channel

Membrane

Na⁺ Na⁺ Na⁺ Na⁺

Na⁺ Na⁺ Cytoplasm Na⁺

Na⁺

Acetylcholine is released and sodium channels are opened.

Figure 11.15

Model of an excitatory synapse. Acetylcholine opens channels for Na⁺ ions in the postsynaptic membrane.

Not all transmissions across a synapse are excitatory. While acetylcholine may act as an excitatory transmitter chemical on one postsynaptic membrane, it may act as an inhibitory transmitter chemical on another. It is believed that many inhibitory transmitter chemicals make the postsynaptic membrane more permeable to potassium. By opening even more potassium gates, the potassium ions on the inside of the neuron follow the concentration gradient and diffuse out of the neuron. The rush of potassium out of the cell increases the number of positive ions on the outside of the cell relative to the number found on the inside of the cell. Such neurons are said to be **hyperpolarized** because the resting membrane is even more negative. More sodium channels must now be opened to achieve depolarization and an action potential. As their name suggests, inhibitory transmitter chemicals prevent postsynaptic neurons from becoming active.

Figure 11.16 shows a model of a typical neural pathway. Transmitter chemicals released from neurons A and B are both excitatory; however, neither neuron is capable of causing sufficient depolarization to initiate an action potential in neuron D. However, when both neurons A and B fire at the same time, a sufficient amount of transmitter chemical is released to cause depolarization of the postsynaptic

membrane. The production of an action potential in neuron D requires the sum of two excitatory neurons. This principle is referred to as **summation.**

The transmitter chemical released from neuron C produces a dramatically different response. Neuron D becomes more negatively charged when neuron C is activated. You may have already concluded that neuron C is inhibitory. But data reveal even more striking information: transmitter chemicals other than acetylcholine must be present. A number of transmitter chemicals such as serotonin, dopamine, gamma-aminobutyric acid (GABA), and glutamic acid have been identified in the central nervous system. Another common transmitter chem-

Hyperpolarized membranes *are much more permeable to potassium than usual. The inside of the nerve cell membrane becomes even more negative.*

Summation *is the effect produced by the accumulation of transmitter chemicals from two or more neurons.*

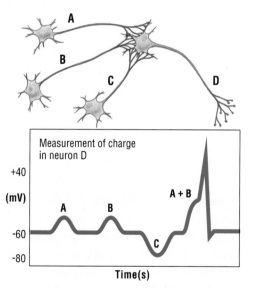

Measurement of charge in neuron D

+40

(mV)

A   B   A + B

-60

C

-80

Time(s)

Figure 11.16

Action potentials must occur simultaneously in A and B to reach threshold level in D.

ical, norepinephrine (noradrenaline) is found in both the central and peripheral nervous systems. To date, all effects of norepinephrine in the peripheral nervous system appear to be excitatory, while those of the central nervous system can be excitatory or inhibitory.

B I O L O G Y   C L I P

The action of many psychoactive drugs can be explained in terms of neurotransmitters. Valium, a tranquilizer, interacts with GABA transmitter-receptor sites on postsynaptic membranes. The greater the number of receptor sites that have been occupied, the more effective is the transmitter chemical. LSD and mescaline, which are hallucinogenic drugs, are thought to interact with the receptor sites of serotonin.

The interaction of excitatory and inhibitory transmitter chemicals is what allows you to throw a ball. As the triceps muscle on the back of your upper arm receives excitatory impulses and contracts, the biceps muscle on the front of your arm receives inhibitory impulses and relaxes. The triceps muscle straightens the arm, while the biceps muscle bends the arm. Inhibitory impulses in your central nervous system are even more important. Sensory information is received by the brain and is prioritized. Much of the less important information is inhibited so that you can devote your attention to the most important sensory information. For example, during a biology lecture, your sensory information should be directed at the sounds coming from your teacher, the visual images that appear on the chalkboard, and the sensations produced as you move your pen across the page. Although your temperature receptors may signal a slight chill in the air, and the pressure receptors in your

skin may provide reassuring information about the fact that you are indeed wearing clothes, the information from these sensory nerves is suppressed.

Various disorders have been associated with transmitter chemicals. Parkinson's disease, characterized by involuntary muscle contractions and tremors, is caused by inadequate production of dopamine. Alzheimer's disease, associated with the deterioration of memory and mental capacity, has been related to decreased production of acetylcholine.

■ REVIEW QUESTIONS   ?

7 What evidence suggests that nerve impulses are not electricity but electrochemical events?

8 Why was the giant squid axon particularly appropriate for nerve research?

9 What is a polarized membrane?

10 What causes the inside of a neuron to become negatively charged?

11 Why does the polarity of a cell membrane reverse during an action potential?

12 What changes take place along a nerve cell membrane as it changes from a resting membrane to an action potential and then into a refractory period?

13 Why do nerve impulses move faster along myelinated nerve fibers?

14 What is the all-or-none response?

15 Use the model of the synapse to explain why nerve impulses move from neuron A to neuron B, but not from neuron B back toward neuron A.

16 Explain the functions of acetylcholine and cholinesterase in the transmission of nerve impulses.

17 Use a synapse model to explain summation.

CHAPTER HIGHLIGHTS

- The nervous system coordinates the activities of other body systems. Nerves enable the body to respond to stimuli and maintain homeostasis.
- Neurons conduct nerve impulses. Sensory neurons carry impulses to the central nervous system. Interneurons carry impulses within the central nervous system. Motor neurons carry impulses from the central nervous system to effectors.
- The neuron is composed of functionally distinct components. Dendrites are projections of cytoplasm that carry impulses toward the cell body. An axon is an extension of cytoplasm that carries nerve impulses away from the cell body. The myelin sheath is a fatty covering over the axon of a nerve cell. The nodes of Ranvier are the regularly occurring gaps between sections of myelin sheath along the axon.
- The nerve impulse results from the movement of ions across the nerve membrane.
- Threshold is the minimum level of a stimulus required to produce a response.
- The all-or-none response of a nerve or muscle fiber indicates that the nerve or muscle responds completely to a stimulus or not at all.
- Synapses are the regions between neurons or between neurons and effectors. Synapses may be excitatory or inhibitory.
- Summation is the effect produced by the accumulation of transmitter chemicals from two or more neurons.
- Somatic nerves lead to skeletal muscle and are under conscious control.

- Autonomic nerves are motor nerves designed to maintain homeostasis. Autonomic nerves are not under conscious control.
- Sympathetic nerves are a division of the autonomic nervous system and prepare the body for stress.
- Parasympathetic nerves are a division of the autonomic nervous system and are designed to return the body to normal resting levels following adjustments to stress.
- The central nervous system consists of the brain and spinal cord. The brain is responsible for the coordination of sensory and motor nerve activity.
- Meninges are protective membranes that surround the brain and spinal cord. Cerebrospinal fluid circulates between the innermost and middle membranes of the brain and spinal cord.
- The cerebrum is the largest and most developed part of the human brain. The cerebrum stores sensory information and initiates voluntary motor activities.
- The corpus callosum is a nerve tract that joins the two cerebral hemispheres.
- The cerebellum is the region of the brain that coordinates muscle movement.
- The pons is the region of the brain that acts as a relay station by sending nerve messages between the cerebellum and the medulla.
- The medulla oblongata is the region of the hindbrain that joins the spinal cord to the cerebellum. The medulla is the site of autonomic nerve control.
- Endorphins belong to a group of chemicals classified as neuropeptides. Containing between 16 and 31 amino acids, endorphins are believed to reduce pain.

APPLYING THE CONCEPTS

1 How did Luigi Galvani's discovery that muscles twitch when stimulated by electrical current lead to the discovery of the ECG and EEG?

2 During World War I, physicians noted a phenomenon called "phantom pains." Soldiers with amputated limbs complained of pain or itching in the missing limb. Use your knowledge of sensory nerves and the central nervous system to explain this phenomenon.

3 Explain the importance of a properly functioning myelin sheath by outlining the pathology of multiple sclerosis.

4 Explain why damage to the gray matter of the brain is permanent, while minor damage to the white matter may only be temporary.

5 Use information that you have gained about threshold levels to explain why some individuals can tolerate more pain than others.

6 Use the diagram below to answer the following question.

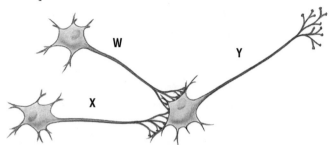

The neurotransmitter released from nerve X causes the postsynaptic membrane of nerve Y to become more permeable to sodium. However, the neurotransmitter released from nerve W causes the postsynaptic membrane of nerve Y to become less permeable to sodium but more permeable to potassium. Why does the stimulation of neuron X produce an action potential in neuron Y, but fail to produce an action potential when neuron X and W are stimulated together?

7 Botulism (a deadly form of food poisoning) and curare (a natural poison) inhibit the action of acetylcholine. What symptoms would you expect to find in someone exposed to botulism or curare? Provide an explanation for the symptoms.

8 Nerve gas inhibits the action of cholinesterase. What symptoms would you expect to find in someone exposed to nerve gas? Provide an explanation for the symptoms.

9 A patient complains of losing his sense of balance. A marked decrease in muscle coordination is also mentioned. On the basis of the symptoms provided, which area of the brain might a physician look to for the cause of the symptoms?

10 In a classic experiment performed by Roger Sperry, two patients viewed the word "cowboy." The first subject acted as a control. The second subject had his corpus callosum severed. Note the right and left hemispheres have been supplied with different stimuli. Predict the results of the experiment and provide an explanation for your predictions.

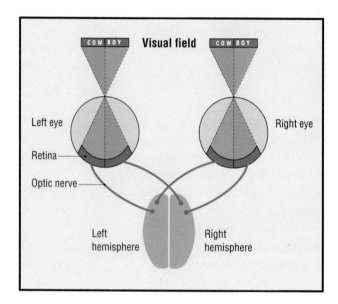

CRITICAL-THINKING QUESTIONS

1 People with Parkinson's disease have low levels of the nerve transmitter dopamine. In 1982, a group of Swedish scientists grafted cells from a patient's adrenal glands into the patient's brain. The adrenal glands produce dopamine. In July 1987, Swedish scientists announced the first transplant of human fetal brain cells into other animals. In September 1987, officials at several hospitals in Mexico City announced the transplant of brain tissue from dead fetuses into patients with Parkinson's disease. Although this research is very new, it would appear that the fetal cells begin producing dopamine. Patients with Parkinson's disease who have received the transplanted tissue have demonstrated remarkable improvement. Comment on the ethics of this research.

2 Research by Dr. Bruce Pomerantz of the University of Toronto has revealed a link between endorphins and acupuncture. Dr. Pomerantz believes that acupuncture needles stimulate the production of pain-blocking endorphins. Although acupuncture is a time-honored technique in the east, it is still considered to be on the fringe of modern western medical practice. Why have western scientists been so reluctant to accept acupuncture?

Unit Three:
Coordination and Regulation in Humans

3 Painkillers are big business. Television commercials tell us that the answer to our headaches, sore backs, or sore muscles is a pill. Should people be encouraged to take pills? What are the implications for individuals who refuse to give their children pills for pain?

4 Aspirin and Tylenol were both marketed long before their action was defined. Although both drugs are generally considered safe, they are not without side effects. Aspirin in high dosages can destroy the protective lining of the stomach. Tylenol in high dosages can cause liver damage. Should the biochemical mechanism of drug action be defined before a drug is marketed? Should drugs that produce side effects be sold over the counter? What tests would you like to see before drugs are approved for sale?

5 The EEG has been used to legally determine death. Although the heart may continue to beat, the cessation of brain activity signals legal death. Ethical problems arise when some brain activity remains despite massive damage. Artificial resuscitators can assume the responsibilities of the medulla oblongata and regulate breathing movements. Feeding tubes can supply food, and catheters can remove wastes when voluntary muscles can no longer be controlled. The question of whether life should be sustained by artificial means has often been raised. Should a machine like the EEG be used to define the end of life?

6 According to the study of craniometry, males were once considered more intelligent than females because they have larger skulls. Outline some of the ethical dangers associated with using skull size to define intelligence.

ENRICHMENT ACTIVITIES

1 Inexpensive galvanic skin-response detectors can be purchased from scientific companies. Although the unit is not a complete lie detector (polygraph), you may wish to test its accuracy.

2 Suggested reading:

- Allport, Susan. *Explorers of the Black Box: The Search for the Cellular Basis of Memory*. New York: W.W. Norton, 1986.

- Gould, Stephen Jay. *Mismeasure of Man*. New York: W.W. Norton, 1981. This book takes a critical view of the pseudoscience of craniometry. Craniometry attempted to link intelligence with skull size and was often inspired by racism. Gould shows how the selective manipulation of data can be used either intentionally or unintentionally to promote untruths.

- Maranto, Gina. "The Mind within the Brain." *Discover*, May 1984, p. 36.

- Montgomery, Geoffrey. "The Mind in Motion." *Discover*, March 1989, p. 58.

- ——. "Molecules of Memory." *Discover*, December 1989, p. 46.

- Penfield, Wilder. *The Mysteries of the Mind: A Critical Study of Consciousness and the Human Mind*, Princeton University Press, 1975.

- Restak, Richard. *The Brain: The Last Frontier*, Warner Books, 1979.

*M*ovement and Support

IMPORTANCE OF THE SKELETON AND MUSCLES

An **exoskeleton** *is an external skeleton.*

An **endoskeleton** *is an internal skeleton.*

Bone marrow *is the tissue located in the central cavity of the long bones.*

The skeletal system performs a number of very important functions. All organisms that require support have some type of skeleton. Some invertebrates, such as insects and crustaceans, rely on chitinous external skeletons **(exoskeletons)** for support; while vertebrates rely on cartilaginous or bony internal skeletons **(endoskeletons)** for support. Skeletons also provide protection for delicate internal organs. Exoskeletons cover all of the internal organs and the muscles. The endoskeletons of vertebrates, however, cover only some of the internal organs. The muscles of vertebrates surround the skeleton and provide for articulated movement. Both types of skeletons serve as shock absorbers for the internal organs. The vertebrate skeleton also produces blood cells, houses some nerve cells, and stores fat in an area called the **bone marrow.** Found within many bones, the marrow is a hub of activity.

An important function of the skeleton is to work in combination with the muscle system. Muscle tissue is specialized for contraction. When muscles move bones

Figure 12.1

Muscles and bones combine to provide the type of complex movements required by athletes.

and cause them to work like levers, movement of body parts occurs, which permits locomotion. Muscles and bones enable you to lift heavy objects, run, jump, and write. Muscle that moves bone is referred to as **skeletal muscle.** However, not all skeletal muscles are involved in locomotion. Your muscle system also allows you to smile, frown, or wink. Other muscles enable you to swallow or draw in air. Muscles also support the internal organs. **Cardiac muscle** causes the pumping of blood by the heart, and **smooth muscle** controls the movement of food along the digestive tract and regulates the diameter of the blood vessels. Because skeletal muscle is regulated consciously, it is referred to as *voluntary muscle.* Smooth muscle and cardiac muscle are usually not controlled consciously, and are therefore referred to as *involuntary muscles.*

SKELETAL STRUCTURE

The human adult skeletal system contains 206 bones, which vary in shape and structure. The vast majority of these bones are contained in the hands and feet. Not surprisingly, these two body parts are capable of the greatest diversity of movement.

The human skeleton has two major divisions: the **axial skeleton** and the **appendicular skeleton.** The axial skeleton includes the 28 bones of the skull, the tiny bones of the middle ear, the backbone, and the 12 pairs of ribs that join with the breastbone. The appendicular skeleton includes the arms, hands, and pectoral girdle as well as the feet, legs, and pelvic girdle.

There are two types of bone: compact bone, which is composed of hard, dense materials, and spongy bone, which is made up of lighter, less dense materials. The long bones consist mostly of dense, compact tissue, while the ribs are largely composed of spongy tissue. The bone is covered by **periosteum,** a thin covering that contains blood vessels, nerves, and bone-forming cells.

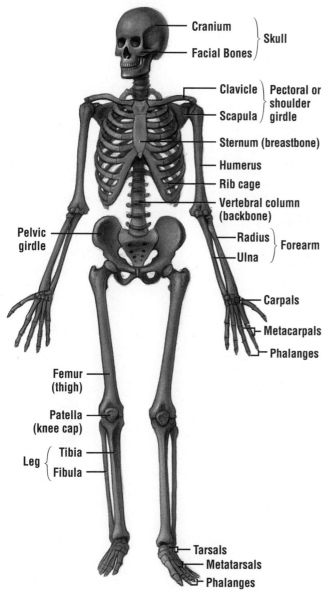

- Cranium ⎫ Skull
- Facial Bones ⎭
- Clavicle ⎫ Pectoral or shoulder girdle
- Scapula ⎭
- Sternum (breastbone)
- Humerus
- Rib cage
- Vertebral column (backbone)
- Radius ⎫ Forearm
- Ulna ⎭
- Carpals
- Metacarpals
- Phalanges
- Pelvic girdle
- Femur (thigh)
- Patella (knee cap)
- Leg ⎧ Tibia
- ⎩ Fibula
- Tarsals
- Metatarsals
- Phalanges

Skeletal muscle *is voluntary muscle that is attached to bones. The muscle is sometimes referred to as striated muscle because of its striped appearance.*

Cardiac muscle *is involuntary muscle found in the heart. This branching, lightly striated muscle is capable of contracting rhythmically.*

Smooth muscle *is involuntary, nonstriated muscle.*

The **axial skeleton** *is the central supporting part of the skeleton and consists of the bones of the skull, backbone, ribs, and breastbone.*

The **appendicular skeleton** *consists of the bones of the upper and lower limbs and their supporting structures.*

Periosteum *is the tissue that covers bone.*

Figure 12.2

The human skeleton, with the axial skeleton shown in green, and the appendicular skeleton shown in purple.

BONE DEVELOPMENT

Bone forms in either cartilage or fibrous connective tissue. In cartilage, bone-forming cells called **osteoblasts** begin to form bone tissue inside the shaft of cartilage. In an embryo, most of the skeleton is cartilage. As the embryo develops, calcium and phosphate ions are deposited in the cartilage. The cartilage cells begin to die and bone cells form canals within the structure. This process is called **ossification.** The canals, known as **Haversian canals,** provide pathways for nerves and blood vessels. Because bone is a living tissue, oxygen and other nutrients are essential for its development. Eventually, the canals grow together, forming the bone marrow.

Long bones continue to grow in length into adulthood as the calcified car-tilage is replaced by bony tissue. The process begins at the middle of the bone and proceeds toward the ends. The only remaining cartilage is found in the joints and at the ends of the long bones, where it forms the **epiphyseal plates.** The epiphyseal plates are often referred to as the *growth plates*. New cartilage cells develop in the epiphyseal plates, but eventually they are overtaken by the bone-producing cells. Bone ceases to grow once the epiphyseal plates become ossified. In humans this occurs usually by age 25.

Bone formation in fibrous connective tissue follows much the same plan. Osteoblasts form splinterlike fragments called *spicules*, which eventually become joined to form spongy bone. Should the matrix continue to thicken, compact bone is formed. Although bone is often

Osteoblasts *are bone-forming cells.*

Ossification *is the process by which bone is formed.*

Haversian canals *are small canals located in bone tissue. The canals are occupied by blood vessels and nerves.*

Epiphyseal plates *contain cartilage and join two sections of bone in youngsters.*

Figure 12.3

Anatomy of the long bone of a mammal.

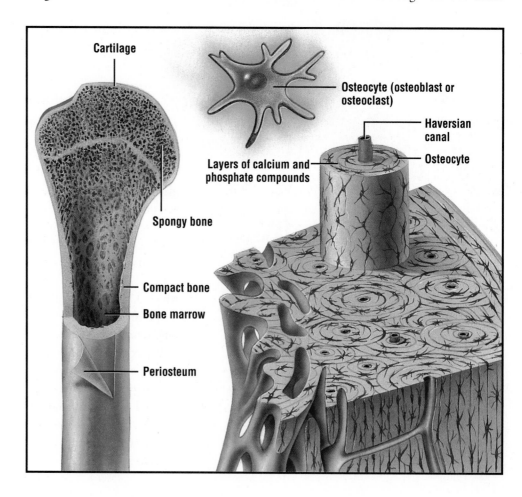

Unit Three:
Coordination and Regulation in Humans

considered to be dead, it is one of the most dynamic tissues in your body. Bone is continually being remodeled by **osteoclasts**, which dissolve bone, and osteoblasts, which prepare new structures. In much the same way as buildings within a city's core are torn down and replaced with newer structures, your bones are continually being refurbished. As it is in a city, the balance between destruction and restoration is a critical one. A rapid loss of bone structure in the vertebral column, legs, and feet is known as *osteoporosis*. The breakdown of bone, especially prevalent in older women, can lead to a partial collapse of the vertebral column. In some

extreme cases the ribs may even come to rest on the pelvic girdle. Although the cause of the disease is not known, many researchers believe that the disease can be linked to decreasing levels of female sex hormones, calcium deficiencies, increased protein consumption, and decreased activity.

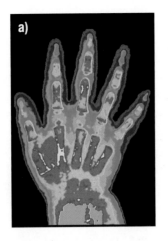 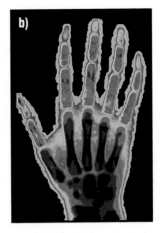 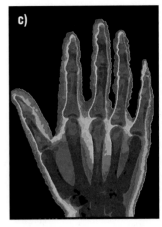

Figure 12.4

X rays of (a) 2 1/2 (b) 6 1/2 and (c) 19-year-old male hand, showing the replacement of cartilage by bone.

Osteoclasts *are cells that dissolve bone.*

DEMONSTRATION
BONE STRUCTURE

Procedure

1 Put on your safety goggles and lab apron.
2 In a 100 mL beaker pour 80 mL of 5 M HCl. Place an end of one of the bones in the HCl and allow the bone to soak for 48 h. (The beaker containing HCl must be labeled and stored in a safe place. Fume hoods are well suited for storing the beaker.)
3 Using tongs, remove the bone from the beaker and rinse the HCl from the bone.
4 Compare the treated end of the bone with the untreated end.
   a) Describe the appearance of the bone soaked in HCl. How has the acid altered the bone?
5 Attempt to twist the ends of each bone.
   b) Describe what happens to each bone.
6 Dispose of the acid after neutralization and thoroughly wash your hands. ■

Objective

To study the structure of spongy bone.

Materials

safety goggles
chicken ribs or leg bones (2)
beaker

lab apron
5 M HCl
tongs

CAUTION: 5 M HCl is very corrosive. Wash all splashes off your skin and clothing thoroughly. If you get any chemical in your eyes, rinse for at least 15 min and inform your teacher.

TYPES OF JOINTS

The point at which two or more bones come together is referred to as a **joint.** Some joints, like those of the skull, are not capable of movement. *Fibrous*, or immovable, joints have no gaps between the bones. As the fetus descends through the birth canal, the bones of the skull shift slightly to permit passage. A small area, often called the "soft spot" (fontanelle), is later replaced by bone and the small spaces between the skull bones become fused.

Other joints, like those of the vertebral column, are capable of restricted movement. Referred to as *cartilaginous* joints because of small bands of cartilage between the bones, these joints can act as shock absorbers in a fall. The cartilaginous disks found between the bony vertebrae compress upon impact.

Still other joints, like those found in your arms and legs, move freely. These are called *synovial* joints. The joining surfaces of the bones are covered by a slippery, smooth cartilage that promotes frictionless movement of the bones. Many of the joining surfaces are covered by a fluid-secreting membrane called the *synovial membrane*. The synovial membrane acts like a capsule lubricating the joint and helping to reduce friction.

Movable joints can also be classified according to the type of movement the joint permits, as shown in Figure 12.6.

> **BIOLOGY CLIP**
> Arthritis is a disease of the joints. In osteoarthritis, the cartilage at the end of the bones wears away. In rheumatoid arthritis, the synovial membrane becomes inflamed and thickens. During the final stages of rheumatoid arthritis, the cartilage between the joints deteriorates.

*A **joint** is the point of contact of two or more bones of the skeleton.*

Figure 12.5

Joints can be classified as (a) immovable, (b) slightly movable, and (c) freely movable.

Figure 12.6

Joints that move can be classified according to the type of movement they permit.

a) Fibrous joint   **b) Cartilaginous joint**   **c) Synovial joint**

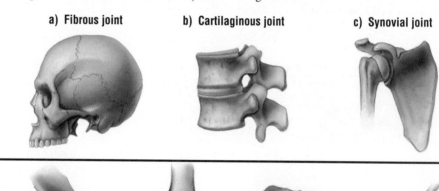

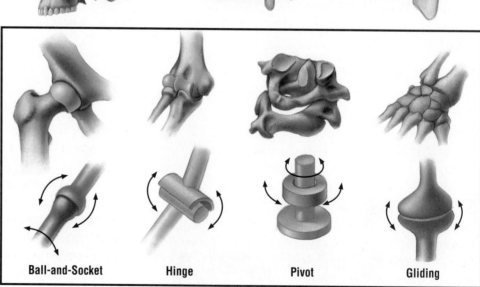

Ball-and-Socket   Hinge   Pivot   Gliding

LIGAMENTS AND TENDONS

Ligaments are bands of connective tissue that join bone to bone. Although ligaments are found in your hands, ankles, wrists, and feet, the most obvious are those of the knees. Seven ligaments hold the leg bones together at the knee. A severe blow to the knee can often sprain or tear the ligaments that hold the femur, the upper bone of the leg, to the tibia and fibula, the lower bones of the leg. Torn ligaments are often attacked by roving white blood cells. White blood cells remove debris that accumulates from day-to-day wear. Unfortunately, the white blood cells can also turn against the torn ligaments, transforming the tissues of the knee to mush.

Tendons join muscle to bone. This remarkably strong connective tissue will often hold even when the bones break. Tendons smaller than the width of a pencil are capable of supporting a load of several thousand kilograms. Wide muscles often taper to small tendons, which attach to small surfaces of bones. **Bursae,** which are the small, fluid-filled sacs located between the bone and tendon, reduce the rubbing of the tendon on bone. An inflammation of the bursae is known as *bursitis*. Baseball pitchers are particularly susceptible to this condition.

FRONTIERS OF TECHNOLOGY: ARTHROSCOPIC SURGERY

Not so long ago a torn cartilage required surgery under general anesthesia, a 10 cm incision, four days of hospital stay, and six weeks of slow recovery. After three months, the individual might be allowed to resume normal activity. An innovative technique called arthroscopic surgery, named after a viewing device called an arthroscope, has dramatically improved the prognosis for people with knee injuries.

The first arthroscope was used in Japan in 1917; however, today's instruments barely resemble the early predecessor. A needlelike tube, less than 2 mm wide, is equipped with a fiber-optic light source. Using only local anesthesia, doctors can insert the needle through a small puncture in the knee. The fiber-optic lens can be linked to a television screen, that provides an internal view of the damaged knee. In addition, the arthroscope is fitted with two thin surgical tools that can snip away unhealthy cartilage. Under most circumstances, hospitalization is not required following the surgery. Normal activity can be resumed in a much shorter time than previously possible.

> **B I O L O G Y   C L I P**
> The shoulder is the most easily dislocated major joint. The joint has little support from underneath. A sharp downward pull can dislodge the joint.

■ REVIEW QUESTIONS   ?

1 How are an exoskeleton and an endoskeleton alike? How are they different?
2 List the functions of an endoskeleton.
3 Differentiate between the axial and appendicular skeleton.
4 Explain the functions of osteoblasts and osteoclasts in maintaining bone growth and structure.
5 Briefly outline how the long bones grow.
6 List three types of joints, and provide an example of each.
7 Differentiate between tendons and ligaments.

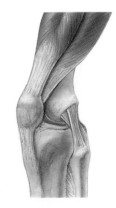

Figure 12.7
The knee is held together by seven ligaments.

Ligaments *are bands of connective tissue that join bone to bone.*

Tendons *are bands of connective tissue that join muscle to bone.*

Bursae *are sacs of fluid found in joints.*

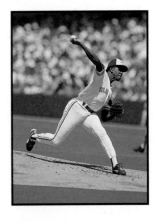

Figure 12.8
Throwing a baseball puts tremendous strain on the shoulder and elbow joints.

RESEARCH IN CANADA

Hip Prostheses

More than 20 000 hips are replaced in Canada annually, and as many as 200 000 in North America alone. Yet little information is available to evaluate the different types of procedures and prostheses being used. Originally, most hip replacements were performed on patients over 65 years of age; today, however, replacements are also being carried out on younger, more active patients. Therefore, the demands placed on the use of the prosthesis are much greater. Although a prosthesis can be replaced, the procedure is much more complicated and involves much higher costs.

Because the prostheses are continually being improved, the long-term effects of the artificial hips are difficult to determine. Currently, there are about 25 different models. An artificial hip should last about 10 years. Although each one is touted as being superior to its predecessor, it could take 10 years to discover if this is in fact true.

The first artificial hip replacement, developed by British orthopedic surgeon Sir John Charnely, was cemented into the hip bone. Today many models are implanted without cement. It would appear that these models provide adequate support and allow for growth of the bone. Dr. Michael Gross of Dalhousie University in Halifax is comparing the function and durability of cemented versus uncemented hip prostheses. Dr. Gross wants to determine whether the implants that do not use cement are as effective as those that use cement.

DR. MICHAEL GROSS

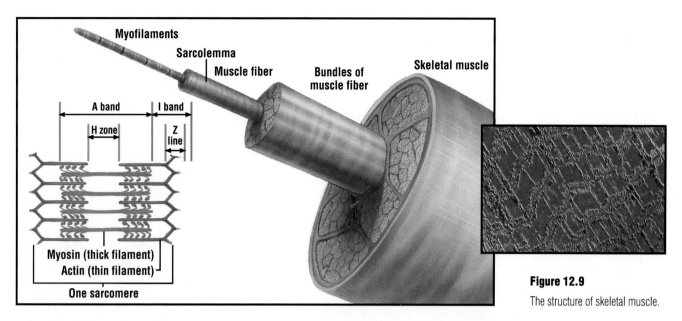

Labels in figure:
Myofilaments
Sarcolemma
Muscle fiber
Bundles of muscle fiber
Skeletal muscle
A band
I band
H zone
Z line
Myosin (thick filament)
Actin (thin filament)
One sarcomere

Figure 12.9

The structure of skeletal muscle.

SKELETAL MUSCLE STRUCTURE

In this chapter only skeletal muscle, the muscle that surrounds the skeleton, will be discussed. Cardiac muscle and smooth muscle were dealt with in previous chapters. If you bend your arm and squeeze your fist together, the muscles in your forearm and the biceps, the large muscle above the elbow, bulge. These muscles are skeletal muscles. Skeletal muscles permit you to move, enable you to smile, and help keep you warm. An estimated 80% of the energy used in the contraction of skeletal muscles is lost as heat. Is it any wonder that you shiver when you are cold?

Skeletal muscle is composed of cells, called fibers, bundled together. Unlike other body cells, which contain one nucleus, each muscle cell has many nuclei. The fibers are enclosed within a membrane called the **sarcolemma.** Within the muscle fibers are tiny **myofilaments** bundled together. Two kinds of myofilaments can be seen under the electron microscope. Thin filaments, called *actin,* and thick filaments, called *myosin,* overlap, producing a striated or striped appearance.

The alternating dark and light bands of the muscle fibers can be explained by examining the arrangement of the myofilaments. The length of the muscle fiber is defined by the Z lines that anchor the actin filaments. The area between the Z lines is referred to as the *sarcomere.* The thick myosin filaments form the darker A bands, while the thinner actin filaments allow more light to penetrate and form the lighter I bands shown in Figure 12.9.

THE SLIDING-FILAMENT THEORY

As the word "theory" in the heading suggests, not everything about muscle contraction is known. The sliding-filament theory provides a working model that helps explain what scientists believe is happening and shows how cell structure is related to function.

Muscles cause movement by shortening. However, the filaments themselves do not contract. Instead, the actin filaments slide over the myosin filaments. Z lines move closer together when muscle fibers contract. As the actin and myosin filaments begin to overlap, the lighter I

The **sarcolemma** *is the delicate sheath that surrounds muscle fibers.*

Myofilaments *are the contractile proteins found within muscle fibers.*

band becomes progressively smaller. But what causes the actin and myosin filaments to slide? It is believed that knoblike projections on the thick myosin filaments form cross-bridges on receptor sites of the thinner actin filaments. A series of cross-bridges attach and detach, drawing the actin filaments inward.

Figure 12.10

Sliding-filament theory, showing one actin and one myosin filament.

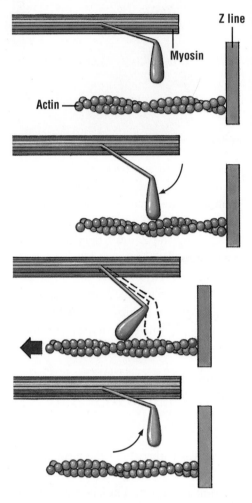

The energy required for muscle contraction and relaxation comes from ATP, adenosine triphosphate. In the absence of ATP, the cross-bridges fail to detach and the muscle becomes rigid. The term *rigor mortis* is used to describe the rigidity of muscles (the inability of muscles to relax) following death. With death, ATP production ceases and skeletal muscle becomes frozen in a fixed position. The condition may last up to 60 h after death.

The release of a transmitter chemical at the junction between the nerve and muscle initiates muscle contraction. Once the transmitter chemical reaches a specialized endoplasmic reticulum, found within the cytoplasm, calcium ions are released. The calcium ions bind to sites along the actin filaments and initiate the formation of cross-bridges with the myosin fibers. It is believed that the release of calcium ions begins the breakdown of ATP by the myosin filaments. The breakdown of ATP causes the filaments to slide over one another. Eventually, the calcium ions are actively taken up and stored once again in the specialized endoplasmic reticulum. The muscle relaxes. When the calcium ions are released once more from the endoplasmic reticulum, the muscle contracts.

Have you ever felt your muscles begin to burn while skiing? Have your muscles ever failed you during a race? No matter how hard you resist, you begin to lose control of your muscles. Muscle fatigue is caused by a lack of energy and the build-up of waste products within your muscles. Unfortunately, very little ATP can be stored in muscle tissues. Ideally, the energy demand is met by aerobic respiration. Glucose is systematically broken down by a series of enzymes found in the cytoplasm and mitochondria of every cell of your body. Glucose is oxidized by oxygen to form ATP, carbon dioxide, and water. A high-energy compound called **creatine phosphate,** found in all muscle cells, ensures that ATP supplies remain high. Creatine phosphate supplies a phosphate to ADP to replenish ATP supplies. If creatine phosphate levels remain high in muscle cells, ATP levels can be maintained.

BIOLOGY CLIP

Although bodybuilders work hard to develop well-defined biceps muscles, it is the triceps that is used most often during sports that require throwing such as baseball, javelin, and shot put.

As long as oxygen can be supplied and cellular respiration can meet the demands of muscle cells, the filaments will continue to be drawn together. However, should energy demand exceed ATP supply, lactic acid begins to accumulate. Lactic acid causes muscle pain and is associated with fatigue. The burning that you feel in your legs while skiing a difficult run, or the pain you feel in your rib muscles after prolonged heavy exercise is due to an accumulation of lactic acid. During this condition, referred to as *oxygen debt*, the fluids surrounding the muscles become acidic, and eventually the muscle fails to contract. The rapid breathing that takes place after heavy exercise is designed to repay the oxygen debt.

INTERACTION OF MUSCLE AND SKELETON

Many muscles work in pairs, causing the limbs to either bend or straighten. Because muscles cannot push bones into the desired position, antagonistic pairs alternate to move the long bones in opposite directions. For example, your arm is bent when the biceps muscle contracts. The shortening of the biceps muscle draws the forearm toward the shoulder. When the biceps muscle relaxes and the triceps muscle contracts, the forearm extends.

The central nervous system ensures that the biceps and triceps do not attempt to pull against each other. Excitatory nerve impulses that cause the triceps to contract are accompanied by inhibitory nerve impulses, which cause the biceps to relax.

Creatine phosphate *is a compound found in muscle cells that releases a phosphate to ADP and helps regenerate ATP supplies in muscle cells.*

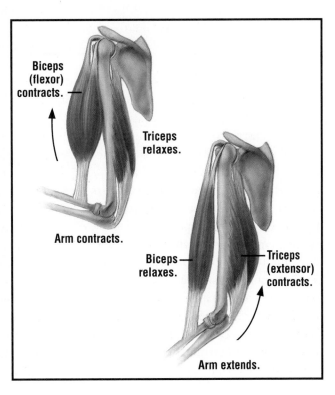

Figure 12.11

The biceps and triceps muscles are antagonistic pairs. The contraction of the biceps muscle causes the arm to bend, while the contraction of the triceps muscle causes the arm to extend.

Biceps (flexor) contracts.

Triceps relaxes.

Arm contracts.

Biceps relaxes.

Triceps (extensor) contracts.

Arm extends.

Figure 12.12

(a) Recording of a muscle twitch that lasts approximately 1 s.
(b) Single muscle twitches approximately 1 s apart. The muscle returns to its original length before succeeding stimuli cause contractions.
(c) Summation of muscle twitches from about six stimulations every second. Following the contraction, the muscle does not have enough time to return to its original length before being stimulated once again. (d) Tetanus resulting from about 20 stimulations per second. The actin and myosin filaments remain overlapped.

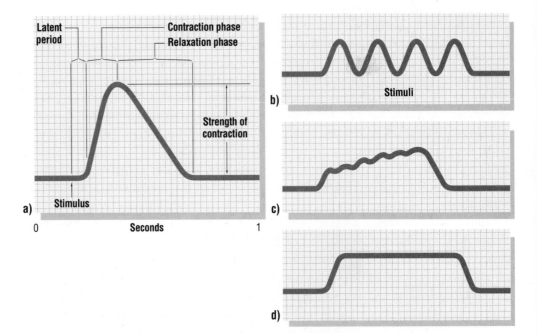

Summation *is the effect produced by the combination of stimuli.*

Tetanus *is the state of constant muscle contraction caused by sustained nerve impulses*

A muscle twitch occurs when a nerve impulse stimulates several muscle cells. A pause between the impulse and the muscle contraction is referred to as the *latent period*. Upon contraction, actin and myosin fibers slide over one another, causing the muscle to shorten. After the contraction phase, the actin and myosin filaments disengage and the muscle begins to relax. Once the relaxation phase is complete, each muscle cell usually returns to its original length. Should a muscle cell be stimulated once again, it will contract with equal force.

An interesting phenomenon occurs when a stimulation happens before the relaxation phase is complete. Predictably, the actin and myosin filaments slide over one another; but because the relaxation has not yet been completed, the overlap is increased and greater muscle shortening can be observed. The sum of the shortening that remains from the first muscle twitch and the shortening produced by the second muscle twitch cre-

ates a greater force of contraction. The strength of the contraction depends on how close the second stimulus is to the first stimulus. The process, shown in Figure 12.12(c), is referred to as **summation.** Occasionally, repeated muscle stimulation prevents any relaxation phase. The state of constant muscle contraction, known as **tetanus,** is shown in Figure 12.12(d).

> **BIOLOGY CLIP**
> Muscle spasms are caused by involuntary contractions of muscles. A pinched nerve is often responsible for the spasm.

■ REVIEW QUESTIONS ?

8 What is the sarcolemma?

9 Name the two myofilaments found in muscle fibers and briefly outline their function.

10 Why does skeletal muscle appear striated, or striped?

11 Why is ATP needed for extended muscular activity?

12 Why is creatine phosphate required for extended muscular activity?

13 Define muscle tetanus.

Unit Three:
Coordination and Regulation in Humans

Objective

To determine how temperature affects muscle contraction.

Materials

ice   large beaker

pen and paper   stopwatch

Procedure

1 Write your name as many times as you can in 2 min. Use the stopwatch to time yourself. In a chart like the one shown, record the number of times that you were able to write your name.

2 Immerse your hand in ice water for as long as you possibly can, but no longer than 1 min. Once again write your name as many times as possible in 2 min. Record the number of times you were able to write your name.

3 Rub your hand until it is warm, and repeat the procedure. Record your observation in a chart like the one below.

a)

Hand condition	Number of signatures
Initial	
After cold-water bath	
After hand warmed	

b) Compare the quality of the signatures.

c) Why does cold water affect the muscles?

d) What other factors, besides cold, would affect the number and quality of the signatures? ■

ATHLETIC INJURIES

As athletes push their bodies to become faster, stronger, and more flexible, injuries inevitably arise. Several factors increase the chances of injury. Cold weather may increase the probability of an injury. Reduced blood flow to exterior muscles of the legs and arms, and the greater muscle tension associated with cold temperatures increase the likelihood of muscle strains and sprains. You should always do warm-up exercises and stretching before beginning any vigorous exercise. This helps reduce muscle tension and increases blood flow and muscle temperature.

Fatigue is another common cause of injuries. The build-up of lactic acid associated with vigorous exercise reduces the elasticity of the muscles. Tired muscles do not always do what you want them to. The problem of athletic injuries has escalated since Canadians have become more preoccupied with physical fitness.

A new category of athletic injuries to children is called the "Nadia syndrome," named after Nadia Comaneci, the 15-year-old gymnast, who won the hearts of spectators at the 1976 Montreal Olympics. This young gymnast inspired the youth of the world to emulate her accomplishments, and created a new market for child stars in athletics. Pushed beyond their body limits, tiny gymnasts, skaters, and dancers have acquired lasting injuries to their growth plates, ligaments, and tendons. A push for greater flexibility by some young gymnasts has produced stress fractures in the back. Young skaters often have stress fractures in their legs. Some young ballet dancers have caused irreparable damage to their growth plates, disrupting the development of their joints.

SOCIAL ISSUE:
Contact Sports

Some sports carry a much higher incidence of injury than others. Contact sports are the most dangerous.

Statement:

Full-contact sports like hockey and football should be restricted.

Point

- In 1904 U.S. President Theodore Roosevelt threatened to outlaw football after 28 deaths were reported for approximately 100 000 players. Although the incidence of injuries for football has dropped considerably since 1904, the prospects for injury remain high.
- The current move to allow NHL players to forego the wearing of helmets provides even greater risks. The disregard for safety will be emulated by younger players and injuries will only increase.

Counterpoint

- Although hockey and football are demanding sports, they are no more dangerous than baseball, which is not designed to be a contact sport. The protection afforded players in hockey and football is designed to reduce the chances of injury. In no other sport has such an effort been made to protect participants.
- The idea that laws or regulations will reduce injury is unfounded. Canadians, of all ages, enjoy participating in hockey and football. As athletic activity increases, injuries increase. The answer is not to decrease the activity of a nation, but to educate the public and provide proper equipment to reduce the incidence of injuries.

Research the issue.
Reflect upon your findings.
Discuss the various viewpoints with others.
Prepare for the class debate.

CHAPTER HIGHLIGHTS

- The skeletal system provides protection and support.
- Bone is composed of compact bone tissue, spongy bone tissue, and marrow, which is located in the central cavity of the bones.
- Periosteum is the tissue that covers bone.
- The axial skeleton is the central supporting part of the skeleton. The appendicular skeleton contains the appendages and their supporting structures.
- Ossification is the process by which bone is formed.
- Ligaments are bands of connective tissue that join bones to each other.

- Tendons are bands of connective tissue that join muscle to bone.
- Bursae are sacs of fluid found in joints.
- Skeletal muscle, smooth muscle, and cardiac muscle are three types of muscle.
- The sarcolemma is the delicate sheath that surrounds muscle fibers.
- Muscles are composed of muscle fibers, which contain myofilaments.
- The movement of bones at the joint is performed by muscles, which work in antagonistic pairs.
- The energy for muscle contraction comes from the breakdown of ATP.

APPLYING THE CONCEPTS

1 Some bone fractures cleave through the bone marrow. Explain why exposing the fat cells held within the bone marrow might prove to be life-threatening.

2 The Achilles tendon attaches the leg's calf muscle (gastronemius) to the heel. Predict what would happen if the Achilles tendon were severed.

3 Why are baseball pitchers susceptible to bursitis?

4 What problems are caused by the tearing of cartilage in your knee?

5 Gout is a genetic disorder characterized by an overproduction of uric acid by the cells of the body. Crystals from the uric acid collect in the joints. Explain why gout causes the joints to become enlarged and deformed.

6 How does the fact that muscles shorten when excited help support the "sliding-filament" theory of muscle contraction? What other evidence should be collected to support the theory?

7 Which diagram depicts a contracted muscle fiber? Explain your answer.

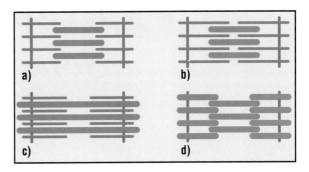

8 Predict what would happen if your biceps and triceps muscles were stimulated simultaneously.

9 Why does the condition of rigor mortis support the theory that ATP is required for muscle relaxation?

CRITICAL-THINKING QUESTIONS

1 Flat feet is a condition in which the arch of the foot has become depressed. The body mass forces the bones of the feet downward, causing the ligaments of the ankle to stretch. Explain why flat feet are more common in people who are overweight. Using the information that you have gained on flat feet, explain why shoes with adequate arch support are especially important for children.

2 About 80% of runners land on the outer part of their foot and roll inward. This action helps absorb the shock. However, for people whose foot bends more than 10°, the action can lead to problems. Examine running shoes that correct for pronation (excessive foot roll). How have running-shoe manufacturers attempted to prevent injuries?

3 Many football injuries occur when a player is blocked from the side unexpectedly. The blow can stretch the ligaments holding the knee in place, and in some cases the cartilage in the joint can be torn. Although some rules have been instituted to prevent certain types of blocking, knee injuries persist. Should below-the-waist blocking be declared illegal?

4 Why do many more knee injuries occur in football than in hockey?

5 Evolutionary biologists suggest that the upright posture of humans is a recent evolutionary adaptation. Considerable muscular effort is exerted in order to keep the back erect. The most mobile portions of the spinal cord, the neck and the lower back, are most susceptible to back injuries because of the amount of muscle development required. Is the human body well-designed to walk upright? Provide your reasons.

ENRICHMENT ACTIVITIES

1 Examine X rays of adults and children in order to determine how the long bones grow.

2 Read about the effects of anabolic steroids on skeletal development.

*S*pecial *Senses*

IMPORTANCE OF SENSORY INFORMATION

Sensory neurons supply the central nervous system with information about the external environment and the quality of our internal environment. Specialized chemoreceptors in the carotid artery provide the central nervous system with information about blood carbon dioxide and oxygen levels. Special osmoregulators in the hypothalamus monitor water concentration in the blood, and highly modified stretch receptors monitor blood pressure in arteries.

Environmental stimuli are conveyed to the central nervous system through sensory nerves. The sounds of thunder, the chill of a cold day, and the smells of food are relayed to the brain by sensory neurons. David Hume, the great Scottish philosopher, once concluded that humans are nothing more than the sum of their experiences. Our experiences, or what some philosophers call reality, exist because of a sensory nervous system. Each of us experiences sensations as highly personal events. Your perception of color is not the same as your friend's. You hear different things while listening to the same song. Indeed, while some individuals enjoy a particular song, others would rather not hear it.

Figure 13.1

The central nervous system processes environmental stimuli. Sound waves can be interpreted as music or noise.

As you learned in the previous chapter, information about the environment is transmitted to the brain along neurons. A neuron that carries visual information from the eye functions in essentially the same manner as one that carries auditory information from the ear. Both are electrochemical impulses. Your ability to differentiate between visual and auditory information depends on the area of the brain that receives the impulse. The brain interprets your reality. Visual information is stored in the posterior portion of the occipital lobe of the cerebrum, while auditory information is stored in the temporal lobe of the cerebrum. The similarities between neurons would be evident if a neuron carrying visual information could be rerouted from the occipital lobe to the temporal lobe or if an auditory neuron could be rerouted from the temporal lobe to the occipital lobe. A bolt of lightning across the sky would be heard, while the crashing thunder that follows would be seen. No doubt, the prospects of hearing lightning and seeing thunder would be confusing.

Clearly, what enables you to distinguish visual and auditory information resides in the sensory receptors. Light-sensitive receptors within the retina of the eye are stimulated by light, not sound. A group of specialized temperature receptors in the skin identify cold, while other ones identify heat. How do different receptors respond to different stimuli? How are different stimuli converted into electrochemical events? How do you identify the intensity of different stimuli? How does the brain interpret stimuli?

BIOLOGY CLIP

Occasionally a sensory receptor can be activated by stimuli that it was not designed to detect. Boxers who receive a blow to the eye often see stars. The pressure of the blow stimulates the visual receptors at the back of the eye, and the blow is interpreted as light. Similarly, a blow near the temporal lobe can often be interpreted as a bell ringing.

WHAT ARE SENSORY RECEPTORS?

A stimulus is a form of energy. Sensory receptors convert one source of energy into another. For example, taste receptors in your tongue convert chemical energy into a nerve action potential, a form of electrical energy. Light receptors of the eye convert light energy into electrical energy, and balance receptors of the inner ear convert gravitational energy and mechanical energy into electrical energy.

Sensory receptors are highly modified ends of sensory neurons. Often, different sensory receptors and connective tissues are grouped within specialized sensory organs, such as the eye or ear. This grouping of different receptors often amplifies the energy of the stimulus to ensure that the stimulus reaches threshold levels. Table 13.1 lists different types of sensory receptors found within the body.

Table 13.1 The Body's Sensory Receptors

Receptor	Stimulus	Information provided
Taste	Chemical	Taste buds identify specific chemicals.
Smell	Chemical	Olfactory cells detect presence of chemicals.
Pressure	Mechanical	Movement of the skin or changes in the body surface.
Proprioceptors	Mechanical	Movement of the limbs.
Balance (ear)	Mechanical	Body movement
Outer ear	Sound	Signals sound waves.
Eye	Light	Signals changes in light intensity, movement, and color.
Thermo-regulators	Heat	Detect the flow of heat.

Sensory receptors are modified ends of sensory neurons that are activated by specific stimuli.

A network of touch, high-temperature, and low-temperature receptors are found all over the skin. A simple experiment indicates that sensations occur in the brain and not the receptor itself. When the transmitter chemical released by the sensory neuron is blocked, the sensation stops. The brain registers and interprets the sensation. This phenomenon is supported by brain-mapping experiments. When the sensory region of the cerebral cortex is excited by mild electrical shock at the appropriate spot, the sensation returns even in the absence of the stimulus.

Despite an incredible collection of specialized sensory receptors, much of your environment remains undetected. What you detect are stimuli relevant to your survival. For example, consider the stimuli from the electromagnetic spectrum. You experience no sensation from radio waves or from infrared or ultraviolet wavelengths. Humans can only detect light of wavelengths between 350 and 800 nm. Your range of hearing, compared with that of many other species, is also limited.

Temperature receptors do not work as thermometers, which detect specific temperatures. Heat and cold receptors are adapted to signal changes in environmental temperatures. The following simple demonstration emphasizes how the nervous system responds to change. Fill three bowls with hot, room-temperature, and cold water respectively. Place your right hand in the cold water and your left hand in the hot water. Allow the hands to adjust to the temperature and then transfer both hands to the bowl that contains room-temperature water. The right hand, which had adjusted to the cold water, now

feels warm. The left hand, which had adjusted to the hot water, now feels cold.

Most animals can tolerate a wide range of temperatures, but are often harmed by rapid temperature changes. For example, a rapid change in temperature of 4 °C will kill some fish. Even people have succumbed to an unexpected plunge in very cold or hot water. This principle can be dramatized by what has come to be known as the "hot frog" experiment. If a frog is placed in a beaker of water above 40 °C, the frog will leap out immediately. However, if the frog is placed in room-temperature water, and the temperature is slowly elevated, it will remain in the beaker. The frog's skin receptors have time to adjust.

Sensory adaptation occurs once the receptor becomes accustomed to the stimulus. The neuron ceases to fire even though the stimulus is still present. This would seem to indicate that the new environmental condition is not dangerous. The same principle of adaptation can be applied to touch receptors in the skin. Generally, the receptors are only stimulated when clothes are put on or taken off. Sensory information assuring you that your clothes are still on your body is usually not required.

Sensory adaptation *occurs once you have adjusted to a change in the environment.*

Figure 13.2
The right hand will adjust to the cold water and the left hand will adjust to the warm water.

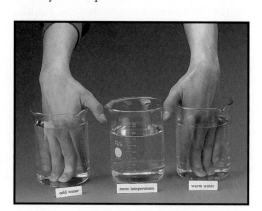

cold water   room temperature   warm water

TASTE AND SMELL

Taste receptors enable you to differentiate between foods and inedible matter. One theory that attempts to explain the extinction of the dinosaur suggests that it was poorly developed taste buds that eventually accounted for the disappearance of the giant reptiles. According to this theory, dinosaurs were not able to identify the bitter taste of the newly evolving poisonous plants.

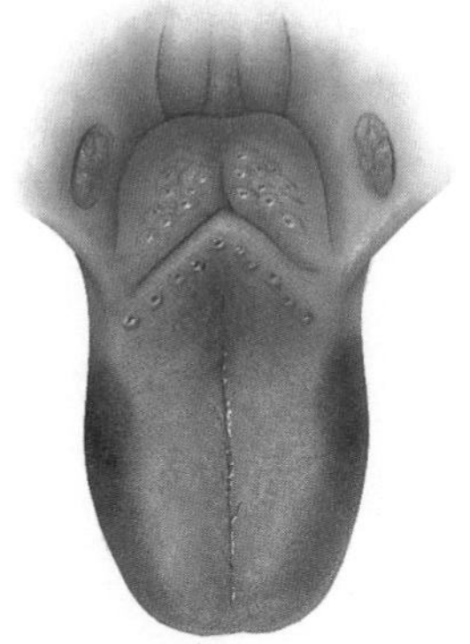

Taste receptors are found in different locations in different species throughout the animal kingdom. For example, octopuses have taste receptors located on their tentacles, crayfish have taste receptors situated on their antennae, and insects have taste receptors on their legs. Human taste receptors are centralized within the taste buds of the tongue. Once dissolved, specific chemicals stimulate the receptors within the taste buds. Scientists believe that four types of taste (salty, sweet, sour, and bitter) can be detected by specialized taste receptors, each of which is located in a specific area of the tongue. Figure 13.3 shows the four regions. Many individuals place an Aspirin near the back of the tongue believing that the bitter-tasting pill will spend less time in the mouth. Why might you suggest placing the pill near the tip of the tongue?

Olfactory cells are located in the nasal cavity (*olfactory* refers to the sense of smell). Airborne chemicals combine with the finely branched receptor ends on olfactory cells to create an action potential. Scientists believe that chemicals with specific geometry gain access to a specific receptor site so that the chemicals combine with complementary receptors. The impulse is carried to the frontal lobe of the brain for interpretation.

Experience tells you that your sense of taste and smell work together. Have you ever noticed that a cold reduces the taste of your food? Clogged nasal passages reduce the effectiveness of the olfactory cells. Since you use both smell and taste receptors to experience food, the diminished taste is actually the result of your reduced capacity to smell the food. The cooperation of the senses of smell and taste is well known to wine tasters. Have you ever noticed how wine tasters smell the wine before sipping it? You may have used the same test on milk that you believe has soured.

Figure 13.3

The human tongue.

Figure 13.4

(a) Specialized hairlike receptor ends in the tongue detect specific chemicals. Axons from the sensory neurons are shown in yellow. (b) Taste buds of a rabbit tongue. The taste bud appears as a bulblike structure.

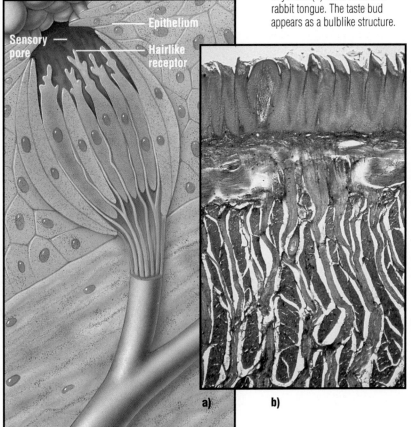

Epithelium

Sensory pore

Hairlike receptor

a)

b)

Olfactory cells demonstrate the phenomenon of sensory adaptation. You may have noticed how a sharp smell tends to disappear after you have been exposed to it for a long period of time. People who live in cities with pulp mills often comment on how they adjust to the odor after a while. Similarly, people who visit the ocean are at first impressed by its characteristic odor. However, the odor seems to fade after a short period of time.

STRUCTURE OF THE EYE

The eye is composed of three separate layers: the sclera, the choroid layer, and the retina. The **sclera** is the outermost layer of the eye. Essentially a protective layer, the white fibrous sclera maintains the eye's shape. The front of the sclera is covered by a clear, bulging **cornea,** which acts as the window to the eye by bending light toward the *pupil*. Like all tis-

*The **sclera** is the outer covering of the eye that supports and protects the eye's inner layers.*

*The **cornea** is a transparent tissue that refracts light toward the pupil.*

*The **aqueous humor** supplies the cornea with nutrients and refracts light.*

*The **choroid layer** is the middle layer of the eye. Pigments prevent scattering of light in the eye by absorbing stray light. Many blood vessels are found in this layer.*

*The **iris** regulates the amount of light entering the eye.*

■ REVIEW QUESTIONS ▨ ?

1 Identify a sensory receptor for each of the following stimuli: chemical energy, mechanical energy, heat energy, light energy, and sound energy.

2 Do sensory receptors identify all environmental stimuli? Give examples to back up your answer.

3 Explain the concept of "sensory adaptation" by using examples of olfactory stimuli and auditory stimuli.

4 Draw a diagram of the various chemoreceptors on the tongue.

5 Why are you less able to taste food when you have a cold?

BIOLOGY CLIP
Despite the wide range of eye colors, the iris contains a single pigment called *melanin.* Eyes appear blue because of a lack of activated melanin. The blue color is produced as light penetrates the clear aqueous humor. Only the shortest wavelength of visible light, blue light, is scattered. With increases in melanin, combinations of other wavelengths of light are reflected and other colors are produced.

You may have noticed that all babies have blue eyes at birth. Within a few months, the melanin migrates toward the surface of the iris and the eyes change color.

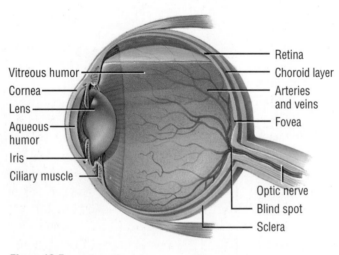
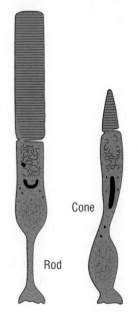

Figure 13.5

Simplified diagram of the human eye.

Labels: Vitreous humor, Cornea, Lens, Aqueous humor, Iris, Ciliary muscle, Retina, Choroid layer, Arteries and veins, Fovea, Optic nerve, Blind spot, Sclera, Cone, Rod

sues, the cornea requires oxygen and nutrients. However, the cornea is not supplied with blood vessels—blood vessels would cloud the transparent cornea. Most of the oxygen is absorbed from gases dissolved in tears. Nutrients are supplied by the **aqueous humor,** a chamber of transparent fluid behind the cornea.

The middle layer of the eye is called the **choroid layer.** Pigmented granules within the layer prevent light that has entered the eye from scattering. Toward the front of the choroid layer is the **iris.** The iris is composed of a thin circular muscle that acts as a diaphragm, controlling the size of the pupil opening.

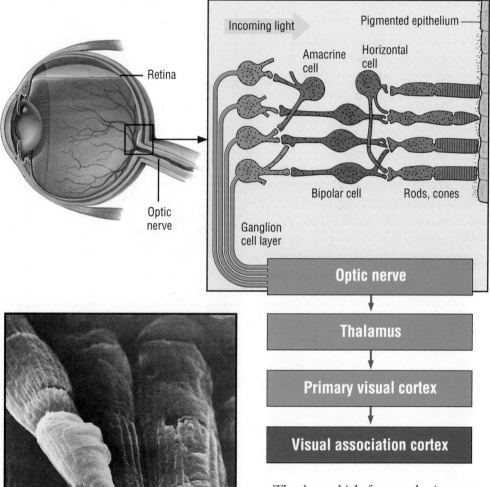

Figure 13.6

The pathway leading from the retina to the brain.

The *lens*, which focuses the image on the retina, is found in the area immediately behind the iris. *Ciliary muscles,* attached to ligaments suspended from the dorsal and ventral ends of the lens, alter the shape of the lens. A large chamber behind the lens, called the *vitreous humor,* contains a cloudy, jellylike material that maintains the shape of the eyeball and permits light transmission to the retina.

The innermost layer of the eye is the **retina,** which is composed of three different layers of cells: light-sensitive cells, bipolar cells, and cells from the optic nerve. The light-sensitive cell layer is positioned next to the choroid layer. Two different types of light-sensitive cells are the rods and cones. The **rods** respond to low-intensity light; the **cones,** which require high-intensity light, identify color. Both rods and cones act as the sensory receptors. Once excited, the nerve message is passed from the rods and cones to the bipolar cells, which, in turn, relay the message to the cells of the optic nerve. The optic nerve carries the impulse to the central nervous system.

In the center of the retina is a tiny depression referred to as the **fovea centralis.** The most sensitive area of the eye, the fovea centralis contains cones packed very close together. When you look at an object, most of the light rays fall on the

fovea centralis. The rods surround the cones. This may explain why you may see an object from the periphery of your visual field without identifying its color. There are no rods and cones in the area in which the optic nerve comes in contact with the retina. Because of this absence of photosensitive cells, this area is appropriately called the **blind spot.**

*The **retina** is the innermost layer of the eye.*

***Rods** are photoreceptors used for viewing in dim light.*

***Cones** are photoreceptors that identify color.*

*The **fovea centralis** is the most sensitive area of the retina and contains only cones.*

*The **blind spot** is the area in which the optic nerve attaches to the retina.*

Table 13.2 Parts of the Eye

Structure	Function
Outer layer:	
Sclera	Supports and protects delicate photocells.
Cornea	Refracts light toward the pupil.
Middle layer:	
Aqueous humor	Supplies cornea with nutrients and refracts light.
Choroid	Contains pigments that prevent scattering of light in the eye by absorbing stray light. Also contains blood vessels.
Iris	Regulates the amount of light entering the eye.
Vitreous humor	Maintains the shape of the eyeball and permits light transmission to the retina.
Lens	Focuses the image on the retina.
Pupil	Functions as the hole in the iris.
Inner layer:	
Retina	Contains photoreceptors.
Rods	Used for viewing in dim light.
Cones	Identify color.
Fovea centralis	Most sensitive area of the retina; contains only cones.
Blind spot	The area in which the optic nerve attaches to the retina.

RESEARCH IN CANADA

Dr. Roseline Godbout from the Cross Cancer Institute at the University of Alberta is investigating gene expression in the developing retina. Dr. Godbout is one of Canada's most promising scientists in a relatively new field: molecular neurobiology.

The retina is a highly specialized neurological network. In much the same way as information is processed in the body's most complicated nerve network—the brain—the retina continuously receives sensory information and adjusts to environmental stimuli during your waking hours.

Genes and the Retina

The conversion of light energy to electrochemical nerve impulses is due to the partnership of two types of cells that make up the retina: the neurons and glial cells. Although both types of cells differ in both structure and function, the neuron and glial cells originate from a common ancestor. Dr. Godbout is interested in determining how cells in the immature retina differentiate to become either glial cells or neurons in the retina. She hopes to be able to identify and characterize genes that regulate cell development in the retina. By identifying the genes in the normal retina, researchers may eventually be able to identify genes that are abnormally expressed in various eye disorders.

DR. ROSELINE GODBOUT

LIGHT AND VISION

The Greek philosopher Democritus first speculated about how the eye worked during the fifth century B.C. In his hypothesis, matter was composed of indivisible particles, which he called atoms. He reasoned that once the atoms touched the eye, they were carried to the soul and therefore could be viewed.

Empedocles, a contemporary of Democritus, proposed a different theory of vision. He believed that matter was composed of four essential elements—earth, air, water, and fire. Vision must be linked to fire because it provided light. According to Empedocles, light radiated from the eye and struck objects, making them visible.

Galen, a Roman physician in the second century A.D., combined Democritus' theory of the eye and the soul with Empedocles' notion that light was emitted from the eye. Galen believed that the optic nerve conducted visual spirits from the brain. The spiritual link between the soul and vision remains today, for example, in the expression "evil eye."

Today scientists accept two complementary theories of light first proposed by Sir Isaac Newton and Christiaan Huygens. Particles of light (photons) travel in waves of various lengths. Light enters the eye as it is reflected or transmitted from objects. In many ways the eye operates by the same principle as a camera. Both camera and eye are equipped with lenses that focus images. The diaphragm of a camera opens and closes to regulate the amount of light entering the camera. The iris of the eye provides an equivalent function.

The image of the camera is focused on a chemical emulsion—the film. Similarly, the image of the eye is focused on the retina.

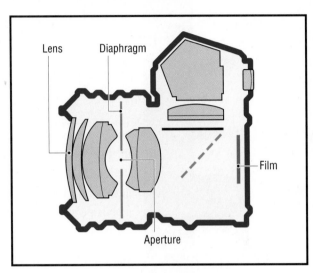

Figure 13.7

Light rays are bent by the lens and an inverted image is projected on the film.

In the 1880s Wilhelm Kuhne, a German physiologist, performed a series of bizarre experiments that underscored the similarities between the actions of the retina and photographic emulsion. Kuhne placed a rabbit in a dark room and fixed its vision on a barred window, the only source of light in the room. Using a dark cloth, Kuhne covered the rabbit's eyes for 10 min and then once again fixed them on the window for 2 min. Immediately, Kuhne decapitated the rabbit and placed its eyes in an alum solution, which fixed the image. The following day, Kuhne examined the eye and found an imprint of the window on the animal's retina. Kuhne repeated the experiments, using progressively more complicated images. Although he was certain that the human retina would behave in much the same manner, he had no experimental proof. In November 1880, Kuhne was presented with the opportunity to test his theory: he was able to secure the head of a criminal who had been beheaded. Following the same procedure as he had with the rabbit, Kuhne was able to fix the criminal's final image.

Afterimages

Have you ever noticed a trailing blue or green line that stays in your vision after you look into a camera flash? What you see is an afterimage. There are two different types of afterimages: positive and negative. The positive afterimage occurs after you look into a bright light and then close your eyes. The image of the light can still be seen even though your eyes are closed. As you read earlier, Wilhelm Kuhne described the positive afterimage as an image printed on the retina. The more dramatic negative afterimage occurs when the eye is exposed to bright colored light for an extended period of time.

Stare at the cross in Figure 13.8 for 30 s, and then stare at a bright white surface. The colors will reverse. The afterimage is believed to be caused by fatigue of that particular type of cone in that area of the retina. The horizontal red cones become fatigued, but the complementary green cones continue to fire. The opposite effect occurs for the vertical bar.

FOCUSING THE IMAGE

As light enters the eye, it is first bent toward the pupil by the cornea. Light waves normally travel in straight lines and slow down when they strike more dense materials like the cornea. The slowing of light by a denser medium causes bending, or refraction. The cornea directs light inward toward the lens, resulting in further bending. Because the biconvex lens is symmetrically thicker in the center than at its outer edges, light is bent to a focal point. An inverted image is projected on the light-sensitive retina.

Ciliary muscles control the shape of the lens, and suspensory ligaments maintain a constant tension. When close objects are viewed, the ciliary muscle contracts, the tension on the ligaments decreases, and the lens becomes thicker. The thicker lens provides additional bending of light for near vision. For objects that are farther away, relaxation of the ciliary muscles increases the tension of the ligaments on the lens, and the lens becomes thinner. The adjustment of the lens to objects near and far is referred to as **accommodation.** Objects 6 m from the viewer need no accommodation.

Figure 13.8

The red bar produces a green afterimage; the green bar produces a red afterimage.

Accommodation reflexes are adjustments made by the lens and pupil for near and distant objects.

Figure 13.9

Distant objects are viewed through a flatter lens. The focal point is farther from the lens and the object appears smaller. Near objects are viewed through a thicker lens. The focal point is closer to the lens and the object appears larger.

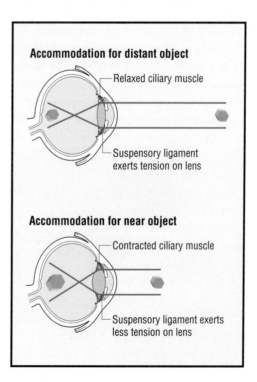

Accommodation for distant object

Relaxed ciliary muscle

Suspensory ligament exerts tension on lens

Accommodation for near object

Contracted ciliary muscle

Suspensory ligament exerts less tension on lens

The importance of the accommodation reflex becomes more pronounced with age. Layers of transparent protein covering the lens increase throughout your life, making the lens harder. As the lens hardens, it loses its flexibility. By the time you reach 40, near-point accommodation has diminished and may begin to hinder reading.

You may have noticed a secondary adjustment during the accommodation reflex. When objects are viewed from a distance, the pupil dilates in an attempt to capture as much light as possible. When objects are viewed close up, the pupil constricts in an attempt to bring the image into sharp focus. Test this for yourself by looking at the print on this page with one eye. Move your head toward the book until the print gets very blurry. Now crook your finger and look through the small opening. Gradually make the opening smaller. The image becomes sharper. Light passes through a small opening and falls on the most sensitive part of the retina, the fovea centralis. The Inuit were aware of this principle when they made eyeglasses for their elders by drilling holes in whalebone. Light passing through the narrow openings resulted in a sharper focus.

> **BIOLOGY CLIP**
> An estimated 10 million molecules of rhodopsin are found in each of the 160 million rods in your eyes.

CHEMISTRY OF VISION

An estimated 160 million rods surround the color-sensitive cones in the center of the retina. The rods contain a light-sensitive pigment called **rhodopsin,** or "visual purple." The cones contain similar pigments, but are less sensitive to light. Rhodopsin is composed of a form of vitamin A and a large protein molecule called opsin. When a single photon, the smallest unit of light, strikes a rhodopsin molecule, it divides into two components: retinene, the pigment portion, and opsin, the protein portion. This division alters the cell membrane of the rods and produces an action potential. Transmitter substances are released from the end plates of the rods, and the nerve message is conducted across synapses to the bipolar cells and the neuron of the optic nerve. For the rods to continue to work, rhodopsin levels must be maintained. Long-term vitamin A deficiency can cause permanent damage of the rods.

The extreme sensitivity of rhodopsin to light creates a problem. In bright light, rhodopsin breaks down faster than it can be restored. The opsins used for color vision are much less sensitive to light and, therefore, operate best with greater light intensity. The fact that images appear as shades of gray during periods of limited light intensity reinforces the fact that only the rods are active. Not surprisingly, the rods are most effective at dusk and dawn.

Color Vision

Vitamin A combines with three different protein opsins in the cones, each of which is sensitive to one of the three primary colors of source light: red, blue, and green. (Do not confuse the primary col-

10 Ask a friend to hold a coin at arm's length not quite above the cup.

11 Observing only the cup, direct the volunteer's hand until it is above the cup. Check your accuracy by asking the volunteer to drop the coin into the cup.

12 Repeat the procedure for 5 trials. Then close the opposite eye and record your results for another 5 separate trials.

j)

Trial	Left eye open	Right eye open
1		
2		
3		
4		
5		

13 Hold a card, with a hole punched in it, at arm's length.

14 In your other hand, hold a pencil at arm's length. Close one of your eyes and attempt to bring the point of the pencil into the hole in the card. Move the pencil and the paper. Repeat the procedure for 5 trials with each eye.

Trial	Left eye open	Right eye open
1		
2		
3		
4		
5		

k) Record your results.

Laboratory Application Questions

1 In an attempt to map the position of the blind spot, a student rotates a book 180° and follows the same steps described in part II. How would rotating the text help the student map the position of the blind spot?

2 In part III of the laboratory you discovered that the pencil moved when you viewed it with a different eye. Offer an explanation for your observation.

3 Horses and cows have eyes on the sides of their head; their visual fields overlap very little. What advantage would this kind of vision have over human vision?

4 Like humans, squirrels have eyes on the front of their face. The visual field of the squirrel's right eye overlaps with the visual field of its left eye. Why does a squirrel need this type of vision?

5 Students place a large piece of cardboard between their eyes while attempting to read two different books at the same time. Will they be able to read two books at once? Provide an explanation. ■

> **B I O L O G Y   C L I P**
> Leonardo da Vinci (1452–1519) speculated about contact lenses long before A.E. Flick fitted the first pair on a patient in 1887. The original contacts, made of glass, were designed to protect the eyes of a man who had cancerous eyelids. The early contact lenses were thick and cumbersome.

STRUCTURE OF THE EAR

The ear is associated with two separate functions: hearing and equilibrium. Sensory cells for both functions are located in the inner ear. These tiny hair cells contain from 30 to 150 cilia, which respond to mechanical stimuli. Movement of the cilia causes the nerve cell to generate an impulse.

> **B I O L O G Y   C L I P**
> The ear ossicles are the smallest bones in the body. They are fully developed at birth.

The ear can be divided into three sections for study: the outer ear, the middle ear, and the inner ear. The outer ear is composed of the **pinna,** the external ear flap, which collects the sound, and the

*The **pinna** is the outer part of the ear. The pinna acts like a funnel, taking sound from a large area and channeling it into a small canal.*

The **auditory canal**
carries sound waves to the
eardrum.

The **tympanic
membrane** is the
eardrum.

Ossicles are tiny bones
that amplify and carry
sound in the middle ear.

The **oval window** receives
sound waves from the
ossicles.

The **eustachian tube** is
an air-filled tube of the
middle ear that equalizes
pressure between the
external and internal ear.

The **vestibule** is a
chamber found at the base
of the semicircular canals
that provides information
about static equilibrium.

Semicircular canals are
fluid-filled structures
within the inner ear that
provide information about
dynamic equilibrium.

The **cochlea** is the coiled
structure of the inner ear
that identifies various
sound waves.

auditory canal, which carries sound to the eardrum. The auditory canal is lined with specialized sweat glands that produce earwax, a substance that traps foreign invading particles and prevents them from entering the ear.

The middle ear begins at the eardrum, or **tympanic membrane,** and extends toward the oval and round windows. The air-filled chamber of the middle ear contains three small bones, called the ear **ossicles,** which include the malleus (the hammer), the incus (the anvil), and the stapes (the stirrup). Sound vibrations that strike the eardrum are first concentrated within the solid malleus and then transmitted to the incus, and finally to the stapes. The stapes strikes the membrane covering the oval window in the inner wall of the middle ear. The amplification of sound is in part met by concentrating the sound energy from the large tympanic membrane to a smaller **oval window.** The surface area of the tympanic membrane is 64 mm^2, while that of the oval window is about 3.2 mm^2.

The **eustachian tube** extends from the middle ear to the air in the mouth and chambers of the nose. Approximately 40 mm in length and 3 mm in diameter, the eustachian tube permits the equaliza-

tion of air pressure. Have you ever noticed how your ears seem to pop when you go up in a plane? The lower pressure on the tympanic membrane can be equalized by reducing air pressure in the eustachian tube. Yawning, swallowing, and chewing gum allow air to leave the middle ear. An ear infection can cause the build-up of fluids in the eustachian tube and create inequalities in air pressure. Discomfort, temporary deafness, and poor balance can result.

The inner ear is made up of three distinct areas: the vestibule and the semicircular canals, which are involved with balance, and the cochlea, which is connected with hearing. The **vestibule,** connected to the middle ear by the oval window, houses two small sacs, the *utricle* and *saccule*, which establish head position (static equilibrium). Three **semicircular canals** are arranged at different angles. The movement of fluid in these canals helps you identify body movement (dynamic equilibrium). The **cochlea** is shaped like a spiralling snail's shell and contains two rows of specialized hair cells that run the length of the inner canal. The hair cells identify and respond to sound waves of different frequencies and intensities.

Table 13.3 Parts of the Ear

Structure	Function
External Ear	
Pinna	Outer part of the external ear. Amplifies sound by funneling it from a large area into the narrower auditory canal.
Auditory canal	Carries sound waves to the tympanic membrane.
Middle Ear	
Ossicles	Tiny bones that amplify and carry sound in the middle ear.
Tympanic membrane	The eardrum.
Oval Window	Receives sound waves from the ossicles.
Eustachian tube	Air-filled tube of the middle ear that equalizes air pressure in the external ear.
Inner Ear	
Vestibule	A chamber at the base of the semicircular canals; concerned with static equilibrium.
Semicircular canals	Fluid-filled structures that provide information concerning dynamic equilibrium.
Cochlea	A coiled tube within the inner ear that identifies sound waves and converts them into nerve impulses.

Unit Three:
Coordination and Regulation in Humans

INVESTIGATION

CAREER CAREER CAREER CAREER

The human body is an amazingly versatile tool. Many careers depend on a knowledge of how the body performs and how it reacts under stress.

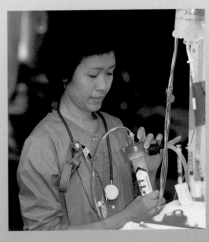

Ophthalmologist/Optometrist/Optician
An ophthalmologist is a physician who specializes in the function, structure, and diseases of the eye. An optometrist examines the eye for defects and prescribes corrective glasses or contact lenses. An optician fits customers with glasses or contact lenses as prescribed by an optometrist or ophthalmologist.

Choreographer
A choreographer should have an intimate knowledge of the workings of the human body, its capabilities and limitations. Routines must be designed to be both aesthetically pleasing and physically impressive, yet within the capabilities of the dancers who will perform them.

Advertising Executive
Good advertising, particularly television advertising, must communicate with the audience on several levels. An advertising executive must know how people perceive and interpret sounds and images in order to create an ad that will sell products.

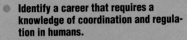

- Identify a career that requires a knowledge of coordination and regulation in humans.
- Investigate and list the features that appeal to you about this career. Make another list of features that you find less attractive about this career.
- Which high-school subjects are required for this career? Is a post-secondary degree required?
- Survey the newspapers in your area for job opportunities in this career.

Nurse
Nurses require not only a thorough knowledge of human anatomy, but also a knowledge of drugs and how to administer them. They must also be sensitive to the psychological needs of their patients. Nurses work in hospitals, in doctors' offices, or in senior citizens' homes.

SOCIAL ISSUE:
Rock Concerts and Hearing Damage

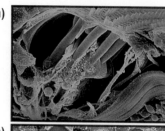

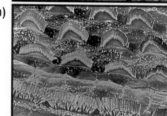

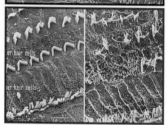

In a recent interview, Pete Townshend stated that after years of playing guitar with the rock group The Who, he had developed partial hearing loss. For concertgoers, the results of listening to loud rock music for almost two hours can include ringing in the ears or impaired hearing for days after.

The most common type of hearing loss is caused by the destruction of the cilia on the hair cells of the cochlea. Although the cilia gradually wear away with aging, high-intensity sounds, such as loud rock music, can literally tear the cilia apart.

The sound level of a normal conversation ranges between 60 and 75 dB, heavy street traffic can reach 80dB, and the crash of thunder usually registers about 100 dB. Pain begins to be felt at 120 dB. At rock concerts, if the audience is close enough, the sound level can register well beyond 130 dB.

(a) Hair cells embedded in the basilar membrane. (b) Normal rows of hair cells in the cochlea. (c) Destruction of hair cells following 24-h exposure to high-intensity rock music, with noise levels approaching 120 dB at 2000 cycles per second.

Statement:

The sound levels at rock concerts should be regulated to protect the audience and performers.

> **Research the issue.**
> **Reflect on your findings.**
> **Discuss the various viewpoints with others.**
> **Prepare for the class debate.**

Point

- The sound levels at rock concerts can register up to 130 dB, well above the level of 85 dB that can cause permanent hearing loss. Governments have a responsibility to protect the health of their citizens by regulating the level of noise to which we are exposed.
- Individuals who knowingly risk hearing damage by attending high-decibel rock concerts are irresponsible, and raise the costs of treatment under our medical insurance plans.

Counterpoint

- Listening to tapes at high volume on a Walkman for extended periods of time is likely to be much more dangerous than occasional exposure to high-decibel levels at rock concerts.

- Earplugs are available to anyone who wants to take extra precautions at a rock concert.

CHAPTER HIGHLIGHTS

- A stimulus is a form of energy detected by the body.
- Sensory receptors are modified ends of sensory neurons that are activated by specific stimuli. Receptors convert one form of energy into another (electrochemical impulses).

- The eye converts light energy into electrical impulses, which are interpreted by the brain.
- Cones, which are specialized nerve cells found in the retina of the eye, contain pigments that are sensitive to light of three different wavelengths.

- The ear detects sound and regulates balance.
- Sound energy is detected by the eardrum and concentrated by the ossicles of the middle ear. The vibrating ossicles push against the oval window of the inner ear, setting up fluid waves that move toward the cochlea. Specialized hair cells in the cochlea detect fluid vibrations.

APPLYING THE CONCEPTS

1 A person steps from a warm shower and feels a chill. Upon stepping into the same room, another person says that the room is warm. What causes the chill?

2 If a frog is moved from a beaker of water at 20°C and placed in a beaker at 40°C, it will leap from the beaker. However, if the temperature is slowly heated from 20 to 40°C, the frog will remain in the beaker. Provide an explanation for the observation.

3 The retina of a chicken is composed of many cones very close together. What advantages and disadvantages would be associated with this type of eye?

4 Why do people often require reading glasses after they reach the age of 40?

5 Indicate how the build-up of fluids in the eustachian tube may lead to temporary hearing loss.

6 A scientist replaces ear ossicles with larger, lightweight bones. Would this procedure improve hearing? Support your answer.

7 When the hearing of a rock musician was tested, the results revealed a general deterioration of hearing as well as total deafness for particular frequencies. Why is the loss of hearing not equal for all frequencies?

CRITICAL-THINKING QUESTIONS

1 One theory suggests that painters use less purple and blue as they age because layers of protein are added to the lens in their eyes and it gradually becomes thicker and more yellow. The yellow tint causes the shorter wavelength to be filtered. How would you test the theory?

2 Should individuals who refuse to wear ear protectors while working around noisy machinery be eligible for medical coverage for the cost of hearing aids? What about rock musicians or other individuals who knowingly play a part in the loss of their hearing?

ENRICHMENT ACTIVITIES

1 Use the diagram below to make a functional model of the ear.

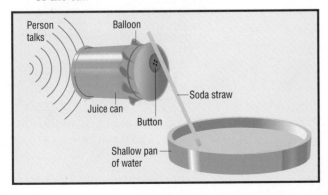

Person talks
Balloon
Soda straw
Juice can
Button
Shallow pan of water

2 Suggested reading:
- Barlow, Robert. "What the Brain Tells the Eye." *Scientific American*, April 1990, p. 90.
- Franklin, Deborah. "Crafting Sound from Silence." *Science News* (October 20, 1984): p. 252.
- Hudspeth, A. "The Hair Cells of the Inner Ear." *Scientific American*, January 1983, p. 58.
- Stryer, L. "The Molecules of Visual Excitation." *Scientific American*, July 1987, p. 42.

*L*ife in the Biosphere

■

U N I T

4

Equilibrium in the Biosphere

IMPORTANCE OF EQUILIBRIUM IN THE BIOSPHERE

One can only imagine how the Apollo astronauts felt when they first set foot on the moon and saw the spectacle of the living earth rising above the lifeless lunar rock. This contrast of life versus desolation must have made them appreciate the uniqueness of living organisms and the narrow range of physical conditions—known only on the earth—within which life can exist.

Think of our planet as a spaceship. Travelling around the sun in a slightly elliptical orbit, the earth carries with it the only forms of life known in the universe. It is a closed system. There is no outside source for life-sustaining raw materials or an interplanetary garbage dump to store wastes. Life is totally dependent on solar energy and the matter available aboard the spaceship earth.

J.E. Lovelock, a British scientist, compares the earth to a living body. The metaphor is referred to as the *Gaia* (pronounced "gay-ah") *hypothesis*, named after the Greek goddess of earth. Although a controversial idea in the

Figure 14.1

Earthrise. A blue and white earth rises over the horizon of the moon.

scientific community, it serves to emphasize that all living things interact with each other and with the nonliving components of our planet. In much the same way that the brain requires oxygen and nutrients from the circulatory system to function properly, each component of the earth's environment must be in a state of balance or equilibrium with every other component. What affects one part affects all parts. The expression "dynamic equilibrium" is used to describe any system in which changes are continuously occurring but whose components have the ability to adjust to these changes without disturbing the entire system.

As the 21st century approaches, ecologists have evidence to suggest that the earth is facing a crisis in which its dynamic equilibrium is being upset. However, scientists have not reached a consensus about the magnitude of the predicament or what can be done. The problems appear to be the result of the activities of a single dominant form of life: humans.

Humans have the ability to understand natural processes and act on this knowledge. By studying some of the well-established principles of ecology, you, as a member of the earth's community, can become a knowledgeable decision-maker. The decisions you make will, in part, determine the future direction of life on this small and fragile planet.

THE BIOSPHERE

There are three basic structural zones of the earth: the lithosphere (land), the hydrosphere (water), and the atmosphere (air). All are visible in Figure 14.1. However, at this distance, no signs of life can be seen. Living organisms are found in all three zones, an area referred to as the **biosphere,** whose limits extend from the ocean depths through the lower atmosphere. Most organisms are confined to a narrow band where the atmosphere meets the surface of the land and water. The regions of the planet that are not within the biosphere, such as the upper atmosphere or the earth's core, are also important because they affect living organisms.

Life forms are referred to as the **biotic,** or living, component. Chemical and geological factors, such as rocks and minerals, and physical factors, such as temperature and weather, are referred to as the **abiotic,** or nonliving, component. It is the interactions within and between the biotic and abiotic components that the ecologist endeavors to understand and explain.

When biologists investigate how a complex organism functions, they must study its various levels of organization. Moving from the simple to the more complex, biologists study individual cells, then tissues, organs and organ systems, and finally the integrated, functioning body. Ecologists investigating the biosphere proceed in a similar manner. Table 14.1 compares the levels of organization of an individual organism with those of the biosphere. By examining its individual parts, ecologists are able to bring together the various data and provide a picture of how the biosphere operates as an integrated unit.

*The **biosphere** is the narrow zone around the earth that harbors life.*

***Biotic** components are the biological or living components of the biosphere.*

***Abiotic** components are the nonliving components of the biosphere. They include chemical and physical factors.*

Table 14.1 Organizational Levels of an Organism and the Biosphere

Organism	Biosphere
Cell	Organism
Tissue	Population
Organ	Community
Organ system	Ecosystem
Complete organism	Biosphere

Ecological studies begin at the organism level. Investigations are designed to determine how the individual interacts with its biotic and abiotic environment. However, an organism does not live in isolation. It tends to group with others of the same species into **populations.** A population influences and is influenced by its immediate environment. When more than one population lives in an area, a **community** of organisms is established.

An **ecosystem,** the functional unit of the biosphere, has both biotic and abiotic components. The physical and chemical environment, as well as the community of organisms, interact with each other in an ecosystem.

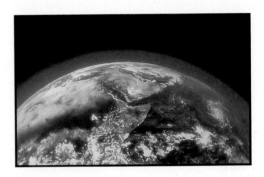

EARTH'S VITAL ATMOSPHERE

The **atmosphere** encircles the earth like a spherical skin. It is held to the earth by gravitational attraction. The atmosphere is composed of gases and contains a vari-

Table 14.2 Gases of the Atmosphere

Name of gas	Approximate percentage
Nitrogen	78.08
Oxygen	20.95
Argon	0.93
Carbon dioxide	0.03
Trace gases (e.g., ozone, methane)	±0.01
Water vapor	variable (0–4)

ety of solid particles such as pollen, dust, spores, bacteria, and viruses. With the exception of water vapor, the percentage composition of the other atmospheric gases is relatively fixed. The amounts of water vapor vary, depending on the temperature and the availability of water.

The atmosphere is described in terms of zones moving upward from the earth's surface to a height of some 900 km,

Figure 14.2

The earth is encircled by an ocean of air.

A **population** *is a group of individuals of the same species occupying a given area at a certain time.*

A **community** *includes the populations of all species that occupy a habitat.*

An **ecosystem** *is a community and its physical and chemical environment.*

The **atmosphere** *is the air that encircles the earth.*

Figure 14.3

Graph showing temperature versus altitude for the zones of the atmosphere.

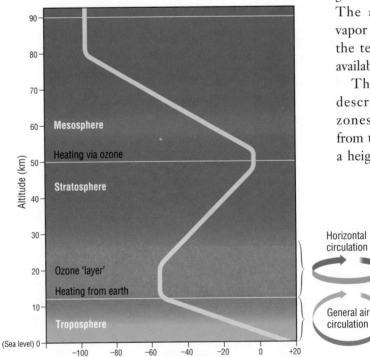

Unit Four:
Life in the Biosphere

where it melds into outer space. The lowest level, the **troposphere,** has a thickness ranging from about 15 km at the equator to 8 km at the poles. This zone contains 80% of the atmosphere's mass along with most of its water vapor and dust. In the higher regions of the troposphere, the temperature drops steadily to about –60°C. Gigantic convection currents arise due to the difference in temperature between the air at the earth's surface and the air in the upper troposphere. These convection currents are the source of winds, turbulence, and weather.

Above the troposphere is the **stratosphere,** which extends to a height of approximately 50 km. The weather in this zone is stable. Here oxygen (O_2) actively absorbs **ultraviolet radiation** from the sun, forming **ozone** (O_3). Ozone is a bluish gas that is spread so thinly that it is barely noticeable. It is possible to catch a strong whiff of this sharp, clean-smelling gas after a thunderstorm. Lightning can split molecules of oxygen, which then combine to form ozone. Because ozone is unstable and reacts quickly with many other chemicals, its smell quickly dissipates after a storm.

Ozone can also form from air pollutants such as gasoline vapors and emissions from automobiles. High levels of ozone have been known to collect over large cities on hot summer days. Although the complete story is not known, high ozone concentrations at lower levels of the atmosphere can be harmful. Prolonged exposure to high concentrations of ozone will damage plants and cause respiratory problems.

The **mesosphere** contains little in the way of atmospheric gases—only traces of water vapor. At its upper limits, temperatures of –113°C are common. The mesosphere represents the limit of what might be called the true atmosphere.

Figure 14.4
The northern lights are caused by charged subatomic particles from the sun interacting with atoms in the upper atmosphere.

The region above the mesosphere is referred to as the upper atmosphere. Of direct significance to life in this region is the **ionosphere,** which absorbs most of the deadly X rays and gamma rays produced by solar radiation. The aurora borealis, or northern lights, is produced in this zone.

Just beyond the limits of the atmosphere is a region sometimes referred to as the **magnetosphere.** Created by the earth's magnetic field, this layer can deflect most of the charged and potentially lethal particles that are emitted by the sun or that come from interstellar space. The Van Allen belts, discovered in 1958, are found at a height of some 950 km. They trap many of these particles in their doughnut-shaped bands.

HOLES IN THE OZONE LAYER

Without the ozone layer the earth would be bathed in ultraviolet radiation from the sun. The damage this form of radiation causes is so severe it is doubtful that plants or animals could survive its effects.

*The **troposphere** is the lowest region of the earth's atmosphere and extends upward about 12 km. Most weather occurs here.*

*The **stratosphere** is the region of the earth's atmosphere found above the troposphere. The ozone layer is found in this region.*

Ultraviolet radiation *is the electromagnetic radiation from the sun and can cause burning of the skin (sunburn) and cellular mutations.*

Ozone *is an inorganic molecule. A layer of ozone found in the stratosphere helps to screen out ultraviolet radiation.*

*The **mesosphere** is the region of the atmosphere found between the stratosphere and the upper atmosphere.*

*The **ionosphere** is a region of the upper atmosphere consisting of layers of ionized gases that produce the northern lights and reflect radio waves.*

*The **magnetosphere** is a region found above the outer atmosphere consisting of magnetic bands caused by the earth's magnetic field.*

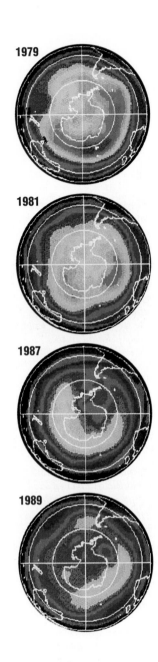

1979

1981

1987

1989

*An **ozone hole** is a region in the ozone layer in which the ozone levels have been considerably reduced and the layer has become very thin.*

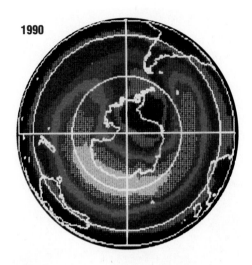

1990

Figure 14.5

Pictures of the antarctic showing levels of ozone from 1979 to 1990. High ozone values are shown in orange/red, while low values are purple. In 1990 the black area inside the purple shows a hole in the ozone layer.

Fortunately, oxygen in the upper atmosphere is converted to ozone, which then screens out ultraviolet (UV) radiation. As a result, nearly 99% of the ultraviolet radiation striking the atmosphere never reaches the earth's surface.

Recent studies indicate that the ozone layer is slowly being depleted. The identification of **ozone holes** above Antarctica and the Canadian Arctic and the concurrent increase in skin cancers and eye problems associated with ultraviolet radiation are raising much concern among ecologists and the general public. Although a number of reasons for this depletion have been suggested, most of the attention has been focused on the release of chlorofluorocarbons (CFCs) into the atmosphere. These odorless compounds have been widely used as propellants in aerosol cans, as coolants in air conditioners, refrigerators, and freezers, and are also the waste products in the

manufacturing of some foam plastics. When the relatively inert CFC compounds diffuse into the atmosphere they are broken down by ultraviolet radiation. This process releases reactive chlorine molecules, each of which is capable of reacting with and breaking down thousands of ozone molecules.

Some scientists have speculated that holes in the ozone layer may have been developing since the beginning of time. They have suggested that we were simply unaware of the holes until recent technological advances allowed us to measure accurately the ozone concentrations in the upper atmosphere.

The rate at which the ozone layer is decreasing has alarmed the world scientific community. Predictions that as much as 60% of the ozone layer will be destroyed by the middle of the next century are not uncommon. Many countries have begun to take action on the problem by either banning or reducing freon-based propellants in aerosol cans, and by modifying the methods of producing plastic foam that produce chlorofluorocarbons. Canada has been heavily involved in international negotiations with countries that produce and release ozone-depleting materials. The Montreal Protocol, an international treaty, was signed in 1987 by over 31 members of the United Nations. Although a total ban on ozone-depleting products has not yet been attained, several governments have set up timetables for cutting emissions by specific percentages over a period of time. The first steps have been taken to correct the ozone problem. Whether the treaty will have a major effect or simply slow down a potentially disastrous situation is not yet known.

> **BIOLOGY CLIP**
> One estimate suggests that a 1% decrease in ozone would cause a 4% increase in skin cancer.

FRONTIERS OF TECHNOLOGY: PROBING THE UPPER ATMOSPHERE

The aurora borealis, or northern lights, is commonly observed in Canada and other northern countries. Since development in the Canadian North began, a variety of electronic communications problems, including total radio blackouts, have been associated with the aurora. This prompted Canadian scientists to study the upper atmosphere. By the 1930s Canada had taken a leading role in upper-atmosphere research.

In the late 1950s, Canada's National Research Council (NRC) began to use balloons to probe the atmosphere. These balloons carry instrumentation to altitudes of up to 40 km. The largest has a volume of 570 000 m³ and carries an instrument package of 1600 kg.

The NRC also supported the development of the Black Brant family of rockets and the Alouette 1515 scientific satellites. The largest, Black Brant X, is a three-stage rocket that can carry a 200 kg payload to an altitude of 700 km. The launch site for these rockets was located at the Churchill Research Range in northern Manitoba. Its three launch pads were used extensively in the 1970s and early 1980s, handling 20 m rockets, each weighing 5000 kg. Significant information on the magnetosphere and ionosphere was obtained. Both the balloon and rocket-based experiments have resulted in a new understanding of the properties of the northern lights and the transfer of solar energy down to the lower atmosphere. Additional information has been provided on the ozone layer, the electromagnetic fields surrounding the earth, cosmic radiation, magnetic storms, and even acid rain.

Through their association with industry, the U.S. National Aeronautics and Space Administration (NASA), the European Space Agency, and joint projects with scientists from many other countries, Canadian scientists are playing a major role in the development of highly sophisticated recording instruments and satellites, keeping Canada in the forefront of atmospheric research.

a)

b)

■ REVIEW QUESTIONS　?

1　In what ways is the earth like a spaceship?

2　What is meant by a closed system?

3　What are the abiotic and biotic components of the biosphere?

4　In what way does a community differ from an ecosystem?

5　Name the levels of organization in the biosphere.

6　Name the specific gases found in the atmosphere. What other materials are found there?

7　State where the ozone layer is located in the atmosphere and explain why it is important.

8　In what ways do the ionosphere and the magnetosphere protect living organisms?

9　What is meant by an ozone hole?

Figure 14.6

(a) A launch of balloons with experimental payloads as part of an arctic ozone hole research project. (b) A rocket leaves the launch pad carrying *Alouette I* spacecraft into orbit.

SOLAR ENERGY AND GLOBAL TEMPERATURES

Because of its unique position in the solar system, earth is the only planet that meets the delicate requirements for sustaining life. At an average distance of 14.5×10^7 km from the sun, the earth receives its primary source of energy in the form of solar radiation. Solar radiation interacts with the oceans, the atmosphere, and the continental masses to help maintain the narrow range of temperatures that can support life forms. Solar radiation is a combination of many forms of energy at different wavelengths, including light, heat, and ultraviolet radiation. Although heat from the earth's core and the tidal effect between the earth, the moon, and the sun do provide some energy to the biosphere, 99% of the earth's energy comes from solar

radiation. Table 14.3 shows what happens to solar energy as it strikes the earth.

Much of the incoming radiation is reflected back out into space by the atmosphere and the earth's surface. The remaining portion is absorbed by the earth, converted to long-wave infrared wavelengths, and re-emitted as heat. Clouds and gases in the atmosphere trap surface heat radiation, reflecting it back to the earth and maintaining a global temperature nearly 33°C warmer than would be possible without an atmosphere.

Not all of the earth's surface is heated equally. Rays of sunlight striking the earth perpendicular to its surface transmit a specific amount of energy per unit area. As shown in Figure 14.8, the same rays hitting the surface at an angle spread the same amount of energy over a larger area. Therefore, each unit area receives less energy with a corresponding temperature drop.

Figure 14.7

Electromagnetic spectrum chart.

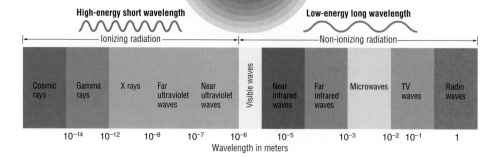

| High-energy short wavelength | | | | | | Low-energy long wavelength | | | | |

| Ionizing radiation | | | | | | Non-ionizing radiation | | | | |

Cosmic rays	Gamma rays	X rays	Far ultraviolet waves	Near ultraviolet waves	Visible waves	Near infrared waves	Far infrared waves	Microwaves	TV waves	Radio waves
	10^{-14}	10^{-12}	10^{-8}	10^{-7}	10^{-6}	10^{-5}		10^{-3}	10^{-2} 10^{-1}	1

Wavelength in meters

Table 14.3 Distribution of Solar Energy

Incoming solar radiation	100%
Reflected by the atmosphere	30%
Absorbed by the atmosphere	45%
Used to drive the water cycle	23%
Used to drive winds and current	<1%
Used for photosynthesis	0.02%

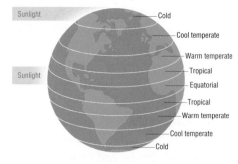

Figure 14.8

All parts of the earth's surface are not heated equally.

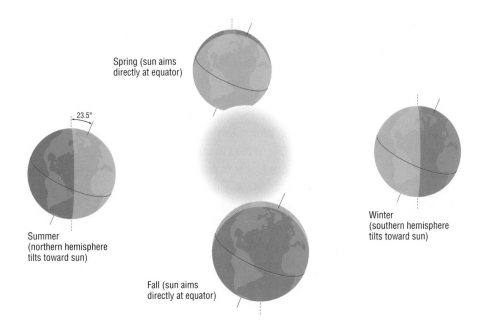

Figure 14.9

Annual variation in the amount of incoming solar radiation.

Spring (sun aims directly at equator)

23.5°

Summer (northern hemisphere tilts toward sun)

Fall (sun aims directly at equator)

Winter (southern hemisphere tilts toward sun)

If the earth's axis were perpendicular to the plane of its orbit around the sun, all areas of the earth would receive light during its 24-hour rotation and the amount of radiation would be the same at equal latitudes. Temperatures would not only become cooler, moving from the equator toward the poles, but the overall climate of any one region would be similar throughout the year; seasons would not exist. However, because the earth's axis is tilted at an angle of 23.5° from the perpendicular, the planet experiences seasons. Seasons are fairly well defined, particularly in the higher latitudes, and are reversed in the northern and southern hemispheres. As seen in Figure 14.9, there is an unequal distribution of incoming solar radiation, with the result that each of the polar regions experiences periods of continuous 24-hour daylight or darkness at some time during the year.

THE ALBEDO EFFECT

The term **albedo** is used to describe the extent to which a material can reflect sunlight. The higher the albedo, the greater the ability to reflect sunlight. Applying this principle to the solar radiation striking the earth, the higher the overall albedo of the earth, the less energy will be absorbed and available for maintaining the earth's global temperature. For example, the albedo of snow cover is extremely high. The presence of snow is a contributing factor to the low temperatures experienced during winter. Snow also delays warming in the spring, even though there is more solar radiation available per unit area. Lighter-colored areas of the earth's surface caused by the chemical composition of rock, the presence of very light-colored sand, or deforestation have a similar effect.

Water vapor added to the atmosphere causes more extensive cloud cover. The high albedo of cloud then causes more of the incoming radiant energy to be reflected directly back into space. An increase of dust in the atmosphere will produce the same effect. This phenomenon has been observed following volcanic activity. The cooling associated with a "nuclear winter," the result of atomic warfare or volcanic ash, increases the albedo of the atmosphere.

Figure 14.10

A column of ash, hot gases, and pulverized rock shoot out of Mt. St. Helens, Washington, during the second eruption in 1980. The increase of dust in the atmosphere produces a cooling effect.

Albedo is a term used to describe the extent to which a surface can reflect light that strikes it. An albedo of 0.08 means that 8% of the light is reflected.

LABORATORY

INVESTIGATING THE ALBEDO EFFECT

Objective

To examine the ability of selected colors and surface conditions to reflect light.

Materials

light source	dissecting pan
2 ring stands	sand
extension clamp	gravel
photocell	soil
voltmeter	water
colored card stock	snow and/or ice
(including black and white)	

Procedure

1 Attach the light source to the top of one of the ring stands. Aim the light down at the bench surface.

2 Attach the photocell unit to the second stand so that it is higher than the light source. Its active surface must face down and have a clear path to the surface of the bench. Attach the voltmeter to the photocell unit so that the photocell, wires, and meter do not interfere with the light source.

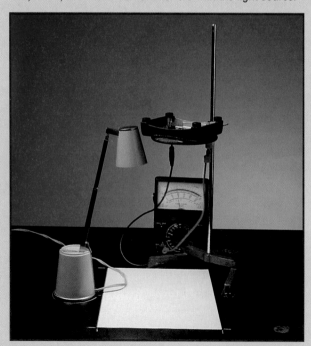

3 Place the sheet of white card stock directly under the light source.

4 Turn on the light source and measure the voltage produced by the photocell. Record the voltage.

   a) Why is it necessary to keep the photocell, wire, and voltmeter out of the direct line of light?

   b) Why does the voltmeter register an electrical change?

5 Repeat the experiment, using the other colored sheets. Remember to control all your variables. Record your data.

   c) What variable must be controlled?

   d) Why are black and white card stock colors used in this experiment?

6 Prepare a table similar to the one below.

Color of surface	Reflected light (in volts)

   e) List the colored surfaces in order from least to greatest reflected light.

7 Set the dissecting pan directly under the light source.

8 Record the reflective value of the pan when empty. Repeat, using a sand surface, a gravel surface, soil, water, and snow and ice.

9 Prepare a table similar to the one in part 6.

   f) Use your data to make comments on the albedo effect of the materials used.

Laboratory Application Questions

1 Design an experiment to measure the effect of color or surface conditions on the absorption of heat. State your hypothesis and the predictions resulting from it.

2 Perform the investigation (approval of instructor required) and draw conclusions based on the collected data.

3 What variable were you unable to control in part 8 of the laboratory procedure? How could you redesign these steps to account for the variable?

4 In order to melt a pile of snow more quickly, a researcher sprayed dye-colored water on it. Which color was probably selected and why? ■

RESEARCH IN CANADA

Dr. Jill Oakes was born in British Columbia and received her Ph.D. from the University of Manitoba. Her research in northern Alaska, Canada, and Greenland has focused on the Inuit peoples, especially the way in which their clothing has helped them withstand the extreme climatic conditions of the arctic environment. With the arrival of northern European, American Alaskan, and southern Canadian influences into the arctic, the various Inuit societies have undergone significant changes. Most Inuit have lost the skills required to produce arctic clothing and many have switched to synthetic materials.

Dr. Oakes has lived with Inuit from two different regions of the arctic and has learned how to make Inuit clothing from the few individuals who still retain the skills. Travelling across the arctic, she has carefully recorded the details of how each item is constructed. Dr. Oakes' investigations led to the discovery of a unique type of parka made by the Inuit of the Belcher Islands in Hudson Bay. Lacking caribou from which to obtain the customary skins used in arctic clothing, the islanders had to rely on the skins and attached feathers of eider ducks. Dr. Oakes is continuing her work at the University of Alberta. She has obtained her commercial flying licence and is in the process of constructing a biplane.

Inuit Clothing

DR. JILL OAKES

BIOGEOCHEMICAL CYCLES

If energy or matter were removed from the biosphere, life would cease to exist. Energy is required to drive both the biotic and abiotic components. All living things depend on abiotic materials to build their body structures and to serve as raw materials during energy-requiring metabolic reactions.

The materials used in building the geological structure of the earth or the bodies of living organisms are limited to those atoms and molecules that make up the planet. In effect, no alternative source of matter is available. Therefore, to maintain the biosphere, matter must be recycled. The elements carbon, hydrogen, oxygen, and nitrogen are the main constituents of the four basic organic compounds found in living organisms: carbohydrates, lipids, proteins, and nucleic acids. When an organism dies and decays, its complex organic molecules are broken down into their basic constituents and once again become available to the biosphere. It is quite possible that atoms such as carbon, hydrogen, oxygen, and nitrogen, which are basic components of your body, could have been part of another organism in the past, possibly a dinosaur. The same atoms will eventually pass from you into the abiotic environment and could once again become part of the biotic world at some future time. This use and reuse of materials of the earth is referred to as a **biogeochemical cycle.**

Table 14.4 The Four Groups of Organic Compounds

Group	Primary Function
Carbohydrates	cell energy
Lipids	energy storage
Proteins	cell structure
Nucleic acids	heredity material

One biochemical process fundamental to life is **photosynthesis.** Photosynthetic organisms combine atmospheric carbon dioxide with water to produce oxygen gas and glucose, a simple sugar molecule. **Chlorophyll** must be present, along with certain enzymes and sunlight (solar radiant energy) for photosynthesis to occur. The balanced chemical equation below (simplified) illustrates the photosynthetic process.

$$6CO_2 + 6H_2O + \text{solar energy} \xrightarrow{\text{enzymes}} C_6H_{12}O_6 + 6O_2$$

carbon dioxide   water   chlorophyll   glucose   oxygen

Solar energy is used to bond carbon, hydrogen, and oxygen atoms into glucose, which is thus made available to other organisms as their primary energy source. Glucose is one of a number of organic molecules produced by photosynthetic

*A **biogeochemical cycle** is the complex, cyclical transfer of nutrients from the environment to an organism and back to the environment.*

***Photosynthesis** is the process by which plants and some bacteria use chlorophyll, a green pigment, to trap sunlight energy. The energy is used to synthesize carbohydrates.*

***Chlorophyll** is the pigment that makes plants green. Chlorophyll traps sunlight energy for photosynthesis.*

Figure 14.11

A simplified pathway of matter as it cycles between living organisms and the nonliving environment.

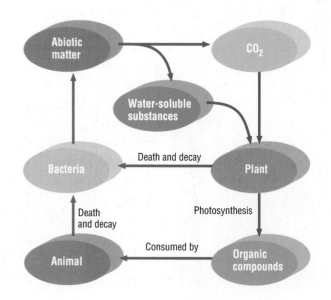

Unit Four:
Life in the Biosphere

organisms. Other molecules, including complex carbohydrates, lipids, proteins, and nucleic acids, are also required by all living organisms. Animals and other non-photosynthetic organisms obtain their organic molecules from the photosynthesizers and, in turn, pass these molecules to other animals. A few primitive organisms are capable of generating organic molecules by means of a process called **chemosynthesis.**

THE HYDROLOGIC CYCLE

Water, a requirement for all living organisms, plays a critical role in maintaining a relatively stable global heat balance. Water is the solvent in which most metabolic reactions take place. It is the major component of a cell's cytoplasm. Many organisms live within water's stable environment, while others depend on water to carry dissolved nutrients to their cells and organs. Table 14.5 lists several ways in which organisms depend on water.

The volume of water in the biosphere in its many forms remains fairly constant. However, the specific amount in any one phase can vary considerably; water is continuously entering and leaving living systems.

Table 14.5 Importance of Water to Organisms

Absorbs and releases heat energy.

Is the medium in which metabolic reactions take place in organisms.

Is an excellent solvent.

Composes about 60% of a cell's mass.

Supplies the hydrogen to organisms during the metabolism of key organic molecules and oxygen atoms for atmospheric oxygen production during photosynthesis.

Is a reactant in some metabolic activities and a product in others.

The pathway of water through the biosphere is called the **hydrologic** or **water cycle.** This cycle is shown in Figure 14.12. Water reaching the earth's surface as precipitation (rain, snow, sleet, hail, or any combination of these forms) can enter a number of pathways. It may remain on the surface as standing water (lakes, swamps, sloughs) or form rivers and streams, which terminate in the oceans, where the bulk of the earth's water reserves are held. Some precipitation seeps into the soil and subsurface

Chemosynthesis *occurs when some organisms use energy from chemicals rather than sunlight to produce glucose.*

The **hydrologic cycle,** *or* **water cycle,** *is the movement of water through the environment—from the atmosphere to earth and back.*

Figure 14.12

Simplified diagram of the water cycle.

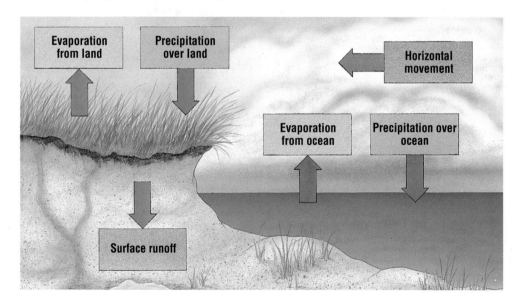

crustal rock to form groundwater. If the rock is permeable, some of the groundwater may seep to the surface, forming individual springs or adding water to existing lakes and streams. The movement of water through rock is slow but measurable. When groundwater is removed from an area during large-scale deep-well drilling, the rock can slowly be depleted of its water. This water may take centuries to be replaced.

By absorbing heat energy from the sun, some of the surface water evaporates as water vapor. It rises upward into the atmosphere until it reaches a point where the temperature is low enough for the water vapor to condense into tiny droplets of liquid water. These droplets are so light that they remain suspended in the atmosphere as clouds, each supported by rising air currents and winds. When conditions are right (i.e., a temperature drop), the droplets come together to form larger drops or sometimes ice crystals. Once the mass of the droplet or ice crystal can no longer be supported by air currents, precipitation occurs. The cycle repeats itself endlessly.

Living organisms also play a vital role in the water cycle. Water enters all organisms through osmosis and is utilized in various ways during their metabolic activities. It would appear then that living things tend to remove water from the environment, thus interfering with the cycle. However, through such processes as **cellular respiration** and the decay of dead organisms, water is released back into the atmosphere. Plants, particularly broadleaf trees and shrubs, play a major role in water recycling through the process of **transpiration.** Where there has been extensive removal of forests by logging or burning, there is less water in the atmosphere and noticeable climatic changes can occur. Surface runoff patterns become disturbed and the water-

holding capability of the soil may be reduced. This helps explain why the destruction of Brazilian rain forests only provides temporary usable land for agriculture.

Acid Deposition and the Water Cycle

Technology is often described as a doubled-edged sword, cutting through problems with one edge, while scarring the environment with the other. Nowhere is this more evident than with the technologies that contribute to acid deposition. Coal-burning plants, metal smelters, and oil refineries provide energy for the industrial world, but at the same time produce oxides of sulfur and nitrogen, listed among the most dangerous of air pollutants.

When fossil fuels and metal ores containing sulfur undergo combustion, the sulfur is released in the form of sulfur dioxide (SO_2), a poisonous gas. Combustion in automobiles, fossil-fuel-burning power plants, and the processing of nitrogen fertilizers produce various nitrous oxides (NO_x). Sulfur and nitrous oxides often enter the atmosphere and combine with water droplets to form acids. On entering the water cycle, the acids return to the surface of the earth in the form of snow or rain. The term *acid rain* is used to describe the movement of sulfur- and nitrogen-containing acids from the atmosphere to the land and water. Acid rain 40 times more acidic than normal rain has been recorded. Such acid precipitation, some as acidic as lemon juice, reacts with marble, metal, mortar, rubber, and even plastics. The impact of acid rain on ecosystems has been well documented. Acid precipitation kills fish, soil bacteria, and both aquatic and terrestrial plants. However, the devastation is rarely uniform in all areas. Some regions are more sensitive to acid rain than others. Alkaline soils neutralize

Cellular respiration *is the process by which living things convert the chemical energy in sugars into the energy used to fuel cellular activities.*

Transpiration *is the loss of water through the leaves of a plant.*

Unit Four:
Life in the Biosphere

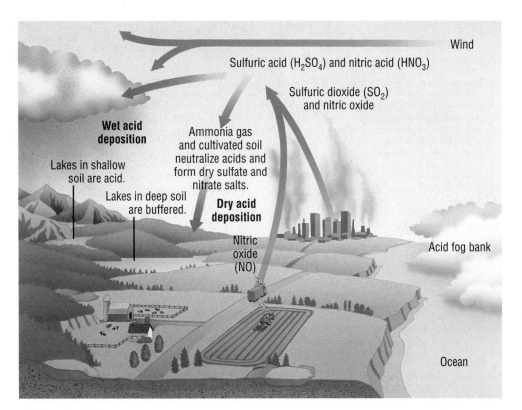

Wind

Sulfuric acid (H_2SO_4) and nitric acid (HNO_3)

Sulfuric dioxide (SO_2)
and nitric oxide

**Wet acid
deposition**

Ammonia gas
and cultivated soil
neutralize acids and
form dry sulfate and
nitrate salts.

Lakes in shallow
soil are acid.

Lakes in deep soil
are buffered.

**Dry acid
deposition**

Nitric
oxide
(NO)

Acid fog bank

Ocean

Figure 14.13

Wet and dry acid deposition.

Figure 14.14

Devastation of forest areas
caused by long-term exposure
to acid rain. Conifers are
particularly susceptible to acid
rain. Scarring caused by acids
makes the trees vulnerable to a
variety of infections.

acids, minimizing the impact of the corrosive acids before runoff carries them to streams and lakes. Unfortunately, soils in much of southeastern Canada have a thin layer of rich soil on top of a solid granite base. Granite offers little in the way of buffering agents that would neutralize the acids.

The sulfur and nitrous oxides released from smokestacks do not always enter the water cycle in the atmosphere. Depending on weather conditions, particles of sulfur and nitrogen compounds may remain airborne and settle out. These dry pollutants then form acids when they combine with moisture. The dew from a lawn, the surface of a lake, or the water inside your respiratory tract are but a few of the potential sites where such acids can form.

One early solution to the problem of acid deposition was to build taller smokestacks. The average height of smokestacks in the 1950s was about 100 m. By the 1980s it was 224 m, with some exceeding 300 m. It was reasoned that if the pollutants could be expelled farther into the upper atmosphere, fewer would fall on any specific area. However, such thinking turned a local problem into an international one. Acid deposition remained suspended in the air and was often carried hundreds of kilometers from the site of the pollution across international borders and to areas less able to buffer the acids.

Technology offers some solutions to the problems that have been created by oxides of sulfur and nitrogen. Scrubbers have been placed in smokestacks to remove harmful emissions, and lime has been added to lakes in an attempt to neutralize acids that have descended from the atmosphere. However, both of the solutions are expensive.

The prospect of improving smelters is equally difficult. Mining companies are already battling to remain operational and compete in a world market. Would tougher legislation result in higher levels of unemployment?

THE CARBON CYCLE

Carbon is the key element in all organic compounds. In the abiotic environment, carbon is part of the carbon dioxide (CO_2) molecule. Some carbon dioxide exists in the atmosphere, but most is dissolved in the oceans. Each year about 50 to 70 billion tonnes of carbon enter the biotic environment through the photosynthetic process.

Some of the organic carbon is released back to the atmosphere as carbon dioxide through cellular respiration, as illustrated in the simplified equation below:

$$\text{C}_6\text{H}_{12}\text{O}_6 + 6\text{O}_2 \xrightarrow{\text{enzymes}} 6\text{H}_2\text{O} + 6\text{CO}_2$$

glucose   oxygen   water   carbon dioxide

Because photosynthesis and cellular respiration are complementary processes, this phase is often called the *carbon-oxygen cycle*. Most of the carbon that remains as part of living organisms is returned to the atmosphere or water as carbon dioxide when wastes and the bodies of dead organisms decay. Under certain conditions the decay process is delayed and the organic matter may be converted into fossil fuels such as coal, petroleum, and natural gas. This carbon is then unavailable to the cycle unless natural or human events allow these fuels to be burned. The burning process releases carbon dioxide to the atmosphere.

In aquatic systems, particularly the oceans, inorganic carbon can be incorporated into carbonate compounds, which form the shells and other hard structures of many organisms. If the carbonates become part of sedimentary rock, the carbon can be trapped for millions of years until geological conditions bring it back to the surface. Natural heat produced by volcanic activity can break down carbonate-containing rocks such as limestone, releasing carbon dioxide.

Figure 14.15

Simplified diagram of the carbon cycle.

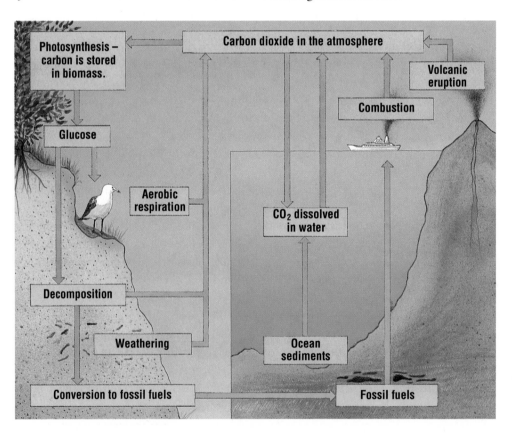

plants to make proteins. This also explains why older lawns lack a rich green color. Chlorophyll, the green pigment of plants, is a protein and therefore requires sufficient levels of nitrates to maintain the green color.

The denitrification process is enhanced when the soil is very acid or waterlogged. Bogs, for example, are deficient in nitrogen. Unique plants, such as sundews and pitcher plants, which are native to bogs, obtain their nitrogen by digesting trapped insects. In an interesting reversal of roles, these plants obtain their protein from animals.

An alternate pathway for the production of nitrates is through **nitrogen-fixing bacteria.** These bacteria grow in nodules on the roots of legumes such as clover or alfalfa. They convert atmospheric nitrogen to nitrates. The bacteria provide these plants with a built-in supply of usable nitrogen, while the plant supplies the nitrogen-fixing bacteria with needed sugars. There is usually much more nitrate produced than required. The excess nitrates move into the soil, providing an alternate source of nitrogen for other plants. The traditional agricultural practice of rotating crops capitalizes on bacterial nitrogen fixation. Plants that are fertilized with manure and/or other decaying matter also take advantage of the nitrogen cycle. Throughout this cycle the mass of nitrogen gas removed from the atmosphere is balanced by the mass being returned.

Have you ever had to pick clover from your lawn by hand? There is an easier way to get rid of it. Because clover contains microbes capable of fixing nitrogen, it has a special advantage over other plants that cannot fix their own nitrogen. This means that clover grows well on lawns that lack nitrogen. Healthy lawns, which are rich in nitrogen, rarely allow clover to establish itself.

THE PHOSPHORUS CYCLE

Phosphorus is a key element in cell membranes, the energy storage molecules, and the calcium phosphate of mammalian bone. (You can learn more about energy storage molecules in the chapter Energy within the Cell.) Environmental phosphorus tends to recycle in two ways: a long-term cycle involving rocks of the earth's crust and a shorter cycle involving living organisms. Phosphorus is usually found in the form of phosphate ions, combined with a variety of elements, as part of the continental rock. Phosphates are soluble and dissolve in water as part of the hydrologic cycle. During this phase phosphates can be absorbed by photosynthetic organisms and passed through the food chains. Erosion carries phosphates from the land to streams and rivers, and then finally to the oceans. The ocean phosphates form sediments that, through geological activities, may be thrust upward and once again become part of the land surface. The overall process can take millions of years.

Figure 14.21

Nitrogen-fixing bacteria grow in root nodules of plants in the legume family.

Figure 14.22

Phosphate cycles through both long and short cycles.

Nitrogen-fixing bacteria *convert atmospheric nitrogen to nitrogen compounds such as ammonia and nitrate.*

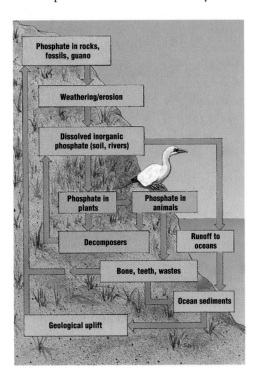

Phosphate in rocks, fossils, guano

Weathering/erosion

Dissolved inorganic phosphate (soil, rivers)

Phosphate in plants

Phosphate in animals

Decomposers

Runoff to oceans

Bone, teeth, wastes

Ocean sediments

Geological uplift

Most of the phosphates used in photosynthesis enter the shorter cycle. This phase is able to meet the high demand for phosphates by all organisms through rapid recycling. Photosynthesis is required for usable phosphates to enter the food chain. Plants are eaten by animals, which, in turn, are eaten by other animals. When organisms decompose, wastes and dead tissue are broken down and the phosphates are released. The phosphates are now available to enter the photosynthesis process again as soluble phosphate ions. During this phase phosphates can be absorbed by photosynthetic organisms and passed through the food chains.

Agriculture and the Nitrogen and Phosphorus Cycles

As crops are harvested, valuable nitrogen and phosphates are removed. The interruption of the nitrogen and phosphate cycles soon depletes the soil. Soil fertilizers restore required nutrients and increase food production. Some estimates suggest that fertilizers containing nitrogen and phosphates can almost double cereal-crop yields. However, fertilizers must be used responsibly. As anyone who has overfertilized a lawn has discovered, more is not necessarily better. Soil bacteria convert ammonia or urea fertilizers into nitrates, but the nitrates in the soil may generate nitric acids. Depending on the amount of buffer, a typical application of between 14 and 23 kg of nitrogen fertilizer per year over 10 years can produce a soil that is 10 times more acidic. This can have devastating effects on food production. Most prairie soils have a neutral pH near 7.0. Should the pH drop to 6.0, some sensitive crops, such as alfalfa and barley, are affected. The effect on soils near the Great Lakes is even more pronounced, as these soils have considerably less

Figure 14.23

Spring runoff carries nitrogen and phosphate fertilizers to streams, promoting aquatic plant growth.

buffering capacity and, therefore, are even more vulnerable to acids. A drop to pH 5.0 will affect almost all commercial crops. The combination of fertilizers and acid deposition only accelerates the destruction of crops.

The accumulation of nitrogen and phosphate fertilizers produces yet another environmental problem. As spring runoff carries decaying plant matter and fertilizer-rich soil to lakes and streams, aquatic plant growth is promoted. Once the aquatic plants die, decomposing bacteria use valuable oxygen from the water to complete the carbon, nitrogen, and phosphate cycles. Because the decomposers flourish in an environment with such an abundant food source, lake oxygen levels drop quickly and fish often begin to die. Unfortunately, this makes the problem worse as decomposers begin to recycle the matter from the dead fish, enabling even more bacteria to flourish and oxygen levels to be reduced even further.

■ REVIEW QUESTIONS ?

16 What is a biogeochemical cycle? Why are these cycles important to living organisms?

17 In terms of biogeochemical cycles, why is photosynthesis critical to the biosphere?

18 Describe the role of plants in the water cycle.

19 What do nitrogen-fixing bacteria and lightning have in common?

20 Why is carbon critical to the biosphere?

21 Where is most of the biosphere's carbon dioxide stored after it is released into the atmosphere?

22 Why are photosynthesis and cellular respiration often called the biotic phase of the carbon cycle?

LABORATORY

RECYCLING PAPER

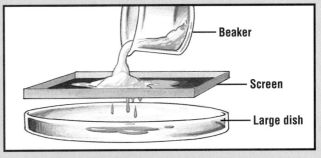

Beaker

Screen

Large dish

Objective

To recycle paper.

Materials

newspapers
large screen
100 mL graduated cylinder
beaker tongs
goggles
blotting paper (or paper towel)

10% starch solution
rolling pin
250 mL beaker
lab apron
hot plate

Procedure

1 Measure 150 mL of the starch solution and pour it into the 250 mL beaker. Using a hot plate, heat the starch solution to boiling.

2 Tear a sheet of newspaper into very small pieces and place it in the boiling starch solution. Remove the beaker from the hot plate and allow the mixture to soak for 10 min.

3 Place the screen over a large dish and pour the solution on the screen. Allow the fluids to collect in the dish.

4 Smooth the pulp evenly over the screen and then place the blotter on top of the screen.

5 Place the screen inside the fold of a number of newspapers and roll with a rolling pin.

Rolling pin

Blotter
Screen
Newspaper

Laboratory Application Questions

1 Why was the starch solution used?
2 Why was the starch solution boiled?
3 Propose a method for recycling paper on a larger scale. ■

IMPORTANCE OF RECYCLING

The renewed environmental awareness of the 1990s has prompted widespread interest in recycling. Nearly 50% of household wastes can be recycled. It takes 95% less energy to reuse an aluminum can than to make a new one. One estimate suggests that if Canadians recycled all of their aluminum food and soft-drink cans, enough metal could be saved to make 145 000 cars. Recycling a tonne of paper saves 11 to 17 trees and uses 50% less energy.

Besides the economic advantages of recycling, there are many environmental benefits as well. Landfill sites take up valuable agricultural land and cause the accumulation of potentially dangerous household wastes. The example most frequently mentioned is disposable diapers. Composed of fibers and plastic, these diapers take several hundred years to decompose. It has been estimated that two million disposable diapers are used in Canada every day and that each child produces one tonne of soiled diapers each year.

SOCIAL ISSUE:
The Greenhouse Effect

Scientists predict that, if global warming trends continue, the average temperature on earth may increase by as much as 5°C by the end of the next century. Global warming could result in changes in weather patterns and growing seasons, the melting of the polar ice cap, rising sea levels, and the flooding of coastal regions.

Statement:

Action must be taken to prevent greenhouse warming.

Point

- The level of greenhouse gases, particularly CO_2, has been on the rise continuously over the past 100 years. This suggests that global warming is more than a distinct possibility and should be prevented.
- The Canadian climate will warm up sufficiently to push the treeline and prairie further north, quickly altering our present ecosystems.
- Computer and mathematical models all suggest that warming is a reality and that within 100 years significant changes to temperatures and ocean levels will take place.

Research the issue.
Reflect on your findings.
Discuss the various viewpoints with others.
Prepare for the class debate.

Counterpoint

- Carbon dioxide is highly soluble in water. Therefore, the oceans will absorb any excess CO_2, and global warming will not be a problem.
- If the climate warms up, it will free areas in Canada from the extremes of winter and put more land to use.
- There are so many variables, including the warming and cooling that have occurred in the recent past, that the "models" cannot predict anything accurately.

CHAPTER HIGHLIGHTS

- Ecology is the study of the interaction between living (biotic) organisms and their nonliving (abiotic) environment.
- Often compared to a spaceship, the earth supports the only known life forms in existence.
- Living organisms are found in a limited region of the earth known as the biosphere.
- The atmosphere is organized in zones. It protects living organisms from excessive radiation from the sun and other sources in space.
- The protective ozone layer appears to be breaking down, resulting in an increase in the amount of harmful ultraviolet radiation reaching the earth.

- Some gases in the atmosphere, such as carbon dioxide, are responsible for trapping heat as it radiates to outer space. This is called the greenhouse effect.
- Greenhouse gases are accumulating in the atmosphere at an ever-increasing rate. Scientists predict that this will cause global warming.
- The tilt of the earth's axis, the physical properties of water, and the albedo of the atmosphere and the earth's surface contribute to the global distribution of heat by causing climate and weather.
- The cycling of matter between the abiotic and biotic environments is called biogeochemical cycling.
- Photosynthesis is the process by which abiotic matter is converted into the organic molecules.

This type of scenario has been repeated in many parts of the world. In Canada the peregrine falcon has become a victim of insecticides. Like the cats on the island north of Borneo the peregrine falcon occupies the uppermost level of the ecological pyramid. Toxins such as DDT and Dieldrin accumulate in the fatty tissues of all consumers. The problem intensifies because the concentrations of toxins become magnified as you move up the food chain. Although relatively low levels of toxins are consumed with the prey, toxins accumulate in predators like the falcon. Because the toxins are soluble in fat but not in water, they are not released with most other waste products. When low levels of the toxin are taken in with each organism consumed, levels in the predator begin to rise. At each stage of the food chain the concentration becomes greater. The higher the trophic level, the greater the concentration of toxins. The process is referred to as **biological amplification.**

The irony of insecticides is that although they were developed to rid the world of harmful insects, they have had a much greater effect on humans. Like other predators at the top rung of the food pyramid, humans are subject to biological amplification. During the 1950s and 1960s, fat-soluble DDT turned up in breast milk and was passed from mothers to their babies. DDT levels were especially high in humans who lived in areas where DDT was used for spraying crops. Even those who ate crops from these areas or consumed the animals that fed on these crops were exposed to DDT. The

> **BIOLOGY CLIP**
> Invented by a graduate student in chemistry in 1873, DDT was almost forgotten until 1939, when the Swiss entomologist, Paul Mueller, rediscovered it. In 1948 Mueller was awarded the Nobel Prize for the discovery of the insect-killing properties of DDT. Since that time nearly two million tonnes of DDT have been used worldwide.

fact that DDT was banned for use in Canada and the United States during the late 1970s has not totally eliminated the problem. Migratory birds like the mallard duck, Canada goose, and peregrine falcon winter in Central America and Mexico, where DDT is still used. A similar scenario exists for the migrating fish of the Atlantic and Pacific Oceans.

Chemical factories, many of which are owned by stockholders from countries that ban DDT and similar chemicals, continue to make these pesticides available for use by the poorer nations of the world. Despite all of the insidious effects of these toxins, the chlorine pesticides still have one alluring feature: they are incredibly cheap to manufacture.

Biological amplification refers to the buildup of toxic chemicals in organisms as tissues containing the chemical move through the food chain.

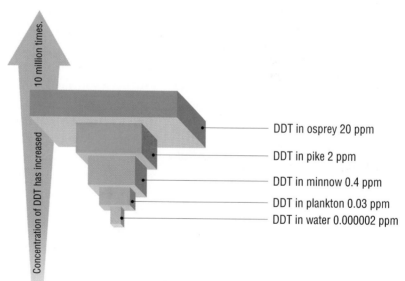

Figure 15.21

Biological amplification. The concentration of DDT increases as you move up the food chain. The greater the number of trophic levels, the greater the amplification. (ppm = parts per million)

DDT in osprey 20 ppm
DDT in pike 2 ppm
DDT in minnow 0.4 ppm
DDT in plankton 0.03 ppm
DDT in water 0.000002 ppm

9 State the first law of thermodynamics.

10 Describe the second law of thermodynamics in your own words.

11 Why is less than 20% of available chemical energy transferred from one organism to another during feeding?

12 Why do energy pyramids have their specific shape?

13 What would be the best source of energy for an omnivore: the plant or animal tissue it feeds on? Explain.

FRONTIERS OF TECHNOLOGY: ARTIFICIAL ECOSYSTEMS

Artificial ecosystems allow ecologists to study food webs while controlling different variables. Biosphere II, the largest of the artificial environments, spans nearly one hectare in the Arizona desert. Many smaller-scale artificial ecosystems exist in Canada, yet the scientific community remains divided on their usefulness. Science writer and broadcaster David Suzuki, for one, suggests that these projects are not worth the money spent on them. By controlling variables, we may get information about interactions within the artificial environment, but how much does that tell us about natural ecosystems? Although scientists have long recognized the value of using models to describe chemical structures or physiological systems, they are divided on the usefulness of models for studying the environment. Can ecosystems devised by humans over a period of a few years approximate natural ecosystems that have evolved over millions of years? Although the artificial ecosystems allow scientists to control variables, can any of their findings be transferred to natural ecosystems?

Research into another type of artificial ecosystem, the indoor environment, has emerged as an important area of study. Many Canadians spend a great deal of time indoors. Downtown office complexes are linked with one another by a series of underground tunnels. Even malls enable shoppers to travel from store to store without going outside. Light, temperature, and moisture are artificially controlled. The scientific community does not agree on the amount of natural light required for a healthy environment. Research shows that workers who receive little natural light are adversely affected.

Pollution has traditionally been seen as a problem affecting outside environments, but studies indicate that indoor environments may pose even greater risks. One research team found that indoor air quality is often much worse than outdoor air. The research team found noxious indoor vapors such as carbon monoxide, nitric oxide, and nitrogen dioxide sometimes exceed the maximum allowable levels for outside air.

Indoor ecologists point out the importance of cleaning air conditioners and heat-exchange systems to prevent the circulation of dust and pollen in buildings. The problems created by office crowding have also been addressed. Have you ever noticed how difficult it can be to breathe if you are working in a crowded classroom? Planting trees and ferns in office buildings and shopping malls to supply more oxygen, and designing buildings with large open areas and skylights are just a few of the newer strategies that attempt to create more livable indoor environments.

SOCIAL ISSUE:
Economics and the Environment

Canada's 1990 Green Plan calls for "responsible sustained development." It recognizes a complex interaction of interests when it calls for global action to ensure that all human activities, including economies, be balanced with environmental protection.

Statement:

Protecting the environment should have priority over economic interests.

Point

- Unless environmental legislation has teeth, it will fail to bring about action. Government initiatives such as Canada's Green Plan of 1990 should make clearer statements about the quality of the air, soil, and water, and ensure that they are put into effect.
- Each country has a responsibility to act for the good of the environment without waiting for other countries to join in. We cannot continue to pass the buck; time is running out. Some sacrifices in terms of jobs and prices are necessary, and Canadians should be willing to make them.
- Improving the environment is expensive. However, less than two weeks' worth of global military spending would pay for a proposed UN plan to minimize clean-water problems in third world countries.

Counterpoint

- Many people blame politicians for short-term goals, but the political system reinforces short-term planning— elections occur every 4 to 5 years.
- Canada should not enact environmental legislation alone. Strict environmental controls could lead to a drop in living standards. Jobs could be lost if companies relocated in developing countries where there are less stringent standards.
- Governments, many of whom have huge debts, cannot afford the enormous amounts of money required to clean up the environment. One estimate states that $774 billion would be needed to reverse negative global trends in soil erosion, to implement reforestation, and to develop renewable energy.

Research the issues.
Reflect on your findings.
Discuss the various viewpoints with others.
Prepare for the class debate.

CHAPTER HIGHLIGHTS

- The energy required for all living organisms originates in solar radiation, which is converted to chemical energy during photosynthesis and stored in the chemical bonds of organic molecules such as glucose.
- Energy is transferred from organism to organism during the feeding process. Therefore, organisms exist in different trophic levels, depending on the number of steps in a food chain they are away from the original source of energy, sunlight.
- Plants belong to the first trophic level and are called producers. Through photosynthesis, they manufacture chemical energy. All animals are called consumers because they must feed on other organisms to obtain energy.
- Primary consumers, called herbivores, eat plant tissues. They are at the second trophic level. Other animals may be carnivores or omnivores, depending on their specific diet.

- A food chain describes a particular feeding (energy) pathway. A food web is a pattern of natural energy flow in which a large number of food chains interlock in an ecosystem.
- Complex food webs are an indication of a stable ecosystem.
- Grazer food chains start with plants that are consumed by herbivores. Decomposer food chains involve the breakdown of wastes and dead tissues from organisms.
- Energy flow in ecosystems must obey the laws of thermodynamics. Therefore, as energy (food) is transferred from one trophic level to the next, over 80% of the original energy is lost.
- For an ecosystem to exist there must be a continuous energy input in the form of sunlight. Energy flows through an ecosystem and is eventually lost.
- Energy flow can be shown graphically in the form of pyramids of numbers, biomass, and energy.

APPLYING THE CONCEPTS

1 In underground caves, where there is permanent darkness, a variety of organisms exist. In terms of energy flow, explain how this can be possible.

2 Based on what you have learned about energy pyramids, comment on the practice of cutting down rain forests to grow grain for cattle.

3 Design complex food webs for a tundra ecosystem and a middle-latitude woodland ecosystem. Reference books can be used to determine the other members of the food web.
   a) Which ecosystem has the greatest biomass? Provide your reasons.
   b) Which ecosystem has the greatest number of organisms? Provide your reasons.
   c) Which ecosystem has the greatest energy requirement? Provide your reasons.
   d) Comparing the tundra and middle-latitude woodland ecosystems, indicate which is more susceptible to environmental pollution. Explain your answer.

4 By law, the cutting of forests must be followed by replanting. Why do some environmentalists object to monoculture replanting programs?

5 Assume that the plant material in a plant–deer–wolf food chain contains a toxic material. Why would the wolf's tissue contain a higher concentration of the toxin than the plant tissue?

6 Provide examples of the two laws of thermodynamics in terms of some common, everyday events.

7 Of the three basic energy pyramids, which best illustrates energy transfer in a food chain? Explain.

8 Provide examples of how technological innovations have altered ecosystems.

9 Assuming an 80% loss of energy across each trophic level, state how much energy would remain at the fourth trophic level if photosynthesis makes available 100 000 kJ of potential energy. Show your reasoning. Construct a properly labelled pyramid to represent this situation. Could a fifth-level organism be added to the chain? Explain.

CRITICAL-THINKING QUESTIONS

1 Assume that a ski resort is proposed in a valley near your favorite vacationing spot. What type of environmental assessment should be done before the ski resort is built? In providing an answer, pick an actual location you are familiar with and give specific examples of studies that you would like to see carried out.

2 Atmospheric warming may cause drought in some parts of the world. Illustrate the impact of drought by drawing an energy pyramid before and after the drought. Explain why the two pyramids are different.

3 Insect-eating plants such as the sundew are commonly found in bogs all across the country. Although referred to as "carnivorous" plants, they are still considered to be members of the first trophic level. Is this the proper trophic level to assign to these plants? Research information on carnivorous plants, then state the trophic level you think is most appropriate. Present the reasoning behind your choice.

4 Some ecologists have stated that to maximize the food available for the earth's exploding human population, we must change our trophic level position. What is the probable reasoning behind this statement? Could any potential biological problems occur if this switch were actually made?

5 A team of biologists notes that the population of white-tail deer is decreasing in a particular area. Hunting pressures have remained stable and the biomass of producers in the area has not changed. What other factors might the biologists consider to determine why the deer population is decreasing?

ENRICHMENT ACTIVITIES

1 Obtain a copy of Canada's Green Plan from Environment Canada.

2 Many businesses have made an attempt to protect Canada's environment. Prepare a visual display explaining some of the initiatives taken by industry.

3 Identify an environmental issue and do a risk-analysis assessment. Show your calculations, identify your assumptions, and justify your conclusions.

CHAPTER 16

Aquatic and Snow Ecosystems

IMPORTANCE OF WATER IN ECOSYSTEMS

Water is the world's greatest natural resource, yet most people undervalue its worth. Gold, diamonds, and emeralds are prized because they are so rare, yet none has the value of water, a chemical so common that it is found almost everywhere you look.

Approximately 70% of your body is composed of water. Water is so vital that most people will die within 48 hours if they cannot replenish their water reserves. By comparison, most people can survive for a week or more without food. No living organism is less than 50% water, and some creatures are as much as 97% water.

Water covers more than two-thirds of our planet, yet only a small portion of it is actually drinkable. Ninety-seven percent of the world's water is salt water. Although unfit for human consumption, the oceans' great reserves of salt water are of tremendous value to all living things. The oceans control the weather patterns on our planet. They also act as a huge reservoir that provides fresh water by evaporation. By far the greatest portion of fresh water is stored as snow and ice,

Figure 16.1

As terrestrial beings we think of the earth in terms of land. However, the vast majority of the earth's surface is covered with water.

Summerlike conditions can be maintained throughout the year because of thermal pollution. Water removed from a lake by industries and used as a coolant is sometimes returned warm enough to prevent freezing over. Many organisms thus continue to thrive through the winter, in turn producing more detritus with its resultant decomposition. Once again an oxygen-depleted hypolimnion may result, with all of its implications. Additional organic matter in the benthic zone causes more rapid eutrophication.

Canadian Lakes

The most productive part of a lake is the littoral zone. Here algae and other green plants take advantage of the sunlight for their photosynthetic activities. The depth of a lake and the extent of the littoral zone are influenced by the type of lake bed in which the water is found. In central Canada into the western Northwest Territories, lakes were typically formed in basins carved in the granite rock of the Canadian Shield during the most recent glaciation. These lakes are usually deep and, due to their granite base, limited in the dissolved natural nutrients (minerals). Most, therefore, are oligotrophic. Although lakes in the Atlantic region are somewhat similar, the underlying rock is more varied, since it originated from ancient mountains. In contrast, lakes in the prairies are found in depressions formed in thick glacial deposits. This varied base is richer in soluble nutrients, making the ecosystems of these lakes more productive. These lakes also tend to be fairly shallow and collect sediments more rapidly than the Shield lakes. They are usually classified as eutrophic lakes.

The largest lakes run in a curve from Great Bear Lake in the Northwest Territories to the Great Lakes. They lie on the boundary between the glacial deposits to the south and west and the Shield rock to the north and east. They are considered to be oligotrophic due to their depth and low water temperatures. The major exception is Lake Erie, which is fairly shallow. Agricultural and industrial pollutants plus human wastes have caused rapid eutrophication of Lake Erie, resulting in noticeable changes in the population and species of organisms. International cooperation between Canada and the United States appears to have reduced the process to a considerable extent.

■ REVIEW QUESTIONS ?

8 Distinguish between the littoral, limnetic, and profundal zones of a lake.

9 What is meant by the term plankton?

10 What is the effect of the fall overturn on a lake?

11 What is unique about arctic lakes?

12 What is the relationship between dissolved oxygen and the temperature of a lake?

13 Why does a thermocline form in the summer?

14 Define algal bloom. What are its effects on a lake?

15 Outline the steps of pond succession through to a mature forest.

16 Explain how waste runoff from a cattle ranch can cause eutrophication.

Figure 16.9

Thermal pollution from industries can produce summerlike conditions in a lake throughout the year.

LABORATORY

ACID RAIN AND AQUATIC ECOSYSTEMS

Objective

To design an experiment to investigate how acid rain will affect an aquatic ecosystem.

Background Information

Instructions will be given for building an acid-rain generator. Suggestions will be provided for an aquatic ecosystem. The experimental design is up to you. Approval should be obtained from the instructor before beginning.

Materials (suggested)

acid-rain generator	glass tubing
2 125-mL flasks	1 M NaHSO$_4$
1 M HCl	aquarium pump
2 rubber (2-hole) stoppers for flasks	lab apron rubber tubing
safety goggles	aquatic ecosystem
field guide for lake and pond organisms	dissecting microscope 2 L pop bottle
graduated cylinder	pH paper

CAUTION: HCl is very corrosive. Avoid skin and eye contact. Wash all splashes off your skin and clothing thoroughly. If you get any chemical in your eyes, rinse for at least 15 min and inform your teacher.

Procedure

Experimental Design

1 Formulate a hypothesis. Most hypotheses are presented in an "if ... then" format (i.e., "If acids are added to an aquatic environment, then ...").
   a) State your hypothesis.
   b) Identify dependent and independent variables.
2 Working in small groups, devise a procedure that allows you to determine the effect of acid rain on an aquatic ecosystem. Remember to accept all opinions of group members and work to a consensus. Attempt to provide a measurement for all data collected.

c) Outline your group's procedure. Your teacher should approve the procedure before you begin.
3 Prepare data tables. Remember: attempt to quantify all possible recordings.
   d) Present your data. Graphs and data tables should be considered.

Constructing an Acid-rain Generator

4 Using a graduated cylinder, add 100 mL of distilled water to flask #2.
5 Arrange the apparatus as shown in the diagram. Connect the rubber tubing.

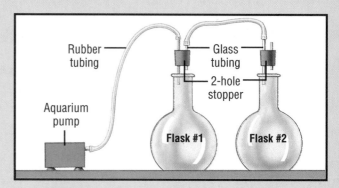

6 Using a graduated cylinder, add 50 mL of NaHSO$_4$ to flask #1. Rinse the graduated cylinder and then follow the same procedure by adding 50 mL of HCl to flask #1.
7 Place stoppers in the flasks and begin your experiment. (It is recommended that you allow the generator to run for at least 20 min.)

Aquatic Ecosystems

8 Fill a 2 L plastic bottle with lake or pond water. Decide whether or not you will aerate the water.
9 Survey the organisms found in the pond water. You may wish to develop a technique for determining the number of organisms.

Aquatic ecosystem

Laboratory Application Questions

1 List the conclusions that you have drawn from the laboratory data.
2 Provide reasons for each of your conclusions.
3 Identify modifications or changes to the procedure that you would incorporate should you wish to carry out the investigation again. ■

Unit Four:
Life in the Biosphere

FRONTIERS OF TECHNOLOGY: ECHO SOUNDING

One of the most important research tools used for investigating the aquatic environment is the echo sounder, or *sonar* device (sonar is an acronym for *so*und *nav*igation and *r*anging). Originally designed to detect submarines in World War II, echo sounding is used by researchers to map the depth and the bottom contours of bodies of water. As well, it is sensitive enough to record the presence and location of fish and large water-dwelling mammals, enabling the researcher to follow their activities and movement under water.

As the name suggests, the echo sounder works on the principle of the echo. Sound is transmitted into the water and the reflected signals are detected by a receiver. The distance of an object is determined by the speed of that particular sound in the water. The results are often displayed on a TV screen.

Modern sonar devices use a variety of sound frequencies, both audible and inaudible. They are also able, to a certain extent, to correct for the variables that affect the speed of sound in water, such as water temperature, density, and salinity. Layers of water having different properties can give an inaccurate reading. However, without echo sounding, knowledge of the physical structure of bodies of water and its effects on the distribution of life forms would be severely limited.

Branches of biology such as marine fisheries management depend on a knowledge of the precise location and boundaries of areas such as Georges Bank, an area with large stocks of fish, located at the edge of the Atlantic continental shelf. In areas of extensive deep water such as the Great Lakes and the oceans, echo sounding has provided vital information on aquatic ecosystems and water circulation patterns.

WINTER IN NORTHERN ECOSYSTEMS

Periods of snow and subzero temperatures characterize ecosystems found in regions that experience a distinct winter season. Much of Canada is included in the arctic or subarctic region of North America.

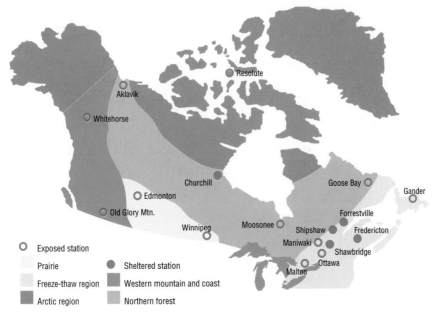

Figure 16.10

Major snow regions of Canada.

Various groups of northern native peoples are so dependent on snow conditions for their survival in winter that they have developed an extensive vocabulary to describe the properties of snow. The Kobuk Valley Inuit of Alaska have devised more than 17 terms to describe different kinds of snow. These terms have now been made part of the language of snow ecology. Snow terminology also uses words from other native groups and from the languages of the countries surrounding the North Pole (i.e., the Russian word *taiga* refers to an area of coniferous forest).

Table 16.3 Some Common Inuit Snow Terms

Term	Pronunciation	Description
Qali	Kall-ee	Snow that collects on trees, i.e., on the branches of spruce and other conifers.
Qamaniq	Com-an-nique	The depression in the snow found around the base of trees, i.e. the snow shadow beneath most conifers.
Anniu	An-nee-you	Falling snow.
Api	Aye-pee	Snow on the ground.
Siqoq	See-cok	Drifting (blowing) snow.
Upsik	Up-sik	Wind-beaten snow.
Pukak	Pew-cak	A crystalline snow layer found at the ground–snow interface.
Siqoqtoaq	See-cok-tow-ak	Sun-crusted snow.
Kimoaqruk	Kee-mow-ak-rook	Drift.
Kaioglaq	Kay—oh-glak	Sharply etched wind-eroded snow surface.

Snow characteristics have influenced the mode and form of transport used by northern natives. For example, different shapes and sizes of snowshoes were constructed based on the type of snow cover

Figure 16.11

Snowshoes can be constructed in different shapes and sizes to suit varying types of snow cover.

found in a specific region. The Inuit sled (komatik) is suitable for travel on the hard-packed snow of the tundra, but useless in the soft snow of the forests to the south. The toboggan was developed for travel in softer snow.

ADAPTATIONS FOR WINTER

For the organisms living in northern ecosystems, winter is the most critical time of the year. These organisms must have some way of dealing with limiting factors of snow, cold, and the reduction in or lack of a food supply, to survive. The problem they face was best described by the Russian ecologist, A.N. Formozov, when he stated, "Snow cover for many species is the most important element of environmental resistance, and the struggle against this particular element is almost beyond the species' ability." Yet even when snow and cold persist for most of the year, many organisms can and do survive. It is the extremes in a winter environment that have the most impact on an organism.

Five key conditions—snow, cold, radiation (of heat), energy (food), and wind, —determine whether an organism can survive in a winter ecosystem. These five factors determine successful adaptation. Adaptations can be anatomical, physiological, or behavioral.

RESPONSE TO SNOW

The classification of animals based on the impact of snow on their behavior was suggested by Formozov in 1946 (see Table 16.4). The Russian word for snow, *chion*, prefixes each of the terms. These terms will be used to describe a number of situations.

Table 16.4 Classification of Animals Based on Their Ecological Relationships with Snow

Classification	Description	Examples
Chionophobes	Animals that avoid snow and winter conditions. Do not inhabit snowy regions.	Pronghorns, many terrestrial birds.
Chioneuphores	Animals that are able to withstand winter conditions.	Deer, voles, small mammals.
Chionophiles	Animals that have adapted to snow. The range of the animal is limited to regions of long, cold winters.	Snowshoe hare, polar bear, caribou, musk-ox.

There are many ways in which animals can adjust to winter conditions. **Migration** is the answer for some. Entire populations of some animals move to regions with a more favorable climate and a plentiful food supply. This is most typical of bird species, some of which migrate thousands of kilometers each year. Many of the waterfowl that breed in the arctic spend the winter along the coast of the Gulf of Mexico.

Some migrations may simply involve a shift in population from an area of one set of winter conditions to that of another. The most spectacular of these migrations is made by the barren-land caribou, which migrate thousands of kilometers. They winter in the boreal forest, which is protected from severe winds, and where lichens, a preferred food, are easily obtained. In the summer the caribou move out into the tundra to reproduce. Elk in the Rocky Mountains usually spend their summers at high altitudes, then, in the autumn, move down into the valleys where there are fewer temperature extremes and greater supplies of food. Deer may simply move from one snowy place to another depending on how easy it is to obtain food. Prior to European settlement, the enormous herds of bison used to migrate the length of the prairies to find available food.

Figure 16.13

A migrating caribou herd numbering 10 000.

Figure 16.12

Animals can be classified according to the impact of snow on their behavior.

Migration *is the movement of organisms between two distant geographic regions.*

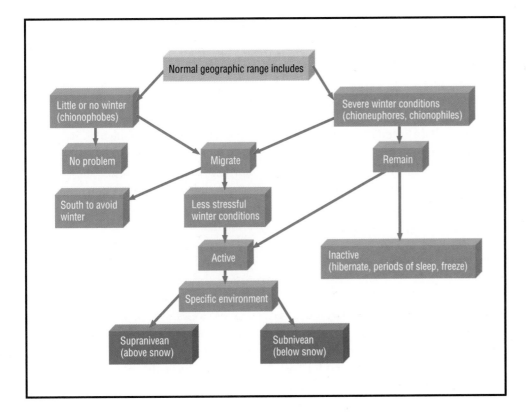

Figure 16.14

Gray squirrel.

Figure 16.15

Moth pupa in cutaway cocoon.

Chionophobes *(snow haters) are animals that avoid snow-covered regions.*

Hibernation *is a dormant (sleep) state in which body temperature and functions are reduced well below normal.*

Supercooling *occurs when a water solution is chilled well below the point at which the solution crystallizes spontaneously into ice.*

Chionophobes

The pronghorn of the southern prairies are nearly unable to cope with snow cover. They rarely attempt to dig through the snow to reach their food and must move to where the food is actually visible. If snow cover persists for an unusual length of time in their range, massive starvation and death can result. Because pronghorn have short legs, walking long distances is a major problem and so they are often forced to use roads for travel. Vehicle traffic then kills many of these unique animals. Their range therefore is limited to regions with little or no snow. Animals that avoid snowy regions are called **chionophobes** (literally, "snow haters").

A number of animals **hibernate** during the most severe part of the winter. Skunks, chipmunks, some ground squirrels, and marmots (groundhogs) enter a period of low metabolic activity (low body temperature, slow heart and breathing rates) that may extend for six months or more. The arctic ground squirrel (Sic-Sic) hibernates for as long as 10 months. These animals enter their hibernation periods at various times before winter. Most of their nesting areas are in burrows below ground, which help protect them from the winter temperatures. Some will come out of hibernation for brief periods during winter to feed on food stores that were built up during the summer. While "sleeping," they rely on stored fat, particularly a high-energy form called brown adipose (fat) tissue, to maintain their body functions.

Although bears enter their dens by early winter, they are not true hibernators. They may sleep for short periods

but with no appreciable change in heartbeat, breathing rate, or body temperature. Some bears have been known to burst from their caves within a few seconds of being disturbed, even in midwinter. In pleasant weather they may roam the areas near their dens. In the Mountain National Parks their activity has often forced the closure of cross-country ski trails. Polar bears are exceptionally well adapted to winter and are active the year round.

A number of insects utilize their life cycle to survive the winter. In the fall they lay their eggs, which then remain dormant until spring conditions trigger their hatching. The adult insects usually do not survive the cold. A few insects overwinter in the pupa stage.

When ice forms, the normal result is increased tissue destruction due to the growth of ice crystals within the cytoplasm of the cells. However, some organisms have been known to survive the cold of winter by freezing solid. Animals that can survive this freezing process include many insects of the arctic and subarctic regions, and animals such as barnacles and mussels of the tidal zone of the northern oceans. Apparently these species can produce an antifreeze mixture of proteins and glycerol that is effective in lowering their freezing point.

The **supercooling** of water also plays a role in organism survival. In this process water can be gradually cooled down to as low as –41°C before it changes to ice. The larvae of *Bracon cephi*, an insect parasite of the Canadian wheatstem sawfly, can survive a temperature of –47°C. It produces an antifreeze—a concentrated glycerol solution—that lowers its freezing point to about −17°C. In addition, through supercooling, the water molecules in the tissues produce ice crystals so slowly that an additional 30°C drop can be reached before tissue damage results.

Recent research has discovered a number of amphibians and reptiles that freeze solid when air and soil temperatures drop below freezing, then thaw in spring and continue on with their lives. Some can even survive several freeze-thaw sequences during the winter. Forest floor frogs, such as the spring peeper, the wood frog, and the striped chorus frog, regularly survive freezing. Young painted turtles, which hatch in late summer, remain in their nests and freeze solid when the ground temperatures fall below −3°C.

Chioneuphores

Animals whose ranges extend into regions where winter snow and cold temperatures are the dominant climatic conditions are called **chioneuphores.** They are capable of surviving in snowy regions, but may also be found in other regions less affected by winter. By remaining in the winter environment they are faced with a reduction in their food supply. Because cold temperatures cause the animals to lose heat more rapidly, they require additional food in order to maintain their normal body temperatures. Energy-intensive activities, such as wading through snow, put additional stress on animals. Some waders, such as the fox, coyote, and wolf, step in their own footprints in the soft snow to reduce the energy requirements of travel.

Animals referred to as *waders*, such as deer, struggle to walk through snow. The deer usually herd together where there is food and protection, trampling down the snow and forming "yards." Moose, on the other hand, are able to walk through deep snow using a gait that involves stepping upward, lifting their hooves above the snow, then "post-holing" their way about. At snow depths greater than 60 cm, however, moose must start wading through the snow, using up precious energy.

Since snow will cover food sources found at the ground level, many of the larger-hoofed animals have the ability and instinct to dig for food, forming craters in the snow. This activity may also expose food sources for other animals. Grouse often depend on food uncovered by deer. However, if the wind has hard-packed the snow or if sleet has formed a thick crust, animals may be unable to dig craters and therefore can starve. The effort required to walk through these snow conditions demands increased energy. The hard-packed surface snow can cut legs, causing blood loss and placing additional stress on the animals.

Animals such as the elk may be forced to change their diet in winter. Often the plants they consume in summer are no longer growing or are covered by deep snow and become unavailable. Elk switch from summer **grazing** to winter **browsing,** feeding on buds, twigs, and the bark of trees and shrubs. Bite marks leave characteristic scars on the trunks of aspen poplar trees where elk have stripped bark in times of low food supplies. Elk, deer, and other animals chew a cud of vegetation prior to digesting the material. Bacteria in the digestive system help the digestive process and act on specific foods. However, the bacteria must change with the diet. This is why, when hay and other foods are brought to deer herds that are trapped by a snowstorm, the starving animals continue to die off—they no longer have "summer-type" bacteria capable of breaking down the food.

The lynx, snowshoe hare, ptarmigan, and ruffed grouse are called *floaters*. They have developed adaptations that enable them to walk or "float" on the snow sur-

Figure 16.16

In winter, snow covers the food sources of deer.

Figure 16.17

Elk winter diet. Bite marks leave scars on aspen trees.

Chioneuphores *are species of animals that can withstand winters with snow and cold temperatures.*

Grazing *is feeding on grass or grasslike vegetation.*

Browsing *is feeding on leaves, twigs, buds, bark, and similar vegetation.*

face. In each case, the feet enlarge in the fall and develop excess hair or feathers. This spreads the overall mass of the animal over a larger foot area, sufficient to keep them from sinking in the snow. If the snow is too soft to support them, hares make trails that they and other animals follow regularly, keeping the energy demands of travel to a minimum.

a)

b)

c)

Figure 16.18

(a) Ruffed grouse.
(b) Willow ptarmigan. (c)Lynx.
(d) Snowshoe or varying hare.

Chionophiles (snow lovers) are animals whose ranges lie within regions of long, cold winters.

d)

The ruffed grouse of the boreal forest has a unique method of surviving the intense cold nights of winter. It will fly directly into the soft snow and wiggle its wings and legs to create a tunnel. Snow then fills the tunnel and completely covers the animal. This behavior allows the birds to take advantage of the warmer temperature in the snow pack. In addition, the food it has eaten, usually frozen buds from trees and shrubs, requires less body heat to thaw and to be digested.

Winter winds pose additional problems for animals. Air moving over the surface of the skin removes heat at a rapid rate. The higher the wind speed, the greater the loss of heat from the skin. The windchill factor, combined with low temperatures, puts extra stress on organisms. Animal herding is one way of reducing the effects of the winds.

Snow can be beneficial to animals that live on its surface. The snowshoe (varying) hare is an excellent example. The winter diet of these animals consists of the buds and branch tips of a variety of shrubs. If the hare population is high, it takes little time for the most nutritious buds to be eaten. Remember, however, that the snowshoe hare is a floater. Each snowfall increases the height of the snow pack. As a result, the hares have access to the upper regions of the shrubs, where a new supply of food awaits them. Eating terminal buds can also trigger lateral branch growth the following summer, providing an even greater food source for the next winter. In addition, the weight of the snow on shrubs and trees can lower the branches, allowing animals to feed.

Chionophiles

Chionophiles are animals whose ranges are usually limited to climatic regions dominated by long, cold winters. Characteristic species include the polar bear, lemmings, musk-ox, barren-ground caribou, arctic fox, arctic hare, and a variety of small rodents and insectivorous mammals. Also referred to as "snow lovers," the chionophiles are so well adapted to winter conditions that they have limited success in southern locations.

- Chionophobes avoid winter if possible; chion-euphores can live in snowy regions; chionophiles must live in regions dominated by long, cold winters.

- To conserve heat, the most northerly mammals tend to be large (reduced surface area–to–volume ratio), are usually covered with a dense winter fur, and have small projecting body structures (e.g., ears).

APPLYING THE CONCEPTS

1 Describe how a water-treatment plant slows down eutrophication.
2 Following a late spring, the ice cover on a lake lasts an extra four weeks past normal breakup. Describe the possible effects on the lake ecosystem.
3 Explain why certain species of animals are used as markers for determining the oxygen levels of lakes. (For example, many leeches are found along the bottoms of lakes with low levels of oxygen; stonefly larvae are indicators of higher levels of oxygen.)
4 How does an increased number of decomposers affect the life of a lake or pond?
5 Explain why eastern lakes are more seriously affected by acid rain than their western counter-parts.
6 After a wind, many of the projections break from snow crystals. Explain how winds affect the insulating value of snow.

7 Does your paper carrier walk across your lawn? Explain why the grass often appears dead along the paper carrier's path, even in early spring. Why does the grass next to the path appear much healthier?
8 A conservationist feeds grain to a herd of starving deer during a particularly difficult winter. Discuss the implications of such practices.
9 Considering the low levels of solar radiation available for heating in the subarctic winter, would a change of coat color from dark to white really be an advantage to an animal such as the snowshoe (varying) hare? Explain.
10 Outline how activities such as cross-country skiing and snowshoeing can affect snow ecosystems.

CRITICAL-THINKING QUESTIONS

1 Discuss the various roles that detritus can play in aquatic ecosystems. Why have some cottage towns insisted that their residents equip cottages with sewage-holding tanks?
2 If humans stopped interfering with lake ecosystems, would eutrophication be stopped? Explain.
3 Thermal pollution has significant effects on Canadian lakes and ponds and the communities associated with them.

a) Why are Canadian standing waters particularly affected?
b) Comment on specific effects of thermal pollution.
4 An early autumn storm dumps 30 cm of heavy, wet snow on a section of the boreal forest prior to leaf fall. What effects, both positive and negative, might the storm have on this section of the forest?

ENRICHMENT ACTIVITIES

1 The barren-land caribou population has dropped significantly over the last 80 to 90 years. Research this subspecies of caribou and discuss the reasons for this negative population growth, the status of the population today, and ways in which the herds can be protected from destruction.
2 Build a small model of a quinzhee. Record the temperature inside and outside of the quinzhee.

CHAPTER 17

*P*opulations and Communities

IMPORTANCE OF POPULATIONS AND COMMUNITIES

*A **population** is a group of individuals of the same species occupying the same area at a given time.*

Twice a year the sounds of the greater snow geese can be heard near Cap-Tourmente, Quebec. Cap-Tourmente is one of the most important staging areas for the geese during their annual migration to their breeding grounds in the high arctic and their wintering areas in the coastal marshes of the eastern United States. The flats along the shoreline turn white as thousands of these birds spend days probing for the roots of aquatic plants. Up to 80 000 geese can be seen feeding at any one time, and visitors from around the world come to view this spectacular scene.

Several questions probably come to mind when you look at this picture of snow geese. Why do they return to Cap-Tourmente? Are the geese endangered? If so, do they require special protection? Is there enough food to support so many geese and the other species of water-fowl living in the same area? To answer these questions and others you must examine the biosphere in terms of populations and communities.

Figure 17.1

Snow geese migration, Cap-Tourmente Wildlife Sanctuary.

A **population** refers to all of the individuals of the same species living in the same place at a certain time. Although individuals from a population can be studied to gain information about life span, food preferences, and reproductive cycle, this tells little about the entire population. Information about competition among members of the same species for food, territory, and reproduction requires a study of populations. Since changes in a population are closely linked with predators, parasites, and food sources, population studies also provide information about relationships among different organisms.

When you investigate how two or more populations interact, you are studying a **community.** A community includes all the species that occupy a given area. The study of a community involves only the organisms, whereas the study of an ecosystem includes both the biotic (living) and abiotic (nonliving) components of a specific area. In spite of this difference between community and ecosystem, it is virtually impossible to examine the structure and activities of any community without some reference to the abiotic factors that may influence its populations.

By comparing population data from a wide variety of organisms, ecologists are able to make a number of generalizations. They have discovered that certain principles govern the growth and stability of populations over time. Analysis of the data allows them to predict what might happen to a species if the environment is altered or a new species is introduced. Ecologists can also use the data to describe how a population interacts with another (interspecific competition) and how individual members of one species relate to each other (intraspecific competition).

In this chapter you will examine the ecological factors that influence the distribution, size, density, and growth of populations.

HABITATS, GEOGRAPHIC RANGE, AND THE ECOLOGICAL NICHE

To understand some of the terms used in population studies, consider the northern flying squirrel (*Glaucomys sabrinus*). The flying squirrel has a flap of loose skin extending between the ankles of its front and hind legs on each side of its body. It is able to glide from tree to tree by extending its front and back legs so that the loose skin forms a kind of parachute. The flying squirrel can make gliding turns and spirals by moving its front legs and thereby changing the shape of the skin flaps.

To learn more about this unique animal, an ecologist would have to know the flying squirrel's **geographic range.** This is a region, usually outlined on a map, where sightings of the animal have occurred. While helpful, the map does not tell you exactly where individual populations are found. The **habitat** of the squirrel is the place where it lives. It is usually determined by the environmental conditions under which the squirrel population has the best chance of survival. For example, a flying squirrel would more likely be spotted in a forest ecosystem than a grassland ecosystem in the same locale. The squirrel's habitat is limited by factors such as vegetation, soil conditions, and climate. In western Canada, its preferred habitat is the northern boreal forest, particularly where the trees are spaced about two to three meters apart. In eastern and Atlantic Canada, its preferred habitat is a mixedwoodland environment dominated by hemlock and yellow birch trees.

Ecologists also need to know the size of the squirrel population and to what extent the population is distributed throughout its range. Changes in the distribution of flying squirrels from season to season would also provide key infor-

Figure 17.2
A flying squirrel leaps from a branch.

*A **community** is made up of the populations of all organisms that occupy an area.*

*A **geographic range** is a region where a given organism is sighted.*

*A **habitat** is the physical area where a species lives.*

Figure 17.3

Geographic range map of the
northern flying squirrel.

DISTRIBUTION OF POPULATIONS

*Ecological niche refers
to the overall role of a
species in its environment.*

*__Clumped distribution__
occurs in aggregates. The
distribution of organisms
is affected by abiotic
factors.*

Clumped

Random

Uniform

Figure 17.4

Population patterns.

*__Sloughs__ are
depressions often filled
with stagnant water.*

Since habitat preference plays an important role in population distribution patterns, ecologists must take great care when determining the population size of any species in a region. Population patterns can be divided into three patterns: clumped, random, and uniform.

A **clumped distribution** occurs when individuals are grouped in patches or aggregations. Organisms are distributed according to a certain environmental factor. For example, in river valleys trees often grow only on the south slopes and grasses dominate the north slopes. The north slope receives direct sunlight, which evaporates moisture from the soil. The drier soils of the north slope cannot support trees. By contrast, the south slope of the valley receives less direct sunlight and the soil tends to hold more water. Figure 17.5 illustrates grassland and woodland communities on opposite sides of the valley. Another example of clumped distribution can be seen in the arrangement of trees around **sloughs** in grassland ecosystems. Some sloughs have trees growing at the water's edge, while others do not. Permanent sloughs have water throughout the summer, whereas temporary sloughs dry up during the summer and do not provide enough water for trees during late fall. As shown in Figure 17.6, the most important limiting factor that affects the distribution of trees along river banks and the edges of sloughs is the presence of water.

mation. Within a habitat, every population occupies an **ecological niche.** This term refers to the population's role in the community, including all the biotic and abiotic factors under which the species can successfully survive and reproduce. Not surprisingly, the niche is not something you can actually see. For example, the niche of a flying squirrel includes its specific feeding habits, its ability to produce a large number of offspring, and its capacity to serve as food for a variety of parasites and predators. The flying squirrel's body wastes enrich the soil, while its use of seeds as a primary food source helps in the distribution of many grasses. The flying squirrel interacts with a number of other organisms and influences several biotic factors in its community.

Objective

To study colonization by plants following the retreat of a glacier.

Procedure

1 Observe diagrams (i) and (ii) below. As a glacier retreats, it leaves a harsh, stony landscape behind.

a) Algae are often found in the snow and ice. Indicate what function they would serve in this ecosystem.

2 As the ice melts, the newly exposed minerals are carried away from the topsoil by streams. Some minerals collect in temporary ponds. Ice immediately below the surface soils also greatly reduces nitrogen fixation. Examine diagram (iii). This shows the first pioneer plants, the mountain avens.

b) Provide a possible explanation for the uneven distribution of the pioneer plants seen in diagram (iii).

c) The mountain avens contain nitrogen-fixing bacteria in their roots. What special advantage do the mountain avens gain from this mutualistic relationship?

3 The next pioneer species, shown in diagrams (iv) and (v), are deciduous shrubs called alders. The alders also benefit from a relationship with nitrogen-fixing bacteria.

d) Describe how the landscape has changed between diagrams (iv) and (v).

4 Eventually, hemlock and cottonwood begin to grow. Examine diagrams (vi) and (vii). Diagram (vi) shows an intermediary phase of succession dominated by cottonwood and hemlock, while (vii) shows a mature forest dominated by spruce trees. The climax community establishes itself approximately 80 years after a glacier has completely retreated.

e) Referring to diagram (vi), speculate why the cottonwood shrubs only begin to colonize the area after the mountain avens are established.

f) Compare diagram (iii) with diagram (vii). Which plant provides the greatest biomass?

Case-Study Application Questions

1 Compare the pioneer vegetation shown in diagram (iii), the intermediary vegetation shown in diagram (iv), and the vegetation of the climax community shown in diagram (vii).

a) Which community would support the greatest number of organisms? Give your reasons.

b) Which community would support the greatest diversity of organisms? Give your reasons.

2 Provide sample food webs for diagrams (iii) and (vii).

3 Briefly explain how the abiotic factors within the community change because of the succession of vegetation. ■

B I O L O G Y C L I P

Increasing acidity in lakes of the Laurentian Shield region can result in "succession in reverse." Fish species lose their ability to reproduce. In an Ontario study all five species found in one lake ceased to reproduce between 1979 (pH = 5.6) and 1982 (pH = 5.1). They did so in sequence from the least to the most acid-tolerant species.

INVESTIGATION

CAREER CAREER CAREER

No living organism is completely independent. We all interact with other organisms and with the abiotic environment. If we are to continue to live on this planet, we must understand and preserve the organisms we share it with.

Fishing boat operator

Fish are an important source of food for people around the world. Overfishing can result in a severe depletion of fish stocks, causing loss of employment and production. Scientific management of fish stocks is necessary if we are to continue to enjoy this valuable resource.

Ski resort operator

Ski hills must be designed not only to provide an exciting ski run for skiers but also to have a minimum impact on the area's ecology. The interactions between the local organisms must be studied and evaluated before construction begins.

Farm manager

On a large farm, a farm manager may be required to supervise up to 100 full-time farm workers. The manager must coordinate production, purchasing, and marketing. Farm managers must keep up to date on new technological developments affecting crops, growing conditions, and plant and animal diseases.

Community planner

Community planners analyze the demographic, economic, and social characteristics of municipalities or cities. They are responsible for setting zoning restrictions, designing transportation routes, and preventing land erosion and pollution. A community planner must attempt to preserve the environment while providing a safe, clean, comfortable area for people to live in.

- Identify a career associated with change in populations and communities.
- Investigate and list the features that appeal to you about this career. Make another list of features that you find less attractive about this career.
- Which high-school subjects are required for this career? Is a post-secondary degree required?
- Survey the newspapers in your area for job opportunities in this career.

SOCIAL ISSUE:
Forest Fires and Ecology

Over 9000 forest fires occur in Canada every year. About 65% of these fires are caused by humans and the remainder by lightning. Fire is an important component of ecosystems. Fire keeps forests from taking over grasslands. Trees such as the black spruce, jack pine, and white birch are all adapted to regenerate after fire. Other species, including the white pine and Douglas fir, require ground that has been prepared by fire in order to reproduce properly. Some trees, such as the jack pine, require fire to reproduce.

Statement:

Naturally occurring forest fires should not be fought unless they endanger lives.

Point

- Each year, more than $250 million is spent fighting forest fires. If these fires were allowed to run their course, the dollar savings would be considerable.

- Some trees, such as the jack pine, require fire to reproduce. Without fire, these species would eventually disappear.

> **Research the issue.**
> **Reflect on your findings.**
> **Discuss the various viewpoints with others.**
> **Prepare for the class debate.**

Counterpoint

- Forest fires annually account for millions of dollars in damage to commercial forests. In addition, many animals and birds are destroyed or displaced by forest fires.
- Although fire causes succession, many other factors in nature ensure change. By stopping fires, succession will not be prevented, only redirected.

CHAPTER HIGHLIGHTS

- A population is a group of individuals of the same species that occupy the same area.
- Community refers to the populations of all organisms that occupy an area.
- An ecological niche is the overall role of a species in its environment.
- Dynamic equilibrium refers to any condition within the biosphere that remains stable within fluctuating limits.
- Carrying capacity refers to the supply of resources (including nutrients, energy, and space) that can sustain a population in an ecosystem.
- Biotic potential is the number of offspring that could be produced by a species under ideal conditions.
- Environmental resistance includes all the factors that tend to reduce population numbers.

- Gause's principle states that no two species can occupy the same ecological niche without one being reduced.
- Interspecific competition involves competition among different species; intraspecific competition involves competition within the ecological niche among members of the same species.
- Symbiosis is a relationship in which two different organisms live in a close association.
- Succession is the slow, orderly, progressive replacement of one community by another during the development of vegetation in any area. A climax community is the final, relatively stable community reached during successional stages.

APPLYING THE CONCEPTS

1 Calculate the rate of change in a moose population using the following data:
1978: 25 moose; area surveyed = 40 ha
1984: 11 moose; area surveyed = 30 ha

2 Plot a population curve from the following data, which were obtained after the introduction of deer mice on an isolated hillside.

   a) Is the curve characteristic of an open or closed population? Why?

   b) Using a line graph, show changes in the population between 1979 and 1990.

   c) Use a red line to indicate the probable position of the carrying capacity of the ecosystem. Why have you placed it where you did?

Date	Numbers
1979	20
1980	20
1981	22
1982	26
1983	25
1984	28
1985	40
1986	80
1987	130
1988	128
1989	133
1990	132

3 Using the data provided above, calculate the growth rate between 1986 and 1987. Identify factors that might account for the accelerated growth rate.

4 Explain how the carrying capacity of an ecosystem is related to the biotic potential of a species and its environmental resistance.

5 Predict what might happen to a population of moose if their numbers exceeded the carrying capacity of their environment. Provide the reasons for your prediction.

6 Design a research study that would supply you with data on the effect of latitude on the population cycle of the common deer mouse.

7 In what ways is Shelford's law a more appropriate principle in population studies than the law of the minimum?

8 Draw a population histogram from the following data on the white-tailed deer.

Age	Males	Females
1	72	75
2	35	33
3	24	25
4	17	15
5	14	11
6	8	9
7	7	6
8	5	5
9	4	3
10	2	3

What information is provided by this histogram? (Note: The few animals over 10 years of age are not included.)

CRITICAL-THINKING QUESTIONS

1 How could habitat preference influence the results of a population density study? Is it possible to account for this factor when designing a population study? Explain clearly.

2 Compare the concept of dynamic equilibrium in an ecosystem with homeostasis in the body of an advanced animal.

ENRICHMENT ACTIVITIES

Suggested reading:
- Gordon, Anita, and David Suzuki. *It's a Matter of Survival*. Toronto: Stoddart Publishing, 1990.

- Mungall, Constance, and Digby McLaren. *Planet under Stress*. Toronto: Oxford University Press, 1991.

FRONTIERS OF TECHNOLOGY: POISON-EATING MICROBES

A toxic wood preservative, called pentachlorophenol, has been known to seep from storage containers, contaminating nearby soil and underground water. The cost of cleaning the soil by traditional means of excavation and incineration usually ranges between $200 and $300 per cubic meter. In some cases, hundreds of tonnes of soil must be excavated. However, a microbe from the genus *Flavobacterium* has proved to be a dramatically economical and efficient alternative to these traditional methods. Tiny microbes dismantle complex toxic penta molecules, leaving nontoxic carbon dioxide, water, and harmless chlorides. The microbes work cheaply, requiring only oxygen and nutrients normally found in the soil. One biotechnology company quotes between $30 and $50 per cubic meter of soil for cleanup using microbes. Another advantage of poison-eating microbes is that the microbe population grows as long as penta molecules remain; once the toxin has been removed, the microbe dies.

Gene-splicing techniques can greatly enhance the ability of microbes to destroy pollutants. Combining genes from different organisms creates a multitude of possibilities. Super-toxin-digesting microbes can be produced by selecting genetic information that promotes destruction of the toxin while accelerating the rate of growth and reproduction of the microbe. Some environmental spills require different kinds of microbes used in tandem. For example, three types of bacteria can be employed to eliminate three components of gasoline: benzene, xylene, and toluene. The combination approach has been used with success in breaking down harmful PCBs (polychlorinated biphenyls), which were once used in hydraulic and electrical systems.

> **BIOLOGY CLIP**
>
> In a type of bacteria known as magnetobacteria, a chain of cytoplasmic, magnetic particles act as compasses. These so-called compasses help animals determine direction.

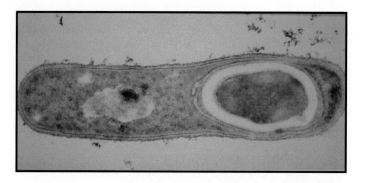

Figure 18.11

Electron micrograph showing an endospore.

■ REVIEW QUESTIONS ?

11 What features are shared by most microorganisms?

12 How are monerans similar to and different from viruses?

13 What feature(s) might cause cyanobacteria to be classified as plants by some biologists?

14 Draw and classify three types of bacteria according to shape.

15 Why is endospore formation important to bacteria?

16 Describe the binary fission process in monerans.

17 What is conjugation in monerans? Why is it important?

CASE STUDY

FOLLOWING AN INFECTION

Objective

To evaluate an experimental design. To analyze data and draw conclusions.

Background Information

Two women buy mascara, a cosmetic applied to the eyelashes. One develops an eye infection while using the mascara. They decide to test the mascara to determine if it might have caused the infection. You will be asked to evaluate the experimental method that the two women used. Later you will be asked to analyze the data they gathered and to draw conclusions.

Procedure

1 Two new containers of the same brand of mascara were purchased. One of the mascara samples was labeled A and the other B.
2 A sterile swab was smeared across the eyelid of one of the women, and then placed in sample B. After the swab was introduced, sample B was closed immediately. The procedure took approximately 20 s. Sample A was opened and then closed after 20 s, but no swab was placed in the sample.
   a) What purpose does sample A serve?
   b) Why did the experimenters swab the eyelid of one of the women?
3 Bacteria populations were sampled every day for the next two weeks. The results are shown in the graph below.

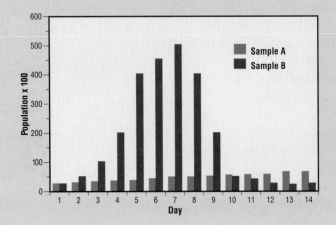

c) Which bacterial population grew the fastest in the mascara?
d) On which day was the population greatest in sample A? In sample B?
e) What was the population size of bacteria found on day 7 in sample A? In sample B ?
f) Where did the bacteria in the samples come from?

Case-Study Application Questions

1 *Conclusion #1:* The experimenters concluded that the mascara had many essential nutrients that promoted the growth of the bacteria. Do you agree with the experimenter's conclusions? Provide reasons that will support your evaluation.
2 *Conclusion #2:* The experimenters concluded that the bacteria that grew on the skin reproduced faster than those found in the air. Do you agree with the experimenter's conclusions? Provide reasons to support your evaluation.
3 The experimenters disagreed on the reason why the population of bacteria began to decline in sample B after day 7. Read both conclusions carefully. You will be asked to evaluate them.
   a) One of the experimenters concluded that the mascara was capable of supporting a maximum number of bacteria. Once the population of bacteria reached 50 000 per gram of mascara, the population began to decline because many of the nutrients had been used up. Do you agree with the experimenter's conclusions? Provide reasons to support your evaluations. What additional information would you want to collect before accepting this experimenter's conclusion?
   b) The second experimenter concluded that the population in sample B decreased because the overpopulated community produced an enormous amount of wastes. Eventually the wastes began to kill many bacteria. As poisonous wastes began to accumulate, the population of bacteria began to decline. Do you agree with the experimenter's conclusions? Provide reasons to support your evaluation. What additional information would you want to collect before accepting this experimenter's conclusion? ■

KINGDOM—PROTISTA

Protists first appeared in the fossil record about 1.5 billion years ago. Although an ancient group, they are more recent than monerans and demonstrate an important evolutionary advancement—a discrete, membrane-bound nucleus. For that reason protists are called eukaryotic cells. In addition, they contain organelles, such as ribosomes, mitochondria, and lysosomes. These structures provide a more efficient method of using available nutrients and carrying out metabolic activities.

Most protists are microscopic, unicellular, and found in both fresh and salt water. Plankton, tiny floating organisms that include protists, are important producers and consumers in aquatic food chains. The animal-like forms, *zooplankton* protists, are heterotrophs. The *phytoplankton* protists are photosynthetic autotrophs. They produce a significant portion of the oxygen supply in the earth's atmosphere and play an important role in the carbon cycle.

Protist Diversity

What distinguishes protists from higher plants and animals? Although protists have a simple biological organization, biologists remain divided on the exact boundaries for this grouping. Although some multicellular protists exist, none form true tissues.

Protists were once considered to be "first animals" and were placed in a phylum called Protozoa. Recently, many biologists have argued that all single-celled eukaryotes should be placed in a separate kingdom—Protista. Today, the kingdom is made up of three distinct groups: plantlike protists, animal-like protists and funguslike protists. While variations are still used, this tri-level division has gained widespread acceptance.

Plantlike Protists

Traditionally, plantlike protists were referred to as algae. The term algae is often used to refer to all simple aquatic "plants." While still widely used, the term has no formal taxonomic significance. In modern classification schemes, algae are divided among three different kingdoms—Monera, Plantae, and Protista.

Table 18.4 Plantlike Protists

Phylum	Representative	Description	
1. Euglenophyta	Euglena	Euglenophytes live mainly in fresh water, and are particularly abundant in stagnant waters. The photosynthetic euglenids can also take in solid food, a trait commonly associated with animals. Euglenids, like many flagellated protists, reproduce asexually.	1.
2. Chrysophyta	Diatoms	This group is found in both fresh and salt water. They contain chlorophyll and are autotrophs. Many are flagellated and encased in shells or skeletons. Diatoms, the most abundant plantlike protist, are not flagellated but are encased in two thin silica valves or shells joined together.	2.
3. Pyrrophyta	Dinoflagellates	Dinoflagellates are autotrophs and contain chlorophyll and red pigments. They are important primary producers and a major component of the oceanic phytoplankton.	3.

Reproduction in protozoans is usually asexual, by fission. Protozoans may also reproduce sexually, which involves the fusion of gametes and the formation of a zygote. Under adverse conditions, protozoans may form resting cells called *cysts*.

Animal-like Protists

These organisms, known as protozoans, are heterotrophs. They must move about to obtain food. Most engulf their food (bacteria and other microbes), and are said to be **holozoic.** Others, called **saprozoic,** absorb predigested or soluble nutrients directly through the cell membranes. The remaining protozoans are either free-living or parasitic.

Holozoic *organisms feed in an animal-like manner.*

Saprozoic *organisms absorb nutrients directly through cell membranes.*

Table 18.5 Animal-Like Protists

	Phylum	Representatives	Descriptions
	1. Sarcodina	Amoeba, foraminiferans, radiolarians	Most sarcodines are free-living forms that thrive in fresh water, salt water, and soil. A few parasitic species are found in the intestines of animals. Many are motile, with adult forms possessing pseudopods ("false feet") for locomotion. Movement results from repeated cytoplasmic extension and retraction of the pseudopods. The amoeba feeds by having its pseudopods flow around and engulf food particles.
	2. Mastigophora	Flagellated protozoans	Flagellates move by means of flagella and are found in both fresh and salt water. Most are parasitic, and cause disease in animals. Reproduction is asexual, by longitudinal fission. There are no reported cases of sexual reproduction. Flagellates also form cysts, which is the way many parasitic forms are spread from host to host.
	3. Ciliophora	Ciliated protozoans	Ciliates, in contrast to flagellates, are considered to be the most complex and advanced of the protozoans. They are characterized by the presence of hairlike structures called cilia. Cilia are synchronized for swimming in free-moving organisms like paramecia. Typically, reproduction is asexual by binary fission; however, the organisms may also reproduce sexually by conjugation.
	4. Sporozoa	Plasmodium	Sporozoans lack means of independent locomotion. Sporozoans, and their relatives, are exclusively parasitic, and depend entirely upon the body fluids of their hosts for movement. They are characterized by a sporelike stage. Sporozoa display complex life cycles.

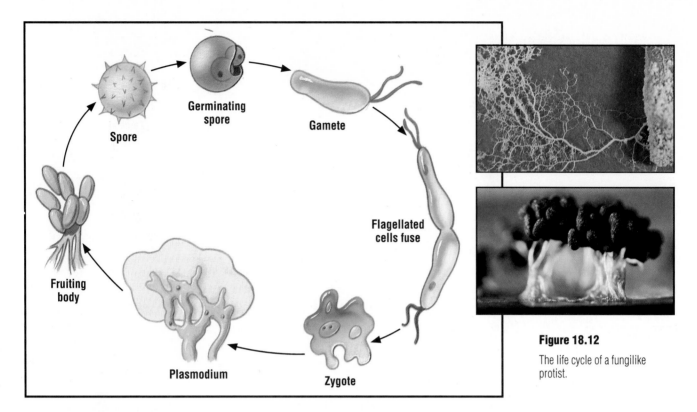

Figure 18.12

The life cycle of a fungilike protist.

Fungilike Protists

Fungilike protists, also referred to as slime molds, are placed in the phylum Gymnomycota. Slime molds prefer cool, shady, moist places and are usually found under fallen leaves or on rotting logs. The name is derived from the slimy trail left behind as the mold moves over the ground.

Categorizing unusual organisms points out the problems associated with classifying members of the kingdom Protista. During some stage of their life cycle, slime molds resemble protozoans and become amoebalike or have flagella. At other times they produce spores much like fungi do.

■ REVIEW QUESTIONS ■ ?

18 How do moneran and protist cells differ?

19 What three groups make up the kingdom Protista?

20 What characteristic differentiates plantlike protists from animal-like protists?

21 Copy and complete the following chart by placing a check mark (√) in the appropriate column.

Phylum	Autotrophic nutrition	Heterotrophic nutrition	Flagellum	Cilia	Sexual reproduction
Euglenophyta					
Chrysophyta					
Pyrrophyta					
Sarcodina					
Mastigophora					
Ciliophora					
Sporozoa					

22 Why are slime molds included in the kingdom Protista?

RESEARCH IN CANADA

Blood Parasites

Much of the research into sporozoan-caused diseases like malaria involves studying nonhuman species such as birds and small mammals. This work is often coordinated in special laboratories or centers that have access to a wide range of expertise, equipment, and information. One such center is the International Reference Centre for Avian Haematozoa (IRCAH), originally located at Memorial University of Newfoundland. As the name avian haematozoa implies, the center specialized in the study of blood parasites of birds.

Since 1969, Dr. G.F. Bennett, a Canadian entomologist born and raised in India, directed the work at the facility, originally established under the sponsorship of the World Health

Organization. The center has accumulated over 120 000 indexed reference collections of bird blood parasites and contains articles representing 99% of the total material on the subject. Bennett and IRCAH have attracted some of the world's leading malariologists to conduct research at IRCAH. Bennett has researched bird-biting insects in Ontario, studied simian (monkey) malaria in Malaysia, and also investigated blood-sucking cattle ticks in Australia. He has personally acted as an experimental host for two species of simian malaria.

Bennett and his colleagues are dedicated to clarifying the connection between bird blood parasites, their vectors, and hosts. This has required a greater understanding of the biology and life cycles of a number of sporozoan diseases. Other projects at IRCAH include studies on environmental impact on birds and the development of biological control agents for insect pests.

DR. G. F. BENNETT

LABORATORY

CULTIVATION AND EXAMINATION OF PROTISTS

Objective

To study the characteristics of protists.

Background Information

Protists are widely distributed in natural environments where water is present. Cultivated pond water and hay infusions provide opportunities to examine protists and other microorganisms.

Materials

hay infusion and/or pond water culture
compound microscope
microscope slides (2)
cover slips (2)
lens paper
paper towel
1.5% methyl cellulose solution
500 mL or larger beakers or jars (2)
eye dropper or dropping pipette
protozoan field guide

Note: If cultures are to be prepared by students, it is important to follow the teacher's instructions carefully.

Procedure

1 Using the dropping pipette, carefully obtain some liquid from the surface of the pond water culture or hay infusion.
2 Place a small drop of this specimen on a slide and cover with a cover slip.
3 Examine the preparation under low-power and high-power magnification. Pay particular attention to the comparative sizes of the microorganisms as well as their shapes, structures, and movements. If the protists are moving too quickly, add a drop of methyl cellulose solution. This will slow their movement without killing them.
4 Obtain a bottom sample and repeat steps 2 and 3.
5 If both cultures are used, repeat the same procedure.
   a) Record your observations by making drawings of the microorganisms seen in both the surface and bottom samples.
   b) What kinds of protists could you identify in the samples observed?

Laboratory Application Questions

1 Were you able to determine any of the cell structures? If so, label them in your drawing.
2 Was there any difference in the numbers and types of organisms found in the surface and bottom specimens? Were there any differences within the two cultures? What might account for this?
3 Which method of locomotion did most protists use? Might this indicate that one type is more efficient than another? Explain.
4 When would low-power magnification be better than high-power magnification? ■

SOCIAL ISSUE:
Microbial Mining

Using microbes to extract desired metals like copper, zinc, and gold from the earth is not a new procedure. The Spanish utilized bacteria to extract copper almost 300 years ago. In 1990 a gold-mining plant went into operation in Colorado to extract gold embedded in iron sulfide ores. It uses Thiobacillus ferroxidans, *a naturally occurring bacterium.*
With modern biotechnology extraction, an ore body is drilled, fractured, and then inoculated with either natural or genetically engineered bacteria to draw out the desired metal.

Statement:

Future microbial mining efforts must undergo thorough assessments to determine the potential risks to living organisms and to the environment.

Point

- Traditional methods employed naturally occurring bacteria and other microbes. Genetically engineered organisms may lead to the production of "super-bugs" with side effects that may be hazardous to life and to the environment.

- The current microbial mining process is slow and mainly effective with low-grade ores. In some cases, it will take decades to extract the same amount of material as with conventional means.

Counterpoint

- Microbial mining is essentially a biological process. Compared with conventional mining techniques, which often use hazardous chemicals to extract metals, microbes are generally safe and environmentally friendly. Genetically engineered microbes may also have additional advantages such as cleaning up hazardous wastes.

- Mining with microbes requires little energy input when compared with conventional mining techniques. It is therefore more cost-effective than traditional means and could cut the cost of precious metal production by up to 40 to 50%.

Research the issue.
Reflect upon your findings.
Discuss the various viewpoints with others.
Prepare for the class debate.

CHAPTER HIGHLIGHTS

- Classification is a process of grouping things based on their similarities. Taxonomy is the science of classifying living things. Living organisms are organized around seven levels of classification within a five-kingdom system.
- The binomial system uses two names for classifying organisms: genus and species.
- Phylogenetic relationships are the basis of modern taxonomy.

- Viruses are unique biological "particles" that consist of a nucleic acid core and a protein capsid.
- Monerans are prokaryotic and have a wide range of metabolic diversity.
- Monerans display a variety of modes of nutrition. They reproduce mainly by binary fission.
- Protista includes three single-celled eukaryote groups: plantlike, animal-like, and fungilike protists.
- Microorganism phylogeny is based largely on indirect evidence and cautious speculation.

APPLYING THE CONCEPTS

1 How was the introduction of classification keys a major contribution to taxonomy?
2 Why is phylogeny sometimes called the foundation of taxonomy?
3 Design a dichotomous key for the animal-like protists shown in Table 18.5.
4 Certain bacteria are credited with establishing the oxygen component of the earth's atmosphere. Explain how the evolution of other life forms might have been different without these bacteria.
5 Identify situations in which a bacteria-free environment is desirable and those situations in which it would not be desirable.

6 Would you expect all students to observe exactly the same shape when they examine a live culture of amoebas under a microscope? Explain.
7 Why are organisms belonging to Monera considered to be more primitive than those belonging to Protista?
8 Unlike the higher plants, the plantlike protists do not have roots, stems, or leaves. Explain why the plantlike protists do not require these features.
9 Explain why some unicellular protists are considered to be complex organisms. Give examples.
10 Provide examples of holozoic and sporozoic protists.

CRITICAL-THINKING QUESTIONS

1 Preventing and controlling bacterial growth is a major concern within the food industry. Suggest some advantages and disadvantages of the following food-preservation methods: freezing, canning, dehydration, pickling, and salting. Can you name some other methods?
2 In prehistoric times, the earth contained little or no free atmospheric oxygen. Using information gained in this chapter, draw a type of bacteria that would suit that environment well. What organelles would be required? Explain.
3 Early scientists strongly believed that species were fixed and unchanging. Is it possible that Linnaeus, by using similar structures as the basis of his classification, was unknowingly supporting the idea of evolutionary relationships? Explain.

4 It has been suggested that sexual reproduction became the dominant type of reproduction among organisms because it had the inherent advantage of variability. This mechanism would have enabled species to adjust to changing conditions. Would you agree or disagree with this reasoning? Explain.
5 Could you accept the hypothesis that viruses were the precursors of life on this planet? Provide reasons for your answer.

ENRICHMENT ACTIVITIES

1 Make collections of local plant leaves or aquatic invertebrates and use classification keys to identify the organisms.
2 Visit a local botanist, plant store owner, or veterinarian to learn more about viral diseases. Prepare a paper and report your findings to the class.

3 Visit or write a food processing plant and/or a sewage treatment plant. Investigate the role of microbes in these facilities and detail the measures taken to safeguard public health.

*F*ungi and Plants

IMPORTANCE OF FUNGI AND PLANTS

Biologists who study fungi are called *mycologists*. Fungi, the decomposers, are part of nature's recycling system. They can break down almost any organic compound including wood, food crops, and plastics. Many fungi benefit humans. Yeast is used to make bread, wine, and beer; penicillin produces antibiotics; *Aspergillus* is used to flavor soft drinks. Other fungi such as mushrooms, morels, and truffles are sought-after food items.

Scientists who study plants are called *botanists*. Plants are important as primary **producers.** Directly or indirectly they nourish almost all other living organisms. They account for much of the world's oxygen and food supply by converting water and carbon dioxide into glucose and oxygen. Plants make soil from bare rock and prevent soil erosion and flooding by holding soil together and regulating runoff. Plants form the basis for a number of industries, and provide essential components for many modern medicines such as Aspirin, morphine, and certain cancer drugs; plant fibers are used to manufacture cotton, rope, paper, and a variety of consumer products.

Producers *are organisms that are capable of making their own food.*

Figure 19.1

The importance of fungi and plants in paving the way for terrestrial organisms cannot be overstated. These two groups have helped shape the biosphere, as it presently exists.

Plant evolution contributed greatly to the establishment of the earth's *biodiversity*. Since the first green shoots appeared on the planet over 400 million years ago, a number of important evolutionary steps have occurred. Most notable have been adaptations to acquire, transport, and conserve water and the evolution of reproductive cycles for survival in terrestrial environments. The ancestors of these early land-dwellers had to make many adjustments for terrestrial living, an existence that was vastly different from the watery existence they left behind.

DIFFERENCES BETWEEN FUNGI AND PLANTS

Fungi were once classified as members of the plant kingdom. Fungi and plants do share certain characteristics: both are eukaryotes with numerous organelles; most have cell walls; often parts of their bodies are anchored in soil; and they are generally stationary and do not move about like animals.

However, enough differences exist between plants and fungi to place them in separate kingdoms. Fungi are not anchored in the ground by roots as are plants, they do not photosynthesize, and in most cases they do not contain cellulose in their cell walls. Lower plants, however, share an additional characteristic with fungi. Both can reproduce by means of spores. The higher plants reproduce by means of seeds.

KINGDOM—FUNGI

Fossils resembling fungi date back about 900 million years. About 300 million years ago, representatives of all modern fungi had evolved. Well before the Cambrian period (570 million years ago), fungi apparently took a different evolutionary pathway than plants. Their predecessors, like all existing fungi, had adapted to a heterotrophic way of life.

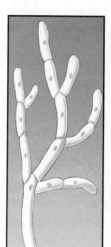

a)   b)   c)

Characteristics of Fungi
- The cells of fungi are eukaryotic.
- Digestion is extracellular (food is digested externally before being absorbed).
- In multicellular forms, food absorption takes place in the **mycelium,** a mesh of microscopic branching filaments. Each filament is called a **hypha** (plural: "hyphae").
- Most hyphae have cell walls that are reinforced with **chitin.**
- All fungi are heterotrophs.
- Reproduction can be asexual, sexual, or both.

Figure 19.2

(a) Hyphae shown with cross-walls.
(b) Hyphae without cross-walls.
(c) Mycelium showing many interlocking strands of hyphae.

Mycelium *is a collective term for the branching filaments that make up the part of a fungus not involved in sexual reproduction.*

Hyphae *are filamentous threads of a fungus.*

Chitin *is a nitrogen-containing polysaccharide, composed of long fibrous molecules, that forms structures of considerable mechanical strength.*

- Fungi are found in dark, warm, moist locations that are rich in organic matter.

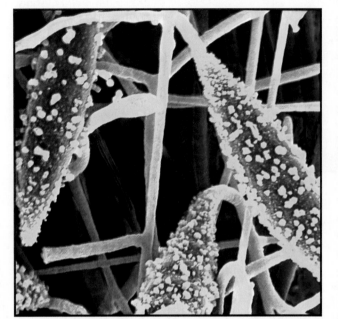

GENERAL LIFE CYCLE

Fungi reproduce both asexually and sexually but always produce spores—small reproductive cells that are dispersed by air currents. Spores have a **haploid chromosome number;** that is, they have one-half of the full complement of chromosomes. Spores have thick, resistant outer coverings to protect them from unfavorable conditions.

Typically, sexual reproduction begins when haploid cells, called *gametes*, from two mating strains undergo cytoplasmic fusion. The gametes are produced in specialized reproductive structures called **gametangia** (singular: "gametangium"). In some fungi the nuclei may not immediately fuse, but will divide independently. In mushrooms, this stage, called the **dikaryotic stage,** may last for generations. When nuclear fusion does occur, a zygote with a **diploid chromosome number,** or full complement of chromosomes, is formed. In most fungi the diploid stage is short-lived. The zygote undergoes nuclear division to form new haploid hyphae.

Asexual reproduction begins with the production of spores in specialized structures called **sporangia** (singular: "sporangium"). Spores have the potential to germinate into new hyphae. Fungi may also reproduce asexually by fragmentation, which is the breaking apart of the hyphae. Single-cell fungi, such as yeast, can reproduce asexually by budding or sexually by forming ascospores.

FUNGAL DIVERSITY

Most fungi, including mushrooms, club fungi, and puffballs, are multicellular, but some simple yeast forms exist as single cells. A recently described fungus from Washington State is claimed to be the world's largest living organism. *Armillaria ostoyae* covers approximately 600 ha and researchers think that still larger fungi may be found.

Fungi are adapted for two main functions: absorption of nutrients and reproduction. Unlike plants, the **vegetative** (nonreproductive) portion is usually below the surface. The only visible parts of a fungus, its reproductive structures, display a wide range of diversity.

CLASSIFICATION OF FUNGI

Classification of fungi may differ slightly, depending on the source. Table 19.1 lists some of the major classes, their common names, and preferred habitats.

*A **vegetative** structure is any part of a fungus or plant that is not involved in sexual reproduction.*

*A **saprophyte** is an organism that obtains its nutrients from dead or nonliving organic matter.*

*A **parasite** is an organism that lives in or on another organism from which it obtains its food.*

Table 19.1 Classification of Fungi

Division—Mastigomycota:	Produce flagellated (motile) spores; cellulose cell walls			
Hyphae with cross walls; **saprophytes;** some **parasites;** mainly aquatic; some terrestrial; sexual and asexual reproduction. Examples: chytrids and water molds				

Division—Amastigomycota:	Produce nonmotile spores; chitin in cell walls			
Class	**Examples**	**Habitat**	**Description**	**Reproduction**
Zygomycetes (zygospore-forming fungi)	Common molds (bread, dung molds)	Mainly terrestrial (soil, decaying plant matter) Some saprophytes	Few single-cell forms Mainly multicellular Cross-walls lacking in hyphae	Asexual spores Sexual reproduction—conjugation, zygospore (thick-walled structures that protect the zygote) formation
Ascomycetes (sac fungi)	Yeast, morels, truffles	Terrestrial and aquatic	Few single-cell forms Mainly multicellular Cross walls in hyphae Many are pathogens of plants	Asexual spores Sexual reproduction—Asci produce ascospores Short dikaryotic stage
Basidiomycetes (club fungi)	Mushrooms, shelf fungi	Mainly terrestrial	Mainly multicellular Cross walls in hyphae Many are pathogens	Asexual spores absent Sexual reproduction Basidia produce basidiospores Long dikaryotic stage

Division—Deuteromycota:	Imperfect fungi in which sexual reproduction is unknown			
Asexual reproduction by spore formation; some resemble sac fungi; others resemble club fungi; mainly terrestrial. Examples: predatory fungi, *Penicillium* and fungi that cause diseases (athlete's foot and yeast infections).				

LABORATORY

MOLD GROWTH ON FOODS

Objective

To culture and observe the germination of mold spores and examine the structure of molds.

Background Information

Molds reproduce by means of spores. Under the proper conditions mold will grow on food and can be observed after several days.

Materials

bread (without mold inhibitor)	hand lens or dissecting
fruit sections,	microscope
potatoes,	compound microscope
rice	forceps
petri dishes, baby food jars,	eye dropper
or plastic bags	paper towels
prepared microscope slides	

Procedure

1 Begin preparing the mold culture by lining the inside of the petri dishes with paper towels. Replace the cover immediately after placing the towel in the dish.
2 Moisten the foods using the eye droppers (do not soak). Allow the foods to stand for 20–30 min. If desired, scatter some dust on the food or wipe food over the bench surface to collect spores.
3 Place one food type on the wet paper towels in each dish. Tape the top and bottom of the dish together and do not open during the growth phase.
4 Store in a cool, dark location for 3–5 d.
5 Make daily observations.
   a) Record your observations in a table.
   b) Measure the diameter of one or two of the larger colonies to determine the growth rate over several days.

6 When the mold turns dark, or different colors become noticeable, examine with a hand lens or dissecting microscope. Identify the following types of hyphae: *sporangiophores*, which develop sporangia at their tips; *stolons*, which run horizontal to the surface on which the mold is growing; and *rhizoids,* which develop at the swellings along the stolons and anchor the fungus to the food source. Do not open the petri dish.

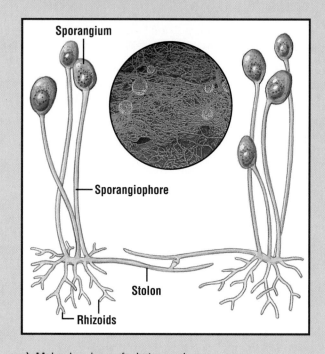

Sporangium
Sporangiophore
Stolon
Rhizoids

c) Make drawings of what you observe.

Note: Mold colors may sometimes indicate different mold types. For example, blue-green mold is *Penicillium*, brown mold is *Aspergillus*, black mold is *Rhizopus stolonifer*, and gray mold is *Mucor*.

7 Examine your slide under the microscope.
   d) Describe and diagram the color and arrangements of the spores.

Laboratory Application Questions

1 Did mold grow to the same extent on all foods? Explain.
2 Did all molds look alike and have the same color?
3 Were other organisms, besides mold, growing in the culture?
4 Why do bakers add mold inhibitors to baked goods?
5 What environmental conditions would you recommend to prevent mold growth on foods? ■

FUNGI SYMBIONTS

Fungi often benefit from symbiotic relationships with other organisms. Two important relationships are those between fungi and algae to form lichens, and between fungi and plant roots to form mycorrhizae. Some lichens that are familiar to many Canadians are the "old man's beard," "reindeer moss," and "British soldier." In the arctic tundra and boreal forests of northern Canada, lichens are an important source of food for reindeer, caribou, and other animals. They can be found on soil, rocks, and trees. Lichens are important in plant succession. By establishing themselves on rocks and in barren areas, lichens help form parent soil material.

Lichens are really "two organisms in one." The algal partner is generally a blue-green or a green alga, and the fungus is usually a sac fungus. In a lichen relationship the fungal mycelium surrounds the algal cells and provides them with carbon dioxide and water for photosynthesis. The mycelium also lends support to the entire organism. The autotrophic algae subsequently share their manufactured food with the fungi. Lichens reproduce asexually by fragmentation.

Mycorrhizae are distinct from lichens. The fungal component of mycorrhizae is usually a club fungus, which typically becomes associated with the roots of trees. They derive their name from the combination of plant structures and fungal hyphae. In this symbiotic relationship, the extensive fungal hyphae help the plant absorb nutrients. In return, the fungi obtain food that is made during photosynthesis. Some plant seeds will not germinate in the absence of the mycorrhizal fungi. For that reason environmental biologists often monitor mycorrhizal growth in contaminated soils.

Lichens—Air-Quality Monitors

Since the Industrial Revolution, researchers have recognized the value of lichens in detecting air pollutants. Unlike plants, which absorb water that has been filtered through the ground, lichens absorb water directly from the air. Thus, they absorb more dissolved toxic substances than do plants. However, as was the case with the radioactive fallout from the Chernobyl nuclear power-plant disaster, pinpointing contaminated areas usually occurs after environmental devastation has taken place. Often, the method is based on which lichens survive and which ones die.

More recently, lichenologists (scientists who study lichens) are attempting to determine the potential of lichens to monitor air quality before they appear damaged. They are hopeful that lichens may be able to measure a wide range of airborne contaminants.

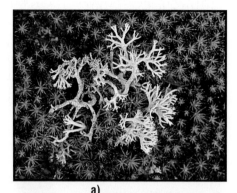

a)

b)

Figure 19.5

Lichens familiar to Canadians are (a) reindeer moss and (b) old man's beard.

■ REVIEW QUESTIONS ?

1 What are some similarities and differences between fungi and plants?
2 Describe the roles of fungi as decomposers and as parasites. Give examples.
3 What are mycelia and hyphae?
4 Refer to table 19.1. How are sac and club fungi different in their reproductive structures?
5 In what ways may fungi be useful or harmful to other living things?
6 Describe the role of fungi in lichens and mycorrhizae.
7 Why are lichens important?

KINGDOM—PLANTAE

The term plant is used to include organisms that share the following characteristics:

- ability to synthesize carbohydrates by photosynthesis;
- presence of cellulose cell walls;
- alternation of generations in their life cycles;
- lack of mobility.

PLANT DIVERSITY

The 300 000 to 500 000 species of identified plants display such diversity that no single species can be cited as a typical example of the kingdom. Consequently, scientists have devised a number of classification schemes for plants. This text recognizes three broad groups—aquatic plants, nonvascular plants, and vascular plants. This classification scheme is based on evolutionary relationships and adaptations that have occurred over millions of years.

As the environment changes, populations have to adapt to their new surroundings if they are to survive. In the case of land plants, the three primary adaptations were the evolution of structures and systems for the transport of water; roots, stems, and leaves; and a method of reproduction that did not depend on water for the dispersal of gametes. As you read ahead, notice how, as plants became more complex in their structure and function, they depend less on water for reproduction.

The stage of a plant that produces gametes is called a **gametophyte;** the stage that produces spores is a **sporophyte.** The evolutionary trend in plant reproduction was from a haploid (gametophyte) dominant plant to a diploid (sporophyte) dominant form. In simple plants like mosses, the sporophyte depends on the gametophyte for water and nutrients. The role is reversed in complex plants like ferns and seed plants.

AQUATIC PLANTS

The simplest of plants live surrounded by water because water satisfies many of their needs. It prevents cells from drying out, lends structural support to the plant, provides nutrients, and accommodates the dispersal of spores and the meeting of sex cells. Most aquatic plants belong to

Gametophyte *refers to a stage in a plant's life cycle in which cells have haploid nuclei. During this stage the sex cells (gametes) are produced.*

Sporophyte *refers to a stage in a plant's life cycle in which cells have diploid nuclei. During this stage spores are produced. This stage arises from the union of two haploid gametophytes.*

Figure 19.6

General life cycles of plants, showing an evolutionary trend from a haploid dominant plant to a diploid dominant form.

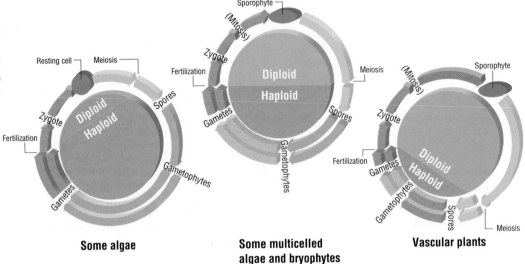

Some algae

Some multicelled algae and bryophytes

Vascular plants

Table 19.2 Classification of Aquatic Plants

	Chlorophyta (green algae)	Phaeophyta (brown algae)	Rhodophyta (red algae)
Examples	*Chlamydomonas* *Ulva* (sea lettuce) *Volvox* *Spirogyra*	*Fucus* (Rockweed) *Plumaria*	*Porphyra (laver)* Dulse Irish Moss
Pigments	Chlorophyll a, b Carotene	Chlorophyll a, c Fucoxanthin	Chlorophyll a, d Carotenes, Phycobilins
Food Storage	Starch	Laminarin, oils	Starch
Cell Wall	Cellulose	Cellulose	Cellulose
Habitat	Mainly fresh water Moist soils, coastal tropical seas	Mainly colder seawater	Mainly warmer seawater Some fresh water
Form	Single-cell to multicellular	Multicellular	Multicellular
Reproduction	Sexual reproduction alternation of generations Asexual by fragmentation or spores	Sexual reproduction alternation of generations Asexual by fragmentation or spores	Sexual reproduction alternation of generations Some vegetative reproduction

one of three divisions, commonly known as green, brown, and red algae, as shown in Table 19.2.

Most aquatic plants are multicellular, but some are unicellular, filamentous, or colonial. The multicelled body is called a **thallus** (plural: "thalli"), and lacks conductive tissue as well as true roots, stems, and leaves.

ADAPTATIONS FOR A TERRESTRIAL LIFESTYLE

Plants such as mosses, liverworts, and ferns, representatives of which still survive today, evolved over time and gave rise to the larger and more complex seed-bearing conifers and flowering plants that you know today.

*A **thallus** is an undifferentiated vegetative plant body without distinct roots, stems, and leaves.*

The two lineages of land plants are separated into nonvascular and vascular plants, as represented by a moss plant and a fir tree. The basis for this division is the presence or absence of **vascular tissue,** or conducting tissue, called **xylem** and **phloem.** The vascular system provides support and the transport of water and nutrients, and tends to develop only in diploid or sporophyte plants.

Nonvascular and vascular plants, while developing along separate evolutionary lines and demonstrating quite different adaptations to terrestrial living, have some striking similarities. Plants in both groups usually have a waxy protective **cuticle,** numerous **stomata** (singular: "stoma"), and complex reproductive structures.

NONVASCULAR "LAND" PLANTS

Nonvascular land plants, or *bryophytes*, are confined to moist habitats because they need water for sexual reproduction to occur. They live close to the ground and are abundant in wetlands, rain forests, and roadside ditches. They consist of three classes—mosses, liverworts, and hornworts. Bryophytes are small plants, generally less than 20 cm tall.

Bryophytes have leaflike, stemlike, and rootlike organs. Most species have *rhizoids*, rootlike filaments that anchor the plant to the substrate. The thin, leaflike structures, usually only one cell layer thick, are located atop a short main stalk. The "leaf" is used to absorb water and minerals and to perform photosynthesis. Water and nutrients move from cell to cell by diffusion.

Mosses, the most common bryophytes, hold large amounts of water. This feature explains their use in potting soils and as absorbants to clean up oil spills. *Sphagnum*, found mainly in peat bogs, is common throughout northern parts of the world including Canada. Dried peat moss has been used for insulation and as a source of fuel for hundreds of years.

Reproduction in mosses shows an interesting relationship between the sporophyte and gametophyte generations. Both generations are contained on the same plant, a condition that is unique to land plants. When moss spores mature, the sporangium breaks open and spores are released. As the spore germinates it grows into a filamentous structure called a **protonema,** which closely resembles the filamentous green algae. The protonema is the young, green gametophyte that forms the familiar mosses, often referred to as "carpets" of moss when widespread. Following several months of growth, the haploid gametophyte produces male and

Figure 19.7

A moss plant showing a ripe spore capsule.

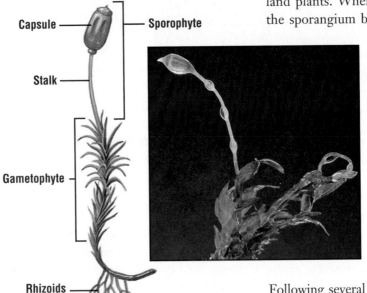

Capsule — Sporophyte

Stalk —

Gametophyte —

Rhizoids —

female sex organs called **antheridia** (singular: "antheridium"), which are sperm-bearing structures, and **archegonia** (singular: "archegonium"), which are egg-bearing structures. After fertilization, the zygote develops into a short-lived sporophyte plant, and grows and is nourished as a single stalk from the female gametophyte.

VASCULAR PLANTS

Vascular plants comprise the division Tracheophyta, which includes the "true" terrestrial plants such as the familiar ferns, herbs, shrubs, trees, and flowering plants. They transport nutrients, water, and minerals via a vascular system. The dominant sporophyte is the generation in which the characteristic vascular tissue, organs, and systems are found. The gametophyte is small, dependent on the sporophyte, and usually short-lived.

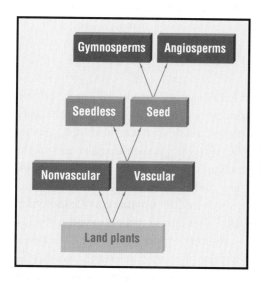

With the exception of ferns, vascular plants have evolved reproductive structures that are free of total dependence on water. The vascular tissue provides a continuous internal conduction system between the roots, stems, and leaves. It also provides structural support, a feature that is essential for the large size and long life exhibited by many land plants. The two main kinds of vascular plants are those that produce seeds and those that do not.

Seedless Plants

About 300 million years ago, seedless plants were the dominant land plants and formed extensive forests in swamplands. These earlier forms formed the basis of today's coal deposits. Most are now extinct, and modern descendants, with the exception of some tropical tree ferns, are miniature versions of the earlier forms. While adapted to life on land, seedless plants are found mainly in humid habitats because their short-lived gametophytes lack vascular tissues. Water is also necessary for fertilization.

a)

Seedless plants include true ferns as well as "fern allies": whisk ferns, club mosses, and horsetails. Ferns are the most common of all seedless plants, with just over 12 000 living species identified. Ferns were so common about 300 million years ago that scientists have named the period the Age of Ferns. All ferns have conducting tissue. The rhizomes of most ferns are underground. Fern leaves are called **fronds.** In temperate climates the fronds die in the winter with a new set

Antheridium *is the sex organ that produces male gametes in mosses and ferns.*

Archegonium *is the sex organ that produces female gametes in mosses, ferns, and gymnosperms.*

Fronds *are the leaves of ferns.*

b)

Figure 19.8

(a) Liverworts lack vascular or conducting tissue and are confined to moist habitats.
(b) *Lycopodium* grows on forest floors and has true roots, stems, and leaves in the sporophyte generation only.

developing in the spring. Unfolding young fronds resemble the heads of violins, and hence are called fiddleheads. Fiddleheads, common throughout the Maritimes, are a delicacy and taste somewhat like asparagus.

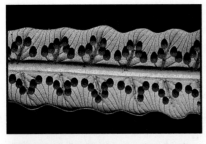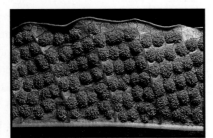

Figure 19.9

The sori found on the under surface of the fern frond produce thousands of dustlike spores.

*A **sorus** is a cluster of sporangia on a fern.*

*A **prothallus** is the gametophyte of mosses, liverworts, and ferns.*

***Gymnosperms** are plants that produce naked seeds.*

***Angiosperms** are plants that produce enclosed seeds.*

On the lower surface of the fronds are clusters of sporangia, called **sori** (singular: "sorus"), that produce thousands of dustlike spores by meiosis. Mature spores are released into the air. A germinating spore develops into a small haploid gametophyte, called a **prothallus.** Antheridia and archegonia develop on the prothallus. As in bryophytes, water is needed for sperm to swim to the nonmotile egg. Following fertilization, the zygote will form a new sporophyte.

Seed Plants

Reproductive adaptations and an improved vascular system largely account for the success of seed plants, the most widely distributed and complex group of plants on earth today. The ancestors of the more than 270 000 known seed plants first appeared about 370 million years ago during the Devonian period. Seed plants have separate male and female gametophytes, as well as roots, stems, and leaves.

A seed contains a plant embryo or a partially developed plant. Many seed plants are free-living, but some are saprophytic or parasitic. Most live on dry land, but others, such as pitcher plants, lilies, and a few trees, prefer wetlands. Seed plants range in size from the very small duckweeds to the giant redwoods and eucalyptus trees, which grow to more than 100 m tall. Vascular tissue allows seed plants to reach heights that are unattainable in the nonvascular world.

Seed type is the main criterion for distinguishing the two major seed-bearing groups, **gymnosperms** and **angiosperms,** terms derived from the Greek *sperma* ("seed"), *gymnos* ("naked"), and *angeion* ("vessel").

Gymnosperms produce unprotected, or naked, seeds in conelike structures, and the angiosperms produce seeds that are enclosed and protected inside the fruit, which is formed by the flower. Seeds ensure the survival of seed plants by resisting desiccation.

GYMNOSPERMS

Gymnosperms include the pines, spruces, junipers, firs, and other cone-bearing plants. Their characteristic thin, needle-like leaves are a special adaptation to the harshness of hot, dry summers, cold winters, and moderate rainfall. The needles are covered by a hard, waxy cuticle that helps the plant to retain moisture.

Another conifer adaptation is the development of roots that extend over a wide surface area rather than penetrate deep into the soil. This feature holds the tree firm even in locations where soil is scanty.

Coniferous forests, besides being a habitat for numerous living organisms, are an important source of raw materials for a variety of commercial and industrial products.

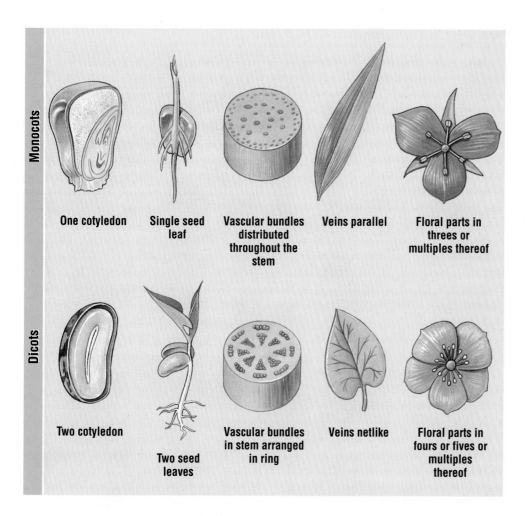

Monocots				
One cotyledon	Single seed leaf	Vascular bundles distributed throughout the stem	Veins parallel	Floral parts in threes or multiples thereof
Dicots				
Two cotyledon	Two seed leaves	Vascular bundles in stem arranged in ring	Veins netlike	Floral parts in fours or fives or multiples thereof

Figure 19.14

The differences between a monocotyledon and dicotyledon.

*A **cotyledon** is a seed leaf that stores food for the germinating seedling. It is the first photosynthetic organ of a young seedling.*

*A **monocotyledon** (monocot) is an angiosperm whose seeds have only one cotyledon or seed leaf.*

*A **dicotyledon** (dicot) is an angiosperm whose seeds have two cotyledons or seed leaves. Most angiosperms are dicots.*

seven cells and eight nuclei. One of the cells contains two haploid nuclei and is called the "endosperm mother cell."

Once the pollen tube has entered the ovule via the micropyle, sperm are released and fertilization occurs. In flowering plants, a unique situation called "double fertilization" takes place. One sperm nucleus fuses with that of the egg, forming a diploid zygote. The zygote will become the new sporophyte plant. Another sperm nucleus fuses with the two nuclei of the endosperm mother cell, forming a triploid nucleus. The endosperm provides nourishment to the developing embryo during the germination of the seed and to the seedling until leaves and photosynthesis take over.

While the ovule is transforming into a seed, the ovary and surrounding structures are developing into a fruit. The fruit assures protection and dispersal of the developing seed.

Angiosperm Diversity

Flowering plants are grouped into two classes: the **monocotyledons** (monocots) and **dicotyledons** (dicots). A **cotyledon** is a seed leaf that stores food for the young sporophyte and becomes the first leaf to appear as the seed germinates. Common monocots include grasses, water lilies, onions, orchids, and crop plants such as wheat, corn, barley, and rye. Dicots, the larger of the two groups, include maples, oaks, cacti, most forest trees, and the majority of flowering plants.

Table 19.3 Classification of Nonvascular & Vascular Plants

Group	Vascular system	Form	Reproduction	Habitat
1. Mosses, hornworts, liverworts	Absent	Simple; no roots, stems, leaves	Nonseed bearing Gametophyte dominant Alternation of generations Water required for sexual reproduction	Mainly moist areas
2. Fern allies: whisk ferns, club moss, horsetails	Simple	True roots, stems, leaves; stems usually grow underground, called rhizomes	Nonseed bearing Sporophyte dominant Water required for sexual reproduction	Mainly terrestrial Moisture required periodically
3. Ferns	Simple	True roots, stems, leaves. Leaves are called fronds.	Nonseed bearing Sporophyte dominant Asexual reproduction from rhizome Water required for sexual reproduction	Mainly terrestrial Moisture required periodically
4. Gymnosperms (Conifers)	Complex	True roots, stems, leaves	Seed bearing Naked seeds Gametophyte reduced to a few cells Water is not required for sexual reproduction	Wide range of terrestrial environments
5. Angiosperms (flowering plants)	Complex	True roots, stems, leaves	Seed bearing Enclosed seeds Gametophyte reduced to a few cells Dependent on pollinating agents for sexual reproduction	Almost all terrestrial environments Some aquatic

■ REVIEW QUESTIONS ■ ?

8 What characteristics are exhibited by all plants?

9 How do nonvascular and vascular plants differ?

10 How are algae and mosses similar? How do they differ?

11 What function does the archegonium and antheridium serve in plants?

12 Draw a diagram and label the main parts of a flower.

13 Describe reproduction in an angiosperm.

14 What are three differences that distinguish monocots from dicots?

15 What are some of the reasons why angiosperms are more widespread than gymnosperms?

*Unit Five:
Diversity of Life*

LABORATORY

GERMINATION OF MONOCOT AND DICOT SEEDS

Objective

Determine whether a germinating seed indicates if a plant is a monocot or dicot.

Materials

2 bean seeds (lima or kidney beans)
2 corn seeds
2 seeds selected from any two of the following: grass, tulips, carrot, poppy
4 beakers or jars (100 ml)
paper towels
hand lens

Procedure

1 Label the four beakers A, B, C, and D. A will be used for the lima beans, B for the corn seeds; C and D will contain the two additional selected seeds.

2 Line each beaker with a double layer of wet paper towels and leave 2 cm of water in the beaker. (It is important that the towels be kept wet throughout the activity.)
   a) Why is it important to keep the paper towels wet?

3 Allow the seeds to soak for several hours before planting.
   b) Why is it necessary to soak the seeds?

4 Position two bean seeds so they are wedged between the wall of the beaker and the wet towels. Repeat for the other seeds in beakers B to D.

5 Predict which seeds are monocots and which are dicots.
   c) Record your predictions.

6 Examine the germinating seeds daily with a hand lens until the root and shoot are established. Allow two weeks for the completed activity.
   d) Record your observations and drawings in a daily log. Be sure to label your drawings.

Laboratory Application Questions

1 What are the differences in the germination of the bean seed and that of the corn grain?

2 Which of the seeds were monocots and which were dicots?

3 Were all of your predictions correct? Why or why not?

4 Which seeds were the first to germinate? Can you explain why some were "late bloomers"?

5 Name some other seeds that are monocots and dicots. ■

Flowering plants can be divided into two main subclasses, (a) dicots and (b) monocots.

RESEARCH IN CANADA

Cytoskeletons

Dr. Faye Murrin of Memorial University in Newfoundland is a cell biologist and mycologist. Her interest in science centers on the microscopic world of the living cell. Using such techniques as fluorescence microscopy, in which fluorescent dyes are used to tag specific molecules, and electron microscopy, which can magnify images up to 200 000 times, Dr. Murrin investigates the functions and mechanisms of the cytoskeleton in cells. The cytoskeleton, a hallmark of eukaryotic cell evolution, is a versatile network of fine protein filaments. The cytoskeleton has a wide range of roles in other organisms as well.

Dr. Murrin uses the fungus *Entomophaga* as a model system in which to study microfilaments and microtubules. As the name implies, this fungus is an insect pathogen. It is useful in elucidating cytoskeleton function during infection. *Entomophaga* is also a potential biological control agent and an alternative to chemical sprays in combating forest pests such as the spruce budworm and hemlock looper.

Dr. Murrin has also set her sights on large fungi and is collaborating on a study of the potential for a wild mushroom harvest as a small-scale industry in Newfoundland, a venture that has proved successful on Canada's west coast.

DR. FAYE MURRIN

PLANT STRUCTURE AND FUNCTION

Photosynthesis, the process by which plants make food, is a cellular process. So how do complex plants, like angiosperms, supply their cells with sunlight, carbon dioxide, water, and minerals? How do they distribute food to all cells? How do they store food? Cells of complex plants, like those of all multicellular organisms, are organized into tissues, organs, and systems.

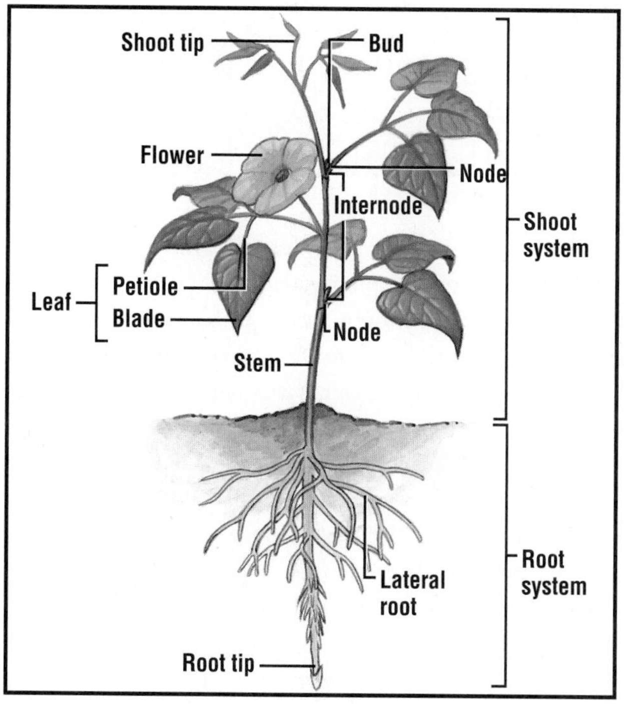

resulting in increases to the diameter of stems and roots.

Unlike animal cells, which have many kinds of cells that undergo division, plants divide only in specific regions, called **meristems.** The meristem at the root and shoot tips is called the *apical meristem.* Meristems that increase a plant's diameter are called *lateral meristems.* Vascular cambium is one lateral meristem that produces additional vascular tissue; cork cambium produces cork. Cells produced by meristematic tissue eventually differentiate into all other plant tissues.

Meristems *are tissue regions of plants where some cells retain the ability to divide repeatedly.*

In seed plants, root and shoot tissues begin to form in the embryo, within the seed. As the seed germinates, cells at the root and shoot tip divide and elongate. This lengthwise growth is called *primary growth.* In young plants, evidence of primary growth is seen when green shoots emerge; mature plants show primary growth in the lengthening of their roots and stems. Many dicots and a few monocots also exhibit *secondary growth.* Secondary growth originates in tissues at sites other than the root and shoot tips,

PLANT TISSUES

Plant tissues are adapted or specialized for functions such as absorption, transport, photosynthesis, reproduction, and storage.

Ground Tissue

The three types of ground tissues are parenchyma, colenchyma, and sclerenchyma. Parenchyma, the most abundant, consists of living cells that make up the bulk of the primary plant body. It functions in healing wounds, regenerating plant parts, and is also involved in photosynthesis, food, and water storage. Plants like cacti, which are adapted for water storage, have large amounts of parenchyma tissue and are called *succulents.*

Collenchyma, also a living tissue, helps strengthen the plant and is specialized for supporting the plant's growth regions. Collenchyma cells have long and thickened cell walls that provide a measure of pliability to certain plant parts such as celery stalks and herbs, which must be flexible to withstand winds.

Figure 19.15

The two main organ systems of plants are the root and shoot systems.

Figure 19.16

Plant tissues: (a) parenchyma, (b) collenchyma, (c) sclerenchyma, (d) xylem tracheids, (e) xylem vessels and fibres, (f) phloem, (g) and (h) epidermis. Photomicrographs (a), (b), (c), and (g) show a cross-sectional view. (d), (e), and (f) are longitudinal sections. (h) is a scanning electron micrograph of the lower surface of a leaf.

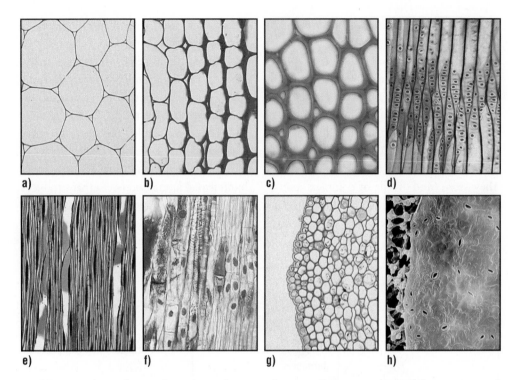

a)   b)   c)   d)

e)   f)   g)   h)

*The **cuticle** is a noncellular layer that covers the epidermis of above-ground plant parts. It prevents water loss.*

Mature sclerenchyma tissue is made up of walls of dead cells. Their long, tapered walls, called fibers, strengthen and support various plant parts. They are particularly adapted to parts where hardness is an advantage such as in the shells of nuts and in cactus spines.

Ground tissue in the center of roots and stems is generally called *pith*. Pith aids in the storage of food and water. Pith consists of spongy parenchyma cells that, in turn, are surrounded by more rigid cells called *cortex*.

Vascular Tissue

Vascular tissue includes xylem and phloem along with some collenchyma and parenchyma cells. The main water-conducting cells in xylem are called *tracheids* and *vessels* shown in Figure 19.16 (d) and (e). Both types are found in flowering plants but gymnosperms have only tracheids. The two types are dead at maturity and are pierced by many recesses, or *pits*. Pits are highly permeable to water. Vessels are usually shorter than tracheids and are joined end to end. Tracheids are

longer than vessels and have tapered overlapping ends.

Phloem tissue transports sugars and other solutes throughout the plant body. Unlike xylem, it is mainly a living tissue. Phloem cells are called *sieve tubes*, because the cells are long and thin, with sieve plates at the end walls and side walls. It is thought that their associated companion cells, which are nucleated, may help to direct activities of the sieve tubes.

Dermal Tissue

The outermost dermal tissue layer of the primary plant body is the *epidermis*. On above-ground parts, the epidermis forms a waxy noncellular covering called the **cuticle.** The cuticle, like animal skin, protects against excessive water loss and infection by microorganisms. Epidermal tissue often contains highly specialized cells such as root hair cells and leaf guard cells. During secondary growth in roots and stems, the epidermis is replaced by *cork*, another dermal tissue. Cork functions in waterproofing the roots and stems.

THE ROOT SYSTEM

Besides anchoring plants, roots also play a role in absorbing water and minerals, and in food storage. Some stored food is used by root cells, but the bulk of it is transported back to the above-ground parts.

The two types of plant roots are *taproots* and *fibrous roots*. If a young root increases in diameter, grows downward, and develops lateral roots, it is called a taproot. Examples of plants with taproots are carrots, sugar beets, dandelions, and pine trees. In monocots like grasses, the primary root is short-lived and is replaced by *adventitious roots* ("adventitious" refers to structures that arise at unusual places). Adventitious roots and their branchings form the fibrous root system. Generally these roots do not penetrate as deeply as taproots do.

Each root has a protective root cap at the tip. Further back are numerous fine root hairs that arise from the epidermal layer. Root hairs increase the root's surface area for absorption.

On the inside of the epidermis of the roots are parenchyma cells that form the root cortex. The cortex acts as storage tissue between the epidermis and vascular cylinder. The inner layer of the cortex is called the *endodermis*. The endodermis forms a waxy substance, *suberin*, which prevents water and minerals from leaving the vascular cylinder. Inside the endodermis, and surrounding the vascular cylinder, is a highly meristematic tissue called the **pericycle.** This tissue forms lateral roots and, in species showing secondary growth, forms the vascular and cork cambia.

Vascular cambium separates the xylem and phloem. When the cambium divides, it produces phloem cells to the outside and xylem cells to the inside. As xylem increases in size, the cambium is displaced outward, causing an increase in the girth (circumference) of the root. This process is also particularly important to secondary growth in stems.

*The **pericycle** is a meristematic tissue consisting of parenchyma cells and sometimes fibers. It is found immediately within the endodermis.*

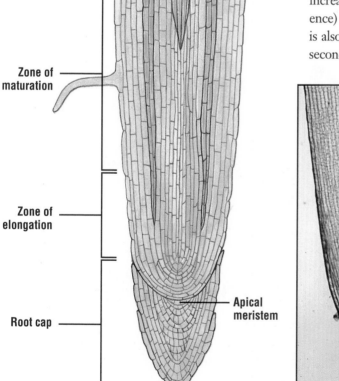

Zone of maturation

Zone of elongation

Root cap

Apical meristem

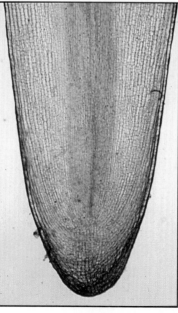

Figure 19.17

A longitudinal section of a generalized root tip.

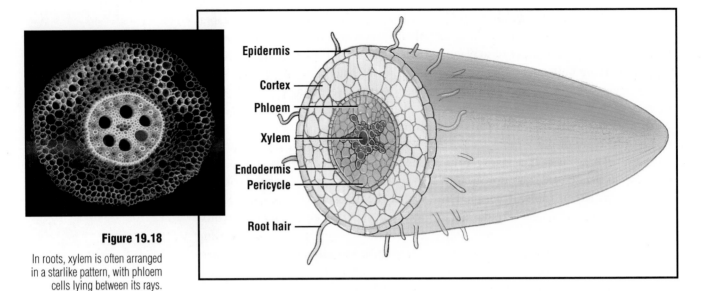

Figure 19.18

In roots, xylem is often arranged in a starlike pattern, with phloem cells lying between its rays.

Labels: Epidermis, Cortex, Phloem, Xylem, Endodermis, Pericycle, Root hair

THE SHOOT SYSTEM

Stem and leaves make up the plant shoot system. Stems provide support for the plant, serve as a transport link between roots and leaves, and act as sites for food storage. Sometimes stems are modified for asexual reproduction. For example, underground white potato tubers and strawberry runners, which reproduce asexually, are stems, not roots.

There are two types of plant stems: *woody stems* and *herbaceous* (nonwoody) *stems*. Herbs contain little or no wood and usually do not grow more than one meter tall. Examples include an assortment of weeds and leafy plants, such as cabbage and lettuce. Examples of woody stems include vines, shrubs, conifers, and dicot trees. Both woody and herbaceous plants undergo secondary growth. However, in comparison to woody stems, the stems of herbs are thin, weak, soft, green, and short-lived.

Figure 19.19

The stem structure of (a) a monocot and (b) a dicot.

Vascular bundles are distributed throughout the stem.

a) Monocot

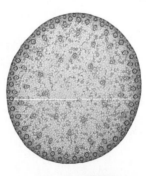

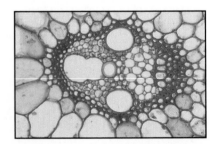

Vascular bundles in the stem are arranged in a ring.

b) Dicot

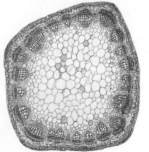

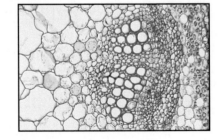

Vertical   Cross-section   Close-up view of cross-section

As in roots, the apical meristem cells in stems differentiate into the primary ground, vascular, and dermal tissues. All stems start as buds. The bud at the shoot tip develops first and is called the *terminal bud*. Inside its bud scales is the shoot tip, the site of primary growth. Primary and secondary growth also occur in other buds, called *lateral*, or *axillary*, buds. Each node on a stem is a region whose cells may give rise to a bud, leaf, or flower. The stem sections between nodes are called *internodes*.

During the first year of growth, woody and herbaceous stems closely resemble each other. However, as growth proceeds, the strength, hardness, and size of a woody stem arise from secondary growth. Cells develop on the outside of the vascular cambium to become secondary phloem, and on the inside secondary xylem develops. The secondary xylem thickens and forms tissue known as *wood*. As the tree ages, additional changes take place. No new ground tissue is added to the tree, and the primary tissue may be displaced by the secondary tissue.

Immediately below the epidermis, a layer of parenchyma cells develops into the cork cambium. The cork produced by this layer eventually replaces the epidermis. Cork is tougher and thicker than the epidermis and offers protection against the environment and injury by insects. As the stem grows, the cork cracks and is renewed.

The outermost layer of a stem or tree trunk is called *bark*. Its tissues are composed of cork, cork cambium, cortex, and phloem. Just beneath the bark, and forming the bulk of the

stem, is secondary xylem, or wood. A thin layer of vascular cambium lies between the bark and wood. If the bark is removed and the phloem destroyed, the tree dies.

In older stems the darker wood at the center is called *heartwood*. The dark color is due to the clogging of cells by waste products—resins, gums, and so on—which prevent liquids from moving up or down. *Sapwood* is the light-colored wood that surrounds the heartwood. This wood conducts water and minerals between roots and leaves. Annually, some sapwood is changed to heartwood.

THE LEAF

The leaf is important because it is the center of photosynthesis and a basic food source for most heterotrophs. Leaves display a variety of adaptations to carry out special functions. Onion bulbs are food-storage leaves attached to a very small stem. Leaves, such as those of the Venus flytrap, are adapted to trapping insects.

Typically, leaves have a network of veins that contain the vascular tissue, which is

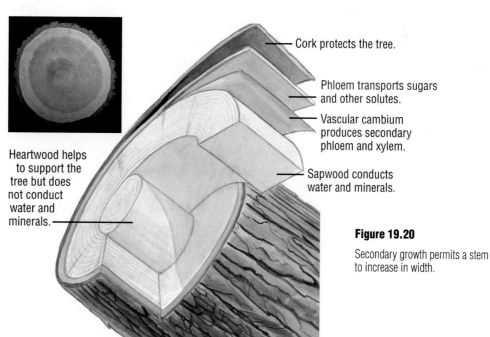

Heartwood helps to support the tree but does not conduct water and minerals.

Cork protects the tree.

Phloem transports sugars and other solutes.

Vascular cambium produces secondary phloem and xylem.

Sapwood conducts water and minerals.

Figure 19.20

Secondary growth permits a stem to increase in width.

Figure 19.21

Cross section of a leaf showing the internal structure.

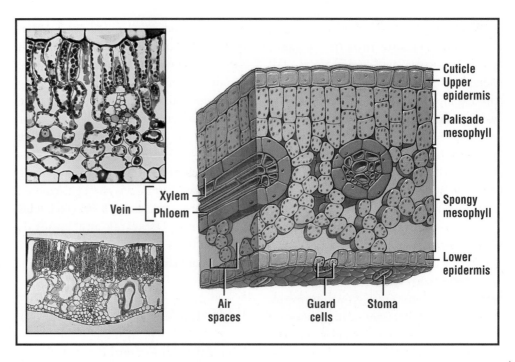

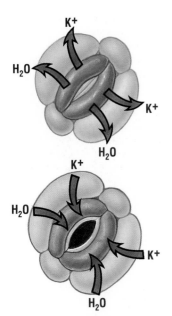

Figure 19.22

The opening and closing of the stoma are regulated by the levels of water and carbon dioxide in the guard cells.

Stomata *are spaces between guard cells in the stems and leaves of plants. They regulate gas exchange and transpiration.*

Guard cells *occur in pairs on the lower epidermis of a leaf or on a stem. They regulate the opening and closing of a stoma.*

Labels on figure: Cuticle; Upper epidermis; Palisade mesophyll; Spongy mesophyll; Lower epidermis; Vein — Xylem, Phloem; Air spaces; Guard cells; Stoma

continuous with the xylem and phloem in the stem. The veins also provide the leaf with support. Each leaf is connected to the stem by a leaf stalk called a *petiole*. The flattened blade is the main body of the leaf. Leaves are positioned along the stem at points called *nodes*. If the leaf is a single blade it is called a *simple leaf*. A blade divided into two or more parts is called a *compound leaf.*

The leaf epidermis is covered by a cuticle and is pierced by tiny pores called **stomata.** Stomata regulate the exchange of water vapor and gases between the leaf and the outside environment. Each stoma is bounded by a pair of sausage-shaped **guard cells.** Water and carbon dioxide levels of the guard cells determine the opening and closing of the stoma. At sunrise, photosynthesis begins in chloroplasts. Unlike most of the surrounding cells of the lower epidermis of the leaf, the guard cells contain many chloroplasts. Photosynthesis lowers the carbon dioxide levels in the guard cells, because carbon dioxide is a reactant used along with water to synthesize sugars. Light from the blue part of the spectrum also triggers active transport of

potassium ions into the guard cells, which in turn alters osmotic pressure. Water moves by osmosis and turgor pressure begins to build in the guard cells. The swollen guard cells change shape when turgid and pull apart, thereby opening the stoma. At nightfall, photosynthesis stops and carbon dioxide levels begin to increase in cells, because carbon dioxide is a by-product of cellular respiration. The buildup of carbon dioxide causes potassium to move out of the guard cells. This is followed by a decrease in turgor pressure and the guard cells collapse closing the stoma.

Between the upper and lower epidermis is a photosynthetic layer called the *mesophyll.* The mesophyll consists of two types of parenchyma cells, called palisade on the upper side, and *spongy mesophyll* on the lower side. Palisade cells are shaped like bricks or columns and are arranged neatly side by side. They contain many chloroplasts. The spongy mesophyll cells are irregular in shape and haphazardly arranged. They have fewer chloroplasts. In the middle of the leaf is xylem and phloem tissue, contained in vascular bundles called *veins.*

TRANSPORT IN PLANTS

About 90% of water loss in plants occurs through stomatal openings in the leaves. This process is called **transpiration.** The height that water must move in tall trees can exceed 100 m. How do plants move water over such great distances and how can they survive losing so much of their moisture?

Several forces are involved. As water is lost from the leaf it is replaced by osmosis, which continuously pulls water into the roots. This creates an internal root pressure that pushes water into the stem. Osmosis continues from one parenchyma cell to another until the xylem is reached. The force of attraction that pulls the water up the narrow xylem tubes is called *capillary action*. While root pressure and capillary action may contribute to water movement over shorter distances, they do not account for water movement to the heights mentioned above.

Researchers believe that the major force behind water movement in plants is the attraction of water molecules for one another. The theory is called the **cohesion-tension,** or *transpiration-cohesion*, theory. Water molecules adhere strongly to the walls of the xylem tubes (attraction of unlike molecules). In addition, the molecular structure of water results in an attraction or cohesion between adjacent water molecules. These opposing forces of cohesion and adhesion provide for continuous water movement and keep the water column intact.

Foods produced by photosynthesis are moved from leaf cells to other plant parts through phloem tissue. This process is called **translocation.** Phloem, like xylem tissue, forms a continuous pipeline from leaves to roots. Although translocation is poorly understood, one hypothesis, called the *pressure-flow hypothesis*, suggests that food flows from the source (where food is made) to a sink (where food is used or stored) because of unequal osmotic pressures at both locations.

RESPONSES IN PLANTS

People rarely think of plants as responding to stimuli. But plant development and growth is controlled by hormones (internal chemical regulators) and external environmental factors like light, gravity, and temperature.

Hormones

Hormones are chemicals produced in one part of a plant that influence growth, cell division, and other activities in another part of the plant. The five hormone groups known to exist in most plants are auxins, gibberellins, cytokinins, abscissic acid, and ethylene.

Auxins are the best-known plant hormones. Although they both stimulate and inhibit plant growth, they are primarily involved in promoting plant cell elongation. Apical bud growth and ripening of fruit is stimulated by auxins. Artificial auxins are widely used.

Gibberellins, like auxins, are best known for their promotion of stem elongation. In some plants they cause the stem to elongate just before the plant flowers. This process, called *bolting*, produces a long stem that holds the flower up to pollinators and the wind. The same effect can be achieved artificially by applying the hormone to the plant stem.

As the name suggests, cytokinins are hormones that stimulate cell division. They also stimulate leaf growth. *Cytokinins* are mostly concentrated in endosperm tissue and young fruit.

Abscissic acid (ABA) is important in promoting the closure of stomata, regulating seed and bud dormancy, and providing resistance to water stress.

Transpiration *is the loss of water from terrestrial plants.*

The **cohesion-tension** *theory provides an explanation for water movement in plants by the attraction between water molecules.*

Translocation *in vascular plants refers to the conduction of organic compounds throughout the plant by way of phloem vessels.*

Figure 19.23

Tropisms. (a) Plant stems generally show positive phototropism as they grow towards the source of light. b) Plant stems usually exhibit negative geotropism, growing against the force of gravity. (c) The tendril of a creeping vine "pulls down" a leaf for support. Growth in response to touch is called thigmotropism.

Ethylene, which exists as a gas, stimulates the ripening of fruits and the conversion of starches and acids into sugars.

TROPISMS

Growth or movement of plants toward or away from an environmental stimulus, is called a tropism. If the response is toward the environmental stimulus, it is called a *positive* tropism. If the movement is away from the stimulus, it is called a *negative* tropism. Responses to environmental stimuli that are independent of the direction of the stimuli are called *nastic movements*. The three main environmental stimuli are light, gravity, and touch. Other tropisms include chemotropism and hydrotropism.

Most plants respond to light to ensure that leaves get plenty of light for photosynthesis. This is called *phototropism*. Experiments have shown that auxins cause cells on the shaded side of shoots to expand and elongate more rapidly than those on the lightened side.

Geotropism is plant growth in response to gravity. Researchers believe that, as in phototropism, the hormone auxin also regulates this process.

Thigmotropism is growth in response to touch. This response is exhibited by many climbing plants such as grapevines and sweet peas. When modified leaves, called *tendrils*, are touched by an object such as a stick or supporting branch, they wind themselves around it and slowly begin to "climb." Quick responses to touch are also exhibited by a number of carnivorous plants including the Venus flytrap. While this response is similar to an animal reflex, it is important to remember that plants do not have muscles or nerves, so the mechanism is quite different.

body for growth and energy supply, and dispose of it as metabolic waste. Animals must coordinate their activities, avoid predators and other hazards, grow, and reproduce. The variety of ways in which they meet these challenges makes animals a diverse and fascinating kingdom.

Scientists divide this complex kingdom into two broad groups: invertebrates and vertebrates. The **invertebrates** do not have a backbone. They are the only large taxonomic group defined by the *lack* of a characteristic (a backbone) rather than by the presence of a common feature. The **vertebrates,** to which humans belong, have a **notochord** for at least a part of their life history.

also come from the ectoderm. Cells from the endoderm will form the lining of the gut in animals. In most animals the mesoderm, a germ layer between the ectoderm and endoderm, gives rise to organs of the circulatory, reproductive, urinary, and muscular systems.

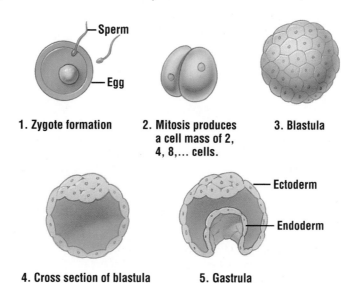

1. Zygote formation
2. Mitosis produces a cell mass of 2, 4, 8,... cells.
3. Blastula
4. Cross section of blastula
5. Gastrula

Invertebrates are multicellular, eukaryotic heterotrophs that do not have a backbone.

Vertebrates are multicellular, eukaryotic heterotrophs that have a notochord at some stage in their life.

Figure 20.2

Gastrula formation in a sea urchin. Many cell divisions form a blastula, a cell mass that resembles a hollow sphere. In turn, one of the edges of the blastula begins to push inward, much like a deflated basketball, to form the gastrula.

DEVELOPMENT

All animals begin life as a single egg cell, which, when fertilized, becomes a *zygote.* The zygote begins a series of mitotic divisions. Each division doubles the number of cells (i.e., into cell masses of 2, 4, 8, 16, 32, and so on). As division continues, the mass of cells develops into a hollow sphere called a **blastula,** which then begins to fold inward, as shown in Figure 20.2. The process by which the blastula folds inward is referred to as **gastrulation.** During this process, cells that will form the outside of the body, the *ectoderm,* become segregated from those that will form the inside of the body, the endoderm. The ectoderm and endoderm are called **germ layers,** because each gives rise to specific tissues in the adult. Cells from the ectoderm will form skin and the nervous system. In more complex animals, feathers, scales, hair, and nails

Complexity and evolutionary development can be studied by examining germ layer development. Lower invertebrates, such as the sponges and jellyfish, have no middle layer of cells; therefore, these animals lack structures, such as the circulatory system, which develop from the mesoderm. With only an ectoderm and endoderm, each cell of the animal is exposed to seawater. Because nutrients are received directly from the seawater, a circulatory system is unnecessary. However, in more complex organisms, as a larger mesoderm layer develops and structures become more elaborate, an internal transport system becomes important. The middle layer of cells is no longer directly connected to the external environment. The transport system must also carry waste products from the mesoderm cells to the external environment. Not surprisingly, the more complex the mesoderm, the more highly developed the transport system.

A notochord is a skeletal rod of connective tissue that runs lengthwise along the dorsal surface and beneath the nerve cord. Notochords are present at some time during vertebrate development.

A gastrula is a stage of embryo development in which germ layers are formed.

A blastula is an early stage of embryo development in which repeated cell divisions have formed a hollow sphere.

A germ layer is a layer of cells in the embryo that gives rise to specific tissues in the adult.

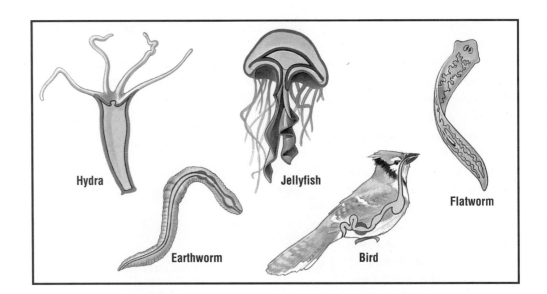

Figure 20.3

Hydra, jellyfish, and flatworms have a saclike gut with a single opening. The earthworm and bird have a tubelike gut with a separate mouth and anus. Regions of the tubelike gut can be specialized for grinding, chemical digestion, and absorption.

BODY CAVITIES

Complexity and evolutionary development can also be studied by surveying the development of the body cavity. All animals with tissues have cells specialized for digestion. In the simplest animals, cells from the endoderm layer form a pouchlike gut with a single opening. The opening acts as both a mouth and anus. More complex animals have a gut with two distinct openings, a mouth and anus. This arrangement enables the one-way movement of food through the gut, and permits greater regional specialization, such as for grinding or chewing, chemical digestion, and absorption.

A body cavity called the **coelom,** located between the gut and the body wall, is found in all vertebrates and many invertebrates. A layer of mesodermal cells, called *epithelial cells,* lines the body cavity and the gut and covers

the internal organs of the body, making up the **peritoneum.** Less complex invertebrates may lack a coelom, or may have an intermediate structure called a **pseudocoelom,** a fluid-filled cavity of variable shape, which has no peritoneum.

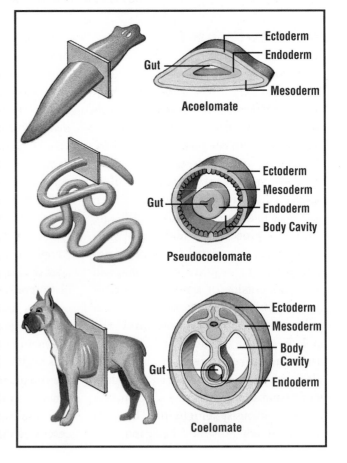

Figure 20.4

Acoelomates, such as flatworms, lack a body cavity. Pseudocoelomates, such as roundworms, do not have a continuous peritoneal lining. Coelomates, such as the higher invertebrates and vertebrates, have a true coelom lined with a continuous peritoneum.

*A **coelom** is a body cavity or space lined with a layer of cells called the peritoneum.*

*The **peritoneum** is a covering membrane that lines the body cavity and covers the internal organs.*

*A **pseudocoelom** is a fluid-filled cavity that lacks the mesodermal lining of a true coelom.*

SYMMETRY

The *symmetry* of an organism gives clues to its complexity and evolutionary development. In the chapter on Organs and Organ Systems, you saw that higher animals, including humans, are symmetrical along the mid-sagittal plane. This body plan is referred to as *bilateral symmetry,* in which the right and left halves of the organism are mirror images of each other. Some animals, however, are *radially symmetric,* or symmetric about a central axis. Figures 20.5 and 20.6 show both types of symmetry.

How is body symmetry related to the speed at which an animal moves and to brain development? In general, animals that display radial symmetry are not highly adapted for movement. For example, the hydra moves little throughout its life, and a jellyfish, even though it floats, is not well suited for rapid motion. Even highly evolved invertebrates such as starfish and sea urchins, which display radial symmetry as adults, are not fast moving. One explanation for the slower movement can be traced to the fact that no one region always leads. Only bilaterally symmetrical animals have a true head region. Because the head, or anterior region, always enters a new environment first, nerve cells tend to concentrate in this area. **Cephalization,** or the concentration of nerve tissue at the anterior end of an animal's body, is an adaptation that enables the rapid processing of stimuli such as food or danger. Not surprisingly, the faster the animal moves, the more important is the immediate processing of environmental information. Imagine the difficulties that would arise if sensory receptors were concentrated near the animal's posterior. Information about an inhospitable environment would only be gained after the entire body of the animal was exposed. At that point, recognition of poisonous chemicals would be of little advantage.

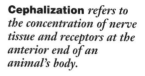

Cephalization *refers to the concentration of nerve tissue and receptors at the anterior end of an animal's body.*

Figure 20.5

Planes of symmetry in a starfish.

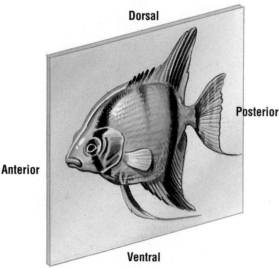

Dorsal

Anterior

Posterior

Ventral

Figure 20.6

Planes of symmetry in a fish.

Zygote development, body cavity arrangement, and symmetry allow scientists to divide animals into about 30 phyla. The phyla, summarized in Table 20.1, demonstrate a wide range, from the least to the most complex organisms. In this chapter, you will see how form and function vary in representative groups, and how these groups are linked phylogenetically.

Table 20.1 Summary of Major Phyla of the Animal Kingdom with Identifying Characteristics

Phylum (common name)	Representative members	Characteristics	Approx. no. of species
1. Porifera (sponges)	Giant sponge vase, redbeard sponge	Sessile, irregular shape, no mouth or digestive cavity; marine and freshwater	5 000
2. Cnidaria	Jellyfish, hydra	Sessile or free-swimming; medusoid form and polyp form in separate life-history stages of the same organism; stinging cells called nematocysts; marine, with a few freshwater	10 000
3. Ctenophora (comb jellies)	Sea gooseberry	Eight vertical rows of "combs" made up of cilia; free-swimming, marine	100
4. Platyhelminthes (flatworms)	Turbellarians, flukes, tapeworms	Free-living in marine or fresh water, or parasitic; body flattened dorsoventrally; mouth, but no anus	19 000
5. Nematoda (roundworms)	Hookworm, pinworm, vinegar eel	Cylindrical, slender, tapered at either end; free-living or parasitic; all habitats	20 000+
6. Rotifera (wheel animals)	Rotifers	Anterior end ringed with cilia, posterior end tapering to a "foot"; mostly freshwater, with some marine	1 500
7. Annelida (annelids)	Earthworms, leeches, polychaetes	Body cylindrical, segmented, "wormlike"	12 000+
8. Mollusca (mollusks)	Snails, clams, squid	Muscular foot; shell present in many forms; all habitats	100 000+
9. Arthropoda (arthropods)	Insects, crabs, mites, ticks, spiders, centipedes	Segmented body, some segments may be fused; jointed appendages; external skeleton; all habitats and modes of life, including parasitism	1 000 000+
10. Echinodermata	Starfish, sea cucumbers, sea urchins	Secondary pentamerous (five-sided) radial symmetry; marine	7 000
11. Chordata (chordates)	Fishes, amphibians, reptiles, birds, mammals	True backbone at some time in life history; all habitats	45 000

INVERTEBRATES

The invertebrates comprise over 95% of described animal species. From microscopic mites in house dust to graceful Portuguese men-of-war drifting in the ocean, from termites digesting a house to leeches ingesting blood, invertebrates are an integral part of life on earth.

ANIMALS WITHOUT TISSUES: PHYLUM PORIFERA

The Porifera, or sponges, are **sessile** animals found on the bottom of oceans and lakes. Sponges were once thought to be plants, because they did not appear to have any of the characteristics of animals;

Sessile *animals remain fixed in one place throughout their adult lives and are not capable of independent movement.*

however, by the early 1800s, they were recognized as the simplest and most primitive of animals. While sponges have some specialized cells, these cells are not arranged into tissues or organs. Without tissue development, the sponge has many limitations. It has no nervous system or brain to coordinate activities. A sponge also lacks muscle tissue, and hence is incapable of independent movement.

Despite their limitations, sponges are highly successful. They have existed for over 500 million years. In part, their success can be attributed to a system of feeding that brings the food to them. Cells called *collar cells*, each specially fitted with a long flagellum, line the inside of the body cavity. The collar cells create currents that move water into and out of the sponge. Water and food enter the sponge through many small pores, and exit via the osculum, or open end, as shown in Figure 20.7.

Most sponges are **hermaphroditic,** producing both male and female sex cells or *gametes;* however, only one type of gamete (egg or sperm) is produced at a time, so that sperm will not fertilize eggs from the same sponge. The fertilized egg develops into a free-swimming **larva,** which soon attaches to the bottom and undergoes **metamorphosis.** A free-swimming stage is common in sessile aquatic invertebrates, providing a means by which they can become distributed to new areas. Sponges can also reproduce asexually, by budding or branching, and have considerable regenerative capabilities. Occasionally a part of the sponge may be broken off. Some of the portions will form new individuals.

TISSUE DEVELOPMENT: PHYLUM CNIDARIA

Cnidarians (from the Greek *knide* meaning "nettle") represent the next higher level of organization of animals. They are radially symmetric and possess true tissues. However, like sponges, cnidarians have only two germ layers: ectoderm and endoderm. A middle layer, called *mesoglea*, is largely composed of a jellylike material. Hydra, jellyfish, sea anemones, and coral colonies belong to this diverse group. Members of this phylum, which consists of about 10 000 species, are found exclusively in aquatic ecosystems. The vast majority reside in marine environments and are common along Canada's three shorelines. Fewer than 50 species can be found in freshwater ecosystems, hydra being one of the most commonly occurring cnidarians in our lakes and ponds.

Cnidarians have two basic body shapes, which occur in different life-history stages of the same organism. The free-swimming *medusa* form, commonly called jellyfish, is generally transparent and has a bell shape. The sessile *polyp* form has a tubelike body with a mouth surrounded by tentacles at the free end.

The evolutionary advances that the cnidarians demonstrate compared with the less complex sponges can be attributed to the development of true tissues. Cnidarians have specialized nerve, muscle, and digestive tissues. A nerve net encircles their body. Some of the free-swimming forms have a ring of nerve cells around the bell, perhaps the primitive beginning of a central nervous system.

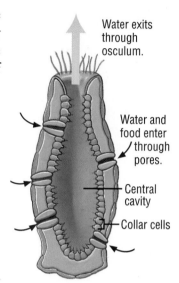

Water exits through osculum.

Water and food enter through pores.

Central cavity

Collar cells

Figure 20.7
The body plan of a simple sponge.

A **hermaphroditic** *organism contains both male and female sex cells or organs.*

A **larva** *is a juvenile form that looks very different from the adult. Tadpoles are larval frogs; caterpillars are larval butterflies or moths.*

Metamorphosis *is a process in which a larval form drastically changes shape to become more like the adult.*

a)

Figure 20.8

(a) Hydra may reproduce by budding.
(b) A typical medusoid jellyfish.

Nematocysts *are stinging capsules that aid in the capture of prey.*

b)

If hydra or sea anemones are touched, they respond by flattening out and pulling their tentacles inward. This type of movement indicates coordination of nerves and muscles, something not seen in the primitive sponge.

Cnidarians capture food by using specialized cells called *cnidocytes*, which contain sacs known as **nematocysts.** The nematocysts contain coiled, hollow, threadlike tubes that can shoot out rapidly and penetrate the skin of a prey or predator, some injecting toxic material. If sufficient nematocysts are triggered, the target can be paralyzed. Thousands of nematocysts can be found on a single tentacle.

Beach-goers may have painful memories of toxin-laden nematocysts. Thousands of the stinging cells can be found along the tentacles of jellyfish and floating bell-shaped colonies. The purple-colored medusa, or *Vellela*, occasionally floats into popular beaches that

dot the maritime shores. Anyone unlucky enough to brush up against the tentacles trips the triggering mechanism on the nematocyst and receives the tiny stinging threads. Although intended for small fish and other marine life, the nematocyst can cause mild swelling in humans. However, the mild irritation caused by these stinging cells is pale in comparison with that of a variety of jellyfish found off the coast of Australia. An Australian variety produces toxins so powerful that they can be life-threatening for humans. The sting of this jellyfish has claimed twice as many victims as have shark attacks in Australia.

Captured food is digested in a saclike cavity that has a single opening. Food enters the same opening from which wastes are expelled. Because the entry to the gut can be opened and closed by muscles, cnidaria can ingest larger animals than can sponges. The more primitive sponge is restricted to prey that can pass through its small pores.

Table 20.2 Sponges and Cnidarians

	Porifera	**Cnidaria**
Representative	Sponge	Hydra, jellyfish, coral colonies
Habitat	Predominantly marine, but also fresh water	Predominantly marine, but also fresh water
Body plan	Asymmetrical, two germ layers	Radial symmetry, bell-shaped medusae, and tube-shaped polyp; two germ layers
Reproduction	Asexual—budding; sexual—egg and sperm	Asexual—budding; sexual—egg and sperm
Transport and digestion	None—incoming current supplies food and excurrent pore carries away wastes	Body cavity functions in transport and digestion
Nervous system	No nerves	Nerve tissues, but no central collection of nerves that would resemble a brain

Unit Five:
Diversity of Life

In most cnidarians, sexual reproduction occurs between separate male and female medusae, and the resulting larvae, after several days of free swimming, attach to the bottom and develop into polyps. The polyps grow and can reproduce asexually, but eventually produce buds that develop into medusae, and the cycle begins again.

Some cnidarians, *Obelia* for example, are aggregations of individuals in colonies. Colonies show evidence of groups of individuals working together in much the same way as groups of cells work together in an individual. Individual polyps specialize in tasks of food gathering, and reproduction.

FRONTIERS OF TECHNOLOGY: DIVING SUBMERSIBLE

Canada's new diving submersible, the Remote Operation Platform for Ocean Science (ROPOS), has revealed a new world of mineral and biological riches on the ocean floor.

The ROPOS has recently explored deep-sea vents, a few hundred kilometers off the B.C. coast and over 2 km deep in the ocean. Deep-sea vents are a phenomenon found on the ocean floor where the earth's crust spreads and molten magma pushes up, releasing massive amounts of heat.

In their exploration of deep-sea vents, scientists have had to cope with severe obstacles such as corrosive water, frigid temperatures (the 2°C ocean water is so cold that it can disable the circuitry on the submersible's underwater navigation system), the inability to send radio waves through water, and the difficulty of getting accurate bearings in the deep sea.

The submersible is able to collect mineral samples and unusual forms of marine life, and insert thermometers, scientific gadgets, and water sampling bottles into holes on the sea floor. All this is controlled by a pilot 3 km away.

Researchers such as biologist Verena Tunnicliffe of the University of Victoria, have had extraordinary glimpses of creatures such as rare clams that have been seen nowhere else on earth—creatures that thrive in scalding, poisonous water. Some of these animals have no mouths, living off the energy extracted from toxic chemicals by bacteria.

BILATERAL SYMMETRY AND THE PRIMITIVE BRAIN: PHYLUM PLATYHELMINTHES

The Platyhelminthes, or flatworms, represent a further increase in complexity. Platyhelminthes are bilaterally symmetrical, and have mesoderm, a tissue not found in less complex forms. They also have true organs, and rudimentary organ systems for digestion and excretion. Flatworms are the most primitive organisms to show cephalization and also contain both free-living and **parasitic** members. Parasitic organisms live in or on the body of a host, from which they obtain their nutrition. The host usually suffers damage in the process.

Adaptations for a Free-Living Lifestyle
The best-known free-living flatworms are the Planaria, which occur in moist or submerged habitats in both marine and fresh water. Most are small, less than 1 cm in length, but some very large terrestrial forms may reach 60 cm in length. Like cnidarians, free-living Platyhelminthes have a "blind" digestive system, with the mouth as the only entrance or exit (see Figure 20.9). Unlike Cnidaria, they have a rudimentary excretory system consisting of a network of fine tubules running

*A **parasite** is an organism that lives in or on the body of another organism, from which it obtains its nutrition.*

throughout the body and opening to the outside via tiny pores.

Figure 20.9

The digestive and nervous systems of a planarian, a typical flatworm.

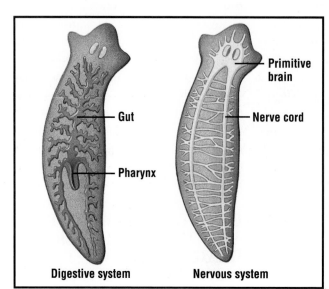

Gut

Pharynx

Digestive system

Primitive brain

Nerve cord

Nervous system

Nervous function is also more complex in Platyhelminthes. A concentration of nerve tissue in the head area resembles a primitive brain. Two nerve cords, each made up of many nerve cells, arise from the brain and run along the ventral side of the body toward the tail (see Figure 20.9). This system coordinates body movements and receives input from simple sensory cells, allowing more complex behavior. Flatworms can avoid light (negative phototaxis), and respond to chemical substances, which enables them to sense and move toward potential food. They can even learn their way out of a simple maze.

Flatworms are hermaphroditic. A reproductive system is present only during the breeding season and degenerates at other times of the year; fertilization occurs after copulation between two individuals. Capsules containing a small number of eggs and thousands of nutritive yolk cells are discharged and fastened to objects in the water. In two to three weeks they hatch into juveniles, which resemble the adults. In addition to sexual reproduction some planarians can regenerate complete worms from small fragments.

Figure 20.10

Flatworms are the simplest animals that display bilateral symmetry.

Adaptations for a Parasitic Lifestyle

Parasitic forms often look quite different from their free-living relatives. Parasites can be found both inside and outside an animal's body. Parasites living inside the body cannot afford to grow too large—no parasite wants to kill its host. Ideally, the parasite wants to extract as much food from the host as possible without killing it. While the largest tapeworm, that of a sperm whale, may reach 30 m, the smallest are only a few millimeters long. Human tapeworms occasionally reach lengths of 7 m.

Many hosts, including humans, maintain a constant internal environment. Parasites living within such a host are highly adapted to specific and reasonably stable environments. Little sensory information is required, and sensory receptors are reduced or absent. In addition, the digestive tract has degenerated or may even be lost. For example, tapeworms feed off the food digested by their host and therefore do not require a digestive system of their own. Other parasites benefit from nutrients carried in the host's blood and similarly do not need a digestive system. Those that feed off the tissues of their host retain a somewhat reduced digestive system. However, what parasites lack in sensory organs and digestive systems, they more than make up for by a highly evolved reproductive system. Parasites are capable of producing hundreds of thousands of eggs. The absence of a digestive system provides more room for eggs. In large part, the success of future generations of parasites depends on a highly developed reproductive system.

Intestinal parasites live and prosper in one of nature's most dangerous environments. These parasites have a modified epidermis called a **tegument,** which protects against the digestive enzymes and immune response of the host. The tegument is folded to increase surface area and to make absorption of nutrients more efficient. Some parasites, such as tapeworms, are protected by a nonliving *cuticle*, which is secreted by the epidermal cells. Like the tegument, the cuticle is designed to protect the parasite.

*The **tegument** is a modified epidermis that protects parasites from the digestive enzymes and immune response of the host.*

THE PSEUDOCOELOMATES: THE ASCHELMINTHES

The Aschelminthes are an aggregation of phyla sharing many characteristics. The most prominent shared characteristic, and the feature that makes the tiny Aschelminthes a bridge between "lower" and "higher" invertebrates, is the pseudocoelom. Aschelminthes are also the simplest organisms to have a complete digestive tract, with a mouth at one end and an anus at the other. The development of an anus means that food can be digested step-by-step as it moves through the system, and the removal of waste does not interfere with the intake of food. This is an important step toward increased complexity.

The Aschelminthe nervous system has an anterior ganglion, or aggregation of nerve cells, and lateral or ventral nerve cords extending along the body. There is no respiratory or circulatory system, due in part to the organism's small size; oxygen and carbon dioxide are exchanged by simple diffusion. The nematodes, a group of considerable importance in global ecol-ogy, will serve as an example of a typical Aschelminthe.

PHYLUM NEMATODA

Most nematodes are free-living, small, and harmless. Published figures for the number of described species of nematode, or roundworms, vary from 20 000 to 80 000, but this likely represents only about 10% of the total number.

Animal, and particularly human, parasites are a minority of nematode species, but can cause major health problems. Trichinosis, for example, can result from eating undercooked pork infected with *Trichinella spiralis*. These worms form cysts in the muscle tissue of the host and remain dormant for years. When the host tissue is eaten, the worms become active again. The adults are small and relatively harmless, and inhabit the digestive tract. However, some 1500 fertile eggs are released per female. The developing young, less than 0.1 mm long, burrow through the intestinal wall and are transported to all organs of the body. They are particularly fond of certain muscles, where they cause excruciating pain, fever, and other problems. Death is possible, but unlikely. Fortunately, North American pork is now relatively free of this parasite, and is subject to stringent government inspection.

Nematode parasites of plants also have far-reaching effects on human society. Potato cyst nematodes, for example, can have debilitating effects on potato crops. These worms are among the most highly specialized and successful plant-parasitic nematodes, perhaps because they origi-

Figure 20.11

An easily recognized annelid, the earthworm.

nated in the same part of the world as the potato—the South American Andes region—and have **coevolved** with their host.

The golden nematode (*Globodera rostochiensis*) is the most common North American potato cyst nematode, and has been subjected to concerted control measures. Confining a population of these worms to one area is very difficult, since the nematodes can be transported by any means that transports soil. The golden nematode is found on the island of Newfoundland, and there are stringent controls in place to keep it from spreading to the mainland. You are not allowed to take any soil-containing materials from the island (including your house plants if you are moving). Cars leaving by ferry are thoroughly washed at the ferry terminal to remove any soil, much to the amusement of tourists not familiar with the reason for their free car wash.

THE SEGMENTED WORMS: PHYLUM ANNELIDA

Although they show superficial similarities to nematodes, annelids are segmented and possess a true coelom. **Segmentation** is an important evolutionary advantage annelids share with humans. Segmentation refers to the repetition of body units that contain similar structures. The advantages of segmentation become more obvious in a discussion of more complex animals such as insects. Segmentation permits greater specialization, as witnessed by the segments of an insect's body: the head, thorax, and abdomen. Insect antennae develop from the head, legs and wings at the thorax, while the abdomen stores reproductive structures and sometimes stingers. The more than 12 000 species

of annelids are divided into three major classes: Oligochaeta, Polychaeta, and Hirudinea.

Oligochaeta

The common earthworm is typical of the class Oligochaeta (meaning "few bristles"). Earthworms are of tremendous importance to agriculture. They burrow in the soil, mixing and churning it, and thereby increasing aeration and drainage. They also help break down organic matter, making it available as nutrients for crop plants.

Oligochaetes are well-adapted for burrowing. They are cylindrical, tapering at either end. Each segment is separated from the next by a septum, or wall, formed by a double layer of peritoneum. A combination of circular muscles, which surround the body, and longitudinal muscles, which run along the length of the body, allows individual segments to contract. The contractions move along the length of the worm in a wave, a process called **peristalsis.** Annelids have no prominent projections, although there are four pairs of chitinous bristles, called *setae*, on each segment. These bristles can be extended to anchor the worm in its burrow, and to assist in locomotion.

Earthworms are nocturnal, feeding on decaying seeds and other plant parts. A mouth and muscular pharynx ingest food, which then moves through an esophagus. The remainder of the digestive tract is the intestine, which ends in an anus at the tip of the last body segment. Another feature of annelids not seen in lower groups is a closed circulatory system. Blood flows in closed vessels, which branch into all parts of the animal. The branches become progressively smaller, until they develop into capillary beds. There is no central heart, but thickened, muscular blood vessels in the anterior region act as pumps to help move the blood.

The annelid's excretory system is very much linked to its segmentation. Primitive paired **nephridia** (singular: "nephridium") are found in each of the approximately 100 segments. The nephridia open into the coelom in one segment, then pass through the septum into the next posterior segment. In that segment, the nephridium ends in an exit pore in the body wall. Waste materials are excreted through this pore to the outside.

the body are scattered cells that appear receptive to chemical, mechanical, and light stimulation.

Earthworms are hermaphroditic, but sperm transfer occurs by copulation. Part way along the body is a prominent swollen segment called the **clitellum,** which contains gland cells. When the eggs are laid, the clitellum secretes a cocoon, which envelops and protects them. The young develop in the cocoon and leave it as fully formed worms.

Nephridia are open-ended tubules that function in excretion.

*The **clitellum** is a smooth and enlarged segment found about one-third of the way along the body of oligochaetes; it secretes a protective covering for the eggs.*

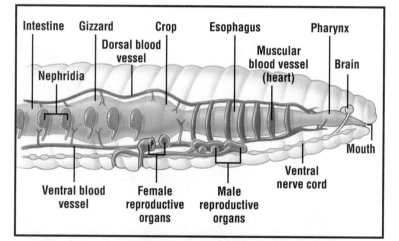

Intestine   Gizzard   Crop   Esophagus   Pharynx
Dorsal blood vessel
Muscular blood vessel (heart)
Brain
Nephridia
Mouth
Ventral blood vessel   Female reproductive organs   Male reproductive organs   Ventral nerve cord

Figure 20.12

The circulatory, excretory, nervous, and reproductive systems of an annelid.

Polychaeta

Polychaetes ("many bristles") are a diverse group of marine worms. Some live on the bottom, beneath stones and shells, some burrow into the bottom, and others build tubes to live in. Some are transparent and free-swimming; a few are even parasitic, living within the coelomic fluid of other polychaetes. Like oligochaetes, polychaetes consist of a series of similar segments; however, polychaetes have

Figure 20.13

A polychaete.

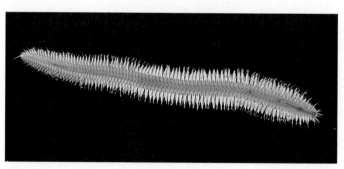

A large, dorsal ganglion is found in the annelid's anterior region. A ventral nerve cord runs the length of the worm, with small ganglia giving rise to nerves in each segment. There is a concentration of sensory cells at the anterior end, and along

flattened, fleshy projections on each side of each segment. These projections, or *parapodia* ("side feet"), function in movement and gas exchange. They are armed with bristles, which provide protection and traction.

Table 20.3 The Worms

	Flatworms	Roundworms	Segmented worms
Representative	Tapeworm, planarian, fluke	Ascaris, hookworm	Earthworm, leech
Habitat	Aquatic and terrestrial; many are parasites.	Aquatic and terrestrial, some are parasites.	Aquatic and terrestrial; few are parasites.
Body Plan	Bilateral symmetry, 3 germ layers, acoelomate	Bilateral symmetry, 3 germ layers, pseudocoelomate	Bilateral symmetry, 3 germ layers, true coelom
Reproduction	Asexual reproduction by fission; sexual reproduction hermaphroditic with cross-fertilization.	Sexual reproduction; separate sexes	Sexual reproduction; hermaphroditic with cross-fertilization
Circulation	None, except diffusion	None, except diffusion	5 pairs of aortic arches and two large blood vessels run along the dorsal and ventral surfaces.
Digestion	Single opening functions as a mouth and anus. Digestive organs present, but reduced in parasitic species.	Separate mouth and anus; digestive organs present.	Separate mouth and anus; digestive organs present.
Nervous system	Primitive brain; two longitudinal nerve cords	Primitive brain; dorsal and ventral nerve cords	More advanced brain; large ventral nerve cord with many peripheral ganglia

Figure 20.14

Feeding leeches. The best known leeches are those used in medicine. The use of leeches for medical purposes goes back to 100 A.D., and was prevalent in the 16th to 19th centuries, when "blood letting" was a common treatment.

Hirudinea

The Hirudinea, or leeches, are found in lakes, slow-moving streams, ponds and marshes, and on moist vegetation in humid environments such as jungles. About 75% of the known species of leeches are blood-sucking, external parasites. The leech attaches to its host and cuts through the skin to create a flow of blood. The leech's salivary glands produce a chemical, *hirudin*, which prevents the host's blood from clotting. Digestion is very slow, and leeches can tolerate long periods (over one year) of fasting between meals.

B I O L O G Y   C L I P

Leeches are still used in some surgical techniques. When severed fingers are reattached, it is impossible to reconnect the small veins that take blood away from the finger. Blood can thus accumulate in the finger and build up dangerous pressure. Leeches have been used successfully to suck out the excess blood with minimum disturbance. In time, the veins grow back normally.

REVIEW QUESTIONS   ?

1 Differentiate between vertebrates and invertebrates.

2 In what ways are sponges more primitive than cnidarians?

3 What is the function of the nematocyst? How does it work?

4 Explain how a concentration of nerve tissue in the anterior portion of an animal's body can be related to bilateral symmetry. Why would the nerve tissue not be concentrated near the posterior end?

5 Why do intestinal parasites, such as tapeworms, have poorly developed digestive systems? Does this provide any evolutionary advantage? Explain.

6 Why do parasitic flatworms have a tegument instead of a normal epidermis?

7 What is segmentation? Why are earthworms considered segmented?

8 How are leeches adapted to feeding on blood meals?

APPLYING THE CONCEPTS

1 How is the development of a true coelom related to the development of a more complex body form?
2 How are the phyla Platyhelminthes, Nematoda, and Annelida fundamentally different despite the fact that all are "wormlike" organisms?
3 List the structural adaptations to a parasitic mode of life, and give examples of each.
4 Why do higher aquatic organisms require gills, while many small forms do not?

5 What factors have enabled the arthropods to become so numerous and diverse?
6 Why have the bony fishes been successful in so many different environments?
7 How did the amniotic egg allow reptiles to dominate the terrestrial environment?
8 Explain why the molting process in birds must take place gradually.
9 How might the upright stance and opposable thumb in humans have influenced technology and cultural development?

CRITICAL-THINKING QUESTIONS

1 In some Cnidaria (e.g., *Obelia*), certain members of a colony are specialized for particular functions (obtaining food, reproduction, etc.). Should the colony be considered an individual organism in its own right? Discuss.
2 Planarians and starfish are very good at regeneration. The ability to regenerate body parts seems to be a very useful ability, yet it is very rare in more complex organisms. Why must this be so?
3 The development of a segmented body, as seen in annelids, is a major step in the evolution of higher organisms. What are the advantages of segmentation?
4 Drawing upon your knowledge of bird flight, indicate what special adaptations humans would need to fly (unassisted by technological devices such as airplanes or hot-air balloons). Draw a picture which illustrates some of the adaptations described.

5 Devise an argument which supports the idea that arthropods, not chordates, are the most highly evolved animals. Now provide the counter-argument.
6 Recently ecologists have been surprised to discover rapid declines in the population of some amphibians. Although many hypotheses have been suggested, no comprehensive explanation has been widely accepted. One hypothesis proposes that amphibians are ecological indicators and that their decline may signal deepening environmental problems. Why might scientists believe that amphibians are ecological indicators? How would you go about testing such a hypothesis?

ENRICHMENT ACTIVITIES

1 There are invertebrates in which the population is primarily female, and reproduction occurs parthenogenetically (i.e., via development of an unfertilized egg), with occasional sexual reproduction. Research this life-history strategy and discuss its advantages and disadvantages.
2 Although the transition from water to land is a major step in animal evolution, some terrestrial animals (e.g., whales) later returned to the sea. Research and discuss the adaptations required to return to an aquatic existence.
3 There has been considerable controversy over the development of game farms and fish farms in some parts of Canada. Research this issue and discuss both the benefits and the potential dangers of such farms.

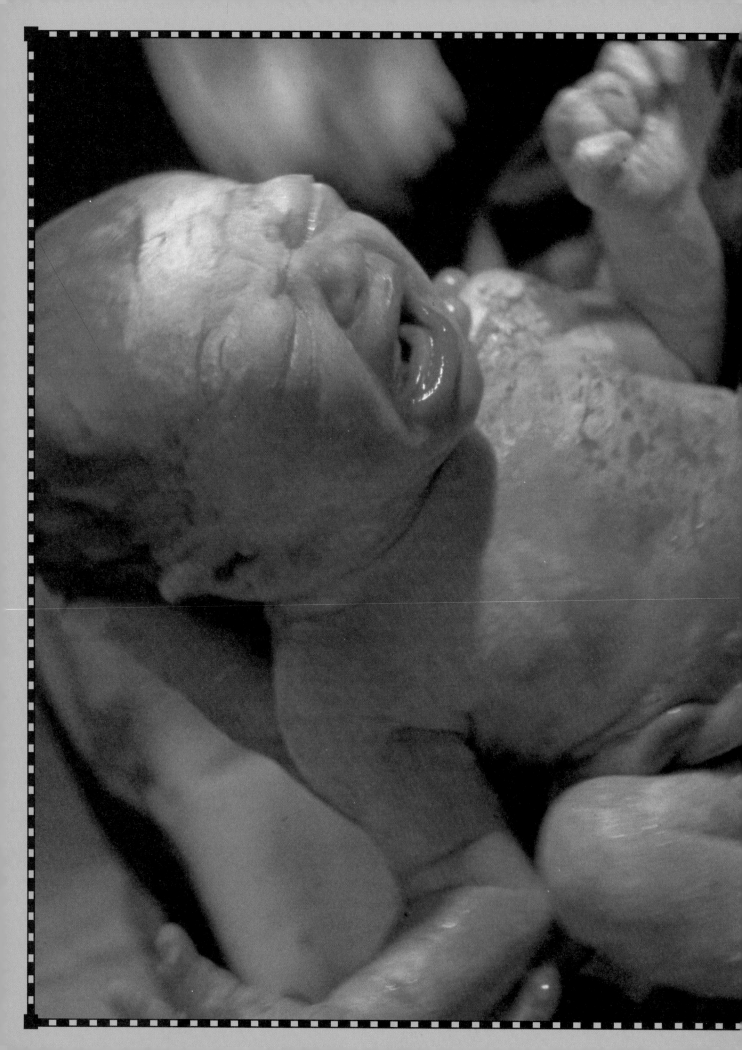

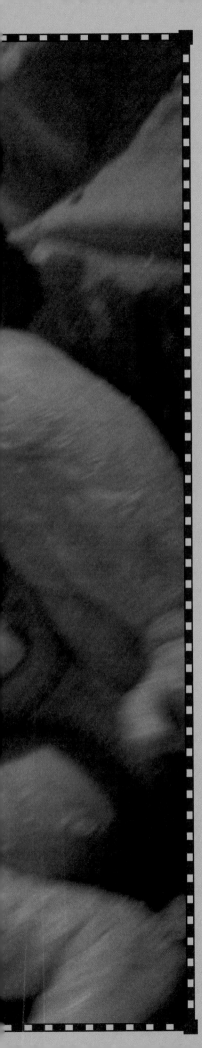

Continuity of Life

.

U N I T

6

CHAPTER 21

*T*he Reproductive System

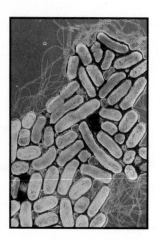

Testes *are the male gonads, or primary reproductive organs. Male sex hormones and sperm are produced in the testes.*

IMPORTANCE OF REPRODUCTION

Reproduction ensures the survival of a species. Sexual reproduction involving the fusion of male and female sex cells creates new gene combinations. The diversity produced by new gene combinations provides a basis for natural selection: only the best-adapted survive. Species survival is based on providing numerous and varied offspring.

Female oysters produce an estimated 115 million eggs for each spawning. Each year, female frogs produce hundreds of thousands of eggs for fertilization. In contrast, human females have 400 000 egg cells, of which only 400 mature throughout the reproductive years—from about age 12 to age 50. According to one source, the greatest number of offspring ever born to one woman was 57. The limited capacity of females to produce sex cells is contrasted with that of males. The average male, beginning at about age 13 to well into his eighties and nineties, can produce as many as one billion sex cells every day.

The human reproductive system involves separate male and female reproductive systems. The male gonads, the **testes** (singular: "testis"), produce male sex

Figure 21.1

From the smallest one-celled organisms to the largest mammals, all living things reproduce, ensuring the survival of the species.

cells called *sperm*. The female gonads, the **ovaries,** produce "eggs." The fusion of a male and a female sex cell, in a process called **fertilization,** produces a **zygote.** The zygote divides many times to form an embryo, which in turn continues to grow into a fetus.

THE MALE REPRODUCTIVE SYSTEM

Male and female sex organs originate in the same area of the body—the abdominal cavity—and are almost indistinguishable until about the third month of embryonic development. At that time, the genes of the sex chromosomes cause differentiation. During the last two months of fetal development, the testes descend through a canal into the **scrotum,** a pouch of skin located below the pelvic region. A thin membrane forms over the canal, thereby preventing the testes from re-entering the abdominal cavity. Occasionally, an injury may cause the rupture of the membrane, producing an inguinal hernia. The hernia can be dangerous because a segment of the small intestine can be forced into the scrotum. The small intestine creates pressure on the testes, and blood flow to either the testes or small intestine may become restricted.

The temperature in the scrotum is a few degrees cooler than that of the abdominal cavity. The cooler temperatures are important, since sperm will not develop at body temperature. Should the testes fail to descend into the scrotum, the male will not be able to produce viable sperm. This makes the male sterile.

Ovaries *are the female gonads, or reproductive organs. Female sex hormones and egg cells are produced in the ovaries.*

Fertilization *occurs when a male and a female sex cell fuse.*

A **zygote** *is the cell resulting from the union of a male and female sex cell.*

The **scrotum** *is the sac that contains the testes.*

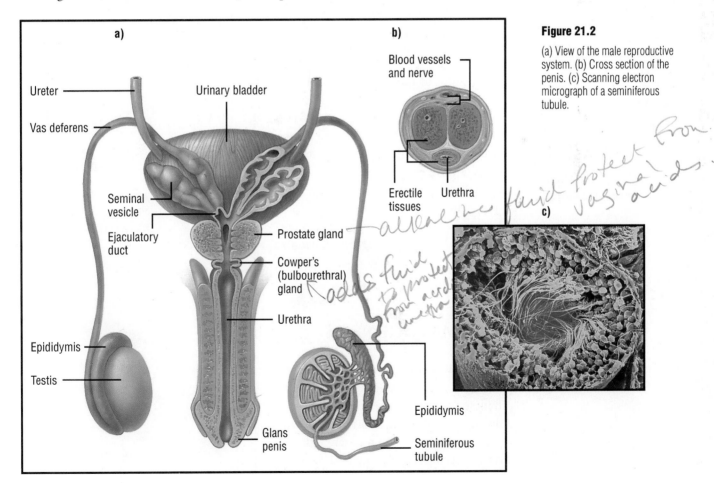

a)
- Ureter
- Urinary bladder
- Vas deferens
- Seminal vesicle
- Ejaculatory duct
- Prostate gland
- Cowper's (bulbourethral) gland
- Urethra
- Epididymis
- Testis
- Glans penis
- Epididymis
- Seminiferous tubule

b)
- Blood vessels and nerve
- Erectile tissues
- Urethra

c)

Figure 21.2

(a) View of the male reproductive system. (b) Cross section of the penis. (c) Scanning electron micrograph of a seminiferous tubule.

A tube called the **vas deferens** carries sperm from the testes to the *urethra*. Any blockage of the vas deferens will prevent the movement of sperm from the testes to the external environment. A surgical procedure, called a vasectomy, can be performed on males as a means of birth control. The *ejaculatory duct* regulates the movement of sperm and fluids, called **semen,** into the urethra, which also serves as a channel for urine. A **sphincter** regulates the voiding of urine from the bladder. Both regulatory functions work independently, and are never open at the same time. At any given time, the urethra conducts either urine or semen, but never both.

BIOLOGY CLIP

The word *testes* comes from the Latin, meaning "witness." It has been suggested that the word *witness* comes from the idea that testes are witness to virility. The Greek word for testes is *orchis.* The orchid flower derives its name from the testes shape.

The **vas deferens** *are tubes that conduct sperm toward the urethra.*

A **sphincter** *is a ring of smooth muscle in the wall of a tubular organ. Contraction of the sphincter closes the opening.*

Semen *(seminal fluid) is a secretion of the male reproductive organs that is composed of sperm and fluids.*

Sertoli cells *nourish sperm cells.*

The **acrosome** *is the cap found on sperm cells. It contains packets of enzymes that permit the sperm cell to penetrate the gelatinous layers surrounding the egg.*

During sexual excitement, the erectile tissue within the penis fills with blood. Stimulation of the *parasympathetic nerve* causes the arteries leading to the penis to dilate, thereby increasing blood flow. As blood moves into the penis, the sinuses swell, compressing the veins that carry blood away from the penis. Any damage to the parasympathetic nerve can cause impotency, in which the penis fails to become erect. (Other causes, such as hormone imbalance and stress, have also been associated with impotency.)

TESTES AND SPERMATOGENESIS

In many ways the sperm cell is an example of mastery in engineering design. Built for motion, the sperm cell is streamlined with only a small amount of cytoplasm surrounding the nucleus. Although reduced cytoplasm is beneficial for a cell that must move, it also presents a problem. Limited cytoplasm means a limited energy reserve. A support cell, the **Sertoli cell,** nourishes the developing sperm cell. Energy-transforming organelles, the *mitochondria*, are located next to the *flagellum*, the organelle that propels the sperm cell. An entry capsule, called the **acrosome,** caps the head of the sperm cell. Filled with special enzymes that dissolve the outer coating surrounding the egg, the acrosome allows the sperm cells to penetrate the egg.

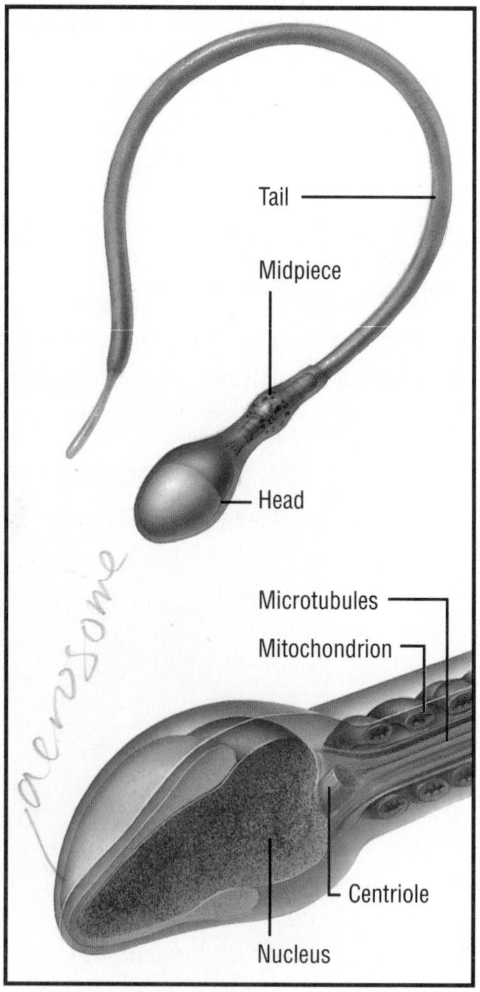

Tail
Midpiece
Head
Microtubules
Mitochondrion
Centriole
Nucleus
acrosome

Figure 21.3

A human sperm cell.

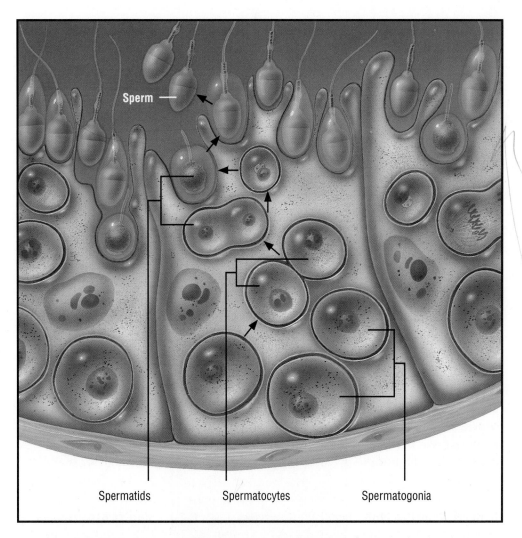

Sperm

Spermatids   Spermatocytes   Spermatogonia

Figure 21.4

Development of sperm cells inside the seminiferous tubule.

The inside of each testis is filled with twisting tubes, called **seminiferous tubules,** that measure approximately 250 m in length. The seminiferous tubules are lined with sperm-producing cells called *spermatogonia*. These immature sperm cells contain a full complement of 46 chromosomes. During a process called *meiosis*, the spermatogonia divide into *spermatocytes*, which contain 23 chromosomes. (Meiosis is described in detail in the chapter Sexual Cell Reproduction.) Human sex cells of both males and females carry 23 chromosomes, which unite at fertilization to restore the original number of 46. Within 9 to 10 weeks, the spermatocytes differentiate into sperm cells. Although sperm cells are produced in the testes, they mature in the **epididymis,** a compact, coiled tube attached to the outer edge of the testis. Sperm cells develop their flagella and begin swimming motions within four days. It is believed that some defective sperm cells are destroyed by the immune system during their time in the epididymis.

> **BIOLOGY CLIP**
> The male sperm cell is dwarfed by the much larger female egg cell. In humans, the egg cell is 100 000 times larger than the sperm cell.

Seminiferous tubules *are coiled ducts found within the testes, where immature sperm cells divide and differentiate.*

The epididymis *is located along the posterior border of the testis and consists of coiled tubules that store sperm cells.*

SEMINAL FLUID

Fluid is secreted by three glands along the vas deferens and the urethra: the seminal vesicle, the prostate gland, and the Cowper's (bulbourethral) gland. Every time a man ejaculates, between 3 to 4 mL of fluid, containing approximately 500 million sperm cells, are released. The fluid, referred to as semen, provides a swimming medium for the flagellated sperm cells. Fluids from the **seminal vesicles** contain fructose and prostaglandins. The fructose provides a source of energy for the sperm cell. Recall that the sperm cell carries little energy reserves in its drastically reduced cytoplasm. Prostaglandins act as a chemical signal in the female system, triggering the rhythmic contraction of smooth muscle along the reproductive tract. It is believed that the contraction of muscles along the female reproductive pathways assists the movement of sperm cells toward the egg. The **prostate gland** secretes an alkaline buffer that protects sperm cells against the acidic environment of the vagina.

Cowper's (bulbourethral) gland secretes mucus-rich fluids prior to ejaculation. It is believed that the fluids protect the sperm cells from the acids found in the urethra associated with the passage of urine. The fluid may also assist sperm movement.

Although sperm cells can exist for many weeks in the epididymis, they have a much reduced life span when they come in contact with the various fluids in the semen. At body temperature, sperm cells will live only 24 to 72 hours. When stored at −100°C, sperm cells have been known to remain viable for many years.

HORMONAL CONTROL OF THE MALE REPRODUCTIVE SYSTEM

Ancient herdsmen discovered that the removal of the testes, known as castration, increased the body mass of their animals, making their meat more tender and savory. The disposition of the castrated males also changed. Steers, which are castrated bulls, tend not to be very aggressive. The castrated animals also lack a sex drive and are sterile.

The male sex hormones—androsterone and **testosterone**—are produced in the *interstitial cells* of the testes. As the name suggests, the interstitial cells are found between the seminiferous cells. Although both hormones carry out many functions, testosterone is the more potent and abundant. Testosterone stimulates spermatogenesis, the process by which spermatogonia divide and differentiate into mature sperm cells. Testosterone also influences the development of secondary male sexual characteristics at puberty, stimulating the maturation of the testes and penis. Testosterone levels have also been associated with sex drive. Evidence comes from ancient times, when eunuchs—males who had had their testes

---

Seminal vesicles *contribute to the seminal fluid (semen), a secretion that contains fructose and prostaglandins.*

The **prostate gland** *contributes to the seminal fluid (semen), a secretion containing buffers that protect sperm cells from the acidic environment of the vagina.*

Cowper's (bulbourethral) gland *contributes a mucus-rich fluid to the seminal fluid (semen).*

Testosterone *is the male sex hormone produced by the interstitial cells of the testes.*

Anabolic steroids *are strength-enhancing drugs.*

Gonadotropic hormones *are produced by the pituitary gland and regulate the functions of the testes and ovaries.*

BIOLOGY CLIP

Boy sopranos are renowned for the beauty of their voices. However, as boys reach puberty, their larynxes begin to change and their voices become lower. During the 17th and 18th centuries, adult male sopranos were very popular. These singers, called *castrati*, had had their testes removed before puberty so that their voices would remain high. In 1878, Pope Leo XVII ended the inhumane practice of castrating boys for the papal choir.

---

removed—were used to guard the harems and households of these rulers. Because they no longer were able to produce testosterone, the eunuchs had a decreased sex drive.

The male sex hormone also promotes the development of facial and body hair, the growth of the larynx, which causes the lowering of the voice, and the strengthening of muscles. In addition, testosterone increases the secretion of body oils and has been linked to the development of acne in males as they reach puberty. Once males adjust to higher levels of testosterone, skin problems decline. The increased oil production can also create body odor. Testosterone, or testosterone-related compounds, are used in the production of **anabolic steroids,** the strength-building drugs often associated with athletes.

The production of sperm and male sex hormones in the testes is controlled by the hypothalamus and the pituitary gland in the brain. Negative-feedback systems ensure that adequate numbers of sperm cells and constant levels of testosterone are maintained. The pituitary gland produces and stores the **gonadotropic hormones,** which regulate the functions of the testes; the male **follicle-stimulating hormone (FSH),** which stimulates the production of sperm cells in the seminiferous tubules; and the male **luteinizing hormone (LH),** which promotes the production of testosterone by the interstitial cells.

At puberty, the hypothalamus secretes the **gonadotropin-releasing hormone (GnRH).** GnRH activates the pituitary gland to secrete and release FSH and LH. The FSH acts directly on the sperm-producing cells of the seminiferous tubules, while LH stimulates the interstitial cells to produce testosterone. In turn, the testosterone itself increases sperm production. Once high levels of

testosterone are detected by the hypothalamus, a negative-feedback system is activated. Decreased GnRH production slows the production and release of LH, leading to less testosterone production. Testosterone levels thus remain in check. The feedback loop for sperm production is not well understood. It is believed that FSH acts on Sertoli cells, which produce a peptide hormone that sends a feedback message to the pituitary, inhibiting production of FSH.

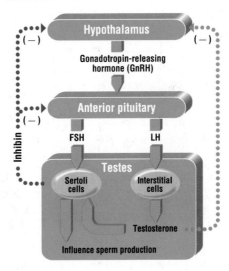

Figure 21.5

Negative feedback regulatory system for FSH and LH hormones. Testosterone inhibits LH production by the pituitary by deactivating the hypothalamus. The hypothalamus will release less GnRH, leading to decreased production of LH. The feedback mechanism for FSH is less understood. It has been suggested that a signalling chemical produced by the Sertoli cells inhibits both GnRH and FSH production.

The **follicle-stimulating hormone (FSH)** *increases sperm production in males.*

The **luteinizing hormone (LH)** *regulates the production of testosterone in males.*

The **gonadotropin-releasing hormone (GnRH)** *is a chemical messenger from the hypothalamus that stimulates secretions of FSH and LH from the pituitary.*

■ **REVIEW QUESTIONS** **?**

1 Name the primary male and female reproductive organs.

2 What would happen if the testes failed to descend into the scrotum?

3 Describe the function of the following structures: Sertoli cells, seminiferous tubules, and epididymis.

4 What is semen? Where is it found? What function does it serve?

5 What is spermatogenesis?

6 Outline the functions of testosterone.

7 How do gonadotropic hormones regulate spermatogenesis and testosterone production?

8 Using examples of LH and testosterone, explain the mechanism of negative feedback.

Objective

To view structures within the testes.

Materials

lens paper   prepared slide of testes (cross section)

light microscope   pencil

Procedure

1 Using lens paper, clean the ocular and all the objective lenses of the microscope. Rotate the revolving nosepiece so that the low-power objective is in place. Position the prepared slide on the stage of the microscope and view the cross section of the testes under low power.

a) Estimate the number of seminiferous tubules seen under low-power magnification.

2 Center the slide on a single seminiferous tubule and rotate the revolving nosepiece to the medium-power objective. Use only the fine adjustment to focus the cells. Locate an interstitial cell and a seminiferous tubule.

b) Estimate the size of the seminiferous tubule and the interstitial cell.

3 Rotate the nosepiece to the high-power objective lens and view the immature sperm cells within the seminiferous tubules.

c) Diagram five different cells viewed in the seminiferous tubule.

d) Would you expect to find mature sperm cells in the seminiferous tubule? Give your reasons. ■

THE FEMALE REPRODUCTIVE SYSTEM

In many ways the female reproductive system is more complicated than that of the male. Once sexual maturity is reached, males continue to produce sperm cells at a somewhat constant rate. By contrast, females follow a complicated sexual cycle, in which one egg matures approximately every month. Hormonal levels fluctuate through the reproductive years that end at **menopause.**

During fetal development, paired ovaries (flattened, olive-shaped organs) form near the kidneys. Like the similarly shaped testes, the ovaries migrate along a canal, but unlike the testes, which come to rest outside of the abdominal cavity, the ovaries remain in the pelvic region.

Egg cells, or *ova* (singular: "ovum"), are found within the ovary. The ovary is also responsible for the production of female sex hormones.

An **oviduct,** or **Fallopian tube** (named after Gabriello Fallopio, a 16th-century Italian anatomist), is found next to each of the ovaries. Once mature, an ovum will enter the oviduct through wide, open ends called *fimbria.* As the ovum is moved along the oviduct by cilia, it goes through its final stages of development. Fertilization of the ovum occurs in the oviduct. However, unless the ovum is fertilized, the cell will deteriorate within 48 hours and die. The paired oviducts join a hollow, inverted, pear-shaped organ called the **uterus,** or **womb.** The length of time required for the fertilized ovum to travel the

Menopause marks the termination of the female reproductive years.

The oviduct, or Fallopian tube, is the passageway through which an ovum moves from the ovary to the uterus, or womb.

CASE STUDY
HORMONE LEVELS DURING THE MENSTRUAL CYCLE

Objective

To investigate how hormone levels regulate the female menstrual cycle.

Procedure

1 Ovarian hormones are regulated by gonadotropic hormones. Study the feedback loop shown in the diagram below.

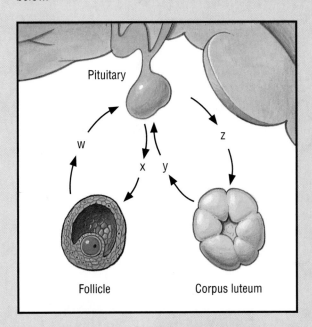

Pituitary

w

z

x   y

Follicle

Corpus luteum

a) Identify as w, x, y, or z, the two gonadotropic hormones represented in the diagram.

b) Identify the ovarian hormones shown in the diagram.

c) Which two hormones exert negative-feedback effects?

2 Body temperatures of two women were monitored during their menstrual cycles. One woman ovulated; the other did not.

d) Graph the data provided. Plot changes in temperature along the *y*-axis (vertical axis) and the days of the menstrual cycle along the *x*-axis (horizontal axis).

e) Assuming this menstrual cycle represents the average 28-day cycle, label the ovulation day on the graph.

Temperature °C		
Days	Ovulation occurs	No ovulation occurs
5	36.4	36.3
10	36.2	35.7
12	36.0	35.8
14	38.4	36.2
16	37.1	36.1
18	36.6	36.0
20	36.8	36.3
22	37.0	36.3
24	37.1	36.4
28	36.6	36.5

f) Describe changes in temperature prior to and during ovulation.

g) Compare body temperatures with and without a functioning corpus luteum.

3 The graph below shows changes in the thickness of the endometrium throughout the female menstrual cycle.

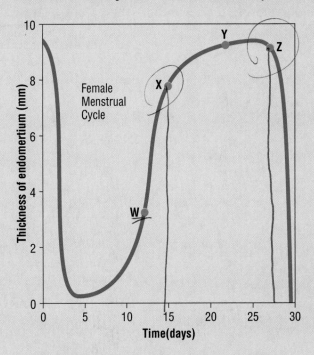

h) Identify the events that occur at times *X* and *Z*.

i) Identify by letter the time when follicle cells produce estrogen.

j) Identify by letter the time when the corpus luteum produces estrogen and progesterone.

4 Levels of gonadotropic hormones are monitored throughout the female reproductive cycle. Levels are recorded in relative units.

k) How does LH affect estrogen and progesterone?

Case-Study Application Questions

1 Explain why birth control pills often contain high concentrations of progesterone and estrogen.

2 Why would a woman not take birth control pills for the entire 28 days of the menstrual cycle? On which days of the menstrual cycle would the pill not be taken? ■

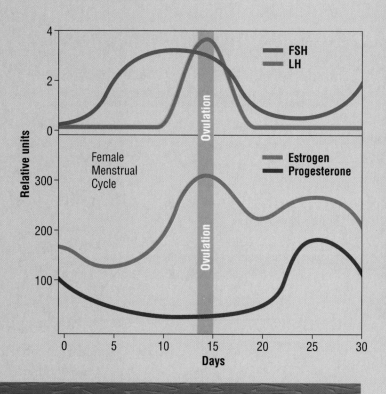

FERTILIZATION AND PREGNANCY

Between 150 million and 300 million sperm cells travel through the cervix into the uterus upon ejaculation during intercourse. However, only a few hundred actually reach the oviducts. Although several sperm become attached to the outer edge of the mature ovum, only a single sperm cell fuses with the ovum.

The fertilized egg, now referred to as a zygote, undergoes many divisions as it travels by way of the oviduct toward the uterus. By the time it reaches the uterus, in about six days, the single fertilized egg cell has been transformed into a cell mass, called a **blastocyst.** Once in the uterus, the blastocyst becomes attached to the wall of the endometrium, a process referred to as **implantation.**

For pregnancy to continue, menstruation cannot occur. Any shedding of the endometrium would mean the dislodging of the embryo from the uterus. However, maintaining the endometrium presents a problem for the hormonal system. High levels of progesterone and

Figure 21.10

Human sperm cell attached to ovum.

Blastocyst *is an early stage of embryo development.*

Implantation *is the attachment of the embryo to the endometrium.*

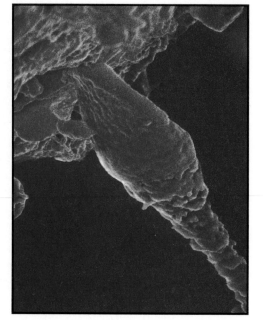

estrogen have a negative-feedback effect on the secretion of gonadotropic hormones. LH levels must remain high to sustain the corpus luteum. Should the corpus luteum deteriorate, the levels of estrogen and progesterone would drop, stimulating uterine contractions and the shedding of the endometrium. For pregnancy to continue, progesterone and estrogen levels must be maintained.

The outer layer of the developing cell mass forms a hormone called **HCG** (human chorionic gonadotropic hormone), which maintains the corpus luteum for the first three months of pregnancy. The functioning corpus luteum continues producing progesterone and estrogen, which in turn maintain the endometrium. The endometrium and embryo thus remain in the uterus. Pregnancy tests identify HCG levels in the urine of women.

Cells from the embryo and endometrium combine to form the placenta, through which materials are exchanged between the mother and developing embryo. At approximately the fourth month of pregnancy, the placenta begins to produce estrogen and progesterone. High levels of progesterone prevent further ovulation. This means that once a woman is pregnant, she cannot become pregnant again during that pregnancy.

PRENATAL DEVELOPMENT

The outer layer of the blastocyst gives rise to two cell membranes. The outer membrane is the **chorion,** which produces HCG. The inner membrane, the **amnion,** develops above the embryo.

The amnion evolves into a fluid-filled sac that insulates the embryo, and later the fetus, protecting it from infection, dehydration, impact, and changes in temperature. By the fourth week of pregnancy the yolk sac, which is a vestigial formation in humans, forms below the embryo.

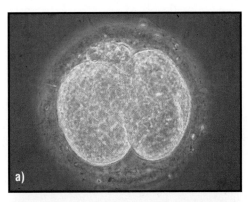

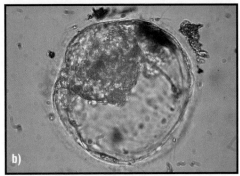

Figure 21.11

(a) Two-cell stage.
(b) Blastocyst after 4 to 6 days.

HCG *is a placental hormone that maintains the corpus luteum.*

The **chorion** *is the outer membrane of a developing embryo.*

The **amnion** *is a fluid-filled embryonic membrane.*

The **placenta** *is the site for the exchange of nutrients and wastes between mother and fetus.*

The **allantois** *is an embryonic membrane.*

> **BIOLOGY CLIP**
> It has been estimated that 1 in 85 births will produce twins, 1 in 7500 will produce triplets, 1 in 650 000 will produce quadruplets, and 1 in 57 000 000 will produce quintuplets.

Cells of the fetus and cells of the endometrium comprise the **placenta.** The placenta is richly supplied with blood vessels. Projections called *chorionic villi* ensure that many blood capillaries of the mother are exposed to a large number of blood capillaries of the fetus. A third membrane, the **allantois,** provides blood vessels in the placenta. However, unlike the chorion and amnion, the allantois does not envelop the fetus. The placenta provides a lifeline between mother and fetus. Nutrients and oxygen diffuse from the mother's blood into the blood of the developing fetus. Wastes diffuse in the opposite direction, moving from the fetus to the mother. The **umbilical cord** connects the embryo with the placenta.

The nine months of pregnancy are divided into three trimesters. The **first trimester** extends from fertilization to the end of the third month. By the second week of development, three germ layers begin to form: the ectoderm, the mesoderm, and the endoderm. Each of the organs shown in Table 21.3 develops from one of the germ layers.

*The **umbilical cord** connects the fetus to the placenta.*

*The **first trimester** extends from conception until the third month of pregnancy.*

Figure 21.12

Formation of the membranes that protect the embryo.

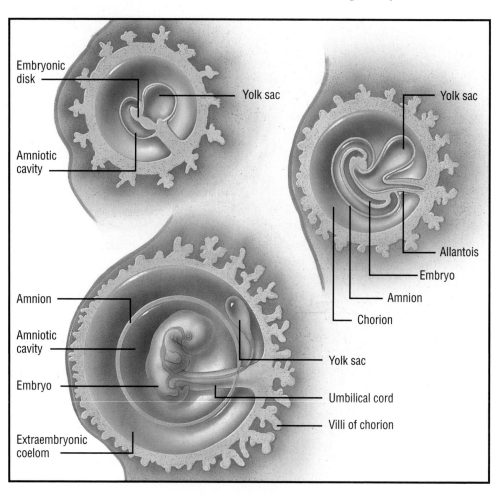

Table 21.3 Organs from Germ Layers

Germ Layer	Organ and Accessory Structures
Ectoderm	Skin, hair, finger nails, sweat glands
	Nervous system, brain, peripheral nerves
	Lens, retina, and cornea
	Inner ear, cochlea, semicircular canals
	Teeth, and inside lining of mouth
Mesoderm	Muscles (skeletal, cardiac, and smooth)
	Blood vessels and blood
	Kidneys and reproductive structures
	Connective tissue, cartilage, and bone
Endoderm	Liver, pancreas, thyroid, parathyroid
	Urinary bladder
	Lining of digestive system
	Lining of the respiratory tract

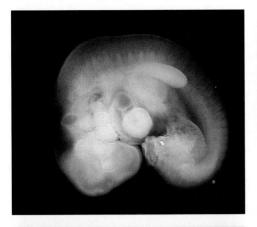

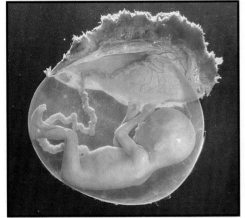
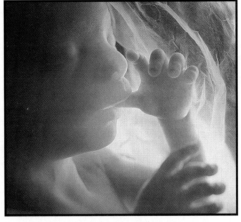

Figure 21.13

Human embryo at four weeks, and the fetus at nine weeks, sixteen weeks, and at eighteen weeks.

By the end of the first month, the 7 mm embryo is 500 times larger than the fertilized egg. The four-chambered heart has formed, a large anterior brain is visible, and limb buds with tiny fingers and toes have developed. By the ninth week, the embryo is referred to as a fetus. Arms and legs begin to move and a sucking reflex is evident.

By the **second trimester,** the 57 mm fetus moves enough to make itself known to the mother. All of its organs have formed and the fetus begins to look more like a human infant. As in other mammals, soft hair begins to cover the entire body. By the sixth month eyelids and eyelashes form. Most of the cartilage that formed the skeleton has been replaced by bone cells. Should the mother go into labor at the end of the second trimester, there is a chance that the 350 mm, 680 g fetus will survive.

During the **third trimester,** the baby grows rapidly. Organ systems have been established during the first two trimesters; all that remains is for the body mass to increase and the organs to become more developed. At birth, the average human infant is approximately 530 mm long and weighs about 3400 g.

BIRTH

Approximately 266 days after implantation, uterine contractions signal the beginning of labor. The cervix thins and begins to dilate. The amniotic membrane is forced into the birth canal. The amniotic membrane often bursts and amniotic fluid lubricates the canal (a process referred to as the breaking of the water). As the cervix dilates, uterine contractions move the baby through the birth canal.

*The **second trimester** extends from the third month to the sixth month of pregnancy.*

*The **third trimester** extends from the seventh month of pregnancy until birth.*

Hormones play a vital role in the birthing process. **Relaxin,** a hormone produced by the placenta prior to labor, causes the ligaments within the pelvis to loosen, providing a more flexible passageway for the baby during delivery. Although the actual mechanism is not completely understood, it is believed that a decreased production of progesterone is crucial to the onset of labor. **Oxytocin,** a hormone from the posterior pituitary gland, causes strong uterine contractions. Prostaglandins, which are also believed to trigger strong uterine contractions, appear in the mother's blood prior to labor.

LACTATION

Breast development is stimulated from the onset of puberty by estrogen and progesterone. During pregnancy, elevated levels of estrogen and progesterone prepare the breasts for milk production. Each breast contains about 20 lobes of glandular tissue, each supplied with a tiny duct that carries fluids toward the nipple. A hormone called **prolactin,** produced by the pituitary gland, is believed to be responsible for stimulating glands within the breast to begin producing fluids. Although small concentrations of prolactin are secreted throughout pregnancy, the levels rise dramatically after birth has occurred. The fact that the rise of prolactin levels coincides with rapid decreases in both estrogen and progesterone levels has led scientists to speculate that the female sex hormones suppress prolactin. Prolactin causes the production of a fluid called *colostrum,* a fluid that closely resembles breast milk. Colostrum contains milk sugar and milk proteins, but lacks the milk fats found in breast milk. A few days after birth, the prolactin stimulates the production of milk.

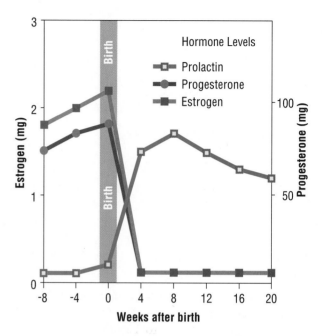

Figure 21.15

A dramatic increase in prolactin is attributed to lowering levels of estrogen and progesterone. Estrogen and progesterone levels drop following childbirth. Measurements for prolactin are in relative amounts.

Although prolactin increases milk production, the milk does not flow easily. Milk produced in the lobes of glandular tissue must be forced into the ducts that lead to the nipple. The suckling action of the newborn stimulates nerve endings in the areola of the breast. Sensory nerves carry information to the pituitary gland, causing the release of oxytocin. The hormonal reflex is completed as oxytocin is carried by the blood to the breasts and uterus. Within the breast, oxytocin causes weak contractions of smooth muscle, forcing milk into the ducts. Within the uterus, oxytocin causes weak contractions of smooth muscle, allowing the uterus to slowly return to its pre-pregnancy size and shape.

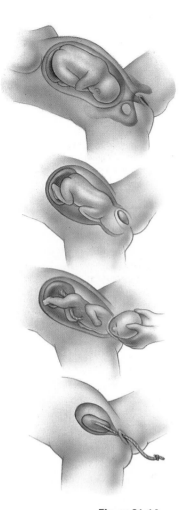

Figure 21.14

Movement of the baby through the birth canal.

Relaxin, *a hormone produced by the placenta prior to labor, causes the ligaments within the pelvis to loosen.*

Oxytocin, *a hormone from the posterior pituitary gland, causes strong uterine contractions.*

Prolactin *is a hormone produced by the pituitary and is associated with milk production.*

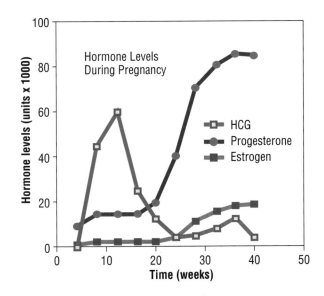

Hormone Levels During Pregnancy

8 During the eighth month of pregnancy, the ovaries of a woman are surgically removed. How will the removal of the ovaries affect the fetus? Explain your answer.

9 Explain why only one corpus luteum may be found in the ovaries of a woman who has given birth to triplets.

CRITICAL-THINKING QUESTIONS

1 Design an experiment to demonstrate the independent functions of the interstitial cells and seminiferous tubules of the testes.

2 Design an experiment to show how female gonadotropic hormones are regulated by ovarian hormones.

3 A 1988 article reported that prenatal testing had been banned in South Korea following a rise in male births. Seoul hospitals noted the change in the ratio of male to female births from 102.5 males for every 100 females, to 117 males for every 100 females. One statistician claimed that the change in ratio could be attributed to the selection of male fetuses following prenatal testing. In India, doctors have been told not to disclose the sex of a child following amniocentesis testing. Consider the implications for a society that attempts to select the sex of unborn children.

4 The embryo is sensitive to drugs, especially during the first trimester. In the late 1950s a drug called thalidomide was introduced in Europe to help prevent morning sickness. Unfortunately, the drug altered the genes regulating limb bud formation in developing embryos. Before the drug could be withdrawn from pharmacies, children were born with lifelong disabilities. Although drugs such as thalidomide are no longer available, a great debate centers around drugs that may have less pronounced effects on the fetus. Some evidence suggests that tranquilizers may cause improper limb bud formation. Even some acne drugs have been linked to facial deformities in newborns, and the antibiotic streptomycin has been associated with hearing problems. Explain why the embryo is so sensitive to drugs.

ENRICHMENT ACTIVITIES

Suggested reading:
- Kolata, Gina. "Operating on the Unborn." *New York Times Magazine*, May 14, 1989, p. 34.
- Krantorvitz, Barbara. "Preemies." *Newsweek*, May 16, 1988.
- Nilson, Leonard. "The First Days of Creation." *Life*, August 1990, p. 26.
- Poole, William. "The First 9 Months of School." *Hippocrates*, July/August 1987, p. 68.
- Vines, Gail. "Test Tube Embryos." *New Scientist*, November 19, 1987, pp. 1–4.
- Weiss, Richard. "Genetic Gender Gap." *Science News*, May 20, 1989, p. 312.
- Wolinsky, Howard. "Transplants from the Unborn." *American Health*, April 1988, p. 47.

Asexual Cell Reproduction

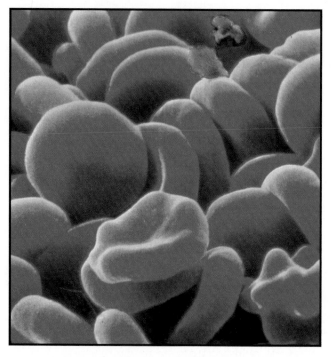

IMPORTANCE OF CELL DIVISION

For evidence of the importance of cell division you need look no further than your own body. The estimated 100 trillion cells that make up your body began from a single fertilized egg. In addition, your body is continually producing new cells. Red blood cells die and are replaced at a rate of one million every second. By the time you complete this biology course you will have totally replaced all of the red blood cells that are currently in your body.

The importance of cell division cannot be underestimated. Have you ever peeled the skin from your back after a sunburn? Imagine what you would look like if new cells did not replace the dead skin cells that had been sloughed off! Without cell division, scratches would never heal and blemishes would never disappear. If your red blood cells did not divide and reproduce, you would die.

Cell division is one of the most studied, yet least understood, areas of biology. Through painstaking hours of observation, scientists have collected a great deal of information about cell division. Yet despite all that has been observed, many questions remain unan-

Figure 22.2

A normal person contains about 25 trillion red blood cells, yet the average life span of a red blood cell is only 20 to 120 days.

Figure 22.1

Early stages of cell division in a frog.

swered. How do cells know when to divide? Why does an egg cell divide so rapidly after fertilization? How does the rate of cell division change (in the formation of calluses, for example)? Why do red blood cells divide at enormous rates, while adult brain cells do not? Why do cancer cells divide at uncontrolled rates?

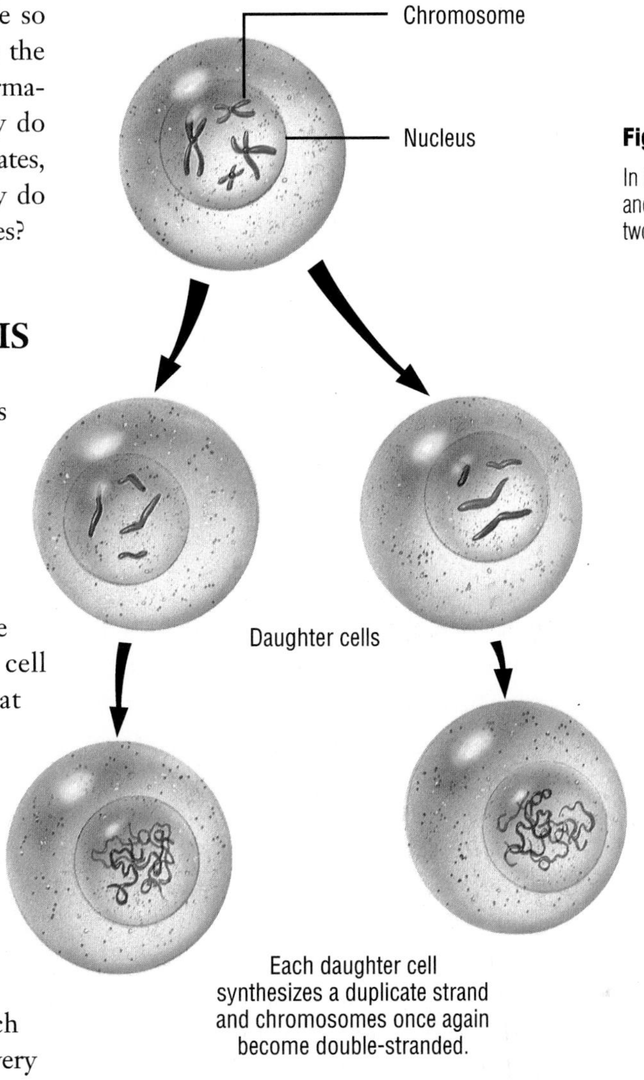

PRINCIPLES OF MITOSIS

The first part of the cell theory states that all living things are made up of one or more cells. It was nearly 40 years after the discovery of the cell that the second part of the modern cell theory was formulated. Scientists concluded that cells are formed from pre-existing cells by cell division. It was earlier believed that some single-cell organisms developed from nonliving objects. The linking of life to reproduction was an important step in the development of modern biology.

Despite great differences in the forms and structures of living things, most cells show remarkable similarities in the manner in which they divide. Cell division occurs in very primitive forms of life, such as bacteria, which split into two identical daughter cells by a process known as **binary fission.** Multicellular organisms follow a similar mechanism. The initial cell, referred to as the mother cell, divides into two identical daughter cells. The division involves two phases: the division of nuclear materials and the division of cytoplasm. During the first phase, the genetic material, located along double-stranded chromosomes, divides and moves to opposite ends of the cell. The second phase of cell division is marked by the separation of cytoplasm, along with its constituent organelles, into equal parts.

Chromosome

Nucleus

Figure 22.3

In binary fission the nucleus and the cytoplasm divide into two equivalent parts.

Daughter cells

Each daughter cell synthesizes a duplicate strand and chromosomes once again become double-stranded.

The process, known as **mitosis,** repeats itself millions of times each day, replacing worn-out cells with new ones.

During mitosis the double-stranded chromosomes separate into single strands. Each of the daughter cells gets one of the single-stranded pairs of genetic material. The daughter cells use the single strands of genetic material to synthesize a complementary strand, and the chromosome again becomes double-stranded. The cell of a cow with 60 chromosomes will divide by mitosis to daughter cells of 60 chromosomes. Each

Binary fission *is a form of asexual reproduction in which one cell divides into two equal cells.*

Mitosis *is a type of cell division in which daughter cells receive the same number of chromosomes as the parent cell.*

cell in your body contains 46 chromosomes; all succeeding generations of cells will also contain 46 chromosomes. The duplication of complementary strands of genetic information ensures that the daughter cells are identical to each other and to the mother cell. The duplication of genetic information also ensures future cell divisions. Each daughter cell is a potential mother cell for the next generation.

Because all cells in the human body are duplicates of the same fertilized egg, all cells have the same genetic information. A muscle cell, for example, has all of the chromosomes of a cell found in your brain, or of a cell found in your heart. However, not all cells in the human body have the same shape or carry out the same functions. One of the most puzzling questions that confronts scientists who study cells is why different cells do different jobs. What makes a brain cell conduct nerve impulses, and what makes a muscle cell contract? How do specialized cells know which genes to use? These questions remain to be answered.

STAGES OF CELL DIVISION

Cell division is often described as taking place in phases. The phases help scientists describe the events of mitosis. Cell division, however, does not pause after each phase—it is a continuous process.

Interphase

Interphase describes the processes of cell activity between cell divisions. Most cells, even the rapidly dividing, immature red blood cells, spend the majority of their time in this phase. During interphase, cells grow, carrying out the chemical activities that sustain life during this period. Cells make structural proteins

that repair damaged parts, move nutrients from one organelle to another, and eliminate wastes. During interphase, cells prepare themselves for mitosis by building proteins. These proteins are used for new cell membranes as well as for constructing enzymes that aid chemical reactions including those that control the duplication of genetic information. The single-stranded chromosomes once again become double strands during interphase. If the cell did not duplicate its chromosomes during interphase, it would never again be able to divide.

Prophase

Prophase is the first true phase of cell division. Under the microscope, the contents of the nucleus become visible as the chromosomes shorten and thicken. In animal cells, a small body in the cytoplasm separates and moves to opposite poles of the cell as the chromosomes become visible. These tiny structures, called **centrioles,** provide attachment for **spindle fibers.** The spindle fibers serve as guide wires for the attachment and movement of the chromosomes during cell division. Small fragments of the spindle fibers, called *asters*, radiate out from the centrioles. The name aster, from the Greek word for "star," is appropriate because the centrioles appear as stars within the cytoplasm. Most plant cells do not have centrioles, but spindle fibers still form and serve a similar purpose. Not surprisingly, plant cells without centrioles do not have asters.

The chromosomes consist of two **chromatids,** held together by a tiny structure called a **centromere.** The centromere helps anchor the chromosomes to the spindle fibers. During prophase, the nuclear membrane appears to fade when viewed under the microscope; in effect, it is dissolving.

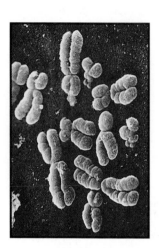

Figure 22.4
Double-stranded chromosomes just before mitosis.

Centrioles *are small protein bodies that are found in the cytoplasm of animal cells.*

Spindle fibers *are protein structures that guide chromosomes during cell division.*

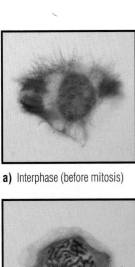

a) Interphase (before mitosis)

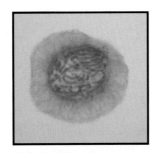

b) Early prophase

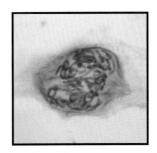

c) Prophase

d) Late prophase

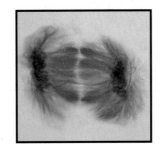

e) Transition to metaphase

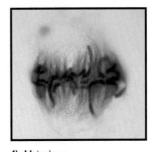

f) Metaphase

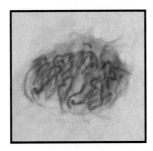

g) Anaphase

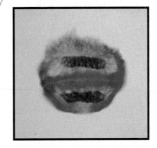

h) Telophase

i) Late telophase

Figure 22.5

Mitosis in a plant cell.

Metaphase

The second phase of cell division is referred to as metaphase. During this phase, chromosomes composed of sister chromatids move toward the center of the cell. This center area is called the *equatorial plate*, because, like the equator of the world, it is an imaginary line that divides the cell and its chromosomes. The chromosomes appear as dark, thick masses that are attached to the spindle fibers. Even though they are most visible at this stage, it is still very difficult to count the number of chromosomes in most cells. Chromatids can become intertwined during metaphase. Occasionally, segments of the chromatids will break apart once anaphase begins.

Anaphase

Anaphase is the third phase of mitosis. During this phase the centromeres divide and the chromatids move to opposite poles of the cell. If mitosis proceeds correctly, the same number of single-stranded chromosomes will be found at each pole.

Telophase

The last phase of cell division is telophase. During this phase the chromosomes reach the opposite poles of the cell, and once again begin to lengthen and intertwine. The spindle fibers dissolve and a nuclear membrane begins to form around each mass of **chromatin.** Telophase is followed by the division of the cytoplasm.

Chromatids *are single strands of a chromosome which remain joined by a centromere.*

Centromeres *are structures that hold chromatids together.*

Chromatin *is the material found in the nucleus and is composed of protein and DNA.*

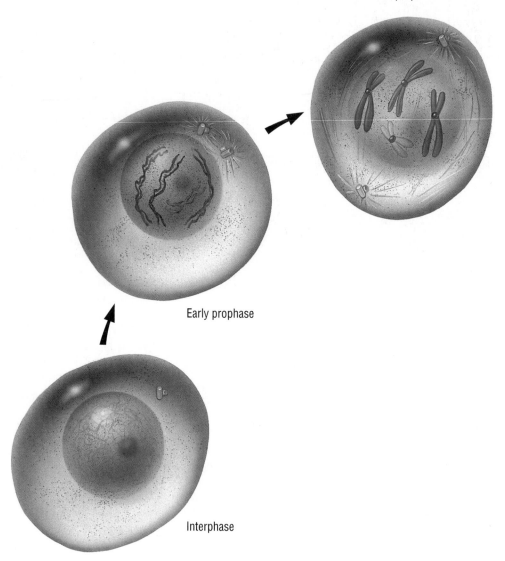

Late prophase

Early prophase

Interphase

Figure 22.6

Mitosis is the nuclear division mechanism that maintains the parental chromosome number in each daughter nucleus. Shown here is a diploid animal cell with pairs of homologous chromosomes derived from two parents.

For the sake of clarity, only two pairs of homologues are shown and the spindle apparatus is simplified. With rare exceptions, the picture is more involved than this, as indicated by the photomicrographs on the previous page.

Cytokinesis *refers to the division of cytoplasm.*

Cytoplasmic Division

Once the chromosomes have moved to opposite poles, the cytoplasm begins to divide. **Cytokinesis,** or the division of the cytoplasm, appears to be quite separate from the nuclear division. The division of the cytoplasm must provide organelles for each of the newly established cells. In plant cells, the separation is accomplished by a cell plate that forms between the two chromatin masses. The cell plate will develop into a new cell wall, eventually sealing the contents of the new cells from each other. Animal cells, however, do not have cell walls to use as barriers. The more elastic animal cells pinch off in the center as the cytoplasm moves to opposite poles.

■ REVIEW QUESTIONS ▮ ?

1 Compare the daughter cells with the original mother cell.
2 Describe the structure and function of the spindle fibers.

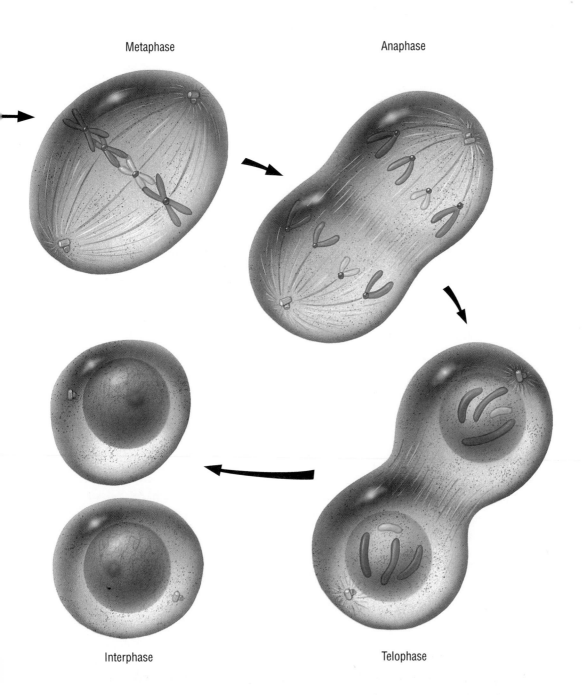

Metaphase

Anaphase

Interphase

Telophase

3 During interphase, what event must occur for the cell to be capable of undergoing future divisions?

4 Describe the events of prophase, metaphase, anaphase, and telophase.

5 Compare and contrast cell division in plant and animal cells.

6 What would happen if you ingested a drug that prevented mitosis?

BIOLOGY CLIP
A giant cell in the bone marrow undergoes cytoplasmic division without nuclear division. Small pieces of cytoplasm pinch off the cell to form blood-clotting platelets.

LABORATORY
MITOSIS

Objectives

To identify cells in the four phases of mitosis and to calculate the rate of cell growth.

Materials

microscope
prepared slides of onion root tip
lens paper
prepared slides of whitefish blastula

Procedure

Observing Dividing Cells

1   Obtain an onion root tip slide and place it on the stage of your microscope. View the slide under low power magnification. Focus using the coarse adjustment.

2   Center the root tip and then rotate the nosepiece to the medium-power objective lens. Focus the image using the fine-adjustment focus. Observe the cells near the root cap. This area is referred to as the *meristematic* region of the root.

3   Move the slide away from the root tip and observe the cells. These are the mature cells of the root.

   a) How do the cells of the meristematic area differ from the mature cells of the root?

4   Return the slide to the meristematic area and center the root tip. Rotate the nosepiece to the high-power objective lens. Use the fine adjustment to focus the image.

5   Locate and observe cells in each of the phases of mitosis. It will be necessary to move the slide to find each of the four phases. Use the pictures on the previous pages as guides.

   b) Draw, label, and title each of the phases. It is important to draw and label only the structures that you can actually see under the microscope.

6   Return your microscope to the low-power objective lens and remove the slide of the onion. Place the slide of the whitefish embryo on the stage and focus, using the coarse adjustment. Repeat the procedure that you followed for the onion and locate dividing animal cells under high-power magnification.

   c) Compare the appearance of the animal cells with that of the plant cells.

Determining the Rate of Cell Division

7   Count 20 adjacent whitefish embryo cells and record whether the cells are in interphase or are dividing.

   d) Cells in interphase   = ___
      Cells actively dividing = ___

   e) Calculate the percentage of cells that are undergoing mitosis:

   $$\frac{\text{Number of cells dividing}}{20} \times 100 = \_\_\_ \text{ \% dividing}$$

8   Repeat the same procedure for the meristematic region of the plant cell.

   f) Cells in interphase   = ___
      Cells actively dividing = ___

   g) Calculate the percentage of cells that are undergoing mitosis:

   $$\frac{\text{Number of cells dividing}}{20} \times 100 = \_\_\_ \text{ \% dividing}$$

   h) Compare the percentage of animal cells that are undergoing mitosis with the percentage of plant cells that are undergoing mitosis.

Laboratory Application Questions

1   Predict what will happen if both sister chromatids move to the same pole during mitosis.

2   A cell with 10 chromosomes undergoes mitosis. Indicate how many chromosomes would be expected in each of the daughter cells.

3   Predict what will happen if a small mass of cells breaks off from a human blastula.

4   Herbicides like 2,4-D and 2,4,5-T stimulate cell division. Why does the stimulation of cell division make these chemicals effective herbicides? ■

CLONING

Cloning is the process in which identical offspring are formed from a single cell or tissue. Because the clone originates from a single parent cell, it is identical to, or nearly identical to, the parent. Although some clones show accidental changes in genetic information, the majority do not show the variation of traits expected from the combination of male and female sex cells. For that reason, cloning is referred to as asexual or nonsexual reproduction.

Figure 22.7

Hydra reproduce asexually by budding. Note the small bud beginning to form on the side of the body.

The word "clone" comes from the Greek word referring to plant cutting. Some plants and animals reproduce naturally by this asexual method. The tiny *Hydra* shown in Figure 22.7 is a good example. The small outgrowth from the parent's body is called a bud. Eventually the bud will break off from the body of the parent and form a separate, but identical, organism. Plants, like strawberries, that reproduce by runners, produce clones. Unlike plants that grow from seeds, these plants have the added advantage of relying on the parent for nutrients. This gives the clone a head start.

Figure 22.8

The strawberry plant reproduces by sending out runners from its main stem.

Fredrick Stewart excited the scientific world in 1958 when he revealed a plant created from a single carrot cell. Stewart's research has had profound economic importance. Most orchids, for example, are produced from clones. Unlike plants that reproduce sexually, cloned plants are identical to their parents. This allows the production of strains of plants with predictable characteristics.

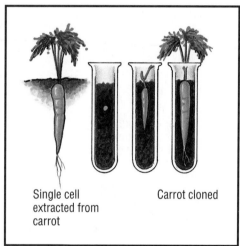

Single cell extracted from carrot

Carrot cloned

Figure 22.9

It is possible for a single cell isolated from a carrot to give rise to an entire carrot plant.

Plant tissue culture and cloning techniques have laid the groundwork for genetic engineering. While carrots, ferns, tobacco, petunias, and lettuce respond well to cloning, grass and legume families do not. So far, scientists do not know why. The secret seems to be hidden somewhere in the genetic makeup of the plant. Each regenerated cell contains the full complement of chromosomes and genes from the parent. Yet some cells specialize and become roots, stems, or leaves.

Figure 22.10

An experiment conducted by Briggs and King placed a nucleus from a frog cell in the blastula stage into an unfertilized, enucleated egg cell.

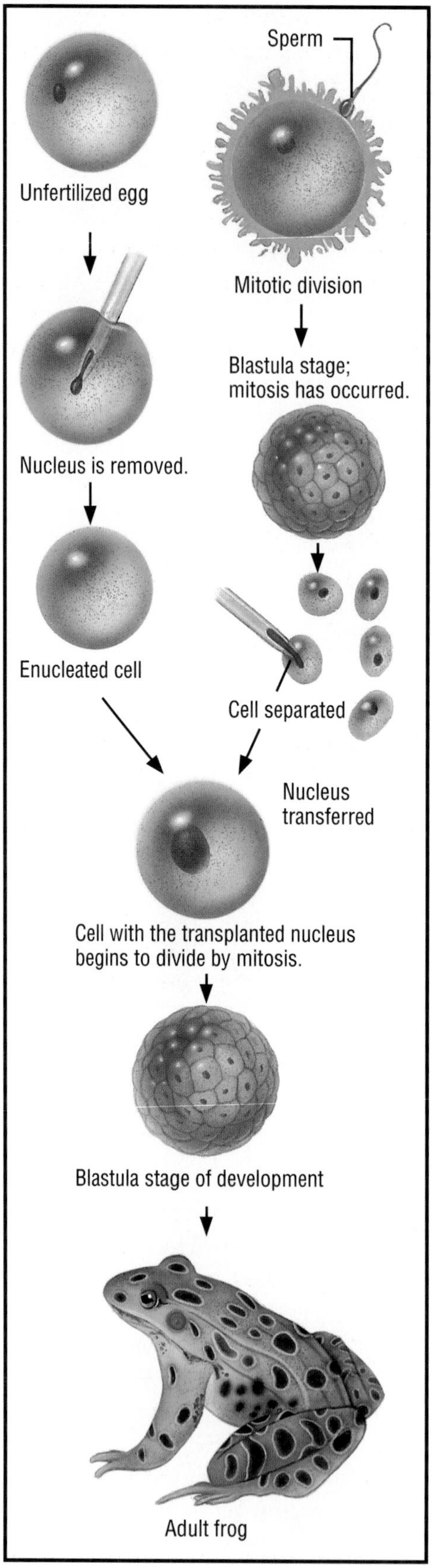

Enucleated cells *do not contain a nucleus.*

Blastula *refers to the early stage of development in which the cells of the dividing embryo form a hollow ball of cells.*

A **totipotent cell** *has the ability to support the development of an organism from egg to adult.*

Sperm

Unfertilized egg

Mitotic division

Nucleus is removed.

Blastula stage; mitosis has occurred.

Enucleated cell

Cell separated

Nucleus transferred

Cell with the transplanted nucleus begins to divide by mitosis.

Blastula stage of development

Adult frog

The cells within the leaf use only certain parts of their DNA, while the root cells use other segments of the DNA. Huge sections of the genetic information remain dormant in specialized plant cells. The trick in cloning plant cells appears to be in delaying specialization or differentiation.

While plant cloning experiments were being conducted, Robert Briggs and Thomas King were busy investigating nuclear transplants in frogs. Working with the common grass frog, the scientists extracted the nucleus from an unfertilized egg cell by inserting a fine glass tube, or *micropipette*, into the cytoplasm. The resulting cell without a nucleus is said to be **enucleated.**

Next, a nucleus from a cell of another frog in the **blastula** stage of development was extracted and inserted into the enucleated cell. The egg cell with the transplanted nucleus began to divide much like any normal fertilized egg cell and eventually grew into an adult frog. Careful analysis proved that the adult was a clone of the frog that donated the nucleus.

However, Briggs and King obtained different results when they transplanted the nucleus of a cell in the later gastrula stage of development. This time the nucleus did not bring the enucleated cell from the single-cell stage to the adult. The cell did not divide. Biologists use the term **totipotent** to describe a nucleus that is able to bring a cell from egg to adult. Something must be missing from the nuclear material of these cells. Since the nucleus of cells in the gastrula stage, unlike the cells of the earlier blastula stage, have begun to specialize, perhaps some regulatory mechanism is turning the genes off.

In another experiment, a scientist took the nuclei from the gut cells of tadpoles of the African clawed toad and inserted them into egg cells whose nuclei had been destroyed by ultraviolet radiation. Although many of these cells failed to develop, some grew into adults. Analysis of the adults confirmed that they were clones. The nuclei of the cells from the gut of the toad remain totipotent much longer than do those of the grass frog.

Scientists have also been able to clone mammal cells. However, successful cloning of mice does not mean that the cloning of humans is just around the corner. Adult mammal cells are not totipotent. However, clones can be obtained by splitting cells of a developing embryo. It would appear that cells must be taken before the eight-cell stage of development to ensure that their nuclei are totipotent. After the eight-cell stage, the cells specialize, and the genetic switch is turned off.

At present, the prospect of cloning humans belongs more to the realm of science fiction than true science, but few scientists would deny that it could ever happen. Genetics has progressed faster than even the most imaginative scientists could have predicted 25 years ago. Who knows what lies in store 25 years from now? Will we ever clone humans? Will we be able to use our clones for organ transplants? Will we be able to clone skin for burn victims? Today's answer is no, but tomorrow's answer may be yes. However, the social and moral consequences of cloning will have to be debated before such technology is embraced wholeheartedly.

FRONTIERS OF TECHNOLOGY: LIVER TISSUE TRANSPLANTS

A revolutionary medical procedure performed on a 21-month-old girl may provide a new direction in transplant surgery. Alyssa Smith has biliary atresia, a common and fatal childhood liver disease.

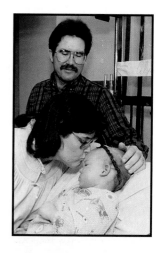

Figure 22.11

Alyssa Smith, shown with her mother and father, received transplanted cells from her mother's liver in a living liver-donor transplant operation.

Figure 22.12

Cloning mammals.

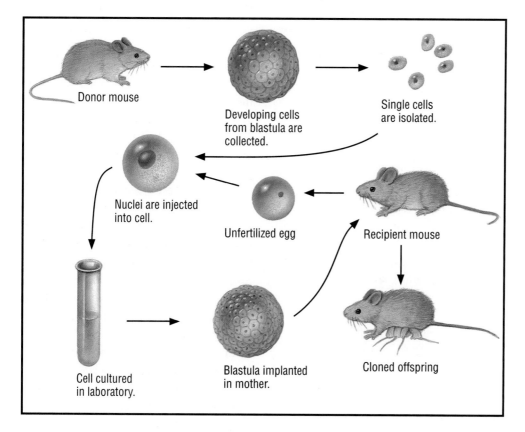

Donor mouse

Developing cells from blastula are collected.

Single cells are isolated.

Nuclei are injected into cell.

Unfertilized egg

Recipient mouse

Cell cultured in laboratory.

Blastula implanted in mother.

Cloned offspring

The liver performs many important functions. It helps maintain constant levels of blood sugar, detoxifies poisons, extracts wastes from the blood, produces bile salts for fat digestion, and stores minerals and vitamins. In this new procedure, doctors removed the left lobe of her mother's liver, and implanted fragments into Alyssa. Because the liver tissue is regenerative, cells from the transplanted liver fragments will continue to divide and grow into a normal liver.

Doctors hope that the transplanted liver will prevent illness as the diseased liver begins to fail. The idea of using a living donor has many benefits, the most obvious being a much wider accessibility to donors.

IDENTICAL TWINS: NATURE'S CLONES

Identical twins originate from a single egg cell. During mitosis, one of the cells breaks free and a second embryo begins to develop. Should the cell masses remain separated, two people with identical gene structures will develop. Identical twins are nature's clones. Barring the possibility of genetic accidents, identical twins are the same sex, have the same blood type, and have similar facial structures.

Fraternal twins originate from two different eggs and are fertilized by different sperm cells. Although both eggs of the fraternal twins carry the same number of chromosomes, they do not have the same genes. Fraternal twins are no more similar than any other sisters or brothers. They merely share the uterus or womb at the same time.

Identical twins provide scientists with an opportunity to study which human characteristics are influenced by genes and which are influenced by environmental factors. In one case, two identical twin

Figure 22.13

The top photograph shows identical twins; the bottom photograph shows fraternal twins.

brothers, who were separated at birth, both gravitated to careers in firefighting. They were both married in the same year and had families of four. Both men became fathers in the same year and lived in similar houses. A coincidence, you say? Perhaps—but other studies show similar findings.

■ REVIEW QUESTIONS ■ ?

7 What is cloning?
8 Discuss the economic importance of plant tissue cloning.
9 What is a nuclear transplant?
10 Define totipotency.
11 Why is it difficult to clone human cells?
12 Why are identical twins often called "nature's clones"?

CELL DEATH AND THE AGING PROCESS

People today are living longer. The number of people over the age of 65 accounts for approximately 10% of our total population. By the year 2030, about 20% of Canada's population will be over the age of 65. The question "Why do cells age?" is becoming central to a growing field of research into the aging process.

Why do some cells in the body divide faster than others? Red blood cells divide at an incredible rate. They live a mere 120 days, while the nerve cells in the adult brain do not divide at all. The giant redwood trees are some of the oldest organisms on earth, yet their oldest living cells are only about 30 years old—the vast majority of the cells in the tree are dead. An understanding of old age must be explored in terms of cell lineages, rather than cell age, since few living things go through life with the same cells.

Research on cells grown in tissue culture seems to indicate that a biological clock regulates the number of cell divisions available to cells. When immature heart cells maintained in tissue culture are frozen, they reveal an internal memory. If a cell has undergone 20 divisions before freezing, it will complete another 30 divisions once it is thawed. However, after the thirtieth division, the cell dies. When a cell is frozen after 10 divisions, it completes another 40 divisions when thawed. The magic number of 50 divisions is maintained, no matter how long the freezing or at what stage the division is suspended.

Not all cells of the body have the same ability to reproduce. Skin cells, blood cells, and the cells that line the digestive tract reproduce more often than do the more specialized muscle cells, nerve cells, and secretory cells. Only two cell types in the human body seem to escape aging: the sperm-producing cells, called spermatocytes, and the cells of a cancerous tumor. Males are capable of producing as many as one billion sperm cells a day from the onset of puberty well into old age. However, once the spermatocyte specializes and becomes a sperm cell, it loses its ability to undergo further divisions. Cancer cells divide at such an accelerated rate that they do not have adequate time for specialization. For example, white blood cells that divide too rapidly are not able to carry out their normal functions. Therefore, people who have leukemia—white blood cell cancer—find themselves with a reduced ability to fight infections. Cells need time to specialize. It would appear that the more specialized a cell is, the less able it is to undergo mitosis. The fertilized egg cell is not a specialized cell; it only begins to differentiate after many divisions. Interestingly, it is at that point that the alarm clock within the cell is turned on. Specialized cells have a limited life span.

Once again, new information raises even more questions. Do all cells age in a similar fashion? Does specialization cause aging? Will an understanding of cancer be linked to the study of aging?

The basic question "What causes aging?" has not yet been answered, but the theory that a clock exists in different cells has provided the groundwork for testing many different hypotheses. One hypothesis suggests that aging is caused by mutations to the genetic messages. The longer the exposure to environmental chemicals, the greater is the probability of damage to the genetic code. However, work with cells grown in tissue cultures does not appear to support this theory. Cells grown in controlled environments still die. There are a finite number of divisions. It has been estimated that a human who did not encounter disease would only live to be about 115. After 115 years, new cells cannot be generated to replace worn-out cells—mitosis has run its course. This estimate seems to be supported by historical records. Even though the average person lives longer today, the maximum longevity of humans has changed little since the days of ancient Rome.

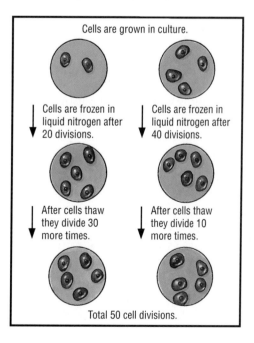

Cells are grown in culture.

Cells are frozen in liquid nitrogen after 20 divisions.

Cells are frozen in liquid nitrogen after 40 divisions.

After cells thaw they divide 30 more times.

After cells thaw they divide 10 more times.

Total 50 cell divisions.

Figure 22.14

Cell division appears to be regulated by a biological clock. For immature heart cells in tissue culture, the total number of divisions does not exceed 50.

A second hypothesis proposes that there are aging genes that shut down chemical reactions within the body. This theory would explain the graying of the hair by a gene message that turns off melanin production. However, the graying of the hair, although associated with aging, does not provide an accurate indicator of age. In some individuals the hair begins to gray in their early twenties or thirties, while some people show only traces of gray hair well into their seventies.

A third theory suggests that aging occurs when cell lineages die out. According to this theory, hair turns gray because the melanin-producing cells no longer divide, and worn-out cells are not replaced. Melanin cells might only live for short periods of time.

If aging is due to a cellular time limit, would decreased activity reduce the wear and tear on tissues? Would inactive people live longer? The answer is no. Evidence from astronauts in weightless environments indicates that more rapid cell deterioration took place. Hospital studies of bedridden patients also support the finding that activity not only helps recovery, but may even reduce aging.

The graph in Figure 22.15 shows some of the organ deterioration that occurs with aging.

ABNORMAL CELL DIVISION: CANCER

Cancer is a broad group of diseases characterized by the rapid, uncontrolled growth of cells. Unlike most diseases, which are associated with cell or tissue death, cancer can be described in terms of too much life. Cancer cells are extremely active, dividing at rates that far exceed those of normal cells.

You began life as a single cell that divided and then specialized to become stomach cells, nerve tissue, and blood. Your cells now only divide to replace damaged cells. A balance between cell destruction and cell replacement maintains a healthy organism. You already know that cells communicate information about the body's needs from one cell to another. The growth of a callus on your hand after hours of gardening is an example of how cellular communication meets the changing needs of the body. In this example, accelerated cell divisions not only replace damaged cells, but by increasing the cell numbers of the protective outer layers of the skin, they also shield the delicate nerve and blood vessels that occupy the skin's inner layer.

Normal cells cannot divide when isolated from one another. Cell-to-cell communication is essential for cellular reproduction. Cancer cells, on the other hand, are capable of reproducing in isolation. A cancer cell growing in an artificial

Figure 22.15

The graph shows the percentage function remaining in two organ systems after the age of 30.

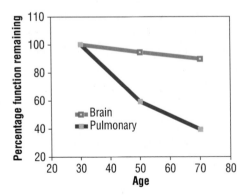

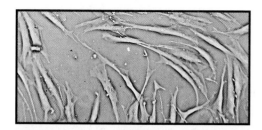

Figure 22.16

Cancerous cells can often be identified by an enlarged nucleus and reduced cytoplasm.

culture is capable of dividing once every 24 h. At this rate of division, a single cancer cell would generate over one billion descendants in a month. Fortunately, cancer cells do not reproduce that quickly in the body of an organism. Such rapid growth would cause a tumor containing trillions of cells, with an estimated mass of 10 kg, within 6 weeks. Even the rapidly dividing cells of the embryo are not capable of such proliferation.

The cancer cell does not totally ignore messages from adjacent cells that regulate its rate of reproduction. Scientists have discovered that some cancers progress at a very slow rate, often stopping for many years, only to become active at a later date. However, all cancer cells demonstrate the ability to reproduce without directions from adjacent cells.

Most normal cells adhere to one another. No doubt this attraction can be associated with the communication that occurs between cells in a given tissue. Kidney, liver, and heart cells, even when isolated from each other in the laboratory, show a remarkable quality of attraction when placed close together. Cancer cells, however, do not adhere to other cancer cells; nor do they stick to normal cells. This ability to separate from other cell masses is what makes cancer growth so dangerous. Cancer cells can dislodge from a tumor and move to another area. This movement, called **metastasis,** makes the source of cancer difficult to locate and control. For example, when Canadian athlete Terry Fox was diagnosed with a rare form of bone cancer, amputation of his leg did not succeed in confining the cancer. During Fox's "Marathon of Hope" run across Canada, cancer was discovered in his lungs.

Another important difference between normal cells and cancer cells is that cancer cells do not change shape as they mature; they are said to lack the ability to differentiate. It has been estimated that there are about 100 different types of cells in the human body. Each cell type has a unique shape that enables it to carry out a specialized function. Since cancer cells do not mature and specialize like normal cells they are inefficient. Therefore, another threat arises because the cancer cells cannot carry out some of the functions of the normal cells.

■ REVIEW QUESTIONS ■ ?

13 Do all the cells of your body divide at the same rate? Explain.

14 What evidence suggests that cells contain a "biological counter"?

15 What evidence supports the theory that the human body is only capable of living about 115 years, even if disease were eliminated?

16 In what ways does cellular communication regulate the division of normal cells?

17 How rapidly can a cancer cell grown in tissue culture divide? Is it likely that the cancer cell could divide that quickly in the human body? Explain.

Figure 22.17

Terry Fox, during his "Marathon of Hope" run.

Metastasis *occurs when a cancer cell breaks free from the tumor and moves into another tissue.*

LABORATORY

IDENTIFICATION OF A CANCER CELL

Objectives

To identify cancerous cells and to recognize the differences between cancerous and noncancerous cells.

Materials

light microscope
lens paper
prepared slide of squamous cell carcinoma

Procedure

1 Clean the microscope lenses with lens paper before beginning. Rotate the revolving nosepiece to the low-power objective.

2 Place the slide of the carcinoma on the stage of the microscope and bring the image into focus using the coarse-adjustment knob.

   a) Locate the dermal and epidermal layers. Draw a line diagram showing the position of the epidermal and dermal cell layers.

   b) Are the cells of the epidermis invading the dermis?

3 Rotate the revolving nosepiece to medium-power magnification and locate a cancerous cell. Use the fine-adjustment focus to bring the image into view. Note how cells of the carcinomas have a much larger nucleus. They appear pink in color and often have an irregular shape.

4 Rotate the nosepiece to high-power magnification, and bring the image into focus. Remember to use only the fine-adjustment focus, and always focus away from the slide.

   c) Estimate the size of a cancerous cell in micrometers (μm).

d) Estimate the size of the nucleus of a cancerous cell in micrometers (μm).

e) Determine the nucleus-to-cytoplasm ratio:

$$\text{Ratio} = \frac{\text{nucleus (μm)}}{\text{cytoplasm (μm)}}$$

5 Locate a normal cell. Using the same procedure as in step 4, determine the nucleus-to-cytoplasm ratio.

   f) Record the nucleus-to-cytoplasm ratio for the normal cell.

6 Compare the cancerous and normal cells in a table similar to the one below.

   g)

	Cell size	Nuclear shape	Nuclear size
Normal cell			
Cancerous cell			

Laboratory Application Questions

1 Cancerous cells are often characterized by a large nucleus. Based on what you know about cancer and cell division, provide an explanation for the enlarged nucleus.

2 Why are malignant (cancerous) tumors a greater threat to life than benign tumors?

3 Provide a hypothesis that explains why the skin is so susceptible to cancer.

4 A scientist finds a group of irregularly shaped cells in an organism. The cells demonstrate little differentiation, and the nuclei in some of the cells stain darker than others.

   a) Based on these findings, would it be logical to conclude that the organism has cancer? Justify your answer.

   b) What additional tests might be required to confirm or disprove the hypothesis that the cells are cancerous? ■

RESEARCH IN CANADA

One of Canada's leading cancer researchers is Dr. Margarida Krause from the University of New Brunswick. Dr. Krause is conducting research on cervical tumors, one of the leading causes of cancer in women. The papilloma viruses have been linked with small tumors on the surface lining of tissues such as the cervix. Dr. Krause's research will provide more information about how viruses interfere with the genetic information to cause cancer.

Dr. Grant McFadden from the Cross Cancer Institute of the University of Alberta is investigating how a normal healthy cell changes into a cancerous cell. Dr. McFadden is studying a virus called Shope fibroma virus which has been linked with tumors in wild animals. It would appear that once the virus infects a cell, the genetic machinery of the host cell shuts down, and the viral genetic information takes over. Dr. McFadden's research will provide important clues to how one type of cancer begins.

Dr. Nancy Simpson from Queen's University in Kingston is currently working on a rare hereditary cancer known as multiple endocrine neoplasia (MEN IIA). Dr. Simpson has traced the cancer gene to the tenth chromosome by using a technique called chromosome markers. The hereditary cancer, which causes tumors of the thyroid, is passed on to half of the children of an affected parent. Dr. Simpson's next stage of research will be carried on by a former student, Dr. Paul Goodfellow, at the University of British Columbia. He is attempting to clone the cancer-causing gene.

Cancer Research

DR. MARGARIDA KRAUSE

DR. NANCY SIMPSON

DR. GRANT MCFADDEN

SOCIAL ISSUE:
Biotechnology and Agriculture

Biotechnology in agriculture includes a variety of techniques designed to increase food production and protect existing strains of plants and trees. Cell fusion techniques can produce hybrid plants that grow faster and have greater resistance to disease. Tissue cloning provides offspring that are identical to the parents—a single parent can be selected to produce offspring with desired traits. Sexual reproduction, by contrast, permits variation, as different genes combine.

Agriculture Canada preserves genetic material for future use in the breeding of new varieties of crops, principally through seed storage. Currently some 82 000 seed samples, mainly barley and wheat, are stored at the Plant Research Centre in Ottawa. Smithfield Experimental Farm in Ontario stores over 1000 varieties of fruit trees, shrubs, and plants to ensure that their genes are preserved after the plants are no longer grown commercially.

Statement:
Tissue cloning and seed banks will provide cheaper and higher-quality food for future generations.

Point

- Tissue cloning provides offspring that are identical to their parents. A single parent can be selected to produce offspring with desired traits. Sexual reproduction, by contrast, permits variation, as different genes combine.

- Biotechnology can provide more food at lower cost. With half the world's population going hungry, we need new sources of cheaper, plentiful food.

Counterpoint

- Biotechnology such as tissue cloning may provide offspring with desirable characteristics, but it also reduces the number of genes that will be expressed. The lack of variation could be disastrous should the climate change or a disease attack a specific gene combination.

- Although biotechnology can provide more food at lower costs, can we trust scientists to control gene combinations? Some random gene combinations may prove less beneficial, other gene combinations could produce new strains of plants with better resistance to disease and better yield.

Research the issue.
Reflect on your findings.
Discuss the various viewpoints with others.
Prepare for the class debate.

CHAPTER HIGHLIGHTS

- Mitosis produces new cells for cell growth and for the replacement of worn-out cells in the body.
- Mitosis involves a series of steps that produce two identical daughter cells. Two divisions occur during mitosis: nuclear division and cytoplasmic division.
- During interphase, genetic material is duplicated.
- Cells can only divide a finite number of times.

- Cloning permits the production of offspring with characteristics identical to those of their parents.
- Totipotency is the ability of a cell to support the development of an egg to the adult form of a multicellular organism.
- Cell division helps us understand aging and cancer.
- Cancer is abnormal cell division.

INVESTIGATION

CAREER CAREER CAREER

Our economic system depends on the production and utilization of living organisms such as plants and animals. To arrange an equitable distribution of these resources among a changing human population, we must be aware of the mechanisms and problems inherent in the continuity of life.

Animal breeder
Animal breeders use techniques such as artificial insemination and fertility drugs to obtain high-quality livestock. Such techniques require the identification of the best breeding period as well as desirable traits.

Family planning counselor
Family planning counselors must be aware of the mechanisms of human reproduction as well as the methods available to slow its rate.

Child care worker
Working with children often involves attending to the special needs of those with genetic disorders such as Down syndrome.

Florist
A florist has to have a constant supply of flowers and plants ready for sale. This involves a detailed knowledge of the plants' different reproductive cycles, growing habits, and resistance to disease.

- Identify a career that requires a knowledge of the continuity of life.
- Investigate and list the features that appeal to you about this career. Make another list of features that you find less attractive about this career.
- Which high-school subjects are required for this career? Is a post-secondary degree required?
- Survey the newspapers in your area for job opportunities in this career.

SOCIAL ISSUE:
Limits on Reproductive Technology

Reproductive technologies include a variety of procedures that have allowed previously childless couples to have children. Debates surrounding two reproductive technologies in particular—surrogate motherhood and "test-tube" fertilization—have received wide media attention in the last several years.

It has been estimated that between 12 and 15% of couples are unable to have children. The demands on scientists to use reproductive technology have been increasing. However, does everyone have the right to produce children? The moral and ethical questions surrounding reproductive technology are far-reaching.

Statement:

Research in reproductive technology should be encouraged.

Point

- A technology like artificial involution was not developed out of any harmful motives. The technology was designed to produce better-quality beef cattle.
- Louise Brown and the other test-tube babies are the very human products of this technology. Clearly, the good outweighs any potential harm: couples who were once childless are now able to have children.

Counterpoint

- Although the technology may not have been developed for harmful purposes, it provides enormous potential for harm. The technology could be exploited to produce humans with selected traits.
- Reproductive technology is advancing faster than moral and ethical questions are being resolved. The fact that the concept of "motherhood" must be redefined is just one example. Scientific research should not determine our moral pathway.

> *Research the issue.*
> *Reflect on your findings.*
> *Discuss the various viewpoints with others.*
> *Prepare for the class debate.*

CHAPTER HIGHLIGHTS

- Meiosis involves the formation of sex cells.
- Cells pass through two divisions in meiosis.
- All gametes produced by meiosis have haploid chromosome numbers.
- Homologous chromosomes are similar in shape, size, and gene arrangement.

- Crossing-over is the exchange of genetic material between chromosomes that occurs during meiosis.
- Sex chromosomes are pairs of chromosomes that determine the sex of an organism.
- Abnormal meiosis, or nondisjunction, produces gametes with irregular chromosome numbers.

APPLYING THE CONCEPTS

1 Explain why sexual reproduction promotes variation.
2 Predict what might happen if the polar body were fertilized by a sperm cell.
3 A microscopic water animal called *Daphnia* reproduces from an unfertilized egg. This form of reproduction is asexual because male gametes are not required. Indicate the sex of the offspring produced.
4 Indicate which of the following body cells would be capable of meiosis. Provide a brief explanation for your answers.
   a) brain cells
   b) fat cells
   c) cells of a zygote
   d) sperm-producing cells of the testes
5 Compare the second meiotic division with mitotic division.

6 King Henry VIII of England beheaded his wives when they did not produce sons. Indicate why a little knowledge of meiosis might have been important for Henry's wives.
7 Explain how it is possible to produce a trisomic female XXX.
8 Abnormal cell division can produce an XYY condition for males. Diagram the nondisjunction that would cause a normal male and female to produce an XYY offspring.
9 A number of important genes are found on the X chromosome. Explain why many biologists suggest that genetic differences account for the fact that more male than female babies die shortly after birth.
10 Why might a physician decide to perform amniocentesis on a pregnant woman who is 45 years of age?

CRITICAL-THINKING QUESTIONS

1 According to one report, the incidence of infertility appears to be increasing in countries like Canada. The sperm count in males has fallen more than 30% in the last half-century and is continuing to fall. Although there is no explanation for this phenomenon at present, environmental pollution is suspected. Suggest other reasons for decreased fertility in males and females.
2 A couple who are unable to have children decide to hire another woman to carry the fetus. The procedure will cost them $10 000. They want absolute legal rights to the child, both while it is in the womb and when it is born. Discuss the ethical implications of hiring surrogate mothers.

3 Advances in modern medicine have increased the number of people who carry genetic disorders. People who might have died at an early age because of their disorder are living longer, marrying and producing children, and thus passing on their defective genes. Nondisjunction disorders could be eliminated by screening prospective sperm and egg cells. Sperm and egg banks could all but eliminate many genetic disorders. Comment on the implications to society of the systematic elimination of genetic disorders in humans.
4 A technique called egg fusion involves the union of one haploid egg cell with another. The zygote contains the full 2*n* chromosome number and is always a female. Discuss the implications for society if this technique were to be employed for humans.

ENRICHMENT ACTIVITIES

1 Contact the department of agriculture in your area and inquire about innovative genetic technology projects. How is farming changing?

2 Research a genetic disorder caused by nondisjunction. How is it detected? What are the physiological effects of the disorder? and psychological?

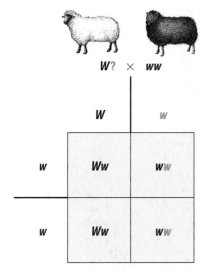

½ of the offspring are black and ½ are white.

All of the offspring are white.

Figure 24.11

A test cross involves crossing an individual of unknown genotype, showing a dominant trait, with a recessive. If any offspring show the recessive trait, then the individual must be hybrid. If all offspring show the dominant trait, then the individual must be homozygous.

TEST CROSS

You have probably heard someone referred to as the "black sheep of the family." Do you know why black sheep are not preferred? Black wool tends to be brittle and is very difficult to dye. How can a sheep rancher avoid getting black sheep? By using a homozygous white ram—the male of the species—a rancher can ensure that all of the flock will have white hair. But how would you know that a ram is homozygous for the white phenotype? Heterozygous white sheep will produce some black wool progeny.

A test cross is often performed to determine the genotype of a dominant phenotype. The test cross is always performed between the unknown genotype and a homozygous recessive genotype. In this case, the homozygous recessive individual would be a black sheep. If one-half of the offspring of the cross are black and the other half are white, then the unknown genotype must be heterozygous white. However, if all the offspring are white, then the unknown genotype must be homozygous dominant.

MULTIPLE ALLELES

For each of the traits studied by Mendel, there were only two possible alleles. The dominant allele controlled the trait. It is possible, however, to have more than two different alleles for one trait. In fact, many traits with multiple alleles exist in nature.

Geneticists who study the tiny fruit fly called *Drosophila* have noted that many different eye colors are possible. The red, or wild-type, is the most common, but apricot, honey, and white colors also exist. Although a fruit fly can only have two different genes at any one time, more than two alleles are possible. A fruit fly may have an allele for wild-type eyes and another for white. Its prospective mate may have an allele for apricot-colored eyes and another for honey-colored eyes. The dominance hierarchy is as follows: wild type is dominant to apricot, is dominant to honey, is dominant to white. In the case of multiple alleles, it is no longer appropriate to use upper- and lower-case letters. Capital letters and superscript numbers are used to express the different genes and their combinations.

Table 24.1 Dominance Hierarchy and Symbols for Eye Color in *Drosophila*

Phenotype	Genotypes	Dominant to
Wild type	$E^1E^1, E^1E^2, E^1E^3, E^1E^4$	Apricot, honey, white
Apricot	E^2E^2, E^2E^3, E^2E^4	Honey, white
Honey	E^3E^3, E^3E^4	White
White	E^4E^4	

The dominance and symbols for eye color in *Drosophila* are shown in Table 24.1

Consider the mating of the following *Drosophila*:

E^1E^4 (wild-type eye color) $\times$ E^2E^3 (apricot eye color)

The phenotypic ratio of the F$_1$ offspring is two wild-type eye color to one apricot eye color to one honey eye color. The Punnett square for this cross is shown in Figure 24.13.

Figure 24.13

A cross between a fruit fly with wild-type eye color and one with apricot eyes.

Figure 24.12

(a) *Drosophila melanogaster*, the fruit fly, is widely used for genetics studies. (b) Wild-type, or red, is the normal, or most common, eye color. It is the dominant allele.

a)

b)

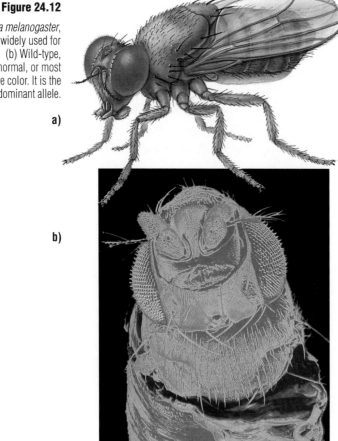

INCOMPLETE DOMINANCE

Prior to Mendel's studies, many scientists believed that hybrids would have a blending of traits. Although Mendel never found any examples of new traits being produced by the combinations of different genes, many do exist in nature. When two genes are equally dominant, they interact to produce a new phenotype. This lack of a dominant gene is known as *incomplete dominance*.

For example, if red snapdragons are crossed with white snapdragons, all of the F$_1$ offspring are pink. The pink color is produced by the interaction of red and white alleles. This type of incomplete dominance is often called *intermediate inheritance*. If the F$_1$ generation is allowed to self-fertilize, the F$_2$ generation produces a surprising ratio of one red to two pink to one white. The Punnett square in Figure 24.14 helps to explain this result.

Another type of incomplete dominance is referred to as *codominance*. In this type of gene interaction, both genes are expressed at the same time. Shorthorn cattle provide an excellent example of codominance. A red bull crossed with a white cow produces a roan calf. The roan calf has intermingled white and red hair. The roan calf would also be produced if a white bull were crossed with a red cow.

$C^R C^R$ = Red
$C^R C^W$ = Pink
$C^W C^W$ = White

Figure 24.14

Color in snapdragons is an example of incomplete dominance. Red-flowering and white-flowering snapdragons combine to produce pink-flowering plants in F_1. The F_2 generation produces one red to two pink to one white.

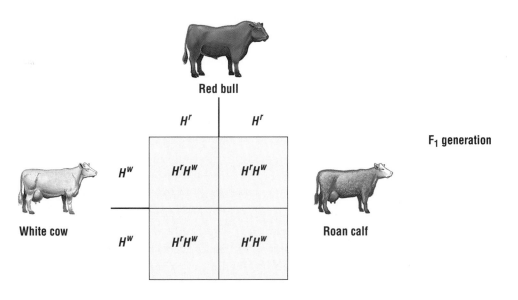

Red bull

White cow

H^r | H^r

H^w | $H^r H^w$ | $H^r H^w$

H^w | $H^r H^w$ | $H^r H^w$

F_1 generation

Roan calf

Figure 24.15

In codominance, the expression of one allele does not mask the expression of the other. In shorthorn cattle, roan calves have intermingled red and white hair.

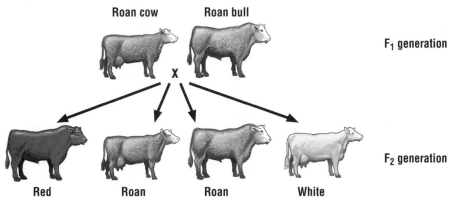

Roan cow   Roan bull

F_1 generation

X

F_2 generation

Red   Roan   Roan   White

REVIEW QUESTIONS ?

5 Multiple alleles control the intensity of pigment in mice. The gene D^1 designates full color, D^2 designates dilute color, and D^3 is deadly when homozygous. The order of dominance is $D^1 > D^2 > D^3$. When a full-color male is mated to a dilute-color female, the offspring are produced in the following ratio: two full color to one dilute to one dead. Indicate the genotypes of the parents.

6 Multiple alleles control the coat color of rabbits. A gray color is produced by the dominant allele C. The C^{cb} allele produces a silver-gray color when present in the homozygous condition, $C^{cb}C^{cb}$, called chinchilla. When C^{cb} is present with a recessive gene, a light silver-gray color is produced. The allele C^b is recessive to both the full-color allele and the chinchilla allele. The C^b allele produces a white color with black extremities. This coloration pattern is called Himalayan. An allele C^a is recessive to all genes. The C^a allele results in a lack of pigment, called albino. The dominance hierarchy is $C > C^{cb} > C^b > C^a$. The table below provides the possible genotypes and phenotypes for coat color in rabbits. Notice that four genotypes are possible for full-color but only one for albino.

Phenotypes	Genotypes
Full color	CC, CC^{ch}, CC^h, CC^a
Chinchilla	$C^{ch}C^{ch}$
Light gray	$C^{ch}C^h, C^{ch}C^a$
Himalaya	C^hC^h, C^hC^a
Albino	C^aC^a

a) Indicate the genotypes and phenotypes of the F_1 generation from the mating of a heterozygous Himalayan-coat rabbit with an albino-coat rabbit.

b) The mating of a full-color rabbit with a light-gray rabbit produces two full-color offspring, one light-gray offspring, and one albino offspring. Indicate the genotypes of the parents.

c) A chinchilla-color rabbit is mated with a light-gray rabbit. The breeder knows that the light-gray rabbit had an albino mother. Indicate the genotypes and phenotypes of the F_1 generation from this mating.

d) A test cross is performed with a light-gray rabbit, and the following offspring are noted: five Himalayan-color rabbits and five light-gray rabbits. Indicate the genotype of the light-gray rabbit.

7 A geneticist notes that crossing a round-shaped radish with a long-shaped radish produces oval-shaped radishes. If oval radishes are crossed with oval radishes, the following phenotypes are noted in the F_2 generation: 100 long, 200 oval, and 100 round radishes. Use symbols to explain the results obtained for the F_1 and F_2 generations.

8 Palomino horses are known to be caused by the interaction of two different genes. The allele C^r in the homozygous condition produces a chestnut, or reddish-color, horse. The allele C^m produces a very pale cream coat color, called cremello, in the homozygous condition. The palomino color is caused by the interaction of both the chestnut and cremello alleles. Indicate the expected ratios in the F_1 generation from mating a palomino with a cremello.

CASE STUDY

A MURDER MYSTERY

Objective

To solve a murder mystery using genetics.

Background Information

There are four different blood types. The alleles for blood type A and B are codominant but are dominant to O.

Phenotypes	Genotypes
Type A	$I^A I^A$, $I^A I^O$
Type B	$I^B I^B$, $I^B I^O$
Type AB	$I^A I^B$
Type O	$I^O I^O$

The rhesus factor is another blood factor that is regulated by genes. The Rh+ gene is dominant to the Rh− gene.

The Case Study

As a bolt of lightning flashed above Black Mourning Castle, a scream echoed from the den of Lord Hooke. When the upstairs maid peered through the door, a freckled arm reached for her neck. Quickly, the maid bolted from the doorway, locked herself in the library, and telephoned the police.

Inspector Holmes arrived to find a frightened maid and the dead body of Lord Hooke. Apparently, the lord had been strangled. The inspector quickly gathered evidence. He noted blood on a letter opener, even though Lord Hooke did not have any cuts or abrasions. The blood sample proved to be type O, Rh negative. The quick-thinking inspector gathered all the members of the family and began taking blood samples. The chart below shows the relatives who were in the castle at the time of Lord Hooke's murder.

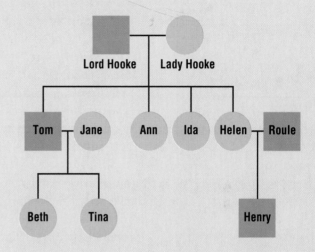

The inspector gathered the following information. Some of the family members were wearing long-sleeved shirts, so the inspector found it difficult to determine whether or not freckles were present on the arms. Note that the gene for freckles is dominant to the gene for no freckles.

The crafty inspector drew the family close together and, while puffing on his pipe, indicated that he had found the murderer. He explained that Lady Hooke had been unfaithful to her husband. One of the heirs to the fortune was not really a *sibling*. The murder was committed to preserve a share of the fortune.

Who was the murderer? State the reasons for your answer. How did the inspector eliminate the other family members? ■

Family member	Blood type	Rh factor	Freckles
Lord Hooke	AB	+	no
Lady Hooke	A	+	no
Helen	A	+	no
Roule	O	+	no
Henry	Refused blood test		?
Ida	A	−	?
Ann	B	+	?
Tom	O	−	no
Jane	A	+	?
Beth	O	−	?
Tina	A	+	yes

SCIENCE AND POLITICS: LYSENKO AND VAVILOV

During the 1940s, Trofim Denisovich Lysenko was considered the father of genetics in the Soviet Union. His theory for explaining genetic inheritance challenged the views of most western geneticists. Lysenko believed that acquired characteristics could be inherited. He maintained that, by exposing wheat to very cold or very arid conditions, he could produce a resistance that would affect genetic structure and be passed on to the next generation. A wheat plant resistant to cold or specially adapted to dry conditions would then arise.

Lysenko's theories were popular in the Soviet Union because he made his arguments fit Marxist philosophy. Joseph Stalin seized the opportunity to add science to Marx's economic and philosophical theories. Great political pressure was applied to silence opponents of Lysenko. Geneticists from the West were accused of fabricating research.

Despite an overwhelming flood of evidence that opposed Lysenko's theories, political pressure ensured that his views dominated Soviet genetics. An academic argument between Lysenko and the noted Soviet geneticist, Nikolai Vavilov, became political when Stalin sided with Lysenko. On August 6, 1940, Vavilov was arrested and subsequently sentenced to death for subversive acts. By his opposition to the theory of the inheritance of acquired characteristics, Vavilov became a threat to communist economic and social philosophy. His sentence was later reduced to ten years' imprisonment.

Lysenko retained a dominant position in Soviet science long after his theories had been discredited. It was not until 1965 that he was removed from his post as director of the Institute of Genetics, a position he had held since 1940.

Figure 24.16

Trofim Denisovich Lysenko (1898–1976) was a Soviet geneticist who believed that acquired characteristics could be inherited.

DIHYBRID CROSSES

Mendel also studied two separate traits with a single cross by following the same procedure he had used for studying single traits. He cross-pollinated pure-breeding plants that produce yellow, round seed coats with pure-breeding plants that produce green, wrinkled seed coats, in order to study the inheritance of two traits. Note that pure-breeding plants always produce identical offspring. The laws of genetics that apply for single-trait inheritance (monohybrid crosses) also apply for two-trait inheritance (*dihybrid* crosses).

Figure 24.17 shows a dihybrid cross. The pure-breeding round coat is indicated by the symbol *RR*, and the pure-breeding wrinkled coat by the recessive alleles *rr*. The pure yellow is indicated by the alleles *YY* and the green by *yy*. The genotype for the pure yellow, round parent is *RRYY*, while the genotype for the green, wrinkled parent is *rryy*. The F$_1$ offspring produced from such a cross are heterozygous for both the yellow and round genotypes.

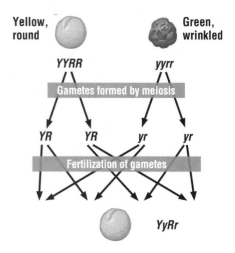

All members of the F$_1$ generation have the same genotype and phenotype.

Figure 24.17

A dihybrid cross between a pure-breeding pea plant with yellow, round seeds and a plant with green, wrinkled seeds.

Consider a cross between a pure-breeding green, round pea plant and a pure-breeding yellow, wrinkled pea plant. Figure 24.18 shows the resulting offspring. Inheritance of the gene for color is not affected by either the wrinkled or round alleles. By doing other crosses, Mendel soon discovered that the genes assort independently. Today, this phenomenon is referred to as *the law of independent assortment*. The genes that govern pea shape are inherited independently of the ones that control pea color.

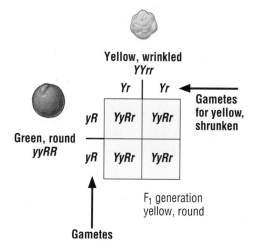

Figure 24.18

The inheritance of the gene for pea color is not affected by the gene for pea shape.

Mendel allowed plants of the F_1 generation to self-fertilize in order to produce an F_2 generation. Each heterozygous, yellow, round plant can produce four different phenotypes. Remember the law of independent assortment, which indicates that if they are located on separate chromosomes, they are inherited independently of each other. As the homologous chromosomes move to opposite poles during meiosis, the yellow gene will segregate with the round and wrinkled genes in equal frequency. This means that the sex cells containing YR will equal the number of sex cells containing yR. Similarly, the green genes will segregate

with round and wrinkled genes in equal frequency. The number of gametes containing Yr will be equal to the number of gametes containing the yr alleles.

Paired Chromosomes

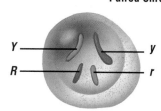

Figure 24.19

Four possible combinations for the gametes of the genotype *YyRr*.

A modified Punnett square (see below) can be used to predict the genotypes and phenotypes of the F_2 generation.

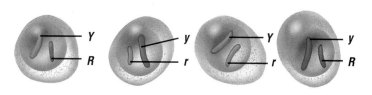

The phenotypes of the F_2 generation shown in the Punnett square in Figure 24.20 are as follows:

9/16 yellow, round   3/16 green, round
3/16 yellow, wrinkled   1/16 green, wrinkled

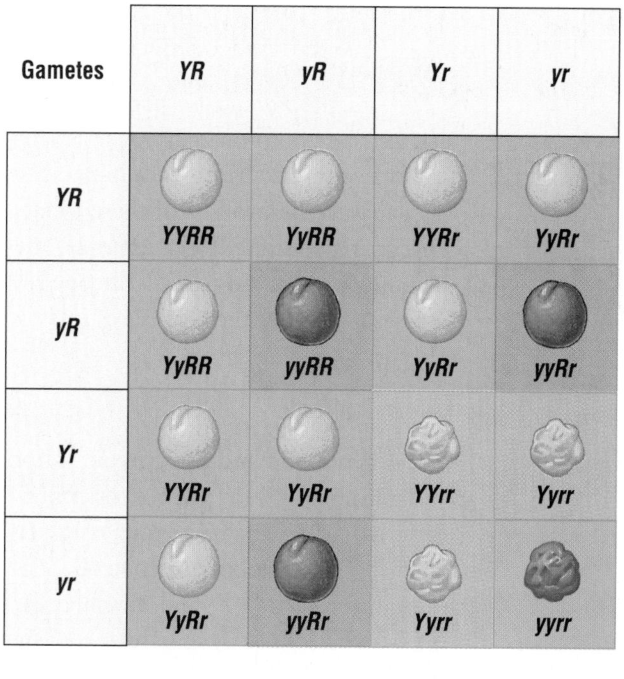

Figure 24.20

A Punnett square that has been modified to find the F_2 generation of a dihybrid cross.

PROBABILITY

Probability is the study of the outcomes of events or occurrences. For example, you can calculate the probability of getting heads when you toss a coin. Because there are only two possibilities—heads and tails—the chances of getting heads is 1/2. Probability can be expressed by the following formula:

$$\text{Probability} = \frac{\text{number of chances for an event}}{\text{number of possible combinations}}$$

Two important rules will help you understand probability:

- *The rule of independent events.* This rule states that chance has no memory. This means that previous events will not affect future events. For example, if you tossed two heads in a row, the probability of tossing heads once again would still be 1/2. Each event or occurrence must be regarded as an individual event.

- *The product rule.* The product rule states that the probability of independent events occurring simultaneously is equal to the product of these events occurring separately. For example, the chances of tossing heads after two tosses is 1/2, but the chances of tossing heads three times in a row is:

$$1/2 \times 1/2 \times 1/2 = 1/8$$

The genotypic and phenotypic ratios are determined by the chance, or probability, of inheriting a certain trait. In humans, free ear lobes are controlled by the dominant allele E, and attached ear lobes by the recessive allele e. The widow's peak hairline is regulated by the dominant allele W, while the straight hairline is controlled by the allele w. Consider the mating of the following genotypes:

EeWw × *EeWw*

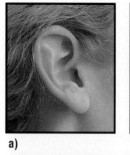
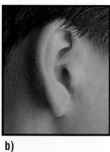

a)   b)

c)

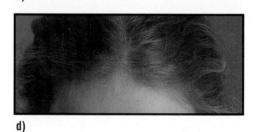

d)

Figure 24.21

The shape of both ear lobes and hairline are inherited characteristics in humans. (a) The free ear lobe is dominant to the attached ear lobe (b), and the widow's peak (c) is dominant to a straight hairline (d).

Dihybrids can be treated as two monohybrids, as shown in Figure 24.22. Isolate the gene for ear lobes and work with it as a monohybrid. The cross between the heterozygous parents can be determined by the F_1 generation. The phenotype probabilities are: free ear lobes 3/4, attached ear lobes 1/4.

Now consider the F_1 generation from the cross between the heterozygous parents for the hairline trait. The phenotype probabilities are: widow's peak 3/4, straight hairline 1/4.

	½ E	½ e
½ E	¼ EE	¼ Ee
½ e	¼ Ee	¼ ee

	½ W	½ w
½ W	¼ WW	¼ Ww
½ w	¼ Ww	¼ ww

Figure 24.22

Punnett squares showing two monohybrid crosses between heterozygous parents for free earlobes and for a widow's peak.

The monohybrids can now be combined to calculate the probabilities of a dihybrid cross. For example, the chances of producing an F_1 offspring from the mating of $EeWw \times EeWw$ who has

- a widow's peak and free ear lobes is: $3/4 \times 3/4$, or $9/16$;
- a straight hairline and free ear lobes is: $1/4 \times 3/4$, or $3/16$;
- a widow's peak and attached ear lobes is: $3/4 \times 1/4$, or $3/16$;
- a straight hairline and attached ear lobes is: $1/4 \times 1/4$, or $1/16$.

The calculation indicates that the dihybrid cross is equivalent to two separate monohybrid crosses. To determine the probability that the first child from the mating of the $EeWw \times EeWw$ parents would be a male with a widow's peak and attached ear lobes, consider each of the separate probabilities:

- The probability of producing a male is $1/2$.
- The probability of widow's peak is $3/4$.
- The probability of attached ear lobes is $1/4$.
- Therefore, the probability of producing a male with a widow's peak and attached ear lobes is $1/2 \times 3/4 \times 1/4$, or $3/32$.

9 In guinea pigs, black coat color (B) is dominant to white (b), and short hair length (S) is dominant to long (s). Indicate the genotypes and phenotypes from the following crosses:

a) Homozygous for black, heterozygous for short-hair guinea pig crossed with a white, long-hair guinea pig.

b) Heterozygous for black and short-hair guinea pig crossed with a white, long-hair guinea pig.

c) Homozygous for black and long-hair crossed with a heterozygous black and short-hair guinea pig.

10 Black coat color (B) in cocker spaniels is dominant to white coat color (b). Solid coat pattern (S) is dominant to spotted pattern (s). The pattern arrangement is located on a different chromosome than the one for color, and its gene segregates independently of the color gene. A male that is black with a solid pattern mates with three females. The mating with female A, which is white, solid, produces four pups: two black, solid, and two white, solid. The mating with female B, which is black, solid, produces a single pup, which is white, spotted. The mating with female C, which is white, spotted, produces four pups: one white, solid; one white, spotted; one black, solid; one black, spotted. Indicate the genotypes of the parents.

11 For human blood type, the alleles for types A and B are codominant, but both are dominant over the type O allele. The Rh factor is separate from the ABO blood group and is located on a separate chromosome. The Rh+ allele is dominant to Rh−. Indicate the possible phenotypes from the mating of a woman, type O, Rh−, with a man, type A, Rh+.

LABORATORY

GENETICS OF CORN

Objective

To investigate the inheritance of traits using corn.

Materials

dihybrid corn ears (sample A, sample B)

Procedure

1 Obtain a sample A corn ear from your instructor. The kernels display two different traits that are located on different chromosomes.

   a) Indicate the two different traits.

   b) Predict the dominant phenotypes.

   c) Predict the recessive phenotypes.

2 Assume that the ear of corn was from the F_2 generation. The parents were pure-breeding homozygous for each of the characteristics. Assign the letters P and p to the alleles for color, and S and s to the alleles for shape. Use the symbols $PPss \times ppSS$ for the parent generation.

   d) Indicate the phenotype of the $PPss$ parent.

   e) Indicate the phenotype of the $ppSS$ parent.

   f) Indicate the expected genotypes and phenotypes of the F_1 generation.

g) Use a Punnett square to show the expected genotypes and the phenotypic ratio of the F_2 generation.

3 Count 100 of the kernels in sequence, and record the actual phenotypes in a table similar to the one below.

Phenotype	Number	Ratio
Dominant genes for color and shape		
Dominant gene for color, but recessive for shape		
Recessive gene for color, but dominant gene for shape		
Recessive genes for color and shape		

4 Obtain sample B. Assume that this ear was produced from a test cross. Count 100 kernels in sequence and record your results.

   h) Indicate the phenotypic ratio of the F_1 generation (the kernel).

   i) Give the phenotype of the unknown parent.

Laboratory Application Questions

1 Why are test crosses important to plant breeders?

2 A dihybrid cross can produce 16 different combinations of alleles. Explain why 100 seeds were counted rather than 16.

3 A dominant allele Su, called starchy, produces kernels of corn that appear smooth. The recessive allele su, called sweet, produces kernels of corn that appear wrinkled. The dominant allele P produces purple kernels, while the recessive p allele produces yellow kernels. A corn plant with starchy, yellow kernels is cross-pollinated with a corn plant with sweet, purple kernels. One hundred kernels from the hybrid are counted, and the following results are obtained: 52 starchy, yellow kernels and 48 starchy, purple kernels. What are the genotypes of the parents and the F_1 generation? ■

SELECTIVE BREEDING

Farmers and ranchers have used **selective breeding** processes to improve domestic varieties of plants and animals. Early farmers identified plants with desirable characteristics. Rust-resistant wheat; sweet, full-kernel corn; and canola, which germinates and grows rapidly in colder climates, have improved food harvests. Selection of specific traits from wild cabbage has produced green and red varieties of cabbage, broccoli, and cauliflower.

You are probably familiar with the term "purebreds." Many dogs and horses are considered to be purebreds, or thoroughbreds. Genotypes of these animals are closely regulated by a process called **inbreeding,** in which similar phenotypes are selected for breeding. The desirable traits vary from breed to breed. For example, Irish setters are bred for their long, narrow facial structure and long, wispy hair. The bull terrier was originally bred for fighting. Quick reflexes and strong jaws were chosen as desirable phenotypes. Some geneticists have complained that inbreeding has caused problems for the general public as well as for the breed itself.

New varieties of plants and animals can be developed by hybridization. This process is the opposite to that of inbreeding. Rather than breed plants or animals with similar traits, the hybridization technique attempts to blend desirable but different traits. Corn has been hybridized extensively. The hybrids tend to be more vigorous than either parent. Figure 24.23 shows the most common method used.

Selective breeding *is the crossing of desired traits from plants or animals to produce offspring with both characteristics.*

Inbreeding *is the process whereby breeding stock is drawn from a limited number of individuals possessing desirable phenotypes.*

Figure 24.23

Hybridization can be used to produce a more vigorous strain of corn.

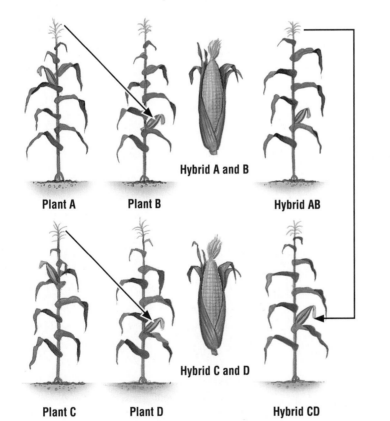

Plant A   Plant B   **Hybrid A and B**   Hybrid AB

Plant C   Plant D   **Hybrid C and D**   Hybrid CD

Hybrid ABCD

RESEARCH IN CANADA

The Plant Breeders

**SIR CHARLES
SAUNDERS
DR. KEITH DOWNEY**

Canada boasts a long history of contributions in plant genetics. One of our first great plant breeders was Sir Charles Saunders. Born in Toronto in 1837, Saunders was noted for developing and introducing Marquis wheat. Marquis is a high-quality, bread-producing wheat that was derived from a cross between the Red Calcutta and Red Fife varieties of wheat. Prior to the Marquis strain, Red Fife was the most common variety of wheat grown in Canada. However, the slower-maturing Red Fife was often damaged by early frosts, especially in Saskatchewan and Alberta. The Marquis variety matures at least a week earlier than Red Fife and also provides better yields.

Keith Downey, formerly of the University of Saskatchewan, and Baldur Stefansson, from the University of Manitoba, are recognized as world leaders in plant genetics. Downey was born in Saskatoon, Saskatchewan, in 1927, and Stefansson was born in Manitoba in 1917.

Their research has helped bring wealth to their respective provinces. Downey and Stefansson developed a rape seed with low levels of erucic acid and glucosinolate. The rape seed was transformed into a high-quality oil-seed crop, renamed and now known worldwide as canola.

DR. BALDUR STEFANSSON

GENE INTERACTION

The traits studied by Mendel are controlled by one gene. However, some traits are regulated by more than one gene. Many of your characteristics are determined by several pairs of independent genes. Skin color, eye color, and height are but a few of your characteristics that are **polygenic.**

Coat color in dogs provides an example of genes that interact. The allele *B* produces black coat color, while the recessive allele *b* produces brown coat color. However, a second gene, located on a separate chromosome, also affects coat color. The allele *W* prevents the formation of pigment, thereby preventing color. The recessive allele *w* does not prevent color. The genotype *wwBb* would be black, but the genotype *WwBb* would appear white. The *W* allele masks the effect of the *B* color gene. Genes that interfere with the expression of other genes are said to be **epistatic.**

Another type of gene interaction, called *complementary interaction*, occurs when two different genotypes interact to produce a phenotype that neither is capable of producing by itself. One of the best examples of complementary interaction can be seen in the combs of chickens. The *R* allele produces a rose comb. Another allele, *P*, located on a different chromosome, produces a pea comb. When the *R* and *P* alleles are both present, they combine to produce a walnut comb. The absence of the rose and pea alleles results in an individual with a single comb.

Polygenic *traits are inherited characteristics that are affected by more than one gene.*

Epistatic *genes mask the expression of other genes.*

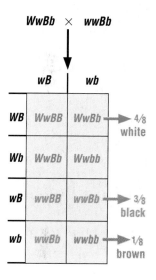

Figure 24.24

A cross between a white dog (*WwBb*) and a black dog (*wwBb*).

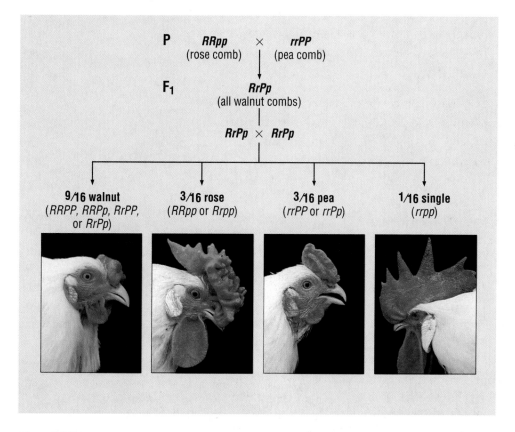

Figure 24.25

Sometimes two gene pairs cooperate to produce a phenotype that neither can produce alone. A cross between a chicken with a rose comb on the crest of its head and one with a pea comb results in a chicken with a walnut comb.

Genes regulate chemical reactions, providing the instructions for cells to build enzymes. Enzymes are protein molecules that speed up chemical reactions that occur in all cells. In some cases, enzymes help molecules combine to form larger molecules, and in other cases, they help larger molecules break down into less complicated molecules. The white clover plant is an excellent example of how genes interact in chemical reactions. The plant's production of cyanide is regulated by two genes. Figure 24.26 summarizes the biochemical pathway.

appears in the homozygous condition, the correct enzyme cannot be produced, and the reaction ends at glucoside. The genotype *GgHh* will produce cyanide, but the genotypes *ggHH* and *GGhh* are not capable of producing cyanide. However, a cross between a *ggHH* and *GGhh* plant will yield offspring capable of producing cyanide.

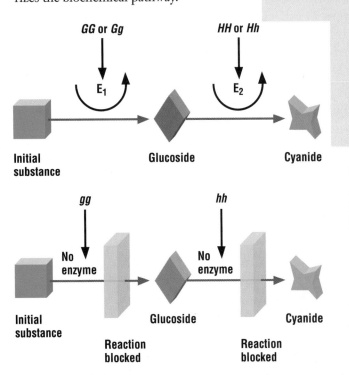

Figure 24.26

The production of cyanide in the white clover plant is regulated by two genes.

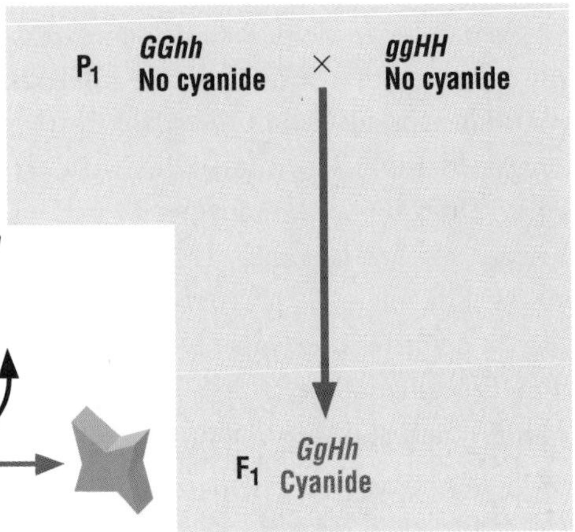

Figure 24.27

A cross between two white clover plants that are not capable of producing cyanide yields offspring that produce cyanide.

ONE GENE, MANY EFFECTS

Pleiotropic genes *are genes that affect many characteristics.*

The first allele, *G*, provides the genetic information for the production of enzyme #1 (E_1). If the recessive mutation, *g*, appears in the homozygous condition, the proper enzyme (E_1) is not produced. Unfortunately, the mutated enzyme will not work, and the reaction is stopped. The *gg* genotype will prevent the formation of the glucoside. Another allele, *H*, is also required. The allele *H* produces enzyme #2 (E_2), which converts glucoside to its final product, cyanide. If the allele *h*

Some genes, called **pleiotropic genes,** affect many different characteristics. Sickle-cell anemia, a blood disorder, is an example of how one gene can cause many devastating symptoms. Hemoglobin, the pigment that carries oxygen, is found in all red blood cells. Normal hemoglobin is produced by the allele Hb^A. Unfortunately, a mutated gene, Hb^S, causes difficulties when it appears in the homozygous condition. Although the hemoglobin pigment produced by the Hb^S gene can still carry oxygen, it creates problems once it gives oxygen to the cells of the body. The oddly shaped hemoglobin molecules begin to interlock with one another. The new arrangement of molecules changes the shape of the red blood cells, which become bent into a sickle shape.

The mutated red blood cells do not pass through the capillaries. Oxygen delivery is halted in the area of the blockage, and normal organ function is impaired. People with sickle-cell anemia can suffer from fatigue and weakness, an enlarged spleen, rheumatism, and pneumonia. Patients often show signs of heart, kidney, lung, and muscle damage.

Marfan's syndrome, an inability to produce normal connective tissue, is also associated with a single gene. Because connective tissue is found in many organs, the symptoms of Marfan's syndrome show up as eye, skeleton, and cardiovascular defects. Many historians and geneticists have speculated that Abraham Lincoln suffered from Marfan's syndrome. A picture of the late U.S. president shows a severe curling of the hands and an infolding of the limbs, characteristic of advanced stages of the disease.

■ REVIEW QUESTIONS ■ ?

12 In mice, the gene *C* causes pigment to be produced, while the recessive gene *c* makes it impossible to produce pigment. Individuals without pigment are albino. Another gene, *B*, located on a different chromosome, causes a chemical reaction with the pigment and produces a black coat color. The recessive gene, *b*, causes an incomplete breakdown of the pigment, and a tan, or light-brown, color is produced. The genes that produce black or tan coat color rely on the gene *C*, which produces pigment, but are independent of it. Indicate the phenotypes of the parents and provide the genotypic and phenotypic ratios of the F_1 generation from the following crosses.

a) *CCBB* × *Ccbb*   **b)** *ccBB* × *CcBb*
c) *CcBb* × *ccbb*   **d)** *CcBb* × *CcBb*

13 The mating of a tan mouse and a black mouse produces many different offspring. The geneticist notices that one of the offspring is albino. Indicate the genotype of the tan parent. How would you determine the genotype of the black parent?

14 The gene *R* produces a rose comb in chickens. An independent gene, *P*, which is located on a different chromosome, produces a pea comb. The absence of the dominant rose comb gene and pea comb gene (*rrpp*) produces birds with single combs. However, when the rose and pea comb genes are present together, they interact to produce a walnut comb (*R_P_*).

Indicate the phenotypes of the parents and give the genotypic and phenotypic ratios of the F_1 generation from the following crosses.

a) *rrPP* × *RRpp*   **b)** *RrPp* × *RRPP*
c) *RrPP* × *rrPP*   **d)** *RrPp* × *RrPp*

Figure 24.28

Abraham Lincoln is believed to have suffered from Marfan's syndrome.

EFFECTS OF ENVIRONMENT ON PHENOTYPE

All genes interact with the environment. At times, it is difficult to identify how much of the phenotype is determined by the genes (nature) and how much is determined by the environment (nurture). Fish of the same species show variable numbers of vertebrae if they develop in water of different temperatures. Primrose plants are red if they are raised at room temperature, but become white when raised at temperatures above 30°C.

Himalayan rabbits are black when raised at low temperatures, but white when raised at high temperatures.

The water buttercup, *Ranunculus aquatilis*, provides another example of how genes can be modified by the environment. The buttercup grows in shallow ponds, with some of its leaves above and others below the water surface. Despite identical genetic information in the leaves above and beneath the water, the phenotypes differ. Leaves found above the water are broad, lobed, and flat, while those found below the water are thin and finely divided.

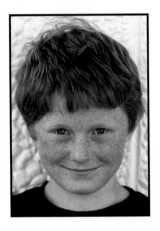

Figure 24.29

The sun may lighten hair and darken freckles.

Figure 24.30

The water buttercup is an example of how variation can be caused by the external environment. The submerged leaves are finely divided, compared with the ones growing in air. This variation can even occur within the same leaf.

Unit Seven:
Heredity

SOCIAL ISSUE:
Genetic Screening

Before the discovery of insulin, many people who had the recessive gene for diabetes in the homozygous condition died before passing on their gene to their offspring. Today genetic screening can tell potential parents if they carry the gene. This information could help a couple decide whether or not they want to have children.

Huntington's chorea is a dominant neurological disorder that only begins to establish itself later in life. The disease is characterized by the rapid deterioration of nerve control, eventually leading to death. Early detection of this disease by genetic screening is possible.

Statement:

Compulsory genetic screening would improve the level of health in our society.

Point

- People are living longer thanks to new medicines. As a result, there are also more people who carry recessive genes for genetic disorders. Screening individuals who plan to have children could cut down on the incidence of these disorders.
- Genetic screening could provide valuable information about worker health and safety. Individuals who are susceptible to industrial chemicals, air pollutants, or nuclear radiation could be identified and then transferred to alternate types of employment.

- Identifying and analyzing human genes could enable doctors to begin treating patients long before the symptoms of disease become serious. This might limit the devastation caused by genetic diseases. The screening could also provide important information used for the diagnosis of diseases related to genetic disorders.

Counterpoint

- Screening individuals who plan to have children is an invasion of privacy. Neither doctors nor the state has the right to interfere in our personal lives to this extent.

- Identification of a genetic disease in an employee could cause early dismissal or end a promising career. All individuals are susceptible to the harmful effects of chemicals, pollution, and nuclear radiation; companies are obligated to provide a safe workplace for all their employees.
- Once genetic screening is done, who gets access to the results? Banks and other financial institutions might refuse to extend loans or give insurance coverage. Companies might only hire workers who can tolerate higher levels of toxic emissions rather than take measures to reduce such emissions.

Research the issue.
Reflect on your findings.
Discuss the various viewpoints with others.
Prepare for the class debate.

CHAPTER HIGHLIGHTS

- Genetics is the study of heredity.
- Gregor Mendel established the principles of genetics by observing the phenotypes arising from the crossing of successive generations of peas.
- The dominant trait will mask the effect of the recessive trait in heterozygous individuals.
- Incomplete dominance occurs when neither gene exerts a dominant influence.
- Hybrids combine characteristics present in both parents.
- The law of segregation states that the two alleles of a gene move to opposite poles during meiosis.

- The principle of independent assortment states that genes for different traits, which are located on separate chromosomes, are organized independently during gamete formation.
- The rules of probability can be applied to genetics.
- Some genetic traits are regulated by more than two possible alleles.
- Dihybrid crosses involve two distinct traits.
- Genes located on one chromosome can affect the expression of other genes that are located on different loci of the same chromosome or on independent chromosomes.

APPLYING THE CONCEPTS

1   Explain why Mendel's choice of the garden pea was especially appropriate.

2   Long stems are dominant over short stems for pea plants. Determine the phenotypic and genotypic ratios of the F_1 offspring from the cross-pollination of a heterozygous long-stem plant with a short-stem plant.

3   Cystic fibrosis is regulated by a recessive allele, c. Explain how two normal parents can produce a child with this disorder.

4   In horses, the trotter characteristic is dominant to the pacer characteristic. A male trotter mates with three different females, and each female produces a foal. The first female, a pacer, gives birth to a foal that is a pacer. The second female, also a pacer, gives birth to a foal that is a trotter. The third female, a trotter, gives birth to a foal that is a pacer. Determine the genotypes of the male, all three females, and the three foals sired.

5   For shorthorn cattle, the mating of a red bull and a white cow produces a calf that is described as roan. Roan is intermingled red and white hair. Many matings between roan bulls and roan cows produce cattle in the following ratio: 1 red, 2 roan, 1 white. Is this a problem of codominance or multiple alleles? Explain your answer.

6   For ABO blood groups, the A and B genes are codominant, but both A and B are dominant over type O. Indicate the blood types possible from the mating of a male who is blood type O with a female of blood type AB. Could a female with blood type AB ever produce a child with blood type AB? Could she ever have a child with blood type O?

7   Thalassemia is a serious human genetic disorder that causes severe anemia. The homozygous condition ($T^m T^m$) leads to severe anemia. People with thalassemia die before sexual maturity. The heterozygous condition ($T^m T^n$) causes a less serious form of anemia. The genotype $T^n T^n$ causes no symptoms of the disease. Indicate the possible genotypes and phenotypes of the offspring if a male with the genotype $T^m T^n$ marries a female of the same genotype.

8   For guinea pigs, black fur is dominant to white fur color. Short hair is dominant to long hair. A guinea pig that is homozygous for white and homozygous for short hair is mated with a guinea pig that is homozygous for black and homozygous for long hair. Indicate the phenotype(s) of the F_1 generation. If two hybrids from the F_1 generation are mated, determine the phenotypic ratio of the F_2 generation.

9 For chickens, the gene for rose comb (R) is dominant to that for single comb (r). The gene for feather-legged (F) is dominant to that for clean-legged (f). Four feather-legged, rose-combed birds mate. Rooster A and hen C produce offspring that are all feather-legged and mostly rose-combed. Rooster A and hen D produce offspring that are feathered and clean, but all have rose combs. Rooster B and hen C produce birds that are feathered and clean. Most of the offspring have rose combs, but some have single combs. Determine the genotypes of the parents.

10 For mice, the allele C produces color. The allele c is an albino. Another allele, B, causes the activation of the pigment and produces black color. The recessive allele, b, causes the incomplete activation of pigment and produces brown color. The alleles C and B are located on separate chromosomes and segregate independently. Determine the F_1 generation from the cross $CcBb \times CcBb$.

CRITICAL-THINKING QUESTIONS

1 Baldness (H^B) is dominant in males but recessive in females. The normal gene (H^n) is dominant in females, but recessive in males. Explain how a bald offspring can be produced from the mating of a normal female with a normal male. Could these parents ever produce a bald girl? Explain your answer.

2 Use the phenotype chart at the right to answer the following questions.
   a) How many children do the parents A and B have?
   b) Indicate the genotypes of the parents.
   c) Give the genotypes of M and N.

Female normal   Female diabetic   Male normal   Male diabetic

3 In Canada, it is illegal to marry your immediate relatives. Using the principles of genetics, explain why inbreeding in humans is discouraged.

ENRICHMENT ACTIVITIES

1 The ability to curl the tongue in a U shape is controlled by the dominant allele T. The recessive allele t results in the inability to curl the tongue. Find an individual who is unable to curl his or her tongue. Test other family members and construct a pedigree chart.

2 Suggested reading:
   • Diamond, Jared. "Curse and Blessing of the Ghetto." *Discover*, March 1991, p. 60. An excellent case study of Tay-Sachs disease.
   • Grady, Denis. "The Ticking of a Time Bomb in Your Genes." *Discover*, June 1987, p. 26. An interesting article on the identification of the gene for Huntington's chorea.

CHAPTER 25

*T*he Source of Heredity

IMPORTANCE OF THE CHROMOSOMAL THEORY

Early scientists believed that hereditary traits were located in the blood. The term "pure bloodline" is still used today and is a reminder of this early misconception. Genes, which are located along the thread like chromosomes found in the nucleus of each cell, are responsible for producing or influencing specific traits in offspring. A description of the structure and location of the gene is essential to an understanding of how it works. Structure provides important clues to function.

During the Dark Ages, strict laws and social pressures prohibited the dissection of corpses; the dead were to be left alone. Yet, in spite of the laws, early physicians and scientists performed dissections secretly in caves. They began to sketch and label different parts of the body, compiling a guide to anatomy in the process. As a composite structure of organs began to appear, theories about function also emerged.

Early geneticists were driven by the same desire for knowledge, but, unlike anatomists, they were not able to see much of the structure they were studying. Science

Figure 25.1

Bobby Hull and his son Brett have starred in the National Hockey League, becoming the first father and son to win the Hart trophy. However, genetic heritage is only part of the equation. Brett Hull has had to work very hard for many years to fulfill his potential as a hockey player.

had to wait for technology to catch up. The development of the light microscope allowed genetic investigations to progress. In this chapter and the next, you will discover how science and technology are intertwined. The light microscope, electron microscope, and biochemical analysis techniques, such as X-ray diffraction and gel electrophoresis, have provided a more complete picture of the mechanisms of gene action.

CYTOLOGY AND GENETICS

Over 2000 years ago, the Greek philosopher Aristotle suggested that heredity could be traced to the power of the male's semen. He believed that hereditary factors from the male outweighed those from the female. Other scientists speculated that the female determined the characteristics of the offspring and that the male gamete merely set events in motion. By 1865, the year in which Mendel published his papers, many of these misconceptions had been cleared up. Nineteenth-century biologists knew that the egg and sperm unite to form a new individual, and it was generally accepted that factors from the egg and sperm were blended in developing the characteristics of the offspring. However, Mendel knew nothing about the structure or location of the hereditary material. He did not know about meiosis, nor did he understand how the genetic code worked. Yet, in spite of the limited knowledge of the day, he set geneticists on the right path. Unfortunately, the lack of complementary information and a comprehensive theory of gene action meant that Mendel's work was interpreted as a mere experiment with garden peas.

About the same time Mendel was conducting his experiments with garden peas,

new techniques in lens grinding were providing better microscopes. A new branch of biology, called cytology, began to flourish. The nucleus was discovered in 1831, just 34 years before Mendel published his results. In 1882, Walter Fleming described the separation of threads within the nucleus during cell division. He called the process *mitosis*. In the same year, Edouard van Benden noticed that sperm and egg cells of roundworms had two chromosomes, but the fertilized egg had four chromosomes. By 1887, August Weisman offered the theory that a special division took place in sex cells. The reduction division is now known as *meiosis*. Weisman had added an important piece to the puzzle of heredity, and in doing so, provided a framework in which Mendel's work could be understood. When scientists rediscovered Mendel's experiments in 1900, the true significance of his work became apparent.

DEVELOPMENT OF THE CHROMOSOMAL THEORY

In 1902, an American biologist, Walter S. Sutton, was studying gamete formation in grasshoppers. At the same time, Theodor Boveri, a German scientist, was studying gamete formation independently of Sutton. Both scientists observed that chromosomes came in pairs, and that they segregated during meiosis. The chromosomes then formed new pairs when the egg and sperm united. The concept of paired, or homologous, chromosomes supported Mendel's two-factor explanation of inheritance. Today, these factors are referred to as genes. One factor, or gene, for each characteristic must come from each sex cell.

The union of two different factors in offspring and the formation of new combinations of factors in succeeding generations could be explained and supported by cellular evidence. The behavior of chromosomes during gamete formation could help explain Mendel's law of segregation and law of independent assortment.

Sutton and Boveri knew that the egg was much larger than the sperm, but that the expression of a trait was not tied to just the male or just the female sex cell. Some structure in both the sperm cell and the egg cell must determine heredity. Sutton and Boveri deduced that Mendel's factors (genes) must be located on the chromosomes. The fact that humans have 46 chromosomes, but thousands of different traits, led Sutton to hypothesize that each chromosome contains many different genes.

CHROMOSOMAL THEORY

The development and refinement of the microscope led to advances in cytology and the union of two previously unrelated fields of study: cell biology and genetics. As you continue reading and exploring genetics, you will discover how other branches of science, such as biochemistry and nuclear physics, have also become integrated with genetics.

The chromosomal theory of inheritance can be summarized as follows:

- Chromosomes carry genes, the units of hereditary structure.
- Paired chromosomes segregate during meiosis. Each sex cell or gamete has half the number of chromosomes found in a **somatic cell.**
- Chromosomes assort independently during meiosis. This means that each gamete receives one of the pairs and that one chromosome has no influence on the movement of a member of another pair.

- Each chromosome contains many different genes.

MORGAN'S EXPERIMENTS

Few people have difficulty distinguishing gender in humans. Mature females look very different from mature males. (Even immature females can be identified on the basis of anatomy.) But can you distinguish whether a blood cell or cheek cell comes from a female or a male? The work of the American geneticist Thomas Hunt Morgan (1866–1945) provided a deeper understanding of gender and the inheritance of some characteristics.

Morgan was one of many geneticists who used the tiny fruit fly, *Drosophila melanogaster*, to study Mendel's principles of inheritance. There are several reasons why the fruit fly is an ideal subject for study. First, the fruit fly reproduces rapidly. Females lay over 100 eggs after mating, and each offspring is capable of mating shortly after leaving the egg. The large number of offspring is ideally suited for genetics, which is based on probability. Because *Drosophila*'s life cycle tends to be only 10 to 15 days, it is possible to study many generations in a short period of time. A second benefit arises from the small size of the *Drosophila*. Many can be housed in a single test tube. A small, solid nutrient at the bottom of the test tube can maintain an entire community. The third and most important quality of *Drosophila* is that males can easily be distinguished from females.

Morgan discovered various mutations in *Drosophila*, noting that some of the mutations seemed to be linked to other traits. Morgan's observations added support to the theory that the genes responsible for the traits were located on the chromosomes.

Somatic cells *are all the cells of an organism except the sex cells.*

GENE LINKAGE AND CROSSING-OVER

It is often said that great science occurs because good questions are asked. Thomas Hunt Morgan, like Mendel, asked great questions. Such questions were raised when Morgan obtained a few odd gene combinations while performing dihybrid crosses with *Drosophila*. The appearance of combinations of recessive traits in very small numbers challenged the principles of Mendelian genetics. The mating of fruit flies that were homozygous for wild-type body coloring and straight wings (*AABB*) with those that had black body color and curved wings (*aabb*) produced offspring that were all heterozygous for both traits (AaBb). Is it possible to get a member of the F_2 generation that demonstrates a recessive phenotype such as curved wings (bb), along with the dominant gene for wild-type body colour? How would you explain it if this combination occurred in only 9 of 300 offspring, as Morgan found? The phenotypic ratios of the offspring could not be explained in terms of a normal 9:3:3:1 ratio. The phenotypes of the F_2 generation could not be explained by normal Mendelian genetics.

By studying the frequencies of other genes located on similar chromosomes, Morgan was able to provide a tentative explanation to these questions. His work with sex-linked traits established that specific genes are located on specific chromosomes. For example, eye color for *Drosophila* is located on an X chromosome. Morgan knew that two or more genes located on nonhomologous chromosomes segregate independently during meiosis. But what about genes located on the same chromosome? Should these genes not segregate together? Figure 25.9 shows what should happen during normal segregation.

Morgan concluded that the gene for wing shape and the gene for body color were located on the same chromosome. The two genes did not segregate independently. Genes located on the same chromosome tend to be transmitted together. These are referred to as **linked genes.**

However, an event called *crossing-over* can provide new combinations. Crossing-over occurs during synapsis of meiosis. This means that a single chromosome can change as it passes from generation to generation. In Figure 25.10, consider the blue-colored chromosome to be the one inherited from the father and the red chromosome to be the one inherited from the mother. Those gametes with chromosomes that have recombined have sections of the chromosome that are maternal (coming from the mother) and other sections that are paternal (coming from the father).

Linked genes *are located on the same chromosome.*

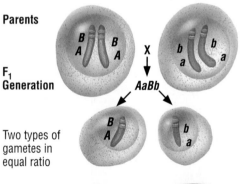

Figure 25.9

Complete linkage. During normal meiosis, the homologous blue and red chromosomes move to opposite poles. One sex cell inherits the *AB* alleles, while the other inherits the *ab* alleles.

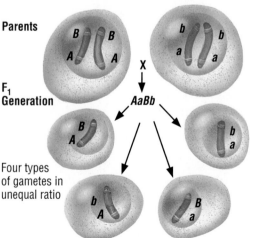

Figure 25.10

Incomplete linkage. Gametes with the gene combination *Ab* and *aB* would not be possible without crossing-over.

Armed with the knowledge of recombination of genes due to crossing-over, we can re-examine the problem that faced Morgan. Consider the mating between fruit flies with wild-type body color and straight wings (*AABB*) and those with black body color and curved wings (*aabb*). If the gene are located on the same chromosome we would expect all members of the F_1 generation to be AaBb. If the F_2 generation AABB, AaBb and aabb are possible. However, Morgan found a few F_2 individuals that were Aabb or aaBb. The new combinations can be explained in terms of crossing-over.

The frequency of crossing-over can be stated as a percentage:

$$\text{Crossover \%} = \frac{\text{number of recombinations}}{\text{total number of offspring}} \times 100\%$$

$$= \frac{18}{300} \times 100\% = 6\%$$

AABB* × *aabb

↓

AaBb

↓

F_2	**282 wild-type body color, straight wings**
Recombinations	**9 black body color, straight wings** **9 wild-type color, curved wings**

Figure 25.11

New combinations can be explained in terms of crossing-over.

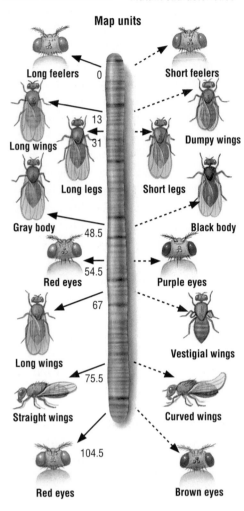

Normal characteristics
Mutant characteristics
Map units
Long feelers
0
Short feelers
Long wings
13
31
Dumpy wings
Long legs
Short legs
Gray body
48.5
Black body
Red eyes
54.5
Purple eyes
67
Long wings
Vestigial wings
75.5
Straight wings
Curved wings
104.5
Red eyes
Brown eyes

Figure 25.12

Gene mapping of chromosome number 2 for *Drosophila melanogaster.* Note that many genes are located on one chromosome.

MAPPING CHROMOSOMES

Gene markers *are often recessive traits that are expressed in the recessive phenotype of an organism. The markers can be used to identify other genes found on the same chromosome.*

Genes located on the same chromosome segregate with each other. A recessive characteristic like white eyes for *Drosophila* can often be used as a **gene marker.** White eyes can be easily observed in the offspring. However, some genes do not code for traits that are readily observable. These genes can be tracked by the use of marker genes located on the same chromosome. The appearance of white eyes in offspring may also provide a key to many other genes located on the same chromosome.

Unfortunately, crossing-over can alter gene linkages along a chromosome. If new segments of DNA are exchanged at the site of the marker, the mapping becomes impossible. By following the offspring from a variety of crossovers, geneticists were able to determine that genes located near each other almost always ended up on the same chromosome. In contrast, genes separated from each other by greater distances were more likely to be affected by crossing-over. This means that a crossover value of 1% indicates that the genes are close to each other, while a value of 12% indicates that the genes are much farther apart. The crossover frequency can help scientists determine the relative position of the genes along the chromosome. The greater the frequency of crossover, the greater the **map distance.** A crossover frequency of 5% means that the two genes are 5 map units apart. A crossover frequency of 16% means that the genes are much farther apart: 16 map units.

The frequency of crossovers can be used to determine gene maps. Consider the following problem. Assume the crossover frequency between genes A and B is 12%, between B and C is 7%, and between A and C is 5%. The fact that map distances are additive allows you to determine the proper sequence of genes. If you assume that gene A is in the middle, then the sum of the distances between B and A, and A and C must equal the distance between B and C. These distances are not equal. Therefore, gene A is not in the middle.

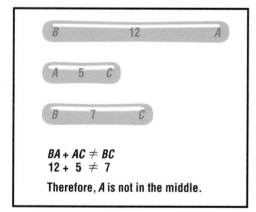

BA + AC ≠ BC
12 + 5 ≠ 7
Therefore, A is not in the middle.

If you assume that gene B is in the middle, then the sum of the distances between A and B, and B and C must equal the distance between A and C. These distances are not equal. Therefore, gene B is not in the middle.

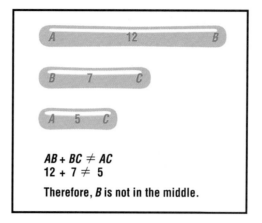

AB + BC ≠ AC
12 + 7 ≠ 5
Therefore, B is not in the middle.

If you assume that C is the middle gene, then the sum of the distances between A and C, and C and B must equal the distance between A and B. These distances are equal. Therefore, gene C is in the middle.

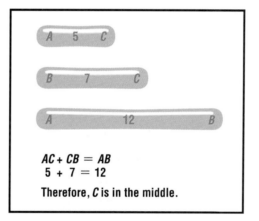

AC + CB = AB
5 + 7 = 12
Therefore, C is in the middle.

The development of special stains for chromosomes has also advanced gene mapping. These stains give the chromosomes distinctive light- and dark-colored bands, which can be used to identify the position of distinctive genes along a particular chromosome.

BIOLOGY CLIP
To date, only 0.1% of all human genes have been mapped.

Map distance *refers to the distance between two genes along the same chromosome.*

Objective

To use crossover frequencies to construct a gene map.

Background Information

A.H. Sturtevant, a student who worked with Thomas Morgan, made the following hypotheses:

Genes are located in a linear series along a chromosome, much like beads on a string. Genes that are closer together will be separated less frequently than those that are farther apart. Crossover frequencies can be used to construct gene maps.

Sturtevant's work with *Drosophila* helped establish techniques for mapping chromosomes.

Procedure

1 Examine the diagram of the chromosome shown. Crossing-over takes place when breaks occur in the chromatids of homologous chromosomes during meiosis. The chromatids break and join with the chromatids of their homologous chromosomes. This causes an exchange of alleles between chromosomes.

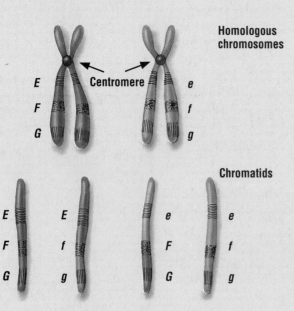

a) Indicate the areas of the chromatids that show crossing-over.

b) According to the diagram, which genes appear farthest apart? (Choose from *EF*, *FG*, or *EG*.)

c) Which alleles have been exchanged?

2 In 1913, Sturtevant used crossover frequencies of *Drosophila* to construct chromosome maps. To determine map distances, he arbitrarily assigned one recombination for every 100 fertilized eggs. For example, genes that had a crossover frequency of 15% were said to be 15 units apart. Genes that had 5% recombinations were much closer: 5 units apart.

d) Using the following data, determine the distance between genes *E* and *F*.

Cross	*EF* × *ef*
Offspring	*EF* + *ef* (from parent) = 94%
Frequency	*Ef* + *eF* (recombination) = 6%

3 Use the data table below to construct a complete gene map.

Cross	Offspring	Frequency
EF × *ef*	*EF* + *ef* (from parent)	94%
	Ef + *eF* (recombination)	6%
EG × *eg*	*EG* + *eg* (from parent)	90%
	Eg + *eG* (recombination)	10%
FG × *fg*	*FG* + *fg* (from parent)	96%
	Fg + *fG* (recombination)	4%

e) What is the distance between genes *E* and *G*?

f) What is the distance between genes *F* and *G*?

Case-Study Application Questions

1 What mathematical evidence indicates that gene *F* must be found between genes *E* and *G*?

2 Draw the gene map to scale. (1 cm = 1 unit)

3 For a series of breeding experiments, a linkage group composed of genes *W, X, Y,* and *Z* was found to show the following gene combinations. (All recombinations are expressed per 100 fertilized eggs.)

Genes	W	X	Y	Z
W	–	5	7	8
X	5	–	2	1
Y	7	2	–	1
Z	8	3	1	–

Construct a gene map. Show the relative positions of each of the genes along the chromosome and indicate distances in gene units.

4 For a series of breeding experiments, a linkage group composed of genes *A, B, C,* and *D* was found to show the following gene combinations. (All recombinations are expressed per 100 fertilized eggs.)

Genes	A	B	C	D
A	–	12	15	4
B	12	–	3	8
C	15	3	–	11
D	4	8	11	–

Construct a gene map. Show the relative positions of each of the genes along the chromosome and indicate distances in gene units. ■

GENE RECOMBINATIONS IN NATURE

Scientists once believed that, with the exception of a few new combinations that might occur because of crossing-over, chromosome structure was fixed. However, in the late 1940s, Barbara McClintock, an American biologist, interpreted her results of experiments with Indian corn and came to a conclusion that would shatter the traditional view of gene arrangement on chromosomes. McClintock suggested that genes can move to a new position. Her theory was dubbed the "jumping gene theory."

Color variation in the kernels of Indian corn led McClintock to hypothesize the existence of transpos-able elements, called **transposons.** The insertion of some genes into a new position along a chromosome inactivated the genes that affect the production of pigment. Despite McClintock's well-documented experiments, the vast majority of the scientific community ignored her work or dismissed her results. The idea of jumping genes challenged the widely accepted theory that genes are fixed in position along the chromosomes. Undaunted by her critics, McClintock continued to gather data on jumping genes. Her contributions to science were finally recognized in 1983, when McClintock was awarded the Nobel Prize for physiology and medicine.

Transposons *are specific segments of DNA that can move along the chromosome.*

Figure 25.13

In the late 1940s, Barbara McClintock's theory of "jumping genes" was greeted with tremendous skepticism by the traditional scientific community.

Transposable genes have since been discovered in other organisms. Bacteria can insert genes randomly along a circular chromosome. The transposon can also move from the chromosome in one bacterium to a chromosome in another bacterium. The inserted genes can even be integrated with a secondary structure of DNA, called a **plasmid,** located in a bacterium. The plasmid is a small ring of genetic material, and can be considered extra DNA.

Some bacteria are capable of a form of sexual reproduction called **conjugation.** During bacterial conjugation, two or more cells fuse, and plasmids are passed from one cell to another. This special property has been exploited by geneticists using revolutionary new technology. But bacterial conjugation has also created some serious consequences for humans. Some disease-causing bacteria have evolved genes that are resistant to certain types of antibiotics, such as penicillin. Under normal conditions, antibiotics interfere with chemical reactions that occur within these harmful microbes. Genes that provide resistance to an antibiotic permit the disease-causing bacteria to continue living and wreaking havoc for the cells of the body.

Plasmids *are small rings of genetic material.*

Conjugation *is a form of sexual reproduction in which genetic material is exchanged between two cells.*

GENE SPLICING TECHNIQUES AND GENE MAPPING

Prior to the 1970s, geneticists studied the inheritance of traits within successive generations of families. Earlier in the chapter, you learned that genes located on the same chromosome are usually inherited together. Recessive genes can be used as markers to follow the inheritance of the linked trait that is being studied; however, linked traits are often disrupted by crossing-over. As described earlier, geneticists have been able to turn the anomaly of crossing-over into an advantage. The closer two genes are on a specific chromosome, the less likely they are to separate during crossing-over. The frequencies of crossing-over can be used to determine gene position. However, the process of following genes during crossover events is tedious and time-consuming. Two recent innovations have increased the speed at which genes can be mapped.

Scientists have been able to fuse the cell of a human with the cell of a mouse. Although the hybrid cell tends to lose most of the human chromosomes during cell replication, all of the rodent chromosomes are maintained. In most cells, a single human chromosome remains, but in each case it appears to be a different chromosome. The production of a human protein by the hybrid cell led geneticists to realize that the genes controlling the production of that protein must exist somewhere on the chromosome that remains in the hybrid cell.

The other major breakthrough in gene mapping technique came on the heels of recombinant DNA research. You will read more about this technology in the next chapter. Special enzymes are used to chop strands of DNA into small segments. Remember that each enzyme

Figure 25.14

Bacteria exchange genetic material during conjugation. The long appendage between the bacteria is called a *pilus*. The pilus will draw the cells closer together and act as a bridge for the exchange of genetic material.

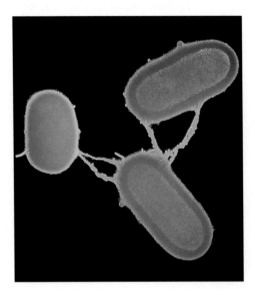

Unit Seven:
Heredity

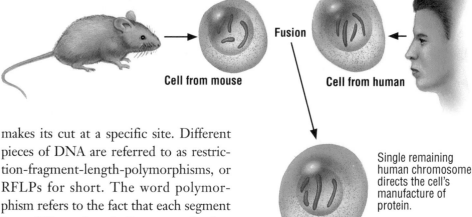

Fusion

Cell from mouse

Cell from human

Single remaining
human chromosome
directs the cell's
manufacture of
protein.

Figure 25.15

The fusion of a human cell with
a mouse cell.

makes its cut at a specific site. Different pieces of DNA are referred to as restriction-fragment-length-polymorphisms, or RFLPs for short. The word polymorphism refers to the fact that each segment has a different length. Because each of us has a distinctive length of RFLP segments, these segments can be used to identify individual people.

Like early cartographers who mapped the seas by using the stars, gene mappers have used RFLP markers to determine the position of genes on a specific site, or locus, on chromosomes. By combining two other existing technologies— radioactive labelling and gel electrophoresis— with the use of restriction enzymes, scientists have been able to significantly increase the rate at which gene maps are made. Radioactive labels can be attached to one end of the DNA. Strands of DNA are then added to four separate test tubes. Each test tube contains specific enzymes that cleave specific bond sites.

Gel electrophoresis was originally developed by Linus Pauling to identify proteins. However, the same principles also apply to strands of DNA. The DNA segments are removed from the test tubes and applied to a thin layer of gel. An electrical field pulls the pieces of DNA through the gel. The smaller pieces of DNA move faster and farther than the larger ones. The final positions of the

> **BIOLOGY CLIP**
> Using biochemical analysis techniques, a team of British scientists has discovered a gene that causes obesity. This gene increases the speed at which food is converted into fat.

DNA segments can be identified by the radioactive label that reacts with the gel to form a distinctive banding pattern. This pattern of banding can be used to determine the sequence of the DNA.

Although technological improvements have greatly increased the speed at which gene maps can be constructed, scientists are far from understanding the complete human **genome.** Of the estimated 100 000 genes that reside on your chromosomes, only about 4500 have been identified. The number mapped changes every day. Even with accelerated techniques, an estimated 1000 person years of effort are still required to map all the genes on your chromosomes. However, some progress has been made. Scientists have already found genes for Huntington's chorea, cystic fibrosis, and Dutchenne's muscular dystrophy. They have also located the genes that make people susceptible to Alzheimer's disease, some forms of schizophrenia, and cancer.

Figure 25.16

A scientist works under ultraviolet light during the preparation of an electrophoresis gel used in DNA separation techniques.

*The **genome** is the complete set of instructions contained within the DNA of an individual.*

RESEARCH IN CANADA

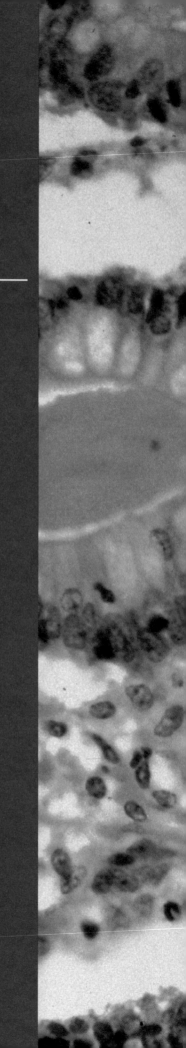

Cystic fibrosis, one of the most serious respiratory disorders, may well be eliminated because of the work of a group of Toronto doctors. Working with a research group from Michigan, Drs. Lap-Chee Tsui, Frank Collins, and Jack Riordin have located the gene responsible for cystic fibrosis. Cystic fibrosis causes secretions of sticky mucus that can block the airways and create digestive problems. It is the most common of all hereditary disorders in Canada, affecting about 1 in 2000 children. One estimate indicates that 1 in every 20 people carry the defect in a heterozygous condition. The recent scientific breakthrough means that carriers can now be identified by gene mapping. Counselling can be provided for couples who might have a child with cystic fibrosis.

The Cystic Fibrosis Gene

Although the discovery of the gene's location does not mean a cure, it does pinpoint the source of the problem. As new techniques become available, it may one day be possible to turn off the defective gene or remove it from the chromosome.

DR. JACK RIORDIN
DR. FRANK COLLINS
DR. LAP-CHEE TSUI

FRONTIERS OF TECHNOLOGY: GENE THERAPY

Over 3500 different genetic diseases have so far been linked to single defective genes. Cystic fibrosis, diabetes, hemophilia, Huntington's chorea, and sickle-cell anemia are but a few of the hereditary diseases known in Canada. Although treatments, such as insulin injections for diabetes, control the disease, no true cure has been found. Medical advances, coupled with modified diet and environmental adjustments, have enabled people with such diseases to continue living productive lives.

Imagine the potential of transforming a defective gene! Although **gene therapy** is in the early stages of development, the prospect of gene correction provides exciting possibilities in the quest to conquer genetic diseases. There are three possible strategies for gene therapy. The first strategy involves *gene insertion*. The normal gene is inserted into position on the chromosome of a diseased cell. Because not every cell uses a particular gene, the insertion can be restricted to those cells in which the gene is active. For example, diabetes occurs when the cells of the pancreas do not produce sufficient amounts of insulin. The normal gene for insulin production need only be inserted into specialized cells within the pancreas. The gene would not have to be inserted into a muscle cell as muscle cells do not produce insulin. A second method involves *gene modification*. The defective gene is modified chemically in an effort to recode the genetic message. This method is much more delicate and requires greater knowledge about the chemical composition of the normal and defective genes. The third technique, *gene surgery*, is the most ambitious: the defective gene is extracted and replaced with a normal gene.

The first attempt at gene therapy dates back to 1980, when Dr. Martin Cline attempted to help two patients who had the genetic disorder called thalassemia (Cooley's anemia). The disorder is characterized by a defective hemoglobin molecule. Hemoglobin, the molecule of red blood cells, is responsible for transporting oxygen to the tissues of the body.

Cline knew that red blood cells were produced in the bone marrow of the long bones. Bone marrow therefore was extracted from the thigh bones, and the defective cells were incubated with normal hemoglobin. It was hoped that the genes from the normal hemoglobin would become embodied within the defective cells. Unfortunately, the procedure failed to produce any significant improvement.

Two years later, the technique was revived. Once again, researchers turned to genetic problems associated with the hemoglobin molecule. Sickle-cell anemia, like thalassemia, is characterized by poor oxygen delivery to the tissues. Sickle-cell anemia affects 1 in 500 people of African heritage in the United States. Red blood cells carrying oxygen appear normal, but once oxygen is released, the hemoglobin turns and folds in an unusual pattern, bending the cell into a sickle shape. Unfortunately, the sickle-shaped cells become lodged in the tiniest blood vessels of the body, the capillaries. The clogged red blood cells impair the transport of oxygen to the tissues and bring about painful swelling.

> **BIOLOGY CLIP**
> It has been estimated that 8% of the human population will show some signs of genetic disorder by age 25.

Gene therapy *is a procedure by which defective genes are replaced with normal genes in order to cure genetic diseases.*

Two genes control the production of hemoglobin during fetal development. As the fetus approaches term, one of the genes is turned off. The gene that remains "on" produces adult hemoglobin. Unfortunately, the gene that is turned on is the very gene that harbours the mutation in people with sickle-cell anemia. If the gene that produces fetal hemoglobin is switched back on and the defective gene turned off, the disorder can be cured. A drug called 5-azacytidine turns the fetal hemoglobin back on; however, this drug is highly toxic, and it is still unclear whether or not the drug turns on other genes as well.

The success of gene therapy has been modest, but its boundaries are extended almost daily. So far, the spotlight of gene therapy has focused on somatic cells in an attempt to cure disease. But what will happen when gene therapy is directed at reproductive cells? Will the ability to select genes for our offspring be far behind? The ethical and moral questions of manipulating life are complex. Is the development of our morality keeping pace with technological advances?

REVIEW QUESTIONS ?

4 List three difficulties that arise when genes are studied in human populations.
5 What are gene crossovers? How does this process affect segregation?
6 What are linked genes?
7 How can gene crossover frequencies be used to construct gene maps?
8 What are some possible benefits of gene therapy?

Figure 25.17

Retroviruses can be used for gene therapy. (a) Genes from the virus are replaced with therapeutic genes. (b) Viral genes are placed in the cell. (c) Genetic information from the virus directs production of proteins which make up the viral shell. The virus assembles (d) and enters the target cell. Genes carried by the virus are spliced into chromosomes of the target cell. Transplanted genes direct the synthesis of needed proteins.

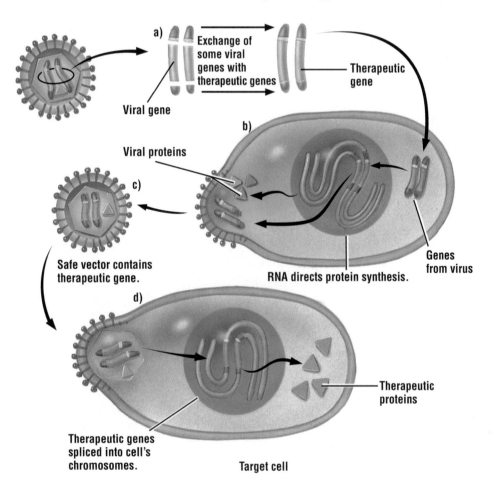

Unit Seven:
Heredity

SOCIAL ISSUE:
The Potential of Gene Therapy

Scientists have located the genes for such hereditary diseases as cystic fibrosis and Huntington's chorea. As new techniques in gene therapy become available, it may soon be possible to switch off defective genes and replace them with normal genes, thereby providing a cure for many genetic diseases. As with most new medical technologies, however, important ethical questions must first be addressed.

Statement:

As a technology, gene therapy is beneficial.

Point

- Gene therapy will eventually provide a cure for most forms of genetic disease, and the cure will be permanent.

- People who once died because of hereditary diseases are being kept alive by modern medicine and their genes are being passed on. Therefore, more and more people will carry genetic disorders unless gene therapy is used. Ultimately, gene therapy will provide a better gene pool.

> **Research the issue.**
> **Reflect on your findings.**
> **Discuss the various viewpoints with others.**
> **Prepare for the class debate.**

Counterpoint

- The question of which genes are "good" or "bad" and which genes should be changed is difficult to decide. Also, gene therapy could be used to do more than just cure diseases. What happens if people decide to use gene therapy to select "desirable" traits for their children?

- The human genome is infinitely complicated. Most characteristics are controlled not by a single gene, but by the interaction of many genes. With such a complicated intertwining of factors, the attempt to control the gene pool would only lead to disaster.

CHAPTER HIGHLIGHTS

- Chromosomes carry genes, the units of hereditary structure.
- Paired chromosomes segregate during meiosis. Each sex cell or gamete has half the number of chromosomes found in a somatic cell.
- Chromosomes assort independently during meiosis. This means that each gamete receives one of the pairs and that one chromosome has no influence on the movement of a non-paired chromosome.
- Each chromosome contains many different genes.
- Sex-linked traits are located on either the X or Y chromosome.

- Barr bodies are small, dark spots of chromatin located in the nuclei of female mammalian cells.
- Genetics can be used to study human populations.
- Crossing-over occurs between homologous chromosomes during meiosis. Small segments of DNA are exchanged, thereby creating new combinations of linked genes.
- Crossover frequencies can be used to determine gene maps.
- Genes do not remain in fixed positions along the chromosomes. Transposons are specific segments of DNA that can move along a chromosome.
- Gene therapy is directed at curing hereditary disorders.

APPLYING THE CONCEPTS

1 In what ways was the development of the chromosomal theory linked with improvements in microscopy?

2 Discuss the contributions made by Walter Sutton, Theodor Boveri, Thomas Morgan, and Barbara McClintock to the development of the modern chromosomal theory of genetics.

3 The gene for wild-type eye color is dominant and sex-linked in *Drosophila*. White eyes are recessive. The mating of a male with wild-type eye color with a female of the same phenotype produces offspring that are 3/4 wild-type eye color and 1/4 white-eyed. Indicate the genotypes of the P_1 and F_1 generations.

4 Use the information from the pedigree chart to answer the following questions.

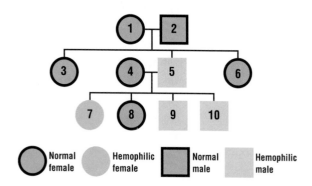

a) State the phenotypes of the P_1 generation.

b) If parents 1 and 2 were to have a fourth child, indicate the probability that the child would have hemophilia.

c) If parents 1 and 2 were to have a second male child, indicate the probability that the boy would have hemophilia.

d) State the genotypes of 4 and 5.

5 The autosomal recessive gene *tra* transforms a female into a phenotypic male when it occurs in the homozygous condition. The females who are transformed into males are sterile. The *tra* gene has no effect in XY males. Determine the F_1 and F_2 generations from the following cross: XX, +/*tra* crossed with XY, *tra/tra*. (Note: the + indicates the normal dominant gene.)

6 Edward Lambert, an Englishman, was born in 1717. Lambert had a skin disorder characterized by very thick skin that was shed periodically. The hairs on his skin were very coarse, like quills, giving him the name "porcupine man." Lambert had six sons, all of whom exhibited the same traits. The trait never appeared in his daughters. In fact, the trait has never been recorded in females. Provide an explanation for the inheritance of the "porcupine trait."

7 A science student postulates that dominant genes occur with greater frequency in human populations than recessive genes. Using the information that you have gathered in this chapter, either support or refute the hypothesis.

8 In 1911, Thomas Morgan collected the following crossover gene frequencies while studying *Drosophila*. Bar-shaped eyes are indicated by the *B* allele, and carnation eyes are indicated by the allele *C*. Fused veins on wings (*FV*) and scalloped wings (*S*) are located on the same chromosome.

Gene combinations	Frequencies of recombinations
FV/B	2.5%
FV/C	3.0%
B/C	5.5%
B/S	5.5%
FV/S	8.0%
C/S	11.0%

Use the crossover frequencies to plot a gene map.

9 Huntington's chorea is a dominant neurological disorder that usually appears when a person is between 35 and 45 years of age. Many people with Huntington's chorea, however, do not show symptoms until they are well into their sixties. Explain why the slow development of the disease has led to increased frequencies in the population.

10 Explain the significance of locating the cystic fibrosis gene.

species, the proportion stays the same in the DNA of all of a given species' cells. That is, the number of adenine molecules is the same as the number of thymine molecules, and the number of guanine molecules is the same as those of cytosine. These observations suggested that the nitrogen bases were arranged in pairs.

Watson's background enabled him to understand the significance of the emerging chemical data, while Crick was better able to appreciate the significance of the X-ray diffraction results. With such data, Watson and Crick developed a three-dimensional model of the DNA molecule, which they presented to the scientific community in 1953. This model, which was visually confirmed in 1969, is still used by scientists today. Some additional information gathered since Watson and Crick's time has been incorporated into it.

Models are very useful tools for scientists. A model airplane looks like the real thing, except that it is much smaller. By studying the replica, one can learn more about how the real plane works. However, the model of a molecule cannot be scaled down for detailed study; molecules are already too small to see. Instead, molecules are made larger to show how the different atoms interact. X-ray diffraction techniques provide a picture that indicates how different chemical bonds interact with each other. Scientists use models as visual devices that help them understand the relationship and interactions of different parts.

Politics and Science

Watson and Crick might not have been credited as the co-discoverers of DNA were it not for politics. The X-ray

diffraction technique developed in England had been used by Maurice Wilkins and Rosalind Franklin to view the DNA molecule. At that time, the American scientist Linus Pauling, a leading investigator in the field, was refused a visa to England to study the X-ray photographs. Pauling, along with others, had been identified by Senator Joseph McCarthy as a communist sympathizer for his support of the anti-nuclear movement. Many scientists believe that the United States passport office may have unknowingly determined the winners in the race for the discovery of the double-helix model of DNA.

The McCarthy era of the early 1950s is considered by many historians as a time of paranoia and repression. Many creative people had their careers stifled or destroyed because of their perceived association with communism. In most cases the charges were unfounded. It is perhaps ironic that, in 1962, Linus Pauling was awarded a Nobel Prize, this time for his dedication to world peace.

STRUCTURE OF DNA

DNA is most often described as a double helix. DNA closely resembles a twisted ladder. The sugar and phosphate molecules form the backbone of the ladder, while the nitrogen bases form the rungs. Nitrogen bases from one spine of the ladder are connected with nitrogen bases from the other spine of the ladder by means of hydrogen bonds. A hydrogen bond is a weak bond that forms between the hydrogen proton on the end of one molecule and the negative charge on the end of another molecule. The backbone

Figure 26.4

James Watson and Francis Crick were awarded the Nobel Prize for physiology/medicine in 1962 for what many people believe to be the most significant discovery of the twentieth century.

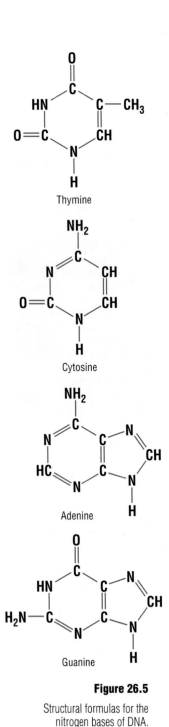

Figure 26.5

Structural formulas for the nitrogen bases of DNA.

Thymine

Cytosine

Adenine

Guanine

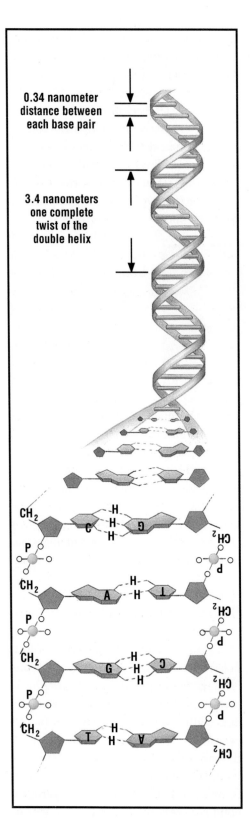

0.34 nanometer distance between each base pair

3.4 nanometers one complete twist of the double helix

Figure 26.6

Representation of DNA molecule. Note the complementary hydrogen bonds.

of the DNA molecule becomes twisted, which makes the molecule look like a winding, spiral staircase. The DNA molecule is made of individual units composed of deoxyribose sugars, phosphates, and nitrogen bases. Each unit is referred to as a nucleotide. The model to the right shows two complementary nucleotides. Notice how the nitrogen base fits into its complementary pair. An adenine molecule always pairs with a thymine molecule, while a guanine molecule always pairs with a cytosine molecule.

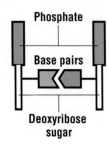

Phosphate

Base pairs

Deoxyribose sugar

■ REVIEW QUESTIONS ■ ?

1 Name two ways in which the DNA molecule is important in the life of a cell.

2 What chemicals make up chromosomes?

3 What are nucleotides?

4 What chemicals are found in a nucleotide?

5 On what basis did some scientists conclude that proteins provided the key to the genetic code?

6 Who were the co-discoverers of the double-helix model of DNA?

7 How did the X-ray diffraction technique provide a clue to the structure of DNA?

8 Which nitrogen base pairs with guanine? Which base pairs with adenine?

REPLICATION OF DNA

DNA is the only molecule known that is capable of duplicating itself. This is accomplished through a process known as **replication.** Replication helps explain how one cell can divide into two identical cells. Each of the daughter cells requires a complete set of genetic information that can only be obtained if the DNA molecule makes an exact duplicate of itself.

During replication, the weak hydrogen bonds that hold the complementary nitrogen bases together are broken. The two edges of the ladder seem to "unzip," as shown in Figure 26.7. The parent strands are conserved (remain intact). Therefore, this process is also known as **semiconservative replication**—semiconservative because only the parent strands are conserved; the original, inte-

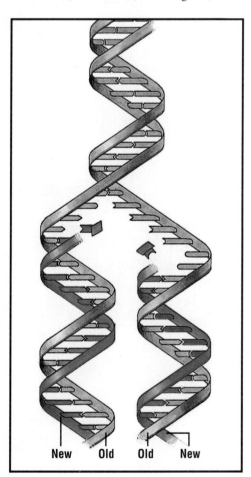

New   Old   Old   New

grated molecule is not. Semiconservative replication produces two "half-old, half-new" strands of DNA. Each parent strand acts as a template, or mold, to which free-floating nucleotides in the cell can attach. The nucleotides attach themselves at their complementary bases: adenine with thymine, and cytosine with guanine. A series of enzymes, called **polymerases,** fuse the free nucleotides together in the complementary chain of DNA.

The free-floating nucleotides in your cells (and those in the cells of other organisms) are derived from the food you eat. A steak supplies you with the muscle cells from a cow. Each of these cells contains the nucleus and the genetic information from the cow. However, the intake of cow DNA does not mean that you will begin to look like a cow! Specialized enzymes in your digestive tract break apart the cow DNA into nucleotides, which you use to make human DNA.

The duplication of DNA in humans is a marvelous feat. It has been estimated that the human genome contains approximately three billion bases. An error rate as infrequent as one per million bases would produce 3000 mistakes, or mutations, in a single division. Imagine what would happen if each cell had 3000 genetic mistakes. Although genetic mistakes do occur during the duplication of genetic information, they are very infrequent.

Environmental factors such as hazardous chemicals or radiation are one source of the genetic mistakes that do occur. These factors can cause uncomplementary nitrogen bases to become paired. Permanent damage is prevented by enzymes that act as proofreaders. These enzymes run along the strands of DNA looking for mismatched pairs. Once a damaged section is detected, it can be repaired. Another enzyme snips the error from the chain and replaces it with the correct nucleotide sequence.

Replication *is the process in which a single strand of nucleotides acts as a template for the formation of a complementary strand. A single strand of DNA can make a complementary strand.*

Semiconservative replication *is the process in which the original strands of DNA remain intact and act as templates for the synthesis of duplicate strands of DNA.*

Polymerases *are enzymes that join individual nucleotides together in complementary strands of DNA.*

Figure 26.7

The DNA molecule unzips, and each strand serves as a blueprint for the synthesis of a complementary strand of DNA.

RESEARCH IN CANADA

With recent advances in reproductive technology and applied genetics, couples who were once unable to have children can now become parents. However, these new techniques raise some very important ethical and legal questions. For example, will frozen embryos be considered property or a life form?

Reproductive Technologies

In 1987, the United States embarked on a massive $3 billion project that will attempt to decipher the entire human genome. It has been predicted that doctors will be able to screen for defective genes by the end of the century. How will this knowledge be applied in Canada? Will couples be allowed to select the genes of their offspring? Who will decide how genetics is to be used? These questions cannot be ignored.

Dr. Patricia Baird, a medical geneticist from the University of British Columbia, heads the Royal Commission on New Reproductive Technologies. The mandate of the commission is to examine how new information is applied to emerging technologies concerned with human reproduction. Who will be able to apply these procedures? Who will be responsible for the expense? How much public money will be spent? Although the commission's mandate extends beyond the confines of science, a knowledge of science and technology is critical to understanding the issues.

DR. PATRICIA BAIRD

LABORATORY

EXPLORING DNA REPLICATION

Objective

To investigate how the double helix of DNA replicates.

Background Information

The DNA molecule is made up of nucleotides, each comprising a deoxyribose sugar, a phosphate, and a nitrogen base. During the replication process, the double strands of DNA separate along the bonds between the nitrogen bases. Each parent strand serves as a template for the arrangement of new nucleotides. An enzyme joins the nucleotides into a complementary strand of DNA. A second enzyme checks the ordering of bases for errors.

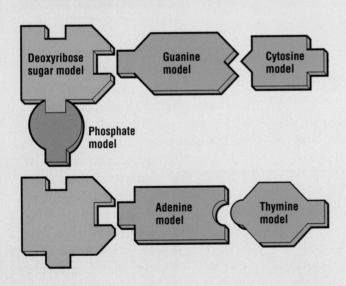

Materials

scissors   transparent tape or glue
toothpicks   blank sheet of paper

Procedure

1   You will be supplied with a page of symbols representing the molecules that make up DNA. Cut out the individual molecules. The toothpicks will be used to represent bonds between the molecules.

a) Why are the adenine and guanine molecules represented by larger shapes than the other two nitrogen bases?

2   Using a toothpick, bond the adenine molecule to a deoxyribose sugar molecule. Then use another toothpick to bond the phosphate molecule to the sugar molecule. Place the phosphate molecule along the left margin of a sheet of paper.

b) What is this structure called?

3   Assemble four different nucleotides, as described in the procedure above. Keeping the phosphate molecules along the left margin of the page, attach each of the nitrogen bases to a different sugar molecule.

4   The phosphate molecules of the DNA bond with two different sugar molecules. Place a second toothpick on the phosphate molecule, and attach a second sugar molecule to the phosphate molecule.

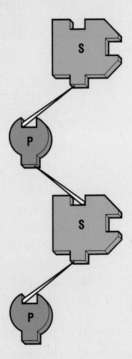

5   Using toothpicks to represent bonds, attach a line of four nucleotides together.

c) Record the genetic code by indicating the letters of the nitrogen bases, beginning from the top of the page.

6   Make a complementary strand of DNA by matching nitrogen bases. The complementary strand should have the phosphate molecules aligned along the right of the page.

d) Record the genetic code of the complementary strand.

7 The hydrogen bonds between the nitrogen bases are easily broken. Separate the two strands so that each is alongside each margin. Then make complementary strands for each of the original strands.

8 Glue each of the strands to a blank sheet of paper.

e) What do you notice about the two strands of DNA?

Laboratory Application Questions

1 Thymine always bonds with adenine. Explain why the thymine used in your model does not bind with guanine.

2 What determines the nitrogen base sequence of DNA in a new strand of DNA?

3 Adenine is a double-chain purine molecule. Purines always bond with pyrimidines. What special problems would be created if the purine adenine joined with guanine? with another purine?

4 A special enzyme scans the DNA helix in search of improper nitrogen base pairings. Using your model, indicate how the enzyme would be able to detect incorrect nitrogen base pairings.

5 Radiation can cleave the bonds between phosphate and sugar molecules, thereby cutting a section of the DNA out of the ladder. Special enzymes have the responsibility of gluing the spliced segment of DNA back into place; however, the segment is not always glued back in the correct place. Explain why it is important to place the broken section of DNA back in its original position. ■

FRONTIERS OF TECHNOLOGY: DNA FINGERPRINTING

In 1986, DNA matching was used to identify a rapist-murderer in Leicester, England. More than 1000 men were tested in three villages near Leicester. The technique, called DNA fingerprinting, was used to free one suspect, and led to the arrest and subsequent confession of another. Later the same year, a rapist in Orange County, Florida, was convicted on the basis of genetic evidence.

The DNA fingerprinting test used in Britain was developed by Alec Jeffreys, a geneticist from the University of Leicester. Jeffreys found that, although long stretches of the DNA molecule are similar from one person to another, particular segments contain unique arrangements of nitrogen bases. Only identical twins share the same nitrogen base arrangements in these sections. These short sequences of DNA appear to have no function. Geneticists believe that the codes are nonsense codes, which seem to repeat in almost a chemical type of stuttering.

In the DNA fingerprinting test, a DNA segment is taken from the semen found in the rape victim and compared with the DNA segment taken from a blood sample of the suspect. The DNA samples are transferred to a nylon sheet, where they are tagged with a radioactive probe that identifies the unique segments of the DNA chain. The nylon sheet is then placed against an X-ray film. Black bands appear where the probes have attached to the segments used to establish identity. A print is then made from the film and used to compare samples.

> **BIOLOGY CLIP**
> Fisheries Canada officials are considering using DNA fingerprinting to identify persons who catch protected species of fish.

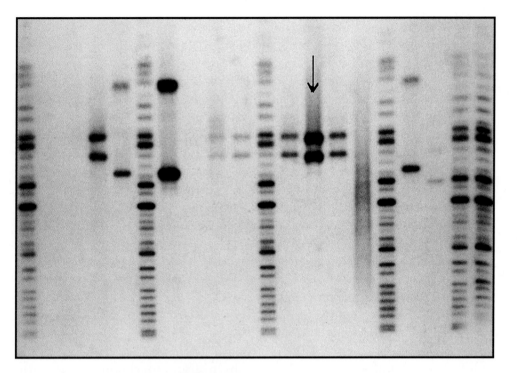

Figure 26.8

DNA profiling can be used to identify criminals.

Restriction enzymes *cut strands of DNA at specific sites.*

A **ligase enzyme** *is a biological glue that permits one section of DNA to be fused to another.*

■ REVIEW QUESTIONS ■ ?

9 What is the significance of DNA replication for your body?

10 Name the nitrogen bases in the DNA molecule, and state which normally pairs with which.

11 Briefly describe the events of DNA replication.

12 Proofreading enzymes scan the strands of DNA to check the nitrogen base pairings. Explain why these enzymes are important.

13 When you eat fish, you take in fish protein and fish DNA; however, you do not assume the characteristics of a fish. Explain why the nucleic acids of a fish do not change your appearance.

14 What is DNA fingerprinting and how does it work?

GENE RECOMBINATIONS IN THE LABORATORY

All cells have enzymes called *endonucleases*, which can cut strands of DNA. Bacteria have special endonucleases called **restriction enzymes,** which cut apart foreign DNA that has invaded the cell. Restriction enzymes act like biological scissors because they snip sites along the DNA molecule. There are over 200 different restriction enzymes, each of which cuts a different segment along the chromosome.

The discovery of the scissor enzyme was preceded by the discovery of another equally important enzyme, the **ligase enzyme,** or the DNA glue. The ligase enzyme is part of the normal DNA repair system. Many things in your environment can damage your DNA. Even sunlight can split chromosomes apart. The ligase enzyme helps restore the fragmented chromosome. Molecular biologists use both enzymes for gene splicing. The restriction enzymes can also be used to snip a recipient chromosome apart. The ligase enzyme permits the DNA snipped from the donor chromosome to be glued into the recipient.

Biologists have recently discovered that restriction enzymes from bacteria can be used to cut genes from many different living things. Herbert Boyer and Stanley Cohen conducted experiments using a common bacterium of the gut, *Escherichia coli* (or *E. coli*). First, they used restriction enzymes to extract a gene from a toad. Next, they used another restriction enzyme to cut the small circular plasmid of an *E. coli* bacterium. The gene from the toad was then inserted into the opening of the plasmid and glued in place. A gene from a toad had thus been incorporated into the genetic material of a bacterium. Boyer and Cohen watched in amazement as the toad DNA was duplicated along the DNA in the plasmid. Even more incredible, the DNA from the toad began providing direction to the cytoplasm of the bacterial cell as if it had always been there. Two organisms that would never exchange genetic material in nature had been joined in a most unusual union. This technique has been called **recombinant DNA**.

The genes from the toad duplicate right along with the genetic information that is native to the bacteria. As the *E. coli* cells conjugate, the toad DNA located in the plasmid is passed to other bacteria. The bacteria that receive the spliced DNA express traits coded for by the new genes just like their conjugate partners. Therefore, the new gene combination can soon be found in many cells as the bacteria reproduce by conjugation.

Imagine blending bits of your own DNA into bacteria cells. These primitive microbes would have segments of your characteristics. Does this sound like science fiction? It is not. Scientists have already placed a number of human genes into bacteria. The gene that produces human insulin, for example, has been spliced into *E. coli* bacteria. The human DNA directs the bacteria to produce human insulin, a hormone essential for regulating blood sugar. People who do

not produce adequate amounts of this hormone have the disease sugar diabetes. Diabetics can regulate their blood sugar by taking an insulin injection. Traditionally, insulin has been extracted from the pancreases of cows and pigs. But cow and pig insulin differs from human insulin, and some patients have allergic reactions to it. Human insulin, produced by *E. coli* bacteria, was first marketed in Canada in 1983.

DNA AND INDUSTRY

Nobody knows when the first human accidentally discovered the intoxicating union of yeast cells with sugar. But once people began to take full advantage of this discovery, biotechnology was born. Biotechnology involves the use of living things—usually but not exclusively microbes—for making products. Today, microbes are a multimillion-dollar business that extends far beyond the traditional boundaries of beer and wines. Modified yeast cells have been employed to produce feed for pigs, fuel for cars, and vaccines for humans. Bacteria have been altered to produce human insulin and growth hormones, as well as glues and artificial sweeteners.

Biotechnology has become a tremendous growth industry. It has been estimated that, by 1995, 40% of Canada's $4 billion pharmaceutical industry will be derived from biotechnology. That figure is expected to increase to 70% by the turn of the century. Canadian biotechnology companies compete with similar companies in Japan, the United Kingdom, Germany, Russia, Australia, and the United States for their share of this promising industry. The Science Council of Canada estimated that biotechnology will be a $186 billion industry by the year 2000. Some economists have predicted

Figure 26.9

Giant fermenter at Alberta Research Park.

Recombinant DNA *is an application of genetic engineering in which genetic information from one organism is spliced into the chromosome of another organism.*

that biotechnology will have a greater impact on the 1990s than the computer chip had on the 1980s.

Table 26.1 Cloned Human Gene Products

Protein	Used in treating
Atrial natriuretic factor	High blood pressure
Erythropoietin	Anemia
Factor VIII	Hemophilia
Factor IX	Hemophilia
Insulin	Diabetes
Interferons	Some cancers, viral infections
Interleukin-2	Cancer
Monoclonal antibodies	Infectious disease
Somatotropin (growth hormone)	Pituitary dwarfism
Tissue plasminogen factor	Heart attack, stroke
Tumor necrosis factor	Cancer

U.S. biotechnology firms like Genetech Inc. and Cetus Corp. have become economic giants virtually overnight. In 1981, Wall Street stockbrokers watched in amazement as stock for a little company, Genetech, skyrocketed from $39 to $89 a share within minutes of trading. Herbert Boyer, who had borrowed $500 to begin the company in 1976, found it had an estimated worth of $80 million only five years later. The trading was spurred by the announcement that interferon, a natural product of the body, had been produced by genetically engineered bacteria. In 1981, there was evidence that interferon could lead to a cure for cancer. Although the drug could be collected by conventional methods, it was very expensive. The promise of mass-producing the drug sent speculators into a frenzy.

NEW, IMPROVED MOUSE

Biotechnology involving recombinant DNA was first directed at placing genes from plants and animals into lower life forms. More recent research has concerned itself with placing genes in higher organisms. In April 1988, researchers used a mouse to combine specific genes customized for cancer research. The mouse contains an oncogene that makes it susceptible to breast cancer. This allows researchers to study which chemicals cause breast cancer in a living model. The mouse can also be used to test cancer-preventing drugs.

E.I. du Pont de Nemours, the company that engineered the new mouse, applied for and received a patent that would prevent other companies from copying their technique. Although other companies have applied for and received patents on recombinant DNA bacteria, no company has received a patent on a mammal. Another group of researchers have placed the genes that carry HIV (the virus that causes AIDS) into a mouse. Like the mouse that contains the oncogenes, this mouse will prove invaluable in determining how HIV destroys the immune response.

This kind of research raises many concerns. Opponents of animal patenting fear that this practice will allow a company to own a life form. Animal-rights supporters are concerned that patents on genetically engineered animals will promote their suffering. Some researchers fear that, by placing patents on mice, companies can control research; only those groups that have enough money or prestige to buy their mice will have an opportunity to study oncogenes. Should companies receive profits for technology that has worldwide benefit? What would happen if the laboratory mice escaped and began to breed with normal mice?

Figure 26.10

Both mice are the same age, but the mouse on the right had the gene for human growth hormone inserted into it.

SOCIAL ISSUE:
Nonhuman Life Forms as Property

Statement:

Companies and research institutes should be forbidden to apply for patents on or restrict access to living organisms.

Research the issue.
Reflect on your findings.
Discuss the various viewpoints with others.
Prepare for the class debate.

Point

- A company should not be allowed to own a life form such as an oil-digesting microbe or to gain exclusive profits from its use. Environmental disasters such as oil spills have such far-reaching effects on water quality, wildlife, native people, and other groups that any helpful technology should be available to all.

- Research institutes should not be able to impede research by other researchers. If they or private companies could have a patent on a mouse, it would not be long before similar patents were sought on more complex life forms.

Counterpoint

- Developing an oil-ingesting microbe is a very expensive process. The patent would protect a company by enabling it to get a return on its investment and encouraging further research. Since the microbe may reduce the dangers imposed by crude-oil spills, the company may be acting for the benefit of both humanity and the environment.
- Profit has motivated many discoveries. An example is the development of antibiotics by drug companies. Few people would choose to do without penicillin or tetracycline, just because a drug company is making a profit on them.

CHAPTER HIGHLIGHTS

- The experiments of Avery and associates demonstrated that DNA was the chemical of heredity.
- Nucleotides of DNA are composed of deoxyribose sugar, phosphates, and four different nitrogen bases.
- The Watson and Crick model of DNA is described as a double helix.
- The Watson and Crick model is based on the fact that specific nitrogen base pairs are constant. Adenine always forms a hydrogen bond with thymine, while guanine always bonds with cytosine.

- DNA, unlike other molecules, is capable of replicating. The double strand of DNA separates, and each of the single strands acts as a template for the alignment of nucleotides and subsequent synthesis of a complementary strand of DNA.
- DNA fingerprinting has many applications, one of which is the identification of individuals.
- Biotechnology is an emerging industrial force, but along with the new technology come moral and ethical questions.

APPLYING THE CONCEPTS

1 What led scientists to speculate that proteins were the hereditary material?
2 Explain how Avery, MacLeod, and McCarty's experiment pointed scientists toward nucleic acids as the chemical of heredity.
3 Science and technology are often referred to as *synergists*. This means that not only do scientific breakthroughs provide information for technologi-

cal applications, but technological advances also spur scientific progress. Explain how X-ray diffraction techniques led to the discovery of DNA.
4 Using DNA as an example, explain why scientists use models.
5 Compare the amount of DNA found inside one of your muscle cells with the DNA found in one of your brain cells.

6 Why are organ transplants more successful between identical twins than between other individuals?

7 A drug holds complementary nitrogen bases with such strength that the DNA molecule is permanently fused in the shape of a double helix. Predict whether or not this drug might prove harmful. Provide your reasons.

8 DNA fingerprinting has been used to identify rapists. Suggest at least two other applications for the DNA matching technique.

9 What follows is a hypothetical situation. Genes that produce chlorophyll in plants are inserted into the chromosomes of cattle. Indicate some of the possible advantages of this procedure.

10 Analysis of chloroplasts and mitochondria reveals that DNA is located within these organelles. Explain how this discovery helps support the theory that these organelles might actually be descendants of individual living creatures?

11 Explain the significance of the following statement to the search for the structure of the genetic material: Structure provides many clues about function.

CRITICAL-THINKING QUESTIONS

1 The European Economic Community (EEC) has ruled that organisms created by biotechnology can be patented. Such patents have been awarded in the United States since 1984. The Supreme Court of Canada ruled that a patent application for a hybrid soybean could not be granted. Although this does not prevent another company from applying for patents in Canada, it does raise an interesting question: Will successful biotechnology companies choose not to locate in Canada unless they are given the same protection as companies in Europe and the United States? Give your opinions on this matter.

2 DNA fingerprinting has had a tremendous impact on law enforcement. Speculate about how DNA fingerprinting will affect criminal trials. Should DNA fingerprinting ever be used to convict a criminal? Why or why not?

3 In December 1988, researchers at Toronto's Hospital for Sick Children transplanted human bone marrow into mice. The mice allow scientists to study human blood cells, immune responses, and genetic blood disorders in a living system. Previously, bone marrow could only be studied in a living culture, not in a functional body. Many supporters of animal rights are appalled by this procedure. They believe that the mice are being exploited. Should animal modeling experiments, like the ones performed at the Hospital for Sick Children, be continued? Give reasons for your answer.

4 Recombinant DNA technology has been described as the 20th century's most powerful technique since the splitting of the atom. Do you agree or disagree with this comparison?

ENRICHMENT ACTIVITIES

1 Isolate some germinating seeds and ask your dentist to treat the seeds with various levels of radiation. Plant the seeds and compare the growth rates of seeds treated with radiation with those of a control group.

2 Using the squash technique, prepare cells from the meristematic region of the germinating roots. Use a stain to make the chromosomes more prominent. Compare the chromosomes of plants treated with radiation with those of a control group.

Protein Synthesis

IMPORTANCE OF PROTEINS

Amino acids *are organic chemicals that contain nitrogen. Amino acids can be linked together to form proteins.*

Proteins *are the structural components of cells.*

Figure 27.1

The spectacular colors of the kingfisher are produced by proteins, the structural components of cells.

Why do organ transplants present medical problems? Why do identical twins look alike? Why does our hair turn white as we age? What causes a baby's blue eyes to turn to hazel as the child ages? To answer these questions we must begin investigating the structural components of cells—the proteins.

The proteins within your body are composed of 20 different **amino acids.** The amino acids can be strung together in an almost endless variety to make different **proteins.** Proteins are the chemicals that make up the structure of cells. Cell membranes and cellular organelles are composed of proteins. Proteins make up the muscle filaments that enable the body to move. Hair and hair color are also produced by proteins. However, proteins are responsible for much more than just appearance. They also produce the enzymes that regulate the speed of chemical reactions within cells. Antibodies—disease-controlling agents—and hormones—chemical messengers produced by certain cells of the body—are all made up of proteins.

a) Record which antibiotic was placed in each section of the petri dish.

5 Invert the petri dish and place it in a bacteriological incubator set at 37°C.

6 Check the petri dishes after 48 h. Measure the growth ring around each disk. Four separate measurements should be taken, as indicated in the diagram below.

Antibiotic	Measurement (mm)				
	#1	#2	#3	#4	Average
1					
2					
3					

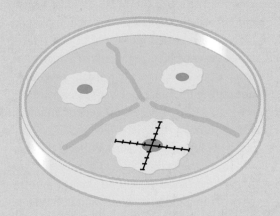

b) Record your data in tabular form as shown.

Laboratory Application Questions

1 On the basis of the experimental evidence, indicate which antibiotic was the most effective.

2 Would the antibiotic that best controlled the bacteria found on your face necessarily be the most effective at controlling the bacteria that causes strep throat? How would you go about testing your hypothesis?

3 What potential problems might be created by placing low dosages of antibiotics in skin creams?

4 Some antibiotics prevent protein synthesis in more advanced cells (eukaryotic cells). Indicate some of the advantages of these antibiotics, and explain why their dosages must be carefully administered. ■

DNA AND MUTATIONS

Mutations are inheritable changes in the genetic material. They can arise from mistakes in DNA replication when one nitrogen base is substituted for another. Cosmic rays, X rays, ultraviolet radiation, and chemicals that alter DNA are referred to as **mutagenic agents.** By changing the arrangement of the nucleotides in the double helix, the mutagen changes the genetic code. The ribosome will read the new code and assemble amino acids according to the new instructions provided. The shift of a single amino acid will, unfortunately, produce a new protein. The new protein has a different chemical structure and, in most cases, is incapable of carrying out the function of the required protein. Without the required protein, cell function is impaired, if not completely destroyed. Although some mutations can, by chance, improve the functioning of the cell, the vast majority of mutations produce adverse effects.

Sometimes the error may arise because of a shortage of a particular type of nucleotide during the replication process. A more plentiful nitrogen base is substituted for the scarce one but, in the process, the genetic code is altered. One of the most common types of errors is created when one nitrogen base is substituted for another. For example, a chemical mutagen called hydroxylamine can modify cytosine so that it pairs with thymine. Normally, adenine forms hydro-

Mutagenic agents *are things that cause changes in the DNA.*

gen bonds with thymine. The hydroxylamine removes a nitrogen group attached to the adenine molecule, making it appear much like the guanine molecule. The guanine then bonds with a thymine-containing nucleotide. Occasionally, X rays will break the backbone of the DNA molecule. Special enzymes will repair the break. Unfortunately, the spliced segment of the DNA ladder is not always placed in the correct section. The misplaced segment of DNA may alter the entire library of genetic information. The impact would be similar to reading the first sentence of this paragraph if the word breaks were moved two letters to the left. "Sometimes the error may arise" would read "Sometim est heerr orm ayar ise."

Figure 27.6

Mutations sometimes occur when spliced sections of DNA are returned in the incorrect position.

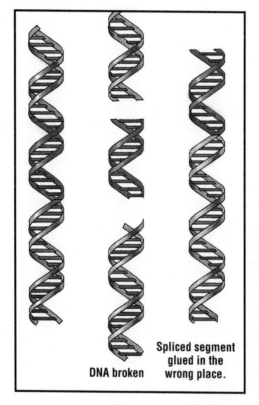

DNA broken

Spliced segment glued in the wrong place.

When mutations occur, they are repeated each time the cell divides. A mutation in an egg or sperm cell will lead to permanent change in the characteristics of an offspring. In the adult, many genes code for characteristics that have already developed. For example, your hands, feet, and toes are already in place. If the gene that controls the production of hair color were altered in a single cell, it would have little effect on your body. However, if this same gene were altered in a fertilized egg, the mutation would be coded into the DNA of every succeeding cell of the body. The single mistake can be repeated billions of times. This is why mutagenic agents are particularly dangerous for pregnant women, especially during the first trimester of pregnancy.

One well-known genetic mutation is a human disorder called sickle-cell anemia. This genetic disorder affects the structure of the oxygen-carrying molecule found in red blood cells. The alteration of a single nitrogen base causes valine to replace glutamate as the sixth amino acid in one of the protein chains. Unfortunately, even this slight change has devastating consequences. The red blood cell assumes a sickle shape and is unable to carry an adequate amount of oxygen. To make matters worse, the sickle-shaped cells clog the small capillaries, starving the body's tissues of oxygen.

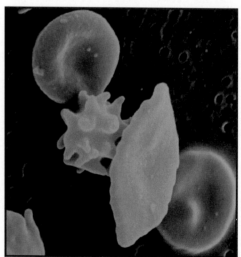

Figure 27.7

Sickle-shaped red blood cells characteristic of sickle-cell anemia. This disease is a hereditary, genetically determined anemia.

Objective

To investigate the genetic properties of HIV.

Background Information

Acquired immune deficiency syndrome, or AIDS, describes a number of disorders associated with the infection of the human immunodeficiency virus, or HIV. Two different types of HIV have been identified. HIV-1 was discovered in 1981, HIV-2 in 1985. HIV invades the very cells whose function is to protect the body from **pathogens,** or disease-causing agents. HIV progressively damages the immune system, predisposing a person to a number of opportunistic infections and malignancies. As of 1993, there was no cure for AIDS. However, advances in microbiology, genetics, and molecular biology, along with improvements in the microscope are a cause for cautious optimism. For the time being, an educated public may prove to be the best defense against the virus.

HIV must be directly transmitted. Unlike the chicken pox and flu viruses, which can be transmitted through the air, HIV must enter the bloodstream. HIV has been found in human body fluids. It is spread primarily through sexual intercourse and by the introduction of blood or blood components into the bloodstream through the sharing of needles or syringes for injection drug use. HIV can also be transmitted to infants during pregnancy or at the time of birth. In rare cases, HIV has been transmitted through the breast milk of an infected mother.

Although HIV is very tiny, the damage it causes can be devastating. The virus contains a control center of RNA, a sister molecule of the more familiar DNA, and an envelope of protein. What makes HIV even more insidious than other viruses is that it attacks the immune system directly. The helper T cells (sometimes referred to as T4 lymphocytes), the cells that act as guards against invading pathogens, are the targets of HIV. Thus, HIV destroys the body's own defenses, rendering it incapable of defeating other invading substances.

The Case Study

HIV attacks the helper T cells. The shape of the protein coat of the virus permits binding to the cell membrane of the helper T cell much like a lock and key. The unique binding site of the T cell is not designed for HIV, but as a port for hormones or other needed chemicals. HIV takes advantage of the contours of the port of the T cell to launch its invasion. The outer membrane of the HIV is compatible with the outer membrane of the T cell. This explains why certain viruses only infect certain cells.

1 View the diagram of the outer membranes of the helper T cell, the skin cell, and the muscle cell. The drawings are models of the cell membranes, not the actual structures.

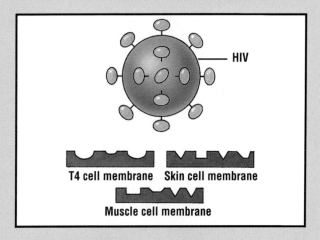

a) Can HIV attach itself to a muscle cell or a cell from the skin?

b) Explain why you cannot get AIDS by shaking hands. (Use the information that you have gained about binding sites.)

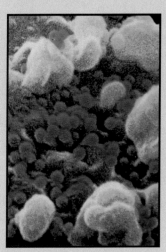

Color-enhanced scanning electron micrograph of the T cell being attacked by HIV particles. The HIV particles appear green in color.

The helper T cell mistakes the virus for a needed substance and engulfs it in a process known as **phagocytosis**. Cells normally engulf large molecules by phagocytosis. The entire virus enters the cell. Once inside, the virus sheds its coat, and the RNA core is set free.

2 View the diagram that shows how HIV enters the helper T cell. The genetic material from HIV is incorporated into the genetic material of the host cell.

c) Most viruses leave their coat on the membrane of the infected cell. Indicate why these viruses are much more easily identified than HIV.

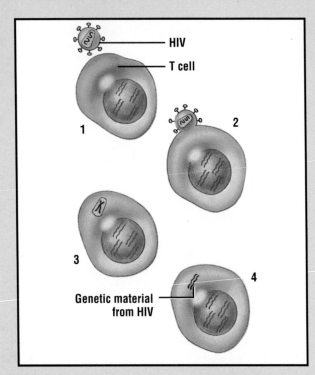

1

2

3

Genetic material
from HIV

4

HIV

T cell

Once inside the cell, the viral RNA behaves in a very special way. Normally, the DNA acts as a blueprint for instructions, and the mRNA molecule reads the hereditary message from it. The process is referred to as *transcription*. Under normal circumstances, the RNA acts as a messenger by carrying the genetic information from the DNA in the nucleus to the cytoplasm.

Once in the cytoplasm, the mRNA provides the ribosome with the correct instruction for the assembly of a protein. Another molecule of RNA, tRNA, carries amino acids to the ribosome. Here the code provided by the mRNA dictates the position of the amino acids. The ribosome links individual amino acids together in a process known as *translation*. Proteins are made from amino acid building blocks.

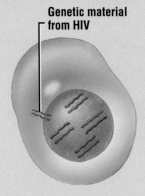

Genetic material
from HIV

Most viruses contain DNA, but HIV is different; that is, its genetic material is RNA. The RNA molecule of HIV is much more than just a messenger. A special enzyme called **reverse transcriptase** allows the genetic message contained in the RNA of the virus to be printed along a strand of DNA.

d) Why is the enzyme referred to as reverse transcriptase?

The newly constructed viral DNA now slips into the nucleus of the infected cell. Here it can splice into healthy DNA. The instructions of HIV are now part of the helper T cell. The virus can remain dormant for many years. The viral DNA becomes part of the human DNA.

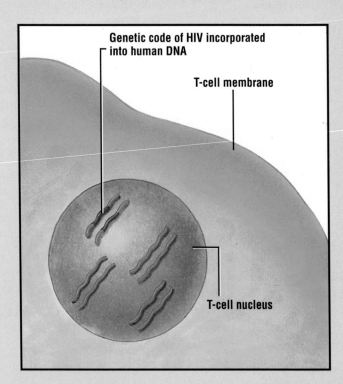

Genetic code of HIV incorporated
into human DNA

T-cell membrane

T-cell nucleus

3 Use the previous diagrams for reference to answer the following two questions.

e) What happens to the viral DNA if the T cell divides?

f) Explain why it is possible for a human to be infected with HIV and not exhibit any of the symptoms of AIDS?

No one knows what activates HIV. Some scientists speculate that it could be other co-infections. Others believe the virus has a gene that acts like a ticking time bomb. Whatever the reason, once activated, the virus reproduces at a furious pace.

Viral DNA stimulates the production of viral RNA, which carries the virus message into the cytoplasm. The transcribed RNA attaches itself to the ribosome and directs it to produce viral proteins. The proteins will serve as coats for the newly released RNA.

The once-healthy helper T cell has been transformed into an HIV factory. Eventually, the overworked and exhausted T cell bursts, and the virus particles are released to infect other T cells.

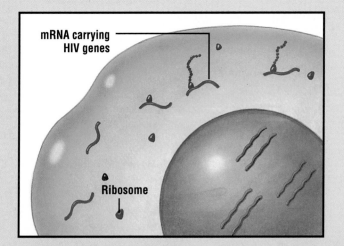

g) Normally, the killer T cells would destroy a cell infected with a virus long before it could become a virus factory. Why are the infected helper T cells not held in check?

HIV creates yet another problem for the helper T cells. Once infected, the outer membrane of the helper T cells changes, causing them to fuse together. This allows the virus to pass from one cell to another without entering the bloodstream. Because antibodies move freely through the fluid portions, HIV avoids them by slipping from one cell to another. The T cell not only produces other HIV viruses, but also protects them from wandering antibodies.

h) Indicate why people infected with HIV most often die of another infection such as pneumonia?

i) David, "the boy in the plastic bubble," suffered from a disorder called "severe combined immunodeficiency syndrome." How does this disorder differ from acquired immune deficiency syndrome?

The challenge presented in finding a cure for AIDS stems from its variable protein coat. Each mutation produces a distinctive coat.

j) Why is it so difficult to destroy a virus that changes shape?

(a) Electron micrograph showing HIV escaping from the cell membrane of a T cell.

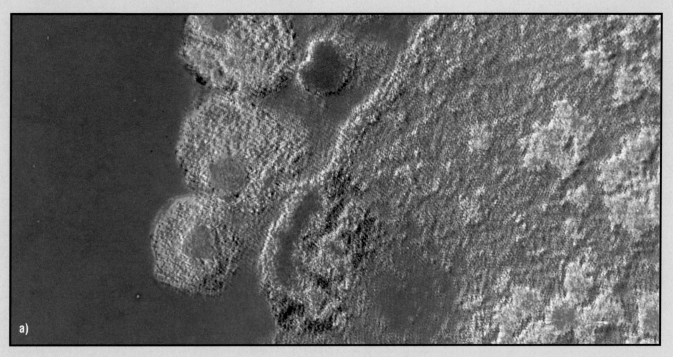

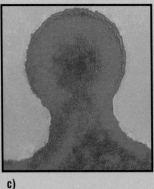

b)   c)

Pictures (b) and (c) show a closer view of the infected T cell as the virus escapes through the cell membrane.

A drug known as Zidovudine (formerly known as AZT) hampers the replication of HIV. Zidovudine has proved somewhat effective if the virus lies dormant. However, for those who already exhibit the symptoms of AIDS, it has limited results. Although Zidovudine may help increase the life span of people infected with HIV, it cannot be viewed as a cure. At $10 000 a year per patient, the cost of the drug is also a major concern.

The testing of vaccines has been underway in both France and the United States. Daniel Zagury of the Université de Pierre et Marie Curie announced that he had injected himself and others with a vaccine developed by Bernard Moss. To date, the vaccine has produced no observable side effects, but hope remains tempered with caution.

Major breakthroughs have come in the form of tests for detecting HIV. Blood collected by the Red Cross in Canada has been screened for the presence of HIV since 1985.

People receiving transfusions or transplants no longer have to worry about acquiring HIV through donated blood. In June 1987, the Du Pont company announced a new test that can detect HIV antibodies. The new test is much cheaper and easier to run. It takes only about five minutes to complete and does not require refrigeration. This breakthrough is especially significant for central Africa, the area with the highest incidence of AIDS cases. Unfortunately, it is also one of the poorest areas in the world. Currently, the United States spends more money testing for AIDS than central African states allocate for their entire health budget.

Case-Study Application Questions

1 Why does the Canadian Red Cross inquire about a person's travel before they accept blood donations?

2 How does transcription for HIV differ from normal cell transcription?

3 Can AIDS be transmitted through either food or beverages? Explain your answer.

4 Can AIDS be contracted by casual contact such as shaking hands, or using the same telephone or toilet seat? Explain your answer.

5 Do tattoos and pierced ears pose any potential risks for infection with AIDS?

6 How is it possible to catch HIV from a person who shows no symptoms associated with AIDS?

7 Should people with AIDS be quarantined? Justify your answer.

8 Should health-care workers such as doctors, dentists, and nurses be screened for HIV? Justify your answer. ■

ONCOGENES: GENE REGULATION AND CANCER

Cancer is characterized by uncontrolled cell division. Cancer cells are often described as cells that are "too alive." But what causes normal cells to become cancerous? Two lines of evidence indicate that cancer results from changes in the genetic code. First, cancer cells often display nitrogen base substitution, or the movement of genetic material from one part of the chromosome to another. A second factor that supports the idea that cancer arises because of alterations of DNA is the fact that many known mutagens are also known to cause cancer. X rays, ultraviolet radiation, and mutagenic chemicals can also induce cancer.

American soldiers took advantage of the North American native peoples' susceptibility to smallpox to drive them from their lands. Blankets, contaminated with the deadly virus, were used in trade.

During World War II, British, American, and Canadian armies were involved in the development of biological weapons. Porton Down in England, Camp Detrick in Maryland, and Suffield in Alberta were designated as research and testing stations. The preferred microbe was the deadly anthrax bacterium, which affects both cattle and humans. The deadly spores of the rod-shaped bacterium live for long periods of time, are highly contagious, and are resistant to many environmental factors. American and British armies had planned to make thousands of anthrax bombs that would disperse the microbes on impact. The war ended before the plan was implemented.

The Japanese army was also engaged in the development of biological weapons during World War II. Pingfan, a small village near Harbin in China, housed over 3000 researchers, technicians, and soldiers who were dedicated to exploiting the disease-causing properties of typhoid fever, anthrax, and cholera.

Almost any disease-causing agent can be exploited for biological weaponry. The microbe, or toxin produced by the microbe, need not be harmful to humans. An enemy could target the destruction of livestock, cereal grains, or bacteria found in the soil to create food shortages and cause economic ruin. Fortunately, few organisms are well-suited for mass destruction. HIV, the virus that causes AIDS, cannot be transmitted through the air or by casual contact. It is transmitted in body fluids such as blood. Therefore, releasing the virus into a city's drinking water would not create an epidemic. The bacterium *Clostridium botulinum* produces one of the most powerful poisons known to humans. It has been estimated that one kilogram of the toxin placed in a typical water reservoir could kill 50 000 people. Sixty percent of the population would die in less than 24 h. However, this microbe would have little effect if released into the air. *Clostridium* cannot live in environments where oxygen is present. The microbe must be cultured in oxygen-free environments, where it is capable of producing the harmful toxin. Even the most dangerous of microbes have some natural controls that have evolved over a period of time. No disease-causing agents can survive if their hosts are eliminated.

Combining genes for weaponry is a frightening prospect. Merging genes that permit rapid reproduction with those that demonstrate a resistance to environmental factors could create a "superbug." Bacteria that carry drug-resistant genes in their plasmids already create problems. The genetic information can be duplicated and passed between microbes during sexual reproduction. Consider the possibility of a microbe that is resistant to a range of antibiotics being introduced to an enemy population. The attacking army might be able to construct a secret drug that is capable of protecting their allies while selectively removing the enemy. By the time the enemy finds an antibiotic for the disease, a significant number would have died and their resistance weakened. The hybrid microbe might provide yet another advantage. It is quite likely that disease-causing microbes could escape detection by both the body's immune system and physicians. For example, *E. coli* bacteria are a normal fauna of the human gut. The body will not mobilize an immune response against this organism. By splicing disease-causing genes into the *E. coli*, the source of the infection would be masked. Most physicians would have little difficulty identifying an anthrax infection, but associating the symptoms of a disease with the *E. coli* microbe would be much more difficult.

CAREER INVESTIGATION

The applications and uses of genetic technology are increasing every day. Biotechnology can be applied to almost everything in our lives. The scope of this field ranges from agriculture to pharmacology, from oil-spill cleanup to outer space exploration.

Forensic scientist

Police are using more sophisticated techniques to determine the identity of criminals. DNA fingerprinting, for example, is a powerful tool in the hands of a trained technician.

- Identify a career associated with biotechnology.
- Investigate and list the features that appeal to you about this career. Make another list of features that you find less attractive about this career.
- Which high-school subjects are required for this career? Is a post-secondary degree required?
- Survey the newspapers in your area for job opportunities in this career.

Astronaut

As well as being physically and mentally fit, an astronaut must possess a wide-ranging scientific knowledge. Astronauts are called on to perform many different experiments in space, some of which deal with biology and biotechnology.

Fertility specialist

Many couples who are unable to have children are turning to fertility specialists. A fertility specialist must determine the nature of the problem and provide a solution. The solution may involve a number of techniques such as hormone treatments or *in vitro* fertilization.

Stockbroker

Many biotechnology companies have started up over the past few years. Biotechnology stocks are a billion-dollar business. Knowledge about which stocks to buy requires scientific knowledge. A stockbroker must be able to predict how well the company will do in order to properly advise clients.

SOCIAL ISSUE:

Biological Warfare

The development of chemical weapons during World War I shows the extent to which science and technology can be exploited for destructive purposes. During that war, chlorine gas, phosgene gas, and mustard gas disabled over one million soldiers.

Advancements in genetics may have made nature an unwilling accomplice in war. Recent biological techniques, responsible for the development of antibiotics, vaccines, and genetically created industrial products, have opened the door for more harmful applications.

Statement:

Research that could be used for biological warfare must be banned.

Point

- In 1985, the United States government reported spending $39 million on defensive biological weapons programs. The programs were directed at developing vaccines for genetically-created microbes. The money would be much better spent on research to develop vaccines for naturally occurring diseases.

- Biological weapons present a real threat to world peace. Unlike nuclear weapons, biological weapons are cheap to produce, portable, and have a built-in time delay. Therefore many smaller countries with militaristic governments are able to build and use them.

Counterpoint

- The money spent on the development of this type of defense cannot be construed as money spent on weapons. The development of vaccines and antibiotics has historically been associated with the growing and culturing of disease-causing organisms.

- The threat posed by biological weapons is exaggerated. It is not likely that humans can construct microbes that are more dangerous than those that nature has taken years to perfect.

> **Research the issue.**
> **Reflect on your findings.**
> **Discuss the various viewpoints with others.**
> **Prepare for the class debate.**

CHAPTER HIGHLIGHTS

- The genetic code is determined by the arrangement of nitrogen bases within the strands of DNA.
- Each gene codes for the production of a specific protein.
- Messenger RNA reads the chemical message inscribed on the DNA and carries the information to the ribosomes located in the cytoplasm. The transfer of the genetic message from the DNA to the tRNA is known as transcription.

- Ribosomes are the site of protein synthesis.
- Transfer RNA molecules carry amino acids to the ribosome for protein synthesis. The tRNA molecules place the amino acids along a strand of mRNA, which is on the ribosome. The mRNA determines the sequencing of the amino acids. The anticodons of tRNA must combine with complementary codons from the mRNA. The process is referred to as translation.

- Gene mutations occur when the nitrogen base sequence in DNA is altered.
- Gene control mechanisms regulate gene expression.
- The operon consists of operator and structural genes.
- Cancer genes, called oncogenes, leave the molecular switches for cell division on.

- The Ames test is used to identify mutagens, and thus potential carcinogens.
- Biological warfare is the exploitation of microbes or the products of living things for military ends. Many ethical questions must be explored as our knowledge of gene action is extended and applied.

APPLYING THE CONCEPTS

1 Why are somatic cell mutations less harmful than germ cell mutations?

2 Explain how Beadle and Tatum's experiments with the bread mold *Neurospora* helped explain protein synthesis.

3 In what ways does mRNA differ from DNA?

4 Suppose that during protein synthesis, nucleotides containing uracil are in poor supply. The uracil is substituted with another nitrogen base to complete the genetic code. How will the protein be affected by the substitution of nitrogen bases?

5 In what ways do the codons and anticodons differ?

6 A scientist discovers a drug that ties up the site at which mRNA attaches to the ribosome. Although mRNA is transcribed in the nucleus, it has no way of attaching itself to the ribosome. Speculate about how the drug will affect the functioning of a cell. If the drug is introduced into a single brain cell, will the organism be destroyed?

7 Outline the advantages of Dr. Ram Mehta's genetic tests using yeast cells over those that use animal testing.

8 Compare protein synthesis in cells with an automobile assembly plant. Match the parts of the auto assembly plant with the correct part of the cell. Provide reasons for each of the matches.

Auto assembly plant:	Cell parts:
1) Corporate headquarters	a) tRNA
2) Master blueprint for the car	b) cytoplasm
3) Entire shop area	c) nucleus
4) Supervisor who carries blueprints	d) DNA
5) Stockperson who brings parts to the assembly worker	e) mRNA
6) Assembly worker	f) amino acid
7) Parts of the automobile	g) ribosome

9 Thalidomide was a drug given to pregnant women in the early 1960s to reduce morning sickness. Unfortunately, the drug caused irreparable damage to the developing fetus. Thalidomide inhibited proper limb formation. Many children were born without arms and legs. Explain how the Ames test could have prevented this tragedy.

10 Recombinant DNA provides a valuable technique for those engaged in biological warfare. The gene that produces the botulism toxin, a deadly food poison, can be extracted from its resident bacteria and placed into a harmless bacterium, called *Escherichia coli*, which is a natural inhabitant of the large intestine of humans. Why would the transfer of the botulism gene to the bacteria found in your gut be so dangerous?

CRITICAL-THINKING QUESTIONS

1 Recombinant DNA has produced human insulin in bacteria. Because millions of people suffer from diabetes, the market for human insulin is enormous, and so are the profits. Because of economic pressures, the nature of scientific research has changed from one of sharing information through publishing to one of patents and secretiveness. Companies are unwilling to release any breakthrough for fear that their ideas might be stolen. Scientific research, at least in some fields, is no longer controlled by researchers, but by investors, who seek the advice of accountants and lawyers.

Should government remove biotechnology from private enterprise? Support your opinion.

2   John Moore suffered from a rare form of leukemia. A team of doctors from UCLA removed his spleen, and he went on to live a fruitful life. However, years after his operation, Moore discovered that the surgeons were cloning cells from his extracted spleen for cancer research. The cells produced significant amounts of a substance called interferon, which has tremendous commercial value. Moore now wants what he believes is his fair share of the money, and he is suing the people who are cloning the cells. Should Moore be entitled to any money? Is the spleen Moore's property? Support your conclusions.

3   *Pseudomonas syringae* is a bacterium found in raindrops and in most ice crystals. Researchers have been able to snip the frost gene from its genetic code, thereby preventing the bacteria from forming ice crystals. The *Pseudomonas* has been aptly named "frost negative." By spraying the bacteria on tomato plants, scientists have been able to reduce frost damage. The bacteria can extend growing seasons, thus increasing crop yields, especially in cold climates. A second version, called "frost positive," has also been developed. The frost-positive strain promotes the development of ice and, not surprisingly, has been eagerly accepted by some ski resorts. When this microbe is sprayed on ski slopes, a longer season and better snow base can be assured. However, environmental groups have raised serious concerns about releasing genetically engineered bacteria into the environment. Could these new microbes gain an unfair advantage over the naturally occurring species? What might happen if the genetically engineered microbes mutate? Could the mutated microbe become a super microbe? Do you think genetically engineered microbes should be introduced into the environment? Support your conclusions.

4   The gene for the growth hormone has been extracted from human chromosomes and implanted into bacteria. The bacteria produce human growth hormone, which can be harvested in relatively large quantities. The production of human growth hormone is invaluable to people with dwarfism. Prior to the development of this hormone, people with dwarfism relied on costly pituitary extracts. Although the prospect of curing dwarfism has met with approval from a majority of the scientific community, some concerns about the potentially vast supply of growth hormone have been raised. How can scientists ensure that the growth hormone produced by these genetically engineered bacteria will not be used by normal individuals who wish to grow a few more centimetres? Do people have the right to choose their own height? Give your opinion and support your conclusions.

ENRICHMENT ACTIVITIES

Suggested reading:

- Barnes, James. "The Ancient Quest." *Canada and the World*, January 1990, pp. 14–16.
- Beardsely, Tim. "Smart Genes." *Scientific American* 265(2) (August 1991).
- DeDuve, Christian. *A Guided Tour of the Living Cell*. New York: Freedman and Company, 1984.
- Eberlee, John. "Biology's Holy Grail." *Canada and the World*, January 1990, pp. 24–25.
- ——. "The Life Molecule." *Canada and the World*, January 1990, pp. 17–19.
- Grady, Denise. "The Ticking Time Bomb." *Discover* 6 (6) (June 1987).

- Mackenzie, R.C., "Designer Genes." *Canada and the World*, January 1990, pp. 20–24.
- Maxson, Linda R., and Charles H. Daughtery. *Genetics: A Human Perspective*. Dubuque, Iowa: Brown, 1986.
- Petruz, Max. "The Birth of Protein Engineering." *New Scientist* 1460 (June 1985).
- Suzuki, David, and Peter Knudson. *Genethics*. Toronto: Stoddart, 1988.
- Suzuki, David, Eileen Thalenburg, and Robert Sinshiner. *Let's Talk about Aids*. Toronto: General Paperbacks, 1987.

Variation and Change

•

TWENTY-EIGHT
Population Genetics

TWENTY-NINE
Adaptation and Change

UNIT

8

Population Genetics

Figure 28.1

Birds like the albatross have developed elaborate courtship behavior.

IMPORTANCE OF VARIATION

Variation among living organisms is not restricted solely to physical appearance. It can also be expressed in an organism's metabolism, fertility, mode of reproduction, behavior, or other measurable characteristics. Although Darwin knew from observation of nature that the majority of species possessed variable phenotypes, he was unable to explain the source of variation or how it is passed on. Ironically, in 1859, the same year in which Darwin published *On the Origin of Species*, Mendel began his genetic research with experiments involving garden peas. This work would eventually lay the foundation for genetics, and demonstrate conclusively that the source of genetic variation among individuals is sexual reproduction.

The modern theory of evolution recognizes that the main source of variation in a population lies in the differences in the genes carried by the chromosomes. Genes determine an organism's appearance, and mutations (permanent genetic changes) can cause new variations to arise. These variations can be passed on from generation to generation.

Certain genotypes may be better equipped than others for survival. Organisms with these genes might be better able to obtain necessary resources such as food and water, or to protect themselves against predators, or they may have higher reproductive potentials. Through sexual reproduction, the genes for these variations would be transmitted to the offspring. Given that such offspring are also more likely to survive, subsequent generations would include an increased frequency of the variant genes. Consequently, there would be natural selection, within the group, of individuals better adapted to prevailing conditions.

Over billions of years the change in genetic makeup of the individual within the species has been a major contributing factor to earth's most valuable resource: biological diversity, or biodiversity.

GENES IN HUMAN POPULATIONS

The principles of genetics, established by studies on plants and fruit flies, can be applied to humans. Human chromosomes, like those of other organisms, are composed of DNA, and undergo mitosis and meiosis. However, the study of human genetics presents some unique problems. Unlike garden peas and *Drosophila*, humans produce few offspring, which makes it difficult to determine the genotypes of both the parents and offspring for any particular trait. Another problem with human genetics is that observing successive generations requires time. *Drosophila*, in contrast, can reproduce every 14 days, and many different generations can be studied within a few months. A third problem is that many human traits, including body size, weight, or even intelligence, are affected by environment as well as by genes.

One of the most common techniques used to study human populations is **population sampling.** In the sampling technique a representative group of individuals within the population is selected, and the trends or frequencies displayed by

the selected group are used as indicators for the entire population.

One example of a trait that can be studied by means of sampling is tongue rolling. The ability to roll the tongue is controlled by a dominant gene. People with two recessive genes cannot roll their tongues. Approximately 65% of the population carries the dominant gene.

Blood type is another example of genes that can easily be studied within a human population. The I^A, I^B, and I^O alleles combine to make different blood types. However, unlike the tongue-rolling trait, in which the dominant allele occurs in greater frequency, in blood types the recessive allele is more common. Type O blood is the most common among North American whites and blacks. An estimated 45% of whites and 49% of blacks contain two recessive I^O alleles. Despite the fact that the I^A and I^B alleles are both dominant to I^O, only 4.0% of the white population and 3.5% of the black population contain both the I^A and I^B alleles. Blood type AB is considered rare.

The recessive Rh negative alleles are found only in 15% of Canadians. However, the frequency of Rh– alleles is much higher along a valley region that borders France and Germany. Some population geneticists have postulated that the origin of the Rh– allele can be traced to that valley.

All of the genes that occur in a population are referred to as the **gene pool.** The gene pool maintains continuity of traits from generation to generation. Although some gene frequencies remain the same over many generations, others change quickly. Geneticists have used gene frequencies to study changes in the human population. Certain gene frequencies have been associated with a particular population of people. For example, red hair is often associated with people of Irish or Scottish ancestry.

Population sampling *is a technique in which gene frequencies for a particular genetic trait are determined in a small sample of the population, and results are applied to the whole population.*

Gene pools *are all of the genes that occur within a specific population.*

CASE STUDY
TRACING THE HEMOPHILIA GENE

Objective

To use pedigree charts to trace the hemophilia gene from Queen Victoria.

Background Information

A pedigree chart provides a means of tracing the inheritance of a particular trait from parents through successive generations of offspring. Hemophilia A is a blood-clotting disorder that occurs in about one in 7000 males. The disorder is associated with a recessive gene located on the X chromosome. The fact that a female must inherit one of the mutated genes from her mother and another of the mutated genes from her father helps explain why this disorder is very rare in females.

Procedure

1 Study the pedigree chart of Queen Victoria and Prince Albert. Note the legend. Males are designated by a square, while females are designated by a circle.

a) Who was Queen Victoria's father?

b) How many children did Queen Victoria and Prince Albert have?

2 Locate Alice of Hesse and Leopold, Duke of Albany, on the pedigree chart.

c) Using the legend, provide the genotypes of both Alice of Hesse and Leopold.

3 Locate the royal family of Russia on the pedigree chart. Alexandra, a descendant of Queen Victoria, married Nikolas II, Czar of Russia. Nikolas and Alexandra had four girls (only Anastasia is shown), and one son, Alexis.

d) Explain why Alexis was the only child with hemophilia.

Case-Study Application Questions

1 Is it possible for a female to be hemophilic? If not, explain why not. If so, identify a male and female from the pedigree chart who would be capable of producing a hemophilic, female offspring.

2 On the basis of probability, calculate the number of Victoria's and Albert's children who would be carriers of the hemophilic trait. ■

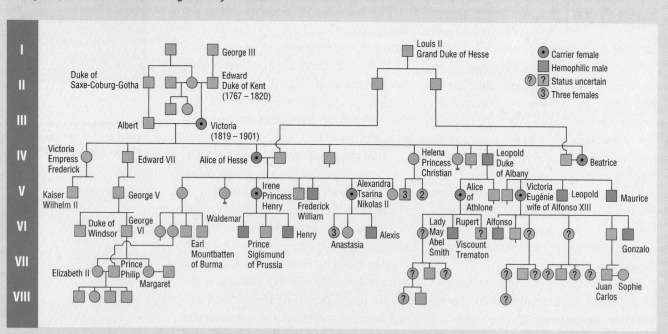

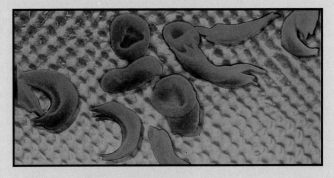

Objective

To investigate two recessive genetic disorders as models for studying evolution in human populations.

Background Information

Human genetic disorders are useful models for studying evolution. Because of their implications for human populations, there is more incentive to study genetic disorders than to study the inheritance and evolution of normal characteristics. Individuals with genetic disorders are often readily identifiable in the larger population. Many disorders are associated with specific populations, reflecting not only differences in lifestyle, but often differences in patterns of genetic inheritance as well. Furthermore, studies of genetic disorders frequently provide evidence of long periods of geographic and genetic isolation within the human population. Examples of disorders that place certain groups and their descendants at a higher risk than others are: cystic fibrosis and European whites; diabetes and Pacific Islanders; sickle-cell anemia and African blacks; Tay-Sachs and eastern European Jews.

Many of the well-known inherited disorders are classified as autosomal (i.e., not sex-linked), recessive disorders. They result from point, or gene, mutations that cause errors in the body's metabolism. The recessive alleles can persist at fairly high frequencies in populations because heterozygous members may still survive and reproduce; only one normal allele is necessary to carry out the specific function.

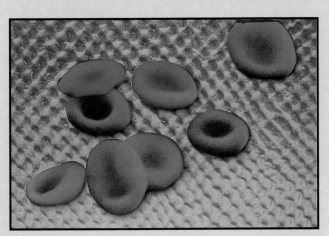

Two extensively studied recessive disorders are sickle-cell anemia and Tay-Sachs disorder. In both cases, the condition is expressed in homozygous, recessive individuals and is usually fatal. The heterozygotes (or "carriers") are usually symptom-free.

Sickle-Cell Anemia

Discovered by a Chicago physician in the early 1900s, this disease gets its name from the sickle-shaped red blood cells found in the blood of those who have the condition. While rare in most human populations, it is common in certain groups of African blacks and in those of African descent. Sickle-cell anemia causes general body pains, loss of appetite, yellowish eyes, a low resistance to infection, and shortness of breath. Death usually occurs in early childhood, but a few cases are known to have survived to adulthood.

It was Dr. Linus Pauling, a noted chemist and double Nobel prizewinner, who uncovered the reason for the presence of sickle cells in individuals with the disorder. By analyzing the hemoglobin molecule from patients' blood, Pauling concluded that the sickle cells occur when one amino acid (valine) is substituted for another (glutamic acid) in one of the four chains of the hemoglobin molecule. He determined that the change in the shape of red blood cells occurs when oxygen levels are low (e.g., at high altitudes, during physical exertion, and so on) and is irreversible.

Since specific amino acids can only be produced from the existing genes within a gene pool, Pauling's work demonstrates the importance of the gene to the production of variation (change) within a population. The shuffling of genes in the process of inheritance provides great variation in gene combinations. The sickle-cell trait, like other genetic traits, is inherited by a simple Mendelian pattern. The genotype of a carrier with the defective gene is written $Hb^A Hb^S$ where Hb is the symbol for hemoglobin and the superscripts A and S represent the genes for normal hemoglobin and the sickle-cell gene respectively.

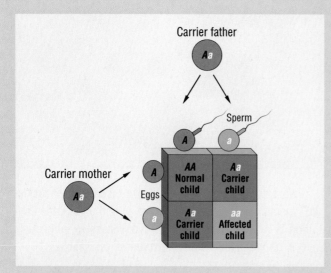

Using the diagram, answer the following questions:

a) Write the genotypes of the offspring from a cross between two carriers of the sickle-cell gene. What is the probability that the couple will have **(i)** a phenotypically normal child? **(ii)** a child with the sickle-cell disorder?

b) What is the probability that a child born of a normal parent and a carrier parent would have a normal child? A child with the sickle-cell disorder?

c) Why would there be no concern about the offspring of two homozygous individuals having the sickle-cell disorder?

d) How does Pauling's discovery provide an answer to the observation that carriers experience mild "sickling" during strenuous exercise?

One biological deduction from this information might be that the recessive allele will eventually disappear from the population through natural selection. However, this has not happened. As many as 40% of the population in certain parts of Africa and 10% of Americans of African descent still carry the trait. What accounts for the frequency of the sickle-cell gene remaining relatively constant over the years?

A review of Example 2: Analysis of Sickle-Cell Anemia Data will shed light on this question and provide additional information for answering the following questions:

e) What selective advantage is afforded a heterozygous individual ($Hb^A Hb^S$) in certain African populations?

f) What might happen to the frequency of the sickle-cell gene if malaria were eliminated in Africa? Why?

g) Estimates suggest that the frequency of the sickle-cell gene in North American blacks has decreased from 22% in the early slavery period to a current value of 10% or less. What might explain this difference?

Tay-Sachs Disorder

Tay-Sachs disorder is also helping scientists understand how genetic disorders evolve, as well as providing insight into precisely how genes (even lethal ones) can persist and spread over the centuries. Tay-Sachs, which occurs in the general population in one out of every 400 000 births, was co-discovered by W. Tay, a British ophthalmologist, and B. Sachs in the 1880s. A notable exception to its occurrence is found in the eastern European Jewish population known as the Ashkenazim. In this group, the Tay-Sachs disorder is 100 times more frequent—about one in 3600 births.

The condition causes a deterioration of the central nervous system and becomes noticeable in infants at about six months after birth. Babies with the disorder lose much of their motor control, have convulsions, and usually die between the ages of two and four years. As with sickle-cell anemia, the genotypes of individuals with Tay-Sachs are homozygous recessive. Carriers of the trait are generally unaffected by the presence of the defective gene.

The cause of the disorder was not determined until 1962, when researchers discovered that people with Tay-Sachs lacked a gene that produces the enzyme ß-N-hexosaminidase A (Hex A). Hex A is required to control excessive accumulation of a fatty substance in body cells, especially nerve cells. Normally, fat is present only in modest levels in cell membranes because it is constantly being broken down by Hex A. A deficiency of the enzyme alters the metabolism of the fat, which then accumulates around the nerve sheath and in time destroys the nerve cells.

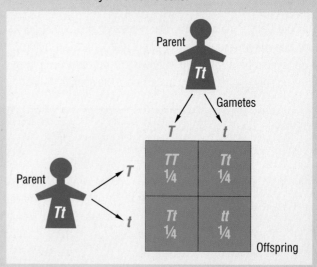

Using the diagram and the preceding text, answer the following questions.

h) Are the odds in this example like those of Mendel's F_2 generation? Under what circumstances would the next generation yield the same odds?

i) Why can scientists be certain that each parent must have had at least one good copy of the gene along with the defective copy?

j) What similarities exist between the biochemical abnormalities that cause sickle-cell anemia and those that cause Tay-Sachs disorder?

k) Explain how scientists recognize that the selection pressure against Tay-Sachs (*tt*) is high?

To understand why there is an excessively high risk of Tay-Sachs disorder among the Ashkenazim, two unique features associated with the group must be examined. One concerns history and lifestyle, and the other relates to a potential selective advantage bestowed by the gene on carriers of the disease.

The lifestyle of the Ashkenazim has been unique among the world's ethnic groups. As a result of certain events over the past several thousand years, including war and various forms of persecution, the population has remained isolated from the general European population. Some reports suggest that this lifestyle has kept intermarriage with other groups down to a mere 15%. Another manifestation of the population's isolation is their susceptibility to 10 other genetic disorders that do not occur to the same extent in other Jews or eastern European non-Jews.

The most plausible hypothesis put forward to explain the high frequency of the Tay-Sachs gene suggests a potential selective advantage conferred upon carriers of the disorder. In 1972, a questionnaire administered to parents of Tay-Sachs children in the United States produced a surprising piece of valuable information. The majority of the children's grandparents who did not emigrate from the old country died of common causes such as stroke or heart attack. Only one of 306 grandparents died of tuberculosis, even though TB was a common killer during their youth. A follow-up study, a decade later, on the distribution of TB and the Tay-Sachs gene within Europe revealed that the Tay-Sachs gene was three times more frequent among the eastern European Jews (9 to 10% of the population were heterozygous) than among other populations of European Jews. If the recessive gene did lend some measure of protection against TB, then the gene should occur more often in areas with a high incidence of TB.

The ghetto conditions in which the eastern European Jews were forced to live is also an environment in which TB thrives. With little or no intermarriage, some investigators believe that the ghetto-bound population was under the strongest pressure to evolve genetic resistance to TB.

l) What are the two selection pressures acting on the eastern European Jewish population?

m) The Tay-Sachs recessive gene has survival value in areas where there is a high incidence of TB. How does natural selection operate in its favor in these areas?

n) What significance can be attached to the fact that eastern European Jews are susceptible to 10 other genetic disorders that are not found in other Jewish and non-Jewish populations?

o) How does this case illustrate that evolution results from the interaction between an organism's genetic makeup and its environment?

Case-Study Application Questions

1 How are the factors in the evolutionary process illustrated in this case study? In your answer consider the ideas of mutation, natural selection, and survival value.

2 What advice should a genetics counsellor give to carriers who are contemplating giving birth to a child?

3 What is the meaning of the statement, "Recessive genetic disorders can be both a blessing and a curse"? ■

FRONTIERS OF TECHNOLOGY: MITOCHONDRIAL DNA AND EVOLUTION

Much is already known about the mitochondrion (the so-called "power plant" of the cell) and its vital importance in cellular respiration. Tens of thousands of these sausage-shaped organelles (roughly the size of bacteria) are present in every eukaryotic cell, from microbes up to the most complex living organisms, including humans. They are particularly abundant in muscle cells and parts of nerve cells, and near the surface of cells that specialize in the transport of nutrients. The role of the mitochondrion (plural: "mitochondria") in the production of ATP, the fuel that provides the energy to power all the body cells' activities, is also well documented.

However, the discovery in the mid-1960s that mitochondria contain their own genetic material shook the scientific community. Scientists had previously believed that all of a cell's structure and activities were under the control and direction of DNA inside the nucleus. This new revelation suggested that mitochondria, by carrying their own genes, maintain a measure of control of their own destiny. Furthermore, because mitochondria are found in all eukaryotic cells, these organelles must somehow figure directly in the evolutionary events that led to the diversity of organisms on earth.

Several other features of the mitochondrion made it a prime candidate for the study of evolutionary relationships. Unlike the nuclear DNA of eukaryotes, mitochondrial DNA (mtDNA) forms small, looping chains, resembling the DNA of viruses and bacteria. The mtDNA is relatively tiny, containing only 17 000 nucleotide pairs (compared with the nearly three billion in nuclear DNA). However, while a cell has only one copy of the nuclear DNA, it can have numerous mitochondria, each with its own mtDNA. The mtDNA also divides and reproduces independently of its host cell.

Recent electron microscope studies and biochemical analyses show that mtDNA contains enough nucleotide pairs to carry the genetic code for 10 to 20 proteins associated with ATP synthesis, the vital energy-transforming function associated with mitochondria. Does this support the idea that the mitochondrion is a "power plant" for the cell?

This recent research and other lines of evidence have led to speculation that mitochondria originally had sufficient genetic material to exist on their own. Lynn Margulis and other evolutionary biologists have expanded this idea into the *symbiotic* or *endosymbiotic hypothesis*. This hypothesis states that at least two organelles, mitochondria and chloroplasts, are the descendants of prokaryotic organisms. The first living cells are thought to have been primitive, anaerobic, one-celled ancestors of some of today's bacteria. These primitive organisms probably obtained their energy from nucleotides such as ATP, likely in a process similar to *fermentation*.

A major event occurred about 1.5 billion years ago. An aerobic organism was captured (either invaded or engulfed) by an anaerobic form. This arrangement was mutually beneficial. The guest (aerobe) was provided protection and increased access to nutrients, while the host (anaerobe) gained an efficient means of respira-

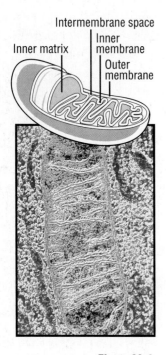

Inner matrix
Intermembrane space
Inner membrane
Outer membrane

Figure 28.4

Transmission electron micrograph of a mitochondrion.

> **BIOLOGY CLIP**
>
> The mitochondria from the male sperm cell do not enter the egg during fertilization. You have mtDNA from your mother but not from your father.

tion. Later, by a now-familiar genetic mechanism, certain guest genes (mtDNA) were transferred to the host nucleus. The host then took over much, but not all, of the control from the guest. Since then, the aerobes have continued to process oxygen both for themselves and their hosts. From this initial symbiotic relationship, or partnership, all other life forms are believed to have evolved. Indeed, the mtDNA in human cells is directly related to that in the ancestral organisms.

Current research on human mtDNA links faulty mitochondrial genes to a growing list of human genetic disorders such as Leber's disease, and less obvious familial illnesses such as Parkinson's disease. Research at the University of California at Berkeley has resulted in the successful cloning of mtDNA fragments from a zebra-like animal that lived a hundred years ago on the steppes of South Africa. The fragments used in the experiment were taken from tissues of a preserved skin in a German museum, and have been used to determine the relationships between and changes in zebras and horses. Some scientists speculate that one day DNA may be retrieved from other animal parts or even mummies to create banks of genes from extinct species. Given that only fragments of mtDNA and no whole intact nuclear DNA has been found, one can only imagine that this pioneering work could one day result in the recreation of extinct species.

SPECIATION

The process by which species originate is called **speciation.** However, it is important to recognize that the origin of species and evolution are not necessarily the same thing. Similarly, natural selection, while the major cause of disruption in genetic equilibrium, does not always lead to speciation. In the peppered moth case discussed in the chapter Adaptation and Change, evolution occurred without the creation of a new species. In the modern sense of the term, a species is a group of similar organisms that can interbreed and produce fertile offspring in their natural environment.

Scientists agree that the number of species today is much greater than it was in the past, even though many species have become extinct. Furthermore, since a species can only arise from existing species, there must be some process or mechanism in which a single species can develop into one or more descendant species.

Two ways in which a new species may arise are through *geographic isolation* and *reproductive (genetic) isolation.* In both instances, populations or parts of a population become isolated and must adapt to conditions in a new environment. Since all environments differ to varying degrees, the "selective" pressures on the populations also vary. It is important to understand that species are not created instantly, but usually evolve over a long period of time.

Geographic Isolation

In geographic isolation, the separation is caused by physical obstacles or barriers such as mountain ranges and bodies of water, or even barriers created by humans. When this happens, gene flow between the isolated group and the main population ceases. Eventually, the groups

Speciation *refers to the formation of a new species.*

Figure 28.5

Mummies may provide scientists with DNA of extinct lineages.

become so different that individuals of one population can no longer interbreed with those of another. Reasons for the differences include different adaptations of populations in the separate environments, the development of different gene frequencies within the separate populations, and different mutations within the populations.

Geographic isolation is used to explain the existence of the 14 species of finches found by Darwin in the Galápagos Islands. The species likely descended from individuals that reached the islands from the mainland of South America. The finches probably arrived by being blown off course in a storm or by getting lost. Finches do not normally fly over great distances. When the finches reached Galápagos, the water that separates the islands acted as a barrier to pre- vent interaction among the separate populations. Over time, this isolation would have resulted in the finches adapting to new conditions of vegetation, food, and so on, characteristic of the different islands. These changed conditions could have caused the populations to evolve in different directions.

Many other examples support the idea of geographic isolation leading to specia- tion. Among these are: the spread of the house sparrow (*Passer domesticus*) in North America; populations of certain gulls of the genus *Larus*, which circle the North Pole and overlap in Great Britain; and variation among turtles on the different islands in the Galápagos.

The question of the "exact" moment at which speciation occurs, or whether, indeed, speciation *has* occurred, cannot be answered precisely. For instance, while

Figure 28.6

Masses of marine iguanas at Espinosa Point, Galápagos. These lizards eat marine algae from rocks.

lions and tigers are markedly different in appearance, and do not interbreed in the wild, they are known to reproduce in captivity (producing "ligers"). A similar situation occurs between the leopard frog (*Rana pipiens*) and the wood frog (*Rana sylvatica*). The interpretations of whether or not speciation has taken place in these cases will require further study.

While the rate of evolution of a new species remains a contentious issue, the widely held view is that speciation is a gradual and lengthy process. There is also ample evidence that in some instances, such as in **polyploidy**, a new species may arise suddenly. Although an unusual event in the animal kingdom, polyploidy is common in plants. It often produces hardier and larger varieties of grains, fruits, and other useful plants.

Polyploids can mate with each other, but not with members of the parent generation, because of different chromosome numbers. In spite of the benefits associated with polyploids, it is not uncommon for them to possess undesirable traits or genes of each parent, which may be passed on to successive generations.

Polyploidy *is a condition in which an organism possesses more than two complete sets of chromosomes.*

Figure 28.7

A liger, also known as a tiglon. The father of this animal was a tiger and the mother was a lion.

Reproductive Isolation

Geographic isolation may also lead to reproductive, or genetic, isolation. Reproductive isolation occurs when organisms in a population can no longer mate and produce offspring, even following the removal of the geographic barriers. Factors that contribute to reproductive isolation include differences in mating habits and courtship patterns, seasonal differences in mating, and the inability of the sperm to fertilize eggs. In some cases, even where fertilization has taken place, the genes and chromosomes are so different that the zygote does not develop or the embryo is prevented from developing normally.

▮ REVIEW QUESTIONS ▮ ?

9 What does the concept of probability enable scientists to predict?

10 What do scientists mean when they say a chi-square value is "significant"?

11 Why are human genetic disorders useful models for studying evolution?

12 What common features are shared by the two genetic disorders in the case study?

13 Why did the discovery that mitochondria contain their own DNA surprise many scientists?

14 What features of the mitochondrion make it useful in the study of evolutionary relationships?

15 What is speciation?

16 Explain two ways in which new species may arise.

SOCIAL ISSUE:
Interbreeding of Plains and Woodland Bison

The North American bison are considered by some biologists to be two different subspecies: the plains bison and the woodland bison. Under natural conditions, the plains bison grazed the prairie regions and the woodland bison occupied lowland meadows and delta regions several hundred kilometers away from the plains bison. In the early 1800s, Canada's plains bison totalled around 60 million. By 1885, they were almost extinct. In 1909, the Canadian government bought plains bison and established Bison Recovery Park in Wainwright, Alberta. In 1922, Canada's largest national park, Wood Buffalo National Park, was established to provide a sanctuary for the woodland bison. To relieve overcrowding in Alberta's Elk Island National Park, plains bison were transported to Wood Buffalo National Park. The resulting hybridization almost destroyed the woodland bison as a separate subspecies.

Statement:

Human beings should not interfere with the natural evolutionary process.

Point

- Distinctive environmental pressures select different genotypes in woodland and plains bison. By placing the subgroups together, the gene pool has been altered. This means that many genes not suited for Wood Buffalo National Park have been reintroduced.
- If left alone, the changes in the two subspecies could become so pronounced that they may develop into two distinct species. Humans should not interfere with such diversification.

Counterpoint

- Recent DNA studies do not support the idea that the two groups of bison are different subspecies. The differences in appearance can be accounted for by a difference in their environments. The heavy fur coat of the more northern bison may be a response to a colder environment.
- The assumption that humans should not interfere with the natural evolutionary process can also be disputed. If we consider humans part of nature, then human interference is but another selective pressure.

Research the issue.
Reflect on your findings.
Discuss the various viewpoints with others.
Prepare for the class debate.

CHAPTER HIGHLIGHTS

- Genetic variation (mutation) is the raw material for evolution.
- All of the genes that occur within a population make up its gene pool.
- The measure of the relative occurrence of genes in a population is called the gene frequency. Evolution occurs when there is a change in the genetic makeup (frequency) of a population.

- The Hardy-Weinberg principle states that the frequency of genes (alleles) in a population stays the same when a population is in genetic equilibrium.
- In nature, populations rarely, if ever, meet the conditions required for genetic equilibrium. Therefore, evolution is recognized as a continuous and ongoing process.

- Three factors that bring about evolutionary change are mutation, genetic drift, and migration. Genetic drift is governed by the laws of probability.

- Two ways in which speciation occurs are through geographic isolation and reproductive isolation.

APPLYING THE CONCEPTS

1 Would it be more correct to say "an organism evolves" or "a species evolves"? Explain.
2 The five conditions of the Hardy-Weinberg principle are rarely met in nature. Yet the theory is still useful for studying "real" populations. How can you account for this apparent contradiction?
3 In a given population of organisms, the dominant allele (p) has a frequency of 0.7, and the recessive allele (q) has a frequency of 0.3. Use the Hardy-Weinberg formula to determine the genotype frequencies within the population.
4 In Tanzania, 4% (0.04) of the population are homozygous sickle-cell anemics (ss) and 32% (0.32) are heterozygotes (Ss). From these data, calculate the proportion of alleles that are s or S.

5 Mutation rates are usually quite low in sexually reproducing organisms, yet mutations are known to be the raw material for evolution. Explain how this is so.
6 A cross between two pea plants, in which tall (T) is dominant to short (t), yields 1000 seeds. Of this number, 550 produce plants that are tall, while 450 produce short plants. Use the chi-square test to determine whether the deviation is the result of chance or some other complicating factor.
7 Describe why long periods of geographic isolation of a small group from other members of a population favor speciation.

CRITICAL-THINKING QUESTIONS

1 How do the genetic disorders discussed in this chapter illustrate the point that evolution involves interactions between an organism's genetic makeup and its environment?
2 The Hardy-Weinberg formula, $p^2 + 2pq + q^2 = 1$, is said to represent all possible genotypes in a population. Verify this statement by determining (a) the frequencies of dominant and recessive alleles, and (b) the number of heterozygotes in a population of 200 pigs in which 72 have the recessive trait. If natural selection removed all of the individuals with the recessive trait, what would be the gene frequencies in the next generation?

ENRICHMENT ACTIVITIES

1 Do library research to learn how mitochondrial DNA investigators are providing insight into the causes of certain illnesses such as Leber's disease and Parkinson's disease.

2 Review the case studies in the chapter Adaptation and Change and in this chapter and develop an argument for the following viewpoint: "Natural selection is no longer in the realm of pure theory—it is now considered an operating principle of biology."

CHAPTER 29

Adaptation and Change

IMPORTANCE OF ADAPTATION

No two organisms are exactly alike—not even identical twins! This observation demonstrates an important feature of all life: diversity. Equally important, but often less apparent, are the similarities shared by all living things. All plants and animals, regardless of size, shape, or level of complexity, share certain characteristics. These include requirements for energy, basic cell structure and function, and **adaptation** to a particular habitat.

All living organisms are "adapted" in the sense that their appearance, behavior, structure, and mode of life make them well-suited to survive in a particular environment. Fish are not generally adapted for flying, and forest birds are not commonly found swimming in the ocean depths. Similarly, pine trees and elms are dryland inhabitants, whereas moss plants and ferns are found in water-laden or damp habitats. In other words, organisms have their own special adaptations and cannot survive or thrive in the habitats of others.

The theory of evolution attempts to explain why living organisms, so similar in their biochemistry and

Figure 29.1

This photograph does not show a twig but the caterpillar of the peppered moth.

Adaptation *is an inherited trait or set of traits that improve the chances of survival and reproduction of organisms.*

Figure 29.2

What similarities or unifying features are most apparent in these organisms? How do they differ? How is each adapted to its habitat?

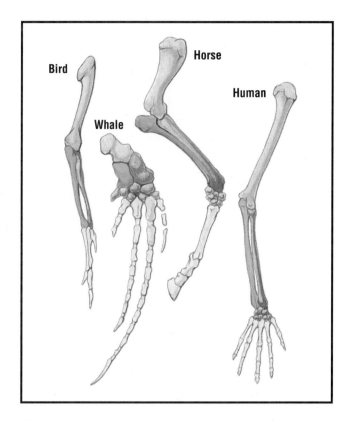

for flying. In contrast to humans and bats, the shorter forelimb and short, flat fingers of seals are adapted to support the flipper, which is used for swimming. The survival value of forelimb adaptations is further demonstrated in most other terrestrial animals, including amphibians, reptiles, and birds.

A number of plants have become adapted to feeding on animals. Perhaps the best known are the Venus flytrap and the pitcher plant. Their leaves have become adapted in different ways to capture prey, usually insects. The leaves of the pitcher plant, for example, are shaped like jars—once the insects enter the "jar," escape is practically impossible. Although these plants photosynthesize, they also require the nutrients that are obtained from the digested remains of their prey. This adaptation likely arose because these plants usually live in bogs and wetlands where the soil is poor in nitrogen-containing substances.

Figure 29.17

The forelimbs of these mammals are adapted to carry out different functions.

*A **pheromone** is a chemical substance produced by an organism that serves as a stimulus to another organism of the same species.*

Structural Adaptations

Most organisms show clear structural adaptations to their environment. Whether these adaptations involve modification to the pentadactyl limb (five digits) in mammals such as humans, seals, and bats, or modification of pitcher plant leaves to capture insects, they represent the most obvious of the three categories of adaptation. But how do these adaptations help the organisms survive and function?

Mammals, including humans, have the same basic limb design. However, in each case the forelimbs have become adapted to do a specific job. In humans, for example, the forelimb (with its five digits) has become particularly well-adapted for grasping and holding things. This feature is important in the human environment for accomplishing everything from eating and caring for the young to providing shelter and developing new technology. The fingers of bats are greatly lengthened to support the wing, which is used

Physiological Adaptations

The example of the pitcher plant illustrates the fact that one type of adaptation frequently depends on other types. While the structural adaptation of the leaf parts is quite apparent, the physiological adaptation that enables the plant to produce chemicals for the digestion of its prey is less obvious, though by no means less crucial for the organism's ability to survive and reproduce.

The production of chemicals such as **pheromones**—chemicals secreted by organisms to influence the behavior of other organisms—also has a physiological basis. In insects, these chemicals can have a number of functions, including serving as sexual attractants and alarm signals. As sexual attractants, the chemicals have reproductive value, whereas the latter adaptation clearly suggests the importance of pheromones as a survival mechanism.

Enzymes are special protein structures that regulate chemical reactions.

A number of physiological adaptations involve specialized **enzymes** that control body functions such as temperature, respiration, digestion, circulation and blood clotting, and muscle and nerve coordination. The secretion of venom by snakes and the production of toxins by certain plants and animals are further examples of physiological adaptations.

Behavioral Adaptations

Behavioral adaptation is a key factor in keeping organisms alive and enabling them to reproduce. Some behavioral adaptations are better known than others. For example, the southward migration of the Canada goose, the hibernation of certain mammals, and the storage of nuts by squirrels are all behaviors involving reactions to the environment. These actions help the organisms adapt to their surroundings.

Have you ever seen the "broken wing" behavior of a bird, which is designed to draw the enemy away from the nest site? Although this behavioral response may signal death for the parent, it increases the likelihood of survival for the offspring.

Another familiar behavioral adaptation can be noticed in cats and dogs. Originally cats were adapted to live in forests, where living or hunting in groups had no advantage. In contrast, dogs in earlier times lived in a grassland environment, where they adapted to hunt cooperatively. This evolutionary history helps explain why cats today tend to be solitary animals while dogs enjoy human companionship.

Most organisms, both plants and animals, respond to stimuli, and may show interesting adaptations in the way they behave. For example, protozoans such as amoebas and hydras respond quickly to touch, and react by moving away. They also react to temperature changes and chemical substances in the water. Depending on whether the experience is stressful or not, the organisms may move away from or toward the stimulus. Similarly, earthworms will react to touch, light, and chemicals in the soil. It is their reaction to light that explains why they burrow into the soil and only come out at night.

Plants also exhibit behavioral adaptations, but instead of being controlled by nerves, as is the case with animals, plants are controlled by chemicals called hormones. Many people are familiar with the behavior of plants that "bend" toward light. The term *tropism* is used to describe the orientation of plants according to some stimulus. In the case of light the orientation is called phototropism.

■ REVIEW QUESTIONS ?

9 How does the modern view of evolution differ from earlier beliefs?

10 What adaptation in the Galápagos finches made the greatest impression on Darwin?

11 What is the significance of the fourth step in Darwin's theory of natural selection?

12 What contribution did Lamarck make to our understanding of the mechanism of evolution?

13 How did Buffon's theory influence Darwin's thinking about evolution?

14 What are the three general types of adaptation?

15 What adaptation did the peppered moth make to the pollution around Manchester in the mid-1800s?

FRONTIERS OF TECHNOLOGY: USING COMPUTERS TO ASSESS ENVIRONMENTAL IMPACT

Much scientific knowledge comes from laboratory experiments in which variables that might affect the results are controlled. This approach has yielded valuable information about many of the components of complex systems. But imagine trying to test in a laboratory all the factors that affect the "health" of a salt marsh, or attempting to enclose a major ecological region to investigate the environmental impact of pollution, acid rain, or ozone depletion.

Supercomputers, the fastest of all computers, provide an approach to scientific investigation that is radically different from traditional observation and experimentation. The combination of computer-enhanced photographic images obtained from spacecraft, satellites, and aircraft, plus improvements in the modelling and simulation capabilities of computers, is playing a vital role in allowing researchers to assess environmental impact on local and global scales.

Supercomputers permit researchers to gather data and test theories on detailed simulations, or models, of physical reality. This modelling provides a framework for comprehending and predicting how nature acts. The supercomputer creates a "virtual reality," in which users participate directly in the world created by the computer. Numerical experiments can be performed and then used with laboratory data to predict what is likely to happen in "real life" situations that cannot be tested directly.

One example of the use of computer simulations to study environmental impact is a project entitled "Designer Wetlands" developed by Ken Pittman at the Visual Environment Laboratory at the North Carolina State University. The program, which puts together general scientific information such as topography, weather, geology, and wildlife, is used to simulate a wetland ecosystem. The computer can actually take the observer on a visual tour of the "designed" wetlands.

The implications of programs of this type are far-reaching. Researchers can quickly see if a given design is feasible and determine its application to wetlands that may be in trouble in the real world.

Figure 29.18

The CRAY X-MP/48 supercomputer consists of four central processors and has a capacity of eight million bits of shared memory.

While supercomputers have allowed us to create and manipulate complex mathematical models, it is important to remember that they are just that—models. They are no better than the set of mathematical relationships on which they are based, and they can never (at least not in the foreseeable future) recreate all of the factors that exist in a real system. They are one more tool in the scientist's repertoire, to be used with caution and care. The best use of such models might well be to direct us toward the best questions to ask, and to help us design experiments that can test the predictions of our models.

CASE STUDY

RESISTANCE TO DDT

Objective

To investigate the connection between pesticides and natural selection.

Background Information

Environmental change introduced by human activities provides good evidence for natural selection. Examples of such activities include the extensive use of drugs and antibiotics to treat certain kinds of pathogens, and the use of pesticides to control diseases such as malaria and yellow fever. Bacteria and insects, like all living organisms, demonstrate variability. Variability results from mutations and chromosomal rearrangements, which are present in a population at very low levels at all times. This variability enables natural selection to take place.

The application of pesticides to target populations favors strains that are resistant to the pesticides. Normally, such strains are uncommon in a population; however, once the environment is changed, the resistant forms have the better chance of survival. The resistant forms will then be reproduced in larger numbers in succeeding generations. Thus, they are spread rapidly throughout the population. As resistance develops, the frequency of application and/or the dosage is increased to maintain levels of control. Increased application and dosages of chemicals further increase the size of the resistant populations. Thus, chemicals that once controlled a pest population are no longer effective.

The Case Study

DDT is a chemical pesticide that has had enormous success in the battle against insect pests. Since it was first used in World War II, to get rid of head lice, it has been effective in the fight against controlling mosquito and other insect populations known to cause diseases such as malaria and yellow fever. Because of its potentially harmful accumulation in the food chain and the possible danger to humans, regulations banning or restricting the use of DDT were implemented by a number of countries in the early 1970s. However, DDT continued to be used in many less developed countries.

Some scientists believe that the resistance displayed by the pests did as much to spell the demise of DDT as did regulations directed at curtailing its use. After 1950, when pesticides came into widespread use, the number of pest species, including insects, that have developed resistant strains has increased dramatically.

Examine the following illustration to determine the effect of repeated sprayings or applications of DDT on a hypothetical population.

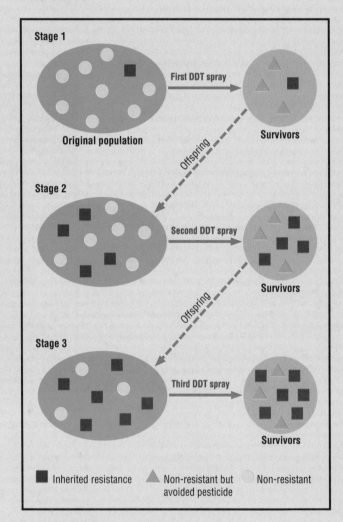

a) Where did the resistant trait come from in the original population?

b) What happens to the nonresistant trait over the three stages of the population? Why?

c) What happens to the resistant trait over the course of evolution in this hypothetical situation? Why?

d) Explain the difference in the selection pressure (the trait natural selection is favoring) in stages 1 and 3.

e) Is this a case of "evolution in action"? Explain.

Unit Eight:
Variation and Change:

A complicating factor in the use of pesticides such as DDT is that some pests develop multiple resistance and cross-resistance. In cross-resistance, the organism develops a resistance to one compound and then achieves resistance to others, usually in the same chemical group. For example, DDT-resistant houseflies tolerate higher levels of a closely related chemical that is also used to control this common household pest. Multiple resistance is much more serious and has wide-ranging consequences. In this type of resistance, pests develop a tolerance for many classes of different compounds. This effect is common in pests that have a strong resistance to DDT. Insects with this factor, called *kdr* (knock-down resistance), are preadapted to synthetic chemicals that are chemically quite different from DDT.

An understanding of the effects of multiple resistance can be illustrated by the World Health Organization's (WHO) 1955 global education program to control the anopheles mosquito (carriers of the malarial parasite). The WHO strategy included a two-stage offensive: (1) spraying DDT and other pesticides inside dwellings where anopheles mosquitoes frequently obtain blood meals from humans, and (2) treating infected individuals with anti-plasmodium drugs to destroy the blood parasites that cause malaria.

For about 15 years, the program showed remarkable success. For example, within a 10-year period in India the annual incidence of malaria was reduced from 100 million cases to 50 000, with a corresponding reduction from three million to 25 in Sri Lanka. By 1970, efforts to control the mosquitoes that carried plasmodia were failing. By 1980, 51 of the known 60 species of anopheles mosquitoes were capable of transmitting the malarial parasite. During the same time as the WHO mosquito spray programs, many countries had implemented extensive agricultural programs. Crops such as cotton, rice, and tobacco were sprayed with pesticides, including many of the same ones used in the malaria-eradication program.

f) What explanation can be given for the early success of the malarial spray program and its eventual failure in later years?

g) What combination of factors most likely contributed to a strong selective force for developing resistant strains in the mosquito population?

h) Because the anopheles mosquito has developed resistance to DDT, what might be the long-term effectiveness of other known synthetic chemicals in controlling malaria? Explain.

Case-Study Application Questions

1 The resistance of mosquitoes to DDT is a good example of natural selection. Explain how this case study supports Darwin's observation that
   a) hereditary variations exist among species
   b) natural selection acts on organisms with these variations.

2 Discuss the following statement: "Natural selection as the cause of evolution has been neither proved nor disproved." ∎

RESEARCH
IN CANADA

Molecular Genetics and Evolution

Dr. Ford Doolittle's work at Dalhousie University in Halifax has focused on applying the recently developed methods of molecular genetics and "genetic engineering" to questions about evolution that have troubled biologists for decades.

In the early 1970s, Lynn Margulis had proposed that chloroplasts and mitochondria—the organelles of photosynthesis and respiration in higher cells—were once free-living bacteria that had become trapped inside the cytoplasm of some larger host cell. Doolittle's laboratory was one of the first to use a molecular approach to provide evidence that at least one type of organelle, chloroplast, was indeed the descendant of ancient bacteria.

Recently, Doolittle's laboratory has focused on archaebacteria. Carl Woese, at the University of Illinois, had shown that these bacteria are as different from the more familiar "eubacteria" like *Escherichia coli* as eubacteria are from humans or plants, and likely branched off from the trunk of the tree of life over three billion years ago. Archaebacteria live in extreme environments such as the bottoms of swamps, salt evaporation ponds, hot springs, and deep-ocean geothermal vents called "black smokers"; some even grow best at temperatures above that of boiling water. Doolittle's laboratory has worked toward developing tools for the genetic engineering of archaebacteria, so that we can both exploit and understand their remarkable and ancient adaptations.

DR. FORD DOOLITTLE

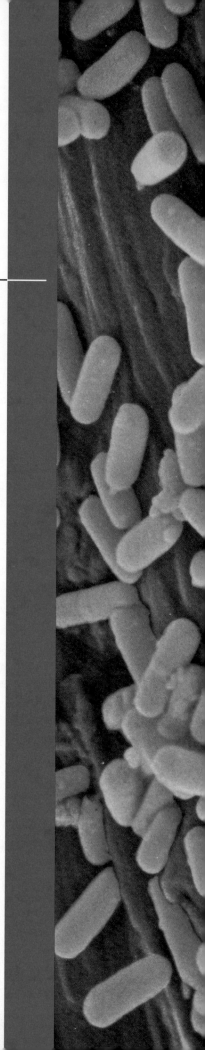

INVESTIGATION
CAREER

With the population of spaceship earth exceeding 6 billion people, our survival depends more and more on an intelligent, well-planned management of the world's resources. A growing number of careers require an awareness of environmental issues. The demand for people with knowledge and skills in environmental science is projected to grow well into the next century.

Architect
An architect must take environmental factors into account when designing a structure. For example, a high-rise building in an earthquake zone must be designed to move with possible earth tremors. In planning any project, architects, developers, and city planners must consider such factors as wind tunnels and access to sunlight for neighboring buildings. A building design must also consider the people who will be working in it. Workers do not want to work in stuffy, airtight boxes.

Retailer
In the 1990s, consumers are demanding "environmentally friendly" products such as phosphate-free detergents or cosmetics that are not tested on animals. Retailers must address consumers' concerns about wasteful packaging by providing products in recyclable containers.

Politician/civil servant
Voters expect their politicians and civil servants to take strong actions to protect the environment. Government officials have to make daily decisions on a wide variety of environmental issues, from water quality and waste disposal sites to the preservation of wilderness areas.

Forestry worker
Canada's vast forests are a valuable, renewable resource that must be handled with care. Forestry workers must understand the impact that the harvesting of trees has on forests and wildlife. The forestry industry also must plan effective reforestation policies so that our forests can be used and enjoyed by future generations.

- Identify a career that requires a knowledge of environmental issues.
- Investigate and list the features that appeal to you about this career. Make another list of features that you find less attractive about this career.
- Which high-school subjects are required for this career? Is a post-secondary degree required?
- Survey the newspapers in your area for job opportunities in this career.

SOCIAL ISSUE:
Pesticides and Evolution

The development of chemical pesticides has led to the evolution of more resistant forms of organisms. Although the controlled and prudent use of pesticides can improve agricultural yields, there remain several concerns about their long-term effects in the ecological chain, including water contamination, which also kills fish; the poisoning of birds and other nontarget organisms; as well as possible harm to humans.

Statement

There should be a worldwide ban on the use of pesticides in agriculture.

Point

- Since the introduction of DDT in 1939, hundreds of insect species have developed resistant strains. As a result, crop damage from insects is now greater than it was before the introduction of chemicals. Some people are concerned that we will run out of new and effective pesticides before the pests run out of resistant strains.

- Nonchemical practices such as crop rotation, burning or plowing of plant refuse, and proper timing for planting were used before chemical pesticides were introduced. These methods worked adequately and kept a healthy balance of nutrients in the soil.

Counterpoint

- Resistance does not always lead to the evolution of new species. Consider the cotton boll weevil and the coddling moth. In both cases, after more than 30 years of exposure to pesticides, there is no evidence that the effectiveness of the pesticide has been reduced or that the organisms have evolved any resistance.

- Alternative methods of pest control are expensive and out of reach for those who will be affected most: underdeveloped and developing nations. An end to pesticide use would mean a decreased quality of life for all.

Research the issue.
Reflect on your findings.
Discuss the various viewpoints with others.
Prepare for the class debate.

CHAPTER HIGHLIGHTS

- Evolution is a major theme in biology. It attempts to explain how living organisms change through time. Earlier beliefs held that organisms were "fixed" and did not change.
- Adaptation refers to the ways in which whole organisms or their individual parts are suited to carrying out the processes of life. Adaptations improve an organism's chances of survival and reproduction.

- Indirect and direct evidence can be used to support the theory of evolution. Some evidence (e.g., that from fossils) is direct, while other evidence (e.g., that from living organisms) is indirect.
- Darwin's theory of evolution by natural selection helped convince the scientific and intellectual communities that evolution was possible.

- Biogeography is the study of the distribution and dispersal of plants and animals.

- Scientific laws are descriptions based on observed events. Theories are attempted explanations based on observations and tested by hypotheses.

APPLYING THE CONCEPTS

1 In what way is the geological time scale both a biological calendar and a framework that describes major geological events?

2 Contrast the views on evolution held by modern scientists with those of earlier proponents.

3 How is the theory of continental drift useful in the study of evolution?

4 How did Darwin use the ideas of Lamarck and Malthus in formulating his theory of evolution?

5 What does the peppered moth case illustrate with respect to the rate of evolutionary change?

6 Using examples, demonstrate from an evolutionary viewpoint that the survival of species is more critical than the survival of specific individuals.

7 How could supercomputers and computer simulation play a useful role in predicting how plants and animals might adapt to a changing environment? Give examples.

8 Give an example of how environmental impact caused by human activity can alter the natural course of evolution.

9 Explain why organisms with a high reproductive capacity can adapt more readily than many organisms that have a lower reproductive capacity.

CRITICAL-THINKING QUESTIONS

1 The wombat, a marsupial mammal found in Australia, and the rabbit, a placental mammal that was brought into Australia by explorers, have many of the same requirements. A biologist studying both animals suggests that rabbits will eventually take over the ranges occupied by wombats because they reproduce at a faster rate. What evidence would you want to gather before accepting this conclusion?

2 Many of Darwin's ideas about evolution came from his observations on the Galápagos Islands. Today, scientists recognize that islands are literally "laboratories of evolution." Discuss the validity of this viewpoint.

3 Scientists suggest that theories provide tentative explanations of the natural world and should never be interpreted as absolute fact. Provide an example that indicates how biological explanations of the natural world would change if a Lamarckian explanation of evolution were endorsed by the scientific community rather than the Darwinian explanation of evolution by natural selection.

4 Adaptations are so numerous that they embrace practically every structure or occurrence in biology. Even when organs have fallen into disuse they often become adapted to a second purpose. Investigate the present function of the pelvic girdle of snakes and the three hinge bones of reptile jaws.

ENRICHMENT ACTIVITIES

1 Read "In the Beginning" by Wallace Raven in *Discover* (October 1990) and write a book report on the research of paleobiologist J. William Schopf.

2 Research the social differences between two designers of the theory of evolution: Darwin and Wallace.

3 How do the creation stories of the North American aboriginals refute the theory that aboriginals migrated to the North American continent?

abdominal cavity: Extends from the diaphragm to the upper part of the pelvis. The stomach, liver, pancreas, small intestine, large intestine, and kidneys are contained in the abdominal cavity.

abiogenesis: A theory that states that nonliving things can be transformed into living things.

abiotic components: The nonliving components of the biosphere. They include chemical and physical factors.

absorption: The movement of fluids in the direction of a diffusion, or osmotic, gradient.

accommodation reflexes: Adjustments made by the lens and pupil for near and distant objects.

acetylcholine: A transmitter chemical released from vesicles in the end plates of neurons. Acetylcholine makes the postsynaptic membranes more permeable to Na⁺ ions.

acids: Substances that release hydrogen ions in solution.

acrosome: The cap found on sperm cells. It contains packets of enzymes that permit the sperm cell to penetrate the gelatinous layers surrounding the egg.

action potentials: Nerve impulses. The reversal of charge across a nerve cell membrane initiates an action potential.

activation energy: The energy required to initiate a chemical reaction.

active site: The area of the enzyme that combines with the substrate.

active transport: The use of cell energy to move materials across a cell membrane against the concentration gradient.

adaptation: An inherited trait or set of traits that improve the chances of survival and reproduction of organisms.

adrenal cortex: The outer region of the adrenal gland. It produces glucocorticoids and mineralocorticoids.

adrenal medulla: Found at the core of the adrenal gland. The adrenal medulla produces epinephrine and norepinephrine.

aerobic respiration: The complete oxidation of glucose in the presence of oxygen.

afferent arteriole: Carries blood to the glomerulus.

agglutination: The clumping of blood cells caused by antigens and antibodies.

albedo: A term used to describe the extent to which a surface can reflect light that strikes it. An albedo of 0.08 means that 8% of the light is reflected.

aldosterone: A hormone produced by the adrenal cortex. It helps regulate water balance in the body by increasing sodium and water reabsorption by the kidneys.

allantois: An embryonic membrane.

alleles: Two or more alternate forms of a gene.

all-or-none response: The all-or-none response of a nerve or muscle fiber means that the nerve or muscle responds completely or not at all to a stimulus.

allosteric activity: The change in the protein enzyme caused by the binding of a molecule to the regulatory site of the enzyme.

alveoli: Blind-ended sacs of the lung. The exchange of gases between the atmosphere and the blood occurs in the alveoli.

amino acids: Organic chemicals that contain nitrogen. Amino acids can be linked together to form proteins.

amniocentesis: A technique used to identify certain genetic defects in a fetus or embryo.

amnion: A fluid-filled embryonic membrane.

amylase: An enzyme that hydrolyzes complex carbohydrates.

amyloplasts: Colorless plastids that store starch.

anabolic steroids: Strength-enhancing drugs.

anabolism: Chemical reactions in which simple chemical substances are combined to form complex chemical structures.

anaerobic bacteria: Bacteria that grow best in environments that have little oxygen.

anaerobic respiration: Takes place in the absence of oxygen.

analogous structures: Similar in function and appearance but not in origin. The wing of an insect and the wing of a bird are analogous structures.

anemia: The reduction in blood oxygen due to low levels of hemoglobin or poor red blood cell production.

aneurysm: A fluid-filled bulge found in the weakened wall of an artery.

angina: Literally means to suffocate. It is often used to describe the chest pain produced by heart attack.

angiosperms: Plants that produce enclosed seeds.

antheridium: The sex organ that produces male gametes in mosses and ferns.

antibodies: Proteins formed within the blood that react with antigens.

anticodons: The three-base codes found in tRNA that pair with the codons of mRNA.

antidiuretic hormone (ADH): Acts on the kidneys to increase water reabsorption.

antigen: A substance, usually protein in nature, that stimulates the formation of antibodies.

aorta: Carries oxygenated blood to the tissues of the body.

appendicular skeleton: Consists of the bones of the upper and lower limbs and their supporting structures.

aqueous humor: Supplies the cornea with nutrients and refracts light.

archegonium: The sex organ that produces female gametes in mosses, ferns, and gymnosperms.

arteries: High-pressure blood vessels that carry blood away from the heart.

arterioles: Fine branches from arteries.

artificial involution: A process by which egg cells are extracted from a donor and placed in a nongenetic mother.

astigmatism: A vision defect caused by the abnormal curvature of the surface of the lens or cornea.

atherosclerosis: A disorder of the blood vessels characterized by the accumulation of cholesterol and other fats along the inside lining.

atmosphere: The air that encircles the earth.

atoms: Small particles of matter. They are composed of smaller subatomic particles: neutrons, protons, and electrons.

ATP (adenosine triphosphate): A compound that stores chemical energy.

atria: Thin-walled heart chambers that receive blood from veins.

atrioventricular (AV) valves: Prevent the backflow of blood from the ventricles into the atria.

auditory canal: Carries sound waves to the eardrum.

autonomic nerves: Motor nerves designed to maintain homeostasis. Autonomic nerves are not under conscious control.

autosomes: Chromosomes not involved with sex determination.

autotrophs: Organisms capable of obtaining their energy from the physical environment and building their required organic molecules.

axial skeleton: The central supporting part of the skeleton; consists of the bones of the skull, backbone, ribs, and breastbone.

axon: An extension of cytoplasm that carries nerve impulses away from the dendrites.

bacteriophage (phage): A category of viruses that infect and destroy bacterial cells.

Barr bodies: Small, dark spots of chromatin located in the nuclei of female mammalian cells.

bases: Substances that release hydroxide ions in solution.

basilar membrane: Anchors the receptor hair cells in the organ of Corti.

B cells: Make antibodies.

benthos: The bottom of any body of water.

bile salts: The components of bile that emulsify fats.

binary fission: A form of asexual reproduction in which one cell divides into two equal cells.

binomial nomenclature: A method of naming organisms by using two names, the genus and species names.

biogeochemical cycle: The complex, cyclical transfer of nutrients from the environment to an organism and back to the environment.

biological amplification: The buildup of toxic chemicals in organisms as tissues containing the chemical move through the food chain.

biomes: Large-scale ecosystems such as tundra, boreal forest, or grassland.

biosphere: The narrow zone around the earth that harbors life.

biotic components: The biological or living components of the biosphere.

biotic potential: The maximum number of offspring that can be produced by a species under ideal conditions.

blastocyst: An early stage of embryo development.

blastula: The early stage of development in which the cells of the dividing embryo form a hollow ball of cells.

blind spot: The area in which the optic nerve attaches to the retina.

blood pressure receptors: Specialized nerve cells that are activated by high blood pressure.

bone marrow: The tissue located in the central cavity of the long bones.

boreal forest: A worldwide forest region found in the upper latitudes that is characterized by coniferous trees.

Bowman's capsule: A cuplike structure that surrounds the glomerulus. The capsule receives filtered fluids from the glomerulus.

brackish: A mixture of fresh and salt water.

bronchial asthma: Characterized by a reversible narrowing of the bronchial passage.

bronchioles: The smallest passageways of the respiratory tract.

bronchitis: An inflammation of the bronchioles.

Brownian motion: The random movement of molecules.

browsing: Feeding on leaves, twigs, buds, bark, and similar vegetation.

buffers: Neutralize excess acid or base, thereby preventing any significant fluctuation in pH values.

bursae: Sacs of fluid found in joints.

camouflage: An adaptation in form, shape, or behavior that better enables an organism to avoid predators.

capillaries: Tiny blood vessels that connect arteries and veins. The site of fluid and gas exchange.

capsid: The protective protein coat of viruses.

carbonic anhydrase: An enzyme found in red blood cells. The enzyme speeds the conversion of CO_2 and H_2O to carbonic acid.

cardiac muscle: Involuntary muscle found in the heart. This branching, lightly striated muscle is capable of contracting rhythmically.

cardiac output: The amount of blood pumped from the heart each minute.

carnivore: An animal that eats other animals in order to obtain food.

carriers: Individuals that are heterozygous.

carrying capacity: The maximum population that can be sustained by a given supply of resources (nutrients, energy, and space).

catabolism: Reactions in which complex chemical structures are broken down into simpler molecules.

catalysts: Chemicals that regulate the rate of chemical reactions without themselves being altered.

cataracts: Occur when the lens or cornea become clouded.

cell fractionation: The process by which cell fragments are separated by centrifugation.

cellular respiration: The process by which living things convert the chemical energy in sugars into the energy used to fuel cellular activities.

cellulose: A plant polysaccharide that makes up plant cell walls.

centrioles: Small protein bodies that are found in the cytoplasm of animal cells.

centromeres: Structures that hold chromatids together.

cephalization: Refers to the concentration of nerve tissue and receptors at the anterior end of an animal's body.

cerebellum: The region of the brain that coordinates muscle movement.

cerebral cortex: The outer lining of the cerebral hemispheres.

cerebrospinal fluid: Fluid that circulates between the innermost and middle membranes of the brain and spinal cord.

cerebrum: The largest and most highly developed part of the human brain. The cerebrum stores sensory information and initiates voluntary motor activities.

cervix: A muscular band that prevents the fetus from prematurely entering the birth canal.

chemical compounds: Formed when two or more elements are joined by chemical bonds.

chemoreceptors: Specialized nerve receptors that are sensitive to specific chemicals.

chemosynthesis: The formation of carbohydrates from energy resulting from the breakdown of inorganic substances rather than from light.

chioneuphores: Animals that can withstand winters with snow and cold temperatures.

chionophiles: Animals whose ranges lie within regions of long, cold winters ("snow lovers").

chionophobes: Animals that avoid snow-covered regions ("snow haters").

chitin: A nitrogen-containing polysaccharide, composed of long fibrous molecules, that forms structures of considerable mechanical strength.

chlorophyll: The pigment that makes plants green. Chlorophyll traps sunlight energy for photosynthesis.

chloroplasts: Organelles that specialize in photosynthesis, and contain the green pigment chlorophyll found in plant cells.

cholinesterase: An enzyme released from vesicles in the end plates of neurons shortly after acetylcholine. Cholinesterase breaks down acetylcholine.

chordae tendinae: Support the AV valves.

chorion: The outer membrane of a developing embryo.

chorionic villus sampling (CVS): A prenatal diagnosis technique that secures cells from the outer membrane of the embryo for analysis.

choroid layer: The middle layer of the eye. Pigments prevent scattering of light in the eye by absorbing stray light. Many blood vessels are found in this layer.

chromatids: Single strands of a chromosome that remain joined by a centromere.

chromatin: The material found in the nucleus. It is composed of protein and DNA.

chromoplasts: Store orange and yellow pigments.

chromosomes: Long threads of genetic material found in the nucleus of cells. Chromosomes are composed of many nucleic acids and proteins.

cilia: Tiny hairlike protein structures found in eukaryotic cells. Cilia sweep foreign debris from the respiratory tract.

cirrhosis of the liver: A chronic inflammation of liver tissue characterized by an increase of nonfunctioning fibrous tissue and fat.

climax community: The final, relatively stable community reached during successional stages.

climax vegetation: The long-enduring steady-state plant community.

clitellum: A smooth and enlarged segment found about one-third of the way along the body of oligochaetes; it secretes a protective covering for the eggs.

closed population: A population in which density changes are the result of natality and mortality with neither food nor wastes being allowed to enter or leave the given environment.

clumped distribution: Occurs in aggregates. The distribution of organisms is affected by abiotic factors.

coagulation: Occurs when the bonds holding a protein molecule are disrupted, causing a permanent change in shape.

cochlea: The coiled structure of the inner ear that identifies various sound waves.

codons: Three-base codes for amino acids.

coelom: A body cavity or space lined with a layer of cells called the peritoneum.

coenzymes: Organic molecules synthesized from vitamins that help enzymes combine with substrate molecules.

coevolution: Occurs when two different species exert selective pressures on each other.

cofactors: Inorganic molecules that help enzymes combine with substrate molecules.

cohesion-tension theory: Provides an explanation for water movement in plants by attraction between water molecules.

collecting duct: Receives urine from a number of nephrons and carries urine to the pelvis.

colon: The largest segment of the large intestine. Water reabsorption occurs in the colon.

commensalism: An association between two organisms in which one benefits and the other is unaffected.

community: Includes the populations of all organisms that occupy an area.

competitive inhibitor: A molecule that has a shape complementary to a specific enzyme, thereby permitting it access to the active site of the enzyme. Inhibitors block chemical reactions.

complementary proteins: Help phagocytotic cells engulf foreign cells.

cones: Photoreceptors that identify color.

conjugation: A form of sexual reproduction in which genetic material is exchanged between two cells.

connective tissue: Provides support and holds various parts of the body together.

consumers: Heterotrophic organisms.

continuity of life: A succession of offspring that share structural similarities with those of their parents.

controls: Standards used to verify a scientific experiment. Controls are often conducted as parallel experiments.

convergent evolution: The development of similar forms from unrelated species due to adaptation to similar environments.

cornea: A transparent tissue that refracts light toward the pupil.

coronary arteries: Supply the cardiac muscle with oxygen and other nutrients.

corpus callosum: A nerve tract that joins the two cerebral hemispheres.

corpus luteum: Made up of the follicle cells of the ovary following ovulation. The corpus luteum secretes estrogen and progesterone.

cortex: The outer layer of the kidney.

cortisol: A hormone that stimulates the conversion of amino acids to glucose by the liver.

cotyledon: A seed leaf that stores food for the germinating seedling. It is the first photosynthetic organ of a young seedling.

counter-current exchange: An anatomical variation of the circulatory system designed to reduce heat loss.

covalent bonds: Formed when electrons are shared between two or more atoms.

Cowper's (bulbourethral) gland: Contributes a mucus-rich fluid to the seminal fluid (semen).

cranial cavity: Surrounded by the skull, which protects the brain, eyes, and inner ear.

creatine phosphate: A compound found in muscle cells that releases a phosphate to ADP and helps regenerate ATP supplies in muscle cells.

crossing-over: The exchange of genetic material between two homologous chromosomes.

cuticle: A noncellular layer that covers the epidermis of above-ground plant parts. It prevents water loss.

cyclic AMP: A secondary chemical messenger that directs the synthesis of protein hormones by ribosomes.

cytokinesis: The division of cytoplasm.

cytoplasm: The area of the protoplasm outside of the nucleus.

Dalton's law of partial pressure: States that each gas in a mixture exerts its own pressure, which is proportional to the total volume.

deamination: The removal of an amino group from an organic compound.

death phase: Phase that marks a constant decline in the population. Mortality exceeds natality.

decomposer food chains: Usually bacteria and fungi; consume wastes and dead tissue from organisms.

decomposers: Bacteria and fungi that break down the remains or wastes of other organisms in the process of obtaining their organic nutrients.

dehydration synthesis: The process by which larger molecules are formed by the removal of water from two smaller molecules.

dendrites: Projections of cytoplasm that carry impulses toward the cell body.

denitrifying bacteria: Soil bacteria that reduce nitrates or nitrites to gaseous nitrogen and some nitrous oxide.

density-dependent factors: Factors arising from population density (e.g., food supply) that affect members of a population.

density-independent factors: Factors that affect members of a population regardless of population density (e.g., flood, fire).

deoxyribonucleic acid (DNA): The carrier of genetic information in cells.

depolarization: Caused by the diffusion of sodium ions into the nerve cell. Excess positive ions are found inside the nerve cell.

detoxify: To remove the effects of a poison.

detritus: Any organic waste from animals and plants.

diabetes mellitus: A genetic disorder characterized by high blood sugar levels.

diapedesis: The process by which white blood cells squeeze through clefts between capillary cells.

diaphragm: A sheet of muscle that separates the organs of the chest cavity from those of the abdominal cavity.

diastole: Refers to heart relaxation.

dichotomous keys: Two-part keys used to identify living things.

dicotyledon (dicot): An angiosperm whose seeds have two cotyledons or seed leaves. Most angiosperms are dicots.

diffusion: The movement of molecules from an area of higher concentration to an area of lower concentration.

dikaryotic state: A stage in the life cycle of fungi when cells contain two haploid nuclei. Each nucleus is derived from a separate parent.

dioecious: Refers to organisms in which male and female gonads are carried by separate individuals.

diploid chromosome number: The full complement of chromosomes. Every cell of the body, with the exception of sex cells, contains a diploid chromosome number.

disaccharides: Formed by the joining of two monosaccharide subunits.

distal tubule: Conducts urine from the loop of Henle to the collecting duct.

dominant genes: Determine the expression of the genetic trait in offspring.

Down syndrome: A trisomic disorder in which a zygote receives three homologous chromosomes for chromosome pair number 21.

duodenum: The first segment of the small intestine.

dynamic equilibrium: Any condition within the biosphere that remains stable within fluctuating limits.

ecological niche: Refers to the overall role of a species in its environment.

ecosystem: A community and its physical and chemical environment.

edema: Tissue swelling caused by decreased osmotic pressure in the capillaries.

efferent arteriole: Carries blood away from the glomerulus to a capillary net.

electrocardiograph: An instrument that monitors the electrical activity of the heart.

electron transport system: A series of progressively stronger electron acceptors. Each time an electron is transferred, energy is released.

elements: Pure substances that cannot be broken down into simpler substances by chemical means. There are 109 different elements.

embolus: A blood clot that dislodges and is carried by the circulatory system to vital organs.

embryo: Refers to the early stages of an animal's development. In humans, the embryo stage lasts until the ninth week of pregnancy.

emphysema: An overinflation of the alveoli. Continued overinflation can lead to the rupture of the alveoli.

endergonic reaction: Reaction that requires the continuous addition of energy. Low-energy reactants are converted into high-energy products.

endocrine hormones: Chemicals secreted by glands directly into the blood.

endocytosis: The process by which particles too large to pass through cell membranes are transported within a cell.

endometrium: The glandular lining of the uterus that prepares the uterus for the embryo.

endorphins: A group of chemicals classified as neuropeptides. Containing between 16 and 31 amino acids, endorphins are believed to reduce pain.

endoscope: An instrument that views the interior of the body.

endoskeleton: An internal skeleton.

endospore: A dormant bacterial cell encapsulated by a thick, resistant cell wall. These forms develop when environmental conditions become unfavorable.

energy systems: Involve energy input, energy conversion, and energy output.

enkephalins: A group of chemicals classified as neuropeptides. Containing five amino acids, the enkephalins are produced by the splitting of larger endorphin chains.

enterokinase: An enzyme of the small intestine that converts trypsinogen to trypsin.

entropy: A measure of non-usable energy within a system.

enucleated cells: Do not contain a nucleus.

environmental resistance: All the factors that tend to reduce population numbers.

enzymes: Special protein catalysts that permit chemical reactions within the body to proceed at low temperatures.

epididymis: A compact, coiled tube located along the posterior border of the testis. Consists of coiled tubules that store sperm cells.

epiglottis: A structure that covers the opening of the trachea (glottis) during swallowing.

epilimnion: The upper layer of water in a lake; it heats up in the summer.

epinephrine: A hormone produced by the adrenal medulla that initiates the flight-or-fight response.

epiphyseal plates: Contain cartilage and join two sections of bone in youngsters.

epistatic genes: Mask the expression of other genes.

epithelial tissue: A covering tissue that protects organs, lines body cavities, and covers the surface of the body.

equilibrium: A condition in which all acting influences are balanced, resulting in a stable condition.

erepsins: Enzymes that complete protein digestion by converting small-chain peptones to amino acids.

erythroblastosis fetalis ("blue baby"): Occurs when the mother's antibodies against Rh+ blood enter the Rh+ blood of her fetus.

erythropoiesis: The process by which red blood cells are made.

esophagus: A tube that carries food from the mouth to the stomach.

estrogen: A female sex hormone.

estuary: The place where rivers enter the ocean.

eukaryotes: Organisms whose cells contain a membrane-bound nucleus.

eukaryotic cells: Cells that have a true nucleus. The nuclear membrane surrounds a well-defined nucleus.

eustachian tube: An air-filled tube of the middle ear that equalizes pressure between the external and internal ear.

eutrophic lakes: Lakes that are shallow and warm, and are rapidly becoming filled in.

eutrophication: The filling in of a lake by organic matter and silt.

evolution: The cumulative changes in characteristics of populations of organisms in successive generations.

exergonic reaction: A reaction that releases energy.

exocytosis: The passage of large molecules through the cell membrane to the outside of the cell.

exoskeleton: A strong protective covering on the outside of the body of many invertebrates. An external skeleton.

experimental variables: Designed to test a hypothesis. Experimental groups test a single variable at a time.

external intercostal muscles: Muscles that raise the rib cage, decreasing pleural pressure.

extracellular fluids (ECF): Occupy the spaces between cells and tissues.

facultative anaerobes: Prefer environments low in oxygen, but can live in environments with reduced oxygen levels.

Fallopian tube: See "oviduct."

farsightedness: Occurs when the image is focused behind the retina.

fats: Lipids composed of glycerol and saturated fatty acids. They are solid at room temperature.

feedback inhibition: The inhibition of an enzyme in a metabolic pathway by the final product of that pathway.

fertilization: Occurs when a male and a female sex cell fuse.

fetus: Refers to the later stages of an unborn offspring's development. In humans, the embryo is called a fetus after the ninth week of development.

filtration: The selective movement of materials through capillary walls by a pressure gradient.

first law of thermodynamics: States that energy can be changed in form but cannot be created or destroyed. It is often referred to as the law of conservation of energy.

first trimester: Extends from conception until the third month of pregnancy.

flow phase: Phase of the menstrual cycle marked by the shedding of the endometrium.

follicles: Structures in the ovary that contain the egg and secrete estrogen.

follicle-stimulating hormone (FSH): A gonadotropin that increases sperm production in males; promotes the development of the follicles in the ovary of the female.

follicular phase: Phase of the menstrual cycle marked by the development of the ovarian follicles prior to ovulation.

food chains: Illustrate a step-by-step sequence of who eats whom in the biosphere.

food web: A series of interlocking food chains representing the transfer of energy through various trophic levels in an ecosystem.

fovea centralis: The most sensitive area of the retina. It contains only cones.

fronds: The leaves of ferns.

gallstones: Crystals of bile salts that form in the gallbladder.

gametangia: Reproductive structures or organs that produce gametes for sexual reproduction.

gametes: Sex cells. They have a haploid chromosome number.

gametogenesis: The formation of sex cells in animals.

gametophyte: Refers to a stage in a plant's life cycle in which cells have haploid nuclei. During this stage, the sex cells (gametes) are produced.

gastrula: A stage of embryo development in which germ layers are formed.

Gause's principle: No two species can occupy the same ecological niche without one being reduced in numbers or being eliminated.

gene markers: Are often recessive traits that are expressed in the recessive phenotype of an organism. The markers can be used to identify other genes found on the same chromosome.

gene pools: All of the genes that occur within a specific population.

genes: Units of instruction located on chromosomes that produce or influence a specific trait in the offspring.

gene therapy: A procedure by which defective genes are replaced with normal genes in order to cure genetic diseases.

genome: The complete set of instructions contained within the DNA of an individual.

genotype: The genes an organism contains.

geographic range: A region where a given organism is sighted.

germ layer: A layer of cells in the embryo that gives rise to specific tissues in the adult.

glaucoma: A disorder of the eye caused by the build-up of fluid in the anterior chamber to the lens.

glomerulus: A high-pressure capillary bed that is surrounded by Bowman's capsule. The glomerulus is the site of filtration.

glucagon: A hormone produced by the pancreas. When blood sugar levels are low, glucagon promotes the conversion of glycogen to glucose.

glycogen: The form of carbohydrate storage in animals. Glycogen is often called the animal starch.

glycolipids: Compounds consisting of specialized sugar molecules attached to lipids.

glycolysis: The process in which ATP is formed by the conversion of glucose into pyruvic acid.

glycoproteins: Compounds consisting of specialized sugar molecules attached to the proteins of the cell membrane. Many distinctive sugar molecules act as signatures that identify specialized cells.

goiter: An enlargement of the thyroid gland.

Golgi apparatus: A protein-packaging organelle.

gonadotropic hormones: Produced by the pituitary gland. Regulate the functions of the testes and ovaries.

gonadotropin-releasing hormone (GnRH): A chemical messenger from the hypothalamus that stimulates secretions of gonadotropins (FSH and LH) from the pituitary.

grana: Green disks stacked together. The disks are part of the thylakoid membrane.

grazer food chain: Originate with plants that are consumed by herbivores (grazers).

grazing: Feeding on grass or grasslike vegetation.

growth curve: A graph used to show the changes to a population over a specific length of time.

growth hormone: Produced by the cells of the anterior pituitary. Prior to puberty, the hormone promotes growth of the long bones.

growth phase: Phase marked by accelerated reproduction by the population. Natality exceeds mortality.

guard cells: Occur in pairs on the lower epidermis of a leaf or on a stem. They regulate the opening and closing of a stoma.

gymnosperms: Plants that produce naked seeds.

habitat: The physical area where a species lives.

haploid chromosome number: One-half of the full complement of chromosomes. Sex cells have haploid chromosome numbers.

Hardy-Weinberg principle: Indicates conditions under which allele and gene frequencies will remain constant from generation to generation.

Haversian canals: Small canals located in bone tissue. The canals are occupied by blood vessels and nerves.

HCG: A placental hormone that maintains the corpus luteum.

helper T cells: Identify antigens.

hemocoel: An open space or body cavity.

hemoglobin: The pigment found in red blood cells.

herbivore: An animal that obtains its food exclusively from plant tissue.

heredity: The passing of traits from parents to offspring.

hermaphroditic: Containing both male and female sex cells or organs.

heterotrophs: Organisms that obtain food and energy from autotrophs or other heterotrophs; they are unable to synthesize organic food molecules from inorganic molecules.

heterozygous: A genotype in which the gene pairs are different.

hibernation: A dormant (sleep) state in which body temperature and functions are reduced well below normal.

holozoic organisms: Feed in an animal-like manner.

homeostasis: A process by which a constant internal environment is maintained despite changes in the external environment.

homologous chromosomes: Chromosomes that are similar in shape, size, and gene arrangement.

homologous structures: Have similar origins but different uses in different species. The front flipper of a dolphin and the forelimb of a dog are homologous structures.

homozygous: A genotype in which both genes of a pair are identical.

hormones: Chemicals released by cells that affect cells in other parts of the body. Only a small amount of a hormone is required to alter cell metabolism.

hosts: Living organisms from which a parasite obtains its food supply.

hybrids: Offspring that differ from their parents in one or more traits. Interspecies hybrids result from the union of two different species.

hydrogen bonds: Formed between a hydrogen proton and the negative end of another molecule.

hydrologic or water cycle: The movement of water through the environment—from the atmosphere to earth and back.

hyperpolarized membranes: Membranes that are much more permeable to potassium than usual. The inside of the nerve cell membrane becomes even more negative.

hypertonic solutions: Solutions in which the concentration of solutes outside the cell is greater than that found inside the cell.

hyphae: Filamentous threads of a fungus.

hypolimnion: The lower layer of a lake; it maintains a constant low temperature.

hypothermia: A condition in which an animal's body core temperature drops to a level that will eventually cause death.

hypothesis: A possible solution to a problem or an explanation of an observed phenomenon.

hypotonic solutions: Solutions in which the concentration of solutes outside the cell is lower than that found inside the cell.

implantation: The attachment of the embryo to the endometrium.

inbreeding: The process whereby breeding stock is drawn from a limited number of individuals possessing desirable phenotypes.

insulin: A hormone produced by the islets of Langerhans in the pancreas. Insulin is secreted when blood sugar levels are high.

intermediary metabolites: The chemicals that form as reactants are converted to products during a series of chemical reactions.

internal intercostal muscles: Muscles that pull the rib cage downward, increasing pleural pressure.

interneurons: Carry impulses within the central nervous system.

interspecific competition: Competition among similar species for a limited resource (e.g., food or space).

interstitial spaces: The spaces between the cells.

intraspecific competition: Competition within an ecological niche between members of the same species.

invertebrates: Multicellular, eukaryotic heterotrophs that do not have a backbone.

***in vitro* fertilization:** Occurs outside of the female's body. *In vitro* is Latin for "in glass."

ionic bonds: Formed when electrons are transferred between two atoms.

ionosphere: A region of the upper atmosphere consisting of layers of ionized gases that produce the northern lights and reflect radio waves.

ions: Atoms that have either lost electrons or gained electrons to become positively or negatively charged.

iris: Regulates the amount of light entering the eye.

isomers: Chemicals that have the same chemical formula but a different arrangement of molecules.

isotonic solutions: Solutions in which the concentration of solute molecules outside the cell is equal to the concentration of solute molecules inside the cell.

isotopes: Elements with the same number of protons but different numbers of neutrons.

jaundice: The yellowish discoloration of the skin and other tissues brought about by the collection of bile salts in the blood.

joint: The point of contact of two or more bones of the skeleton.

karyotypes: Pictures of chromosomes arranged in homologous pairs.

killer T cells: Puncture the cell membranes of cells infected with foreign invaders, thereby killing the cell and the invader.

Klinefelter syndrome: A trisomic disorder in which a male carries an XXY condition.

K-selected populations: Populations found where environmental conditions are stable. These populations are characterized by intense intraspecific competition.

lag phase: The adjustment period prior to accelerated reproduction by the population.

larva: A juvenile form that looks very different from the adult. Tadpoles are larval frogs; caterpillars are larval butterflies or moths.

lateral line: A line of sensory cells along each side of a fish's body.

larynx: The voice box.

law of the minimum: Of the number of essential substances required for growth, the one with the minimum concentration is the controlling factor.

ligaments: Bands of connective tissue that join bone to bone.

ligase enzyme: A biological glue that permits one section of DNA to be fused to another.

limnetic zone: The open water area of a lake.

linked genes: Genes that are located on the same chromosome.

lipases: Lipid-digesting enzymes.

littoral zone: The edge around a lake or pond where the water is shallow enough to permit the growth of aquatic vegetation.

luteal phase: Phase of the menstrual cycle characterized by the formation of the corpus luteum following ovulation.

luteinizing hormone (LH): A gonadotropin that promotes ovulation and the formation of the corpus luteum in females; regulates the production of testosterone in males.

lymph: The fluid found outside capillaries. Most often, the lymph contains some small proteins that have leaked through capillary walls.

lymph nodes: Contain white blood cells that filter lymph.

lymphocytes: Antibody-producing white blood cells.

lymphokine: A protein produced by the T cells that acts as a chemical messenger between cells.

macrophages: Phagocytotic white blood cells found in lymph nodes or in the blood (in bone marrow, spleen, and liver).

magnetosphere: A region found above the outer atmosphere consisting of magnetic bands caused by the earth's magnetic field.

Malphigian tubules: Tubelike structures that are involved in excretion.

map distance: The distance between two genes along the same chromosome.

matrix: The noncellular material secreted by the cells of connective tissue.

medulla: The area inside of the cortex in the kidney.

medulla oblongata: The region of the hindbrain that joins the spinal cord to the cerebellum. The medulla is the site of autonomic nerve control.

meiosis: The two-stage cell division in which the chromosome number of the parental cell is reduced by half. The process by which sex cells are formed.

memory T cells: Retain information about the geometry of antigens.

meninges: Protective membranes that surround the brain and spinal cord.

menopause: Marks the termination of the female reproductive years.

menstruation: The shedding of the endometrium.

meristem: Tissue region of plants where some cells retain the ability to divide repeatedly.

mesosphere: The region of the atmosphere found between the stratosphere and the upper atmosphere.

metabolism: The sum of all chemical reactions that occur within the cells.

metamorphosis: A process in which a larval form drastically changes shape to become more like the adult.

metastasis: Occurs when a cancer cell breaks free from the tumor and moves into another tissue.

microclimate: The temperature and moisture that occur at the ground level in a plant community.

micropipette: A thin glass rod that can be used to extract urine from the nephron.

microvilli: Infoldings of the cell membrane.

migration: The movement of organisms between two distant geographic regions.

mimicry: A form of camouflage that involves developing a similar color pattern, shape, or behavior that has provided another organism with some survival advantage.

mitochondria (singular: mitochondrion): Organelles that specialize in aerobic respiration.

mitosis: A type of cell division in which daughter cells receive the same number of chromosomes as the parent cell.

molecules: Units of matter that consist of two or more atoms of the same or different elements bonded together.

Monera: A kingdom of organisms without a nucleus.

monocotyledon (monocot): An angiosperm whose seeds have only one cotyledon or seed leaf.

monoculture: Involves growing a single species of plant to the exclusion of others.

monohybrid cross: Involves one gene pair of contrasting traits.

monosaccharides: Single sugar units. Monosaccharides usually include carbon, hydrogen, and oxygen in a ratio of 1:2:1.

monosomy: The presence of a single chromosome in place of a homologous pair.

motor neurons: Carry impulses from the central nervous system to effectors.

murmurs: Caused by faulty heart valves, which permit the backflow of blood into one of the heart chambers.

mutagenic agents: Things that cause changes in the DNA.

mutations: Arise when the DNA within a chromosome is altered. Most mutations change the appearance of the organism.

mutualism: A relationship in which two different organisms living together both benefit from each other.

mycelium: A collective term for the branching filaments that make up the part of a fungus not involved in sexual reproduction.

myelin sheath: A fatty covering over the axon of a nerve cell.

myofilaments: The contractile proteins found within muscle fibers.

myogenic muscle tissue: Contracts without external nerve stimulation.

NAD+: A hydrogen acceptor important for electron transport systems in cellular respiration. NADH is the reduced form of NAD+.

NADP+: A hydrogen acceptor important for electron transport systems in photosynthesis. NADPH is the reduced form of NADP+.

nearsightedness: Occurs when the image is focused in front of the retina.

negative feedback system: A control system designed to prevent chemical imbalances in the body. The body responds to changes in the external or internal environment. Once the effect is detected, receptors are activated and the response is inhibited, thereby maintaining homeostasis.

nematocyst: Stinging capsule that aids in the capture of prey.

nephridia: Open-ended tubules that function in excretion.

nephrons: The functional units of the kidneys.

neurilemma: The delicate membrane that surrounds the axon of nerve cells.

neurons: Cells that conduct nerve impulses.

neutralization: Occurs when the pH is brought to 7, i.e., the $[H^+] = [OH^-]$.

nitrogen cycle: The cycling of nitrogen between organisms and the environment.

nitrogen fixation: The conversion of nitrogen gas (N_2) into nitrates (NO_3) and ammonium ions (NH^+_4), which can then be used by plants.

nitrogen-fixing bacteria: Bacteria that convert atmospheric nitrogen to nitrogen compounds such as ammonia and nitrate.

nodes of Ranvier: The regularly occurring gaps between sections of myelin sheath along the axon.

norepinephrine: A hormone produced by the adrenal medulla that initiates the flight-or-fight response.

notochord: A skeletal rod of connective tissue that runs lengthwise along the dorsal surface and beneath the nerve cord. Notochords are present at some time during vertebrate development.

nuclear imaging: Technique that uses radioisotopes to view organs and tissues of the body.

nuclear magnetic resonance (NMR): Technique that employs magnetic fields and radio waves to determine the behavior of molecules in soft tissue.

nucleolus: A small spherical structure located inside the nucleus.

nucleotides: The building blocks of nucleic acids. Nucleotides are composed of a ribose sugar, phosphate, and a nitrogen base.

nucleus: The control center for the cell. Contains hereditary information.

nutrients: Chemicals that provide nourishment. Nutrients provide energy or are assimilated to form protoplasmic structures.

obligate aerobes: Organisms that require oxygen for respiration.

obligate anaerobes: Organisms that conduct respiration processes in the absence of oxygen.

oils: Lipids composed of glycerol and unsaturated fatty acids. They are liquid at room temperature.

olfactory lobes: Areas of the brain that detect smell.

oligotrophic lakes: Lakes that are cold and deep, and have only begun the process of eutrophication.

omnivore: An organism that eats both animals and plants.

oncogenes: Cancer-causing genes.

ootids: Unfertilized egg cells.

open population: A population in which density changes result from the interaction of natality, mortality, immigration, and emigration.

operculum: A bony plate covering the gill chamber.

organic molecules: Compounds that contain carbon.

organ of Corti: The primary sound receptor in the cochlea.

organs: Structures composed of different tissues specialized to carry out a specific function.

organ systems: Groups of organs that have related functions. Organ systems often interact.

osmotic pressure: The pressure exerted on the wall of a semi-permeable membrane resulting from differences in solute concentration.

ossicles: Tiny bones that amplify and carry sound in the middle ear.

ossification: The process by which bone is formed.

osteoblasts: Bone-forming cells.

osteoclasts: Cells that dissolve bone.

otoliths: Tiny stones of calcium carbonate found within the saccule and utricle. The tiny stones are embedded in a gelatinous coating. Gravity causes the otoliths to slide downward as the head is lowered or raised.

oval window: Receives sound waves from the ossicles.

ovaries: The female gonads, or reproductive organs. Female sex hormones and egg cells are produced in the ovaries.

oviduct (Fallopian tube): The passageway through which an ovum moves from the ovary to the uterus, or womb.

ovulation: The release of the egg from the follicle held within the ovary.

oxidation: Occurs when an atom or molecule loses electrons.

oxytocin: A hormone from the posterior pituitary gland that causes strong uterine contractions.

ozone: An inorganic molecule. A layer of ozone found in the stratosphere helps to screen out ultraviolet radiation.

ozone hole: A region in the ozone layer in which the ozone levels have been considerably reduced and the layer has become very thin.

Pangaea: A large supercontinent that existed approximately 225 million years ago.

parasite: An organism that lives in or on another organism, from which it obtains its food.

parasympathetic nerves: A division of the autonomic nervous system. These nerves are designed to return the body to normal resting levels following adjustments to stress.

pathogens: Disease-causing agents.

pelvis: The area in which the kidney joins the ureters.

pepsin: A protein-digesting enzyme produced by the cells of the stomach.

peptide bonds: Bonds that join amino acids.

pericardium: A saclike membrane that protects the heart.

pericycle: A meristematic tissue consisting of parenchyma cells and sometimes fibers. It is found immediately within the endodermis.

periosteum: The tissue that covers bone.

peristalsis: The rhythmic, wavelike contraction of smooth muscle that moves food along the gastrointestinal tract.

peritoneum: A covering membrane that lines the body cavity and covers the internal organs.

phagocytosis: A form of endocytosis in which cells engulf large molecules and incorporate them into the cytoplasm.

pharynx: The passage for both food and air.

phenotype: The observable traits of an organism that arise because of the interaction between genes and the environment.

pheromone: A chemical substance produced by an organism that serves as a stimulus to another organism of the same species.

phloem: A vascular tissue that transports food, synthesized in the leaves, throughout the plant.

phospholipids: The major components of cell membranes in plants and animals. Phospholipids have a phosphate molecule attached to the glycerol backbone, making the molecule polar.

phosphorylation: The addition of one or more phosphate groups to a molecule.

photolysis: The splitting of water by means of light energy.

photoreceptor: A sensory receptor that detects light energy.

photosynthesis: The process by which plants and some bacteria use chlorophyll, a green pigment, to trap sunlight energy. The energy is used to synthesize carbohydrates.

photosystems: Light-trapping units composed of pigments within the thylakoid membranes.

pH scale: Used to measure how acidic or basic a solution is.

phylogeny: The history of the evolution of a species or a group of organisms.

pinna: The outer part of the ear. The pinna acts like a funnel, taking sound from a large area and channeling it into a small canal.

pinocytosis: A form of endocytosis in which liquid droplets are engulfed by cells.

pioneer communities: The first species to appear during succession.

placenta: An organ made from the cells of the baby and the cells of the mother. It is the site of nutrient and waste exchange between mother and baby.

placoid scales: Toothlike scales composed of dentine, within a pulp cavity.

plasma: The fluid portion of the blood.

plasmids: Small rings of genetic material.

plastids: Organelles that function as factories for the production of sugars or as storehouses for starch and some pigments.

pleiotropic genes: Affect many characteristics.

polar bodies: Formed during meiosis. These cells contain all the genetic information of a haploid egg cell but lack sufficient cytoplasm to survive.

polarized membranes: Charged membranes. Polarization is caused by the unequal distribution of positively charged ions.

polar molecules: Molecules that have positive and negative ends.

polygenic traits: Inherited characteristics that are affected by more than one gene.

polymerases: Enzymes that join individual nucleotides together in complementary strands of DNA.

polymers: Molecules composed of from three to several million subunits. Many polymers contain repeating subunits.

polypeptides: Proteins composed of amino acids and joined by peptide bonds

polyploidy: A condition in which an organism possesses more than two complete sets of chromosomes.

polysaccharides: Composed of many single sugar subunits.

pons: The region of the brain that acts as a relay station by sending nerve messages between the cerebellum and the medulla.

population: A group of individuals of the same species occupying a given area at a certain time.

population density: The number of organisms per unit of space.

population sampling: A technique in which gene frequencies for a particular genetic trait are determined in a small sample of the population, and results are applied to the whole population.

population size: The number of organisms at a certain time.

postsynaptic neurons: The neurons that carry impulses away from the synapse.

precursor activity: The activation of the last enzyme in a metabolic pathway by the initial reactant.

prehensile: Capable of grasping.

presynaptic neurons: The neurons that carry impulses to the synapse.

primary succession: The occupation, by plant life, of an area not previously covered by vegetation.

producers: Organisms that are capable of making their own food.

profundal zone: The region of a lake to which light cannot penetrate.

progesterone: A female sex hormone.

prokaryotes: Single-celled organisms without a nucleus.

prokaryotic cells: Primitive cells that do not have a true nucleus or a nuclear membrane.

prolactin: A hormone produced by the pituitary that is associated with milk production.

prostaglandins: Hormones that have a pronounced effect in a small localized area.

prostate gland: Contributes to the seminal fluid (semen), a secretion containing buffers that protect sperm cells from the acidic environment of the vagina.

protein denaturation: Occurs when the bonds holding a protein molecule are disrupted by physical or chemical means, causing a temporary change in shape.

protein hormones: Composed of chains of amino acids. This group includes insulin, growth hormone, and epinephrine.

proteins: The structural components of cells.

prothallus: The gametophyte of mosses, liverworts, and ferns.

Protista: A kingdom originally proposed for all unicellular organisms such as the amoeba. More recently, some simple multicellular algae have been added to the kingdom.

protonoma: The young gametophyte of a moss in the early stages after the germination of the spore.

protoplasm: All the material within a cell. The protoplasm is composed of the nucleus and the cytoplasm.

proximal tubule: A section of the nephron joining Bowman's capsule with the loop of Henle. The proximal tubule is found within the cortex of the kidney.

pseudocoelom: A fluid-filled cavity that lacks the mesodermal lining of a true coelom.

pukak: A layer of open space containing ice crystals at the base of a snow pack.

pulmonary artery: Carries deoxygenated blood from the heart to the lungs.

pulmonary circulatory system: Carries deoxygenated blood to the lungs and oxygenated blood back to the heart.

pulmonary veins: Carry oxygenated blood from the lung to the heart.

pulse: Caused by blood being pumped through an artery.

Punnett square: A chart used by geneticists to show the possible combinations of alleles in offspring.

pus: Substance formed when white blood cells engulf and destroy invading microbes. The white blood cell is also destroyed in the process. The remaining protein fragments are known as pus.

pyramid of biomass: An energy pyramid based on the dry mass of the tissue of organisms at each trophic level.

pyramid of energy: A pyramid drawn on the basis of the energy produced (as heat) at each trophic level.

pyramid of numbers: An energy pyramid based on the numbers of organisms at each trophic level.

qali: Snow that collects on the branches of trees.

qamaniq ("snow shadow"): The depression in the snow found around the base of trees, particularly conifers.

radioisotopes: Unstable chemicals that emit bursts of energy as they break down.

random distributions: Are arbitrary and appear to be unaffected by biotic factors.

rate of change: The change in a population over a period of time.

receptor sites: Act as ports along cell membranes. Nutrients and other needed materials fit into specialized areas along cell membranes.

recessive genes: Genes that are overruled by dominant genes, which determine the genetic trait.

recombinant DNA: An application of genetic engineering in which genetic information from one organism is spliced into the chromosome of another organism.

reduction: Occurs when an atom or molecule gains electrons.

refractory period: The recovery time required before a neuron can produce another action potential.

regulator genes: Genes that control the production of repressor proteins, which switch off structural genes.

relaxin: A hormone produced by the placenta prior to labor that causes the ligaments within the pelvis to loosen.

rennin: An enzyme that coagulates milk proteins.

replication: The process in which a single strand of nucleotides acts as a template for the formation of a complementary strand. A single strand of DNA can make a complementary strand.

repolarization: A process in which the original polarity of the nerve membrane is restored. Excess positive ions are found outside the nerve membrane.

respiration: The chemical process in which nutrients are broken down to provide energy.

resting membranes: Maintain a steady charge difference across the cell membrane. These membranes are not being stimulated.

restriction enzymes: Enzymes that cut strands of DNA at specific sites.

retina: The innermost layer of the eye.

reverse transcriptase: An enzyme that allows the genetic message from the RNA of a virus to be transcribed into the DNA.

rhodopsin (visual purple): The pigment found in the rods of the eye.

ribonucleic acid (RNA): A single-stranded nucleic acid used to translate the information of DNA into protein structure.

rods: Photoreceptors used for viewing in dim light.

r-selected population: Population that undergoes many changes, many of which cannot be predicted. These populations are characterized by a high birth rate and a short life span.

saccule: Responsible, with the utricle, for static equilibrium.

saprophytes: Organisms that obtain nutrients from dead and decomposing nutrients.

saprozoic organisms: Absorb nutrients directly through cell membranes.

sarcolemma: The delicate sheath that surrounds muscle fibers.

sclera: The outer covering of the eye that supports and protects the eye's inner layers.

scrotum: The sac that contains the testes.

second law of thermodynamics: States that, during an energy transformation, some of the energy produced, usually in the form of heat, is lost from the system.

second trimester: Extends from the third month to the sixth month of pregnancy.

secondary succession: Occurs in an area that was previously covered by vegetation and still has some soil.

secretin: A hormone that stimulates pancreatic and bile secretions.

segmentation: Refers to the repetition of body units that contain some similar structures.

segregation: The separation of paired genes during meiosis.

selective breeding: The crossing of desired traits from plants or animals to produce offspring with both characteristics.

selectively permeable membranes: Membranes that allow some molecules to pass through the membrane, but prevent other molecules from penetrating the barrier.

semen (seminal fluid): A secretion of the male reproductive organs that is composed of sperm and fluids.

semicircular canals: Fluid-filled structures within the inner ear that provide information about dynamic equilibrium.

semiconservative replication: The process in which the original strands of DNA remain intact and act as templates for the synthesis of duplicate strands of DNA.

semilunar valves: Prevent the backflow of blood from arteries into the ventricles. (Semilunar means "half-moon.")

seminal vesicles: A gland located along the vas deferens that contributes to the seminal fluid (semen), a secretion that contains fructose and prostaglandins.

seminiferous tubules: Coiled ducts found within the testes, where immature sperm cells divide and differentiate.

sensory adaptation: Occurs once you have adjusted to a change in the environment.

sensory neurons: Carry impulses to the central nervous system.

sensory receptors: Modified ends of sensory neurons that are activated by specific stimuli.

seral stages: Specific stages in succession identified by the dominant species present.

Sertoli cells: Nourish sperm cells.

sessile animals: Remain fixed in one place throughout their adult lives and are not capable of independent movement.

sex chromosomes: Pairs of chromosomes that determine the sex of an individual.

sex-linked traits: Traits that are controlled by genes located on the sex chromosomes.

Shelford's law of tolerance: Too little or too much of an essential factor can be harmful to an organism.

sinoatrial node: The heart's pacemaker.

skeletal muscle: Voluntary muscle that is attached to bones. The muscle is sometimes referred to as striated muscle because of its striped appearance.

sloughs: Depressions often filled with stagnant water.

smooth muscle: Involuntary, non-striated muscle.

solutes: Molecules that are dissolved in water. Salt and sugars are common solutes.

somatic cells: All the cells of an organism except the sex cells.

somatic nerves: Nerves that lead to skeletal muscle. Somatic nerves are under conscious control.

sorus: A cluster of sporangia on a fern.

speciation: The formation of a new species.

species: A group of organisms that look alike and can interbreed under natural conditions to produce fertile offspring.

sphincter: A ring of smooth muscle in the wall of a tubular organ. Contraction of the sphincter closes the opening.

sphygmomanometer: A device used to measure blood pressure.

spindle fibers: Protein structures that guide chromosomes during cell division.

sporangia: Reproductive structures in which spores are produced by two-staged nuclear division.

sporophyte: Refers to a stage in a plant's life cycle in which cells have diploid nuclei. During this stage spores are produced. This stage arises from the union of two haploid gametophytes.

starch: A plant storage carbohydrate.

stationary phase: Phase that marks equilibrium between natality and mortality.

steroid hormones: Made from cholesterol. This group includes male and female sex hormones and cortisol.

stomata: Pores in the epidermis of plants, particularly in leaves. Stomata permit the exchange of gases between the plant and atmospheric air.

stratosphere: The region of the earth's atmosphere found above the troposphere. The ozone layer is found in this region.

stroke volume: The quantity of blood pumped with each beat of the heart.

stroma: The gel-like substance containing proteins that surrounds the grana.

structural genes: Genes that direct the synthesis of proteins.

subnivean: Means "beneath a snow cover."

substrate molecules: Molecules that attach to enzymes.

succession: The slow, orderly, progressive replacement of one community by another during the development of vegetation in any area.

summation: The effect produced by the accumulation of transmitter chemicals from two or more neurons.

supercooling: Occurs when a water solution is chilled well below the point at which the solution crystallizes spontaneously into ice.

suppressor T cells: Turn off the immune system.

surrogate: Means "substitute." A nongenetic mother is a surrogate mother.

symbiosis: A relationship in which two different organisms live in a close association.

sympathetic nerves: A division of the autonomic nervous system. These nerves prepare the body for stress.

synapses: The regions between neurons or between neurons and effectors.

synapsis: The pairing of homologous chromosomes.

systemic circulatory system: Carries oxygenated blood to the tissues of the body and deoxygenated blood back to the heart.

systole: Refers to heart contraction.

target tissues: Tissues that have specific receptor sites that bind with the hormones.

taxa: Categories used to classify organisms.

taxonomy: The science of classification according to the inferred (presumed) relationship among organisms.

T cells: Produced in the lymph nodes. Different types of T cells regulate an immune response.

tegument: A modified epidermis that protects parasites from the digestive enzymes and immune response of the host.

tendons: Bands of connective tissue that join muscle to bone.

testes: The male gonads, or primary reproductive organs. Male sex hormones and sperm are produced in the testes.

testosterone: The male sex hormone produced by the interstitial cells of the testes.

tetanus: The state of constant muscle contraction caused by sustained nerve impulses.

tetrads: Contain four chromatids.

thallus: An undifferentiated vegetative plant body without distinct roots, stems, and leaves.

thermocline: The zone separating the epilimnion from the hypolimnion.

third trimester: Extends from the seventh month of pregnancy until birth.

thoracic cavity: The chest cavity. Heart and lungs are contained in the thoracic cavity.

threshold level: The minimum level of a stimulus required to produce a response.

threshold level: In terms of reabsorption, the maximum amount of material that can be moved across the nephron.

thrombus: A blood clot that forms within a blood vessel.

thylakoid membranes: Specialized cell membranes found in chloroplasts.

thyroxine: A hormone secreted by the thyroid gland that regulates the rate of body metabolism.

tissues: Groups of similarly shaped cells that work together to carry out a similar function.

totipotent cell: A cell that has the ability to support the development of an organism from egg to adult.

trachea: The windpipe.

tracheal system: A branching series of tubes that conducts air to the cells of insects.

transcriptase: An enzyme that allows the genetic message from the RNA of a virus to be transcribed into the DNA.

transcription: The process by which the genetic code is transferred from the DNA molecule to the messenger RNA molecule.

translation: The process by which proteins are synthesized using the DNA instructions encoded in the mRNA.

translocation: Refers to the conduction of organic compounds throughout cascular plants by way of phloem vessels.

transpiration: The loss of water through the leaves of a plant.

transposons: Specific segments of DNA that can move along the chromosome.

triglycerides: Lipids composed of glycerol and three fatty acids.

trisomy: The presence of three homologous chromosomes in every cell of an organism.

trophic level: The number of energy transfers an organism is from the original solar energy entering an ecosystem.

troposphere: The lowest region of the earth's atmosphere. It extends upward about 12 km. Most weather occurs here.

trypsin: A protein-digesting enzyme.

tundra: The most northerly major life zone. Characterized by low precipitation, low temperatures, permafrost, and a lack of trees.

turgor pressure: The pressure exerted by water against the cell membrane and the nonliving cell walls of plant cells.

Turner's syndrome: A monosomic disorder in which a female has a single X chromosome.

tympanic membrane: The eardrum.

ulcer: A lesion along the surface of an organ.

ultraviolet radiation: The electromagnetic radiation from the sun. It can cause burning of the skin (sunburn) and cellular mutations.

umbilical cord: Connects the fetus to the placenta.

uniform distributions: Are orderly and appear to be affected by competition.

urea: A nitrogen waste formed from two molecules of ammonia and one molecule of carbon dioxide.

ureters: Tubes that conduct urine from the kidneys to the bladder.

urethra: A tube that carries urine from the bladder to the exterior of the body.

uric acid: A waste product formed from the breakdown of nucleic acids.

uterus (womb): The female organ in which the fertilized ovum normally becomes embedded and in which the embryo and fetus develop.

utricle: Responsible, with the saccule, for static equilibrium.

vagus nerve: A major cranial nerve that is part of the parasympathetic nervous system.

varicose veins: Distended veins.

vascular tissue: Tissue that transports water and the products of photosynthesis throughout a plant.

vas deferens: Tubes that conduct sperm toward the urethra.

vasoconstriction: The narrowing of a blood vessel. Less blood goes to the tissues when the arterioles constrict.

vasodilation: The widening of the diameter of the blood vessel. More blood moves to tissues when arterioles dilate.

vegetative structure: Any part of a fungi or plant that is not involved in sexual reproduction.

veins: Carry blood back to the heart.

venae cavae: Carry deoxygenated blood back to the heart.

ventricles: Muscular, thick-walled heart chambers that pump blood through the arteries.

venules: Small veins.

vertebrates: Multicellular, eukaryotic heterotrophs that have a backbone at some stage in their life.

vestibule: A chamber found at the base of the semicircular canals that provides information about static equilibrium.

villi: Small fingerlike projections that extend into the small intestine. Villi increase surface area for absorption.

virus: A microscopic particle capable of reproducing only within living cells.

waxes: Long-chain lipids that are insoluble in water.

womb: See "uterus."

xylem: A vascular tissue in plants that carries water and minerals to the leaves.

zero population growth: Occurs when the population of a species shows no increase or decrease in size over time.

zoology: The study of animal life.

zygote: The cell resulting from the union of a male and female sex cell, until it divides.

and lysosomes, 36, 47
Arthropoda, 485–89
Arthroscopic surgery, 275
Artificial body parts, 118, 120
Artificial involution, 563
Aschelminthes, 477
Aspirin, 269
Aster, 534
Asthma. *See* Bronchial asthma
Astigmatism, 26, 296
Atherosclerosis, 64, 142
Athletes
    and heart rate, 153–54
    and injuries, 281
    and natural painkillers, 265
    and steroid use, 71
Atlantic Ocean, 686
Atmosphere, 311
    composition of, 186, 312
    defined, 312
    and solar radiation, 316–17
    research, 315
    zones of, 312–13
Atom
    description of, 51–52
    discovery of, 51
Atomic mass, 52
Atomic number, 52
Atrioventricular (AV)
    node, 149
    valve, 146–47, 152
Atrium, 146, 152–53
Auditory canal, 300
Aurora borealis. *See* Northern lights
Autonomic nerve, 241, 256
Autonomic nervous system, 241, 256
Autosome, 555, 558, 599
Autotroph
    defined, 74, 335
    See also Producer
Autumn, and lakes, 364
Auxin, 463
Avalanche, 377
Avery, Oswald, 620
Aves (birds), 498–500
Avogadro's hypothesis, 196
Axial skeleton, 271
Axon, 242–43, 252
AZT (azidothymidine), 185, 646

B

Bacilli, 427
Bacon, Roger, 22
Bacteria, 426–29
    anaerobic, 328
    and antibiotics, 640–41
    and binary fission, 533
    and chemosynthesis, 336
    and conjugation, 610
    denitrifying, 328

in lakes, 361, 362
in large intestine, 130
nitrogen-fixing, 329, 336
and restriction enzyme, 629
Bacteriophages, 424
Baird, Patricia, 626
Baldness, 595
Banting, Frederick, 232
Bark, 461
Baroreceptor, 156
Barr bodies, 601
    and gender, 603
Base, 55–57
Basilar membrane, 302, 306
Basophil, 168, 169
B-cell lymphocyte, 168, 176, 178ff.
Beadle, George, 635–36
Beagle, HMS, 688
Bear, 337, 370
Beaumont, William, 132–33
Bee
    and pollination, 407, 550
    wing, 48
Benedict's solution, 61
Bennett, G.F., 434
Benthic detritus feeders, 361
Benthos, 361
Berger, Hans, 246
Berstein, Julius, 247
Beryllium, 54
Best, Charles, 232
Best-fit graph, 393ff.
Biceps, 279
Bilateral symmetry, 471
Bile pigment, 129
Bile salt, 129
Binary fission, 428, 533
Binomial nomenclature, 421
Biogeochemical cycle, 320, 335
Biogeography, defined, 684
Biological amplification, 351
Biological warfare, 649, 651, 653
Biomass, 346
Biome, defined, 357–58
Biosphere, defined, 311
Biosphere II, 352
Biotechnology, 548, 630–31
Biotic, defined, 311
Biotic potential, 396
Birds, 498–500
Birth canal, 525–26
Birth control pills, 517
Bison, 348, 369, 676
Biuret test, 69
Bivalvia, 483
Black spruce, 360
Bladder, 205, 206
Blastocyst, 522, 523
Blastula, 469, 540
Blind spot, 290, 298
Blood
    and age, 171, 172

artificial, 32, 174
and carbon dioxide transport, 197–98
clotting, 168, 170–71
color, 166
composition of, 165ff.
and kidney, 205ff.
oxygen, 143, 147, 149, 167
and oxygen transport, 197
pH, 57
reservoir, 144, 154
tissue, 105
transfusion, 171–72
type, 171–74, 581, 585, 659, 663
Blood flow. *See* Circulation, blood
Blood pressure, 143
    in aorta, 154
    and arteriolar dilation, 155
    in capillaries, 158–59
    and cardiac output, 154–55
    high, 156
    and kidney, 212–13
    low, 156
    measuring, 154
    and posture, 157
    receptors, 156
    regulation of, 156
Blood sugar, 223, 229ff.
Bloom, 364
"Blue baby." *See* Erythroblastosis fetalis
Blueberry, 555
Blushing, 142
Body cavities, 470
Bog, 329
Bolting, 463
Bond
    chemical, 53–55
    covalent, 55
    hydrogen, 55
    ionic, 54
Bone, 271
Bone development, 272–73
Bone marrow, 166, 270, 272
Bony fishes (Osteichthyes), 495–96
Book lung, 487
Boreal forest
    biome, 357, 358
    and global warming, 326
Borg, Bjorn, 153
Botanist, defined, 438
Boveri, Theodor, 597
Bowman's capsule, 206ff.
Boyd, Norman, 65
Boyer, Herbert, 630, 631
Boyle, Robert, 301
Bracon cephi, 370–71
Bradycardia, 150
Bradykinin, 159
Brain. *See also* specific parts of brain
    and autonomic nerves, 256-57
    and cranial nerve, 256
    gray matter in, 243
    hemispheres, 259–60

Nephridia, 479
Nephron, 206ff.
Nerve
    autonomic, 241, 256
    cell, 104
    somatic, 241, 256
Nerve impulse, 243
    and brain, 251
    vs. electricity, 247
    movement of, 250
Nerve tissue, 106
Nervous system
    autonomic, 256
    central, 241, 257ff.
    divisions of, 241
    and endocrine system, 223, 226–27,
        228–29, 260
    peripheral, 241
Neurilemma, 243
Neuron
    and axon, 242–43
    defined, 241
    and dendron, 242
    hyperpolarized, 253
    and ions, 247ff.
    motor, 241ff.
    postsynaptic, 252
    presynaptic, 252
    sensory, 241ff.
    and threshold level, 251
Neurospora crasa, 635–36
Neutralization, 57
Neutral solution, defined, 56
Neutrophil, 168, 169, 175
Newfoundland, 686
Newton, Sir Isaac, 292
Nitrates, 328
Nitrogen, 328–29
    in atmosphere, 186
    base, 636ff.
    and biogeochemical cycle, 320
    partial pressure of, 196
Nitrogen cycle, 328–29
Nitrogen fixation, defined, 328
Nitrogen-fixing bacteria, 329
Nitroglycerine, 147
Nitrous oxide, 322, 325
Noble gases, 53–54
Node of Ranvier, 243
Nodes (leaves), 462
Nondisjunction, 556ff.
Nonvascular plants, 446–47
Noradrenaline. *See* Norepinephrine
Norepinephrine, 228–29, 254, 255
Northern lights (aurora borealis), 313, 315
"Notched" trait, 602
Notochord, 469
Nova Scotia, 686
Nuclear imaging, 108–9
Nuclear magnetic resonance (NMR), 109
"Nuclear winter," 317
Nucleic acid, 70

composition of, 619
    function of, 320
Nucleolus, defined, 30
Nucleotide, 70, 619
Nucleus, 51
    defined, 30
    discovery of, 28
Nutrient, defined, 122

Oakes, Jill, 319
Obesity, and disease, 63
Objective lens, 24
Obligate aerobe, 428
Obligate anaerobe, 428
Observation, and scientific experimentation,
    21
Occipital lobe, 260
Oceanic islands, 686
Oceans
    and carbon cycle, 324, 325
    importance of, 356
    and phosphorus cycle, 329
    and photosynthesis, 86
Octopod, 484, 485
Ocular lens, 24
Oil, 62
Oil-immersion lens, 30
Oil spills, 485
Oleic acid, 63
Olfactory cell, 287
Olfactory lobe, 259
Oligochaeta, 478–79
Oligotrophic lake, 360
Olympics
    and performance-enhancing drugs, 71
    and gender verification, 603
Omnivore, 337
On the Origin of Species, 658, 688
Oncogene, 646–48
"One gene, one protein" hypothesis, 635–36
Oocyte, 516–17, 554
Oogenesis, 554
Ootid, 554
Operculum, 495
Opiates, 265. *See also* Drugs
Opium, 265
Opossum, 501
Opsin, 294
Optic nerve, 289, 290
Orchid, 539
Organelles, defined, 33
Organic compound(s)
    classes of, 58
    defined, 57
Organ of Corti, 302
Organs
    defined, 106
    monitoring, 108–9
    summary of, 107

Organ system
    defined, 106
    summary of, 106–7
Organ transplant, rejection of, 179
Osmoreceptor, 210
Osmosis, 42–43
    and water cycle, 322
Osmotic pressure, 158–59, 210–12
Ossicle, 299ff.
Ossification, 272
Osteichthyes, 495–96
Osteoblast, 272
Osteoclast, 273
Osteoporosis, 273
Otolith, 302–3
Outer ear, 299–300, 301
Oval window, 300, 302
Ovary (female gonad), 509, 514ff.
Ovary, flower, 451, 572
Oviduct. *See* Fallopian tube
Ovulation, 516–17
Ovule, 451
Ovum (egg cell), 514ff., 554
Owl, 290
Oxidation, 92–93
Oxidation-reduction reaction, 86, 87
Oxygen
    and aerobic respiration, 94–96
    in atmosphere, 186
    and biogeochemical cycle, 320
    and capillaries, 143
    and diffusion, 42, 147
    importance of, 186
    and lakes, 359, 360, 361, 362
    and nitrate formation, 328
    and oceans, 86
    and ozone formation, 313, 314
    partial pressure of, 196
    and photosynthesis, 87, 90
    receptor, 191–92
    and solubility in water, 363
    transport, 197
Oxygen debt, 279
Oxyhemoglobin, 197
Oxytocin, 226, 526
Ozone, formation of, 313, 314
Ozone layer
    holes in, 313–14
    and scientific research, 315

Pacific Ocean, 686
Paleoherbs, 450
Palomino horses, 580
Pancreas, 127–28, 223, 230
Pangaea, 684–85
Pap test, 515
Parapodia, 479
Parasite, 406, 428, 475, 476
Parasitism, 406

and predation, 397
rate of change in, 391
r-selected, 399
sampling, 659
size, 390
Population growth
curves, 393ff.
and density–dependent factors, 398–99
and density–independent factors, 398–99
human, 402–3
patterns, 392–97
Porifera, 472–73
Porpoise brain, 261
Postganglionic nerve, 256
Potassium ion, 247ff.
Potato cyst nematode, 477–78
Prebus, Albert, 31, 32
Precursor activity, 81
Predation, 337, 397, 404–5
Preganglionic nerve, 256
Pregnancy, 522–23
and alcohol, 529
ectopic, 515
trimesters of, 524–25
Prehensile, 502
Prenatal development, 523–24
Prenatal testing, 531, 560
Pressure-flow hypothesis, 463
Prey, 337
Price, Gerald, 47
Primary colors, 294–95
Primary growth, 457
Primary succession, 410
Primrose, 592
Prince Edward Island, 686
Prion, 425
Probability, 584–85, 665ff.
Producer, defined, 335, 438
Product rule, 584
Profundal zone, 361
Progesterone, 517, 519–20, 521–23, 526
Prokaryote, 422
Prokaryotic cell, 30
Prolactin, 226–27, 526–27
Pronghorn, 370
Prophase
meiosis, 552, 553
mitosis, 534
Prosimians, 502
Prostaglandin, 236, 512, 526
Prostate gland, 512, 520
Proteins, 320
and active transport, 45
in cell membrane, 40–41
and chromosome, 619
and coagulation, 68
complementary, 175
and deamination, 205
defined, 35, 634
denaturation, 68
and diet, 66
as energy source, 92, 96

identification of, 69
and immune response, 175
and mitosis, 534
and osmosis, 42–43
primary, 67
quaternary, 68
secondary, 67
tertiary, 67
Protein hormone, 224–25
Protein synthesis
and antibiotics, 640–41
and DNA, 636–39
and transcription, 637
and translation, 638–39
Prothallus, 448
Prothrombin, 171
Protista, 336, 422, 431–35
Protium, 52
Proton, 51–52, 92
Protonema, 446
Protoplasm, defined, 29
Protozoans, 432
Proximal tubule, 206
Pruitt, William O., Jr., 382
Pseudocoelom, 470
Puberty
female, 516, 519, 554
male, 512, 513
Pukak, 377, 381
Pulmonary artery, 147
Pulmonary vein, 146
Pulse, 141, 154
Punnett square, 575–76
Pupil, 288, 293, 294
Pus, 36, 168
Pyramids, ecological. *See* Ecological
pyramids
Pyruvic acid, 94, 95

Qali, 379
Qaminq, 379
Quinzhee, 381

Rabbit, coat color of, 580, 592
Radial keratotomy, 297
Radial symmetry, 471
Radioactive dating, 681
Radioactive labelling, 611
Radioactive tracing, 224
Radioisotope, 39, 53, 108
Radionuclide, 108–9
Radio telemetry, 409
Rain forest, 349–50
Random distribution, 389
Rape seed, 588

Ras, 648
Rate of change, in population, 391
Ray, 495
Reabsorption, in kidney, 208–9
Reaction rate, and reflex arc, 245–46
Recapitulation, theory of, 683
Receptor, 243
Receptor site, 177
Recessive genes, 573
Recombinant DNA, 630
Recycling, 331
Red blood cell (erythrocyte)
and blood oxygen, 167
deficiency, 167
and hemoglobin, 165–66
and oxygen transport, 165–66
reproduction of, 166
and sickle-cell anemia, 642
Redi, Francesco, 22, 27
Reduction division, 552
Red-winged blackbird, 360
Reed, Austin, 408
Reflex, 243
Reflex arc, 206, 243, 245, 252
Reforestation, 465
Refractory period, 249
Relaxin, 526
Remora, 407
Remote Operation Platform for Ocean
Science (ROPOS), 475
Renal erythropoietic factor (REF), 167
Renin, 213
Rennin, 125
Repolarization, 248ff.
Reproductive isolation, 675
Reproductive system
female, 514ff.
male, 509–13
Reproductive technology, 563–64, 566, 626
Reptilia, 497–98
Resistance (pesticides), 696–97, 700
Resolving power, 23
Respiratory distress syndrome, 189
Respiratory system
disorders of, 193–94, 200, 201
parts of, 187–89
Resting membrane, 247, 248
Restriction enzyme, 629
Rete mirabile, 373
Retina, 290, 291, 293, 296
Retinene, 294
Reverse transcriptase, 644
Reynolds, James, 265
RFLP markers, and gene mapping, 611
R group, 66, 68
Rh blood type, 173–74, 581, 585
Rheumatoid arthritis, 274
Rhizoid, 446
Rhodopsin, 294, 295
Rib cage, 190
Ribosome, 35, 641
Rigor mortis, 278

UNIT OPENING PHOTOS: Unit 1 Main photo: Freeman Patterson/Masterfile; Sidebar: Dr. Bryan Eyden/Science Photo Library; Unit 2 Main photo: CNRI/Science Photo Library; Sidebar: Manfred Kage/Science Photo Library; Unit 3 Main photo: Tom Sanders/Masterfile; Sidebar: Mel Di Giacomo, Bob Masini/The Image Bank; Unit 4 Main photo: Tom Van Sant/Geosphere Project, Santa Monica/Science Photo Library/Masterfile; Sidebar: Bill Brooks/Masterfile; Unit 5 Main Photo: Janet Foster/ Masterfile; Sidebar: G. Eikerts; Unit 6 Main photo: Michael Melford/The Image Bank; Sidebar: Steve Satushek/The Image Bank; Unit 7 Main photo: Oxford Scientific Films/Animals Animals/Earth Scenes; Sidebar: Science Photo Library/Photo Researchers; Unit 8 Main photo: Comstock/Miller Comstock; Sidebar: Telegraph Color Library/Masterfile; **ONE: 1.1** CNRI/Science Photo Library; **20** M.I.Walker/Photo Researchers; **21** Allen Haslinger/Masterfile; **1.2A** Courtesy of the Burndy Library, Cambridge, MA; **1.2B** Dr. Jeremy Burgess/Science Photo Library; **1.2C** Courtesy of the Burndy Library, Cambridge, MA; **1.3** Martyn Lengden; **1.4** Martyn Lengden; **25** Top, Kam Yu; Bottom left, Martyn Lengden; Bottom right, Kam Yu; **26** Kam Yu; **1.5** Dave Mazierski; **1.6** Kam Yu; **1.7** Dr. Jeremy Burgess/Science Photo Library; **1.8** Martyn Lengden after C. Starr and R. Taggart, *Biology 5ed.*, Wadsworth Publishing Company, 1989; **1.9** Dave Mazierski; **1.10** Martyn Lengden; **1.11** Secchi-Lecaque-Roussel-Uclaf/CNRI/Science Photo Library; **1.12** Comstock/Miller Comstock; **1.13** Dr. Jeremy Burgess/Science Photo Library; **32** Top left, Canapress Photo Service; Bottom left, Dr. Thomas Chang/McGill University; Right, Dr. Brian Eyden/Science Photo Library; **1.14** Dave Mazierski; **1.15** Dave Mazierski; **1.16A** Dr. Gopal Murti/Science Photo Library; **1.16B** Dave Mazierski; **1.17A** Dave Mazierski; **1.17B** CNRI/Science Photo Library; **1.18** Dave Mazierski; **1.19** Dr. B.H. Satir/Albert Einstein College of Medicine; **1.20A** W.P. Wergin, courtesy of E.H. Newcomb, University of Wisconsin/Biological Photo Service; **1.20B** Dave Mazierski; **1.21** Dave Mazierski; **38** Kam Yu; **1.22** Kam Yu; **40** Kam Yu; **1.23** Dave Mazierski; **1.24** Dr. E.R. Degginger; **42** Kam Yu; **1.25A** Kam Yu after C. Starr and R. Taggart, *Biology 5ed.*, Wadsworth Publishing Company, 1989; **1.25B** Dennis Kunkel/Phototake/First Light; **1.26** Photographs courtesy of Thomas Eisner; **1.27A** Dave Mazierski; **1.27B** Leonard Lessin; **1.28** Kam Yu; **1.29** Photographs courtesy of M.M. Perry from M.M. Perry and A.B. Gilbert (1979), *Journal of Cell Science*, Vol.39, pp.257-272; **TWO: 2.1** Oxford Molecular Biophysics Laboratory/Science Photo Library/Masterfile; **51** Science Photo Library/Masterfile; **2.2** Bob Chambers/Miller Comstock; **2.3** Kam Yu; **2.4** Kam Yu; **2.5** Kam Yu; **2.6** Kam Yu; **2.7** Kam Yu; **2.8** Kam Yu; **2.9** Martyn Lengden after C. Starr and R. Taggart, *Biology 5ed.*, Wadsworth Publishing Company, 1989; **2.10** Martyn Lengden; **2.11** Martyn Lengden; **2.12** Martyn Lengden; **2.13** Martyn Lengden; **2.14A** Carolina Biological Supply Co.; **2.14B** Martyn Lengden after C. Starr and R. Taggart, *Biology 5ed.*, Wadsworth Publishing Company, 1989; **61** Kam Yu; **2.15** Martyn Lengden; **2.16** Martyn Lengden; **65** Left, U.S. National Cancer Institute/Science Photo Library; Bottom right, Dr. Leslie Levin; Top right, Dianne Goettler; **66** Martyn Lengden; **2.17** Martyn Lengden; **2.18** Martyn Lengden; **2.19** Martyn Lengden; **2.20** Laboratory of Molecular Biology, MRC/Science Photo Library; **2.21** Laboratory of Molecular Biology, MRC/Science Photo Library; **70** Top, Martyn Lengden; **2.22** Martyn Lengden; **73** Martyn Lengden; **THREE: 3.1** Adam Hart-Davis/Science Photo Library; **75** Mark Tomalty/Masterfile; **3.2** Martyn Lengden; **3.3** Martyn Lengden; **3.4** Margo Stahl; **3.5** Martyn Lengden; **3.6** Martyn Lengden; **3.7** Martyn Lengden; **3.9** Martyn Lengden after art by Victor Royer in C. Starr and R. Taggart, *Biology 5ed.*, Wadsworth Publishing Company, 1989; **3.11** Photograph courtesy of Mike Connolly; **3.12** Martyn Lengden; **3.13** Martyn Lengden; **82** Left, Dr. Arthur Lesk, MRC Laboratory of Molecular Biology/Science Photo Library; Right, Tom Walker, Photo Journalist; **3.14** Martyn Lengden; **3.15** Martyn Lengden; **3.16** Martyn Lengden; **3.17** Martyn Lengden; **3.19** Margo Stahl; **3.20A** D.Cavagnaro/Peter Arnold, Inc.; **3.20B** Dave Mazierski; **3.20C** Dr. Jeremy Burgess/Science Photo Library; **3.20D** Dave Mazierski; **3.20E** Dr. Kenneth R. Miller/Science Photo Library; **3.21** Martyn Lengden after art by Victor Royer in C. Starr and R. Taggart, *Biology 5ed.*, Wadsworth Publishing Company, 1989; **3.22** Kam Yu; **3.23** Martyn Lengden; **91** Kam Yu; **3.24A** Kam Yu; **3.24B** Richard Siemens; **3.25** Martyn Lengden; **3.26** Canapress Photo Services/Pictor Uniphoto; **3.27** Martyn Lengden; **3.28** First Light; **3.29** Martyn Lengden; **3.30** Liz Nyman; **3.31** Martyn Lengden; **97** Left, Grant Johnson/Miller Comstock; Center, Harvey Pincis/Science Photo Library; Top right, Lena Rooraid/Photo Edit; Bottom right, Schmid-Langsfeld/The Image Bank; **99** Martyn Lengden; **100** Margo Stahl; **102** Margo Stahl; **FOUR: 4.1** CNRI/Science Photo Library; **104** CNRI/Science Photo Library; **4.2** Manfred Kage/Science Photo Library; **4.3** Dave Mazierski; **4.4** CNRI/Science Photo Library; **4.5** Ohio-Nuclear Corporation/Science Photo Library; **4.6** CNRI/Science Photo Library; **4.7** Mehau Kulyk/Science Photo Library; **4.8** Kam Yu; **4.9** Kam Yu; **111-116** Steve Gilbert; **4.10** Martyn Lengden; **118** Peter Menzel; **119** Left, Ontario Wheelchair Sport Association; Right, R. Lee Kirby/Dalhousie University; **FIVE: 5.1** Larry Lefever from Grant Heilman/Miller Comstock; **122** Greg Biss/Masterfile; **5.2** Dave Mazierski; **5.3** Kam Yu; **5.4** Martyn Lengden; **5.5** Martyn Lengden; **5.6** Martyn Lengden; **5.7A/B** IMS Creative Communications; **5.7C** Manfred Kage/Science Photo Library; **5.7D** Dr. R.F.R. Schiller/Science Photo Library; **5.8** Kam Yu; **5.9** Dave Mazierski; **5.10A** Dave Mazierski; **5.10B** Manfred Kage/Science Photo Library; **5.11** From *TISSUES AND ORGANS: A Text-Atlas of Scanning Electron Microscopy* by Richard G. Kessel and Randy H. Kardon. Copyright © 1979 by W.H. Freeman and Company. Reprinted by permission. **134** Top left, Service de Polycopie et de Photographie Université de Montréal; Bottom left, Dr. Gilles Caillé/Université de Montréal; Right, CNRI/Science Photo Library; **136** Martyn Lengden; **137** Martyn Lengden; **SIX: 6.1** Lennart Nilsson from *Behold Man*, © 1974 by Albert Bonniers Forlag and Little, Brown and Company, Boston; **138** Dr. E.R. Degginger; **139** Phototake/First Light; **6.2** Courtesy of the Burndy Library, Cambridge, MA; **6.3** Dave Mazierski; **6.4** Dave Mazierski; **6.5** Kam Yu; **6.6** Dave Mazierski; **6.7** Dr. Jeremy Burgess/Science Photo Library; **6.8** Dave Mazierski; **6.11** Dave Mazierski; **6.12** Dave Mazierski; **6.13** Cardio-Thoracic Center, Freeman Hospital, Newcastle-Upon-Tyne/Science Photo Library; **6.14** Dave Mazierski; **6.15** Dave Mazierski; **6.16** Martyn Lengden; **6.17** Martyn Lengden; **153** Left, Lennart Nilsson from *Behold Man*; Top right, Dr. Stuart Connolly/McMaster University; Bottom right, Dr. Gerald Stronik/Dalhousie University; **6.18** Dave Mazierski; **6.19** Sheila Terry/Science Photo Library; **6.20** Martyn Lengden. **6.21** Dave Mazierski; **6.22** Martyn Lengden; **6.23** Martyn Lengden; **6.24** Kam Yu; **6.25** Kam Yu; **6.26** Dave Mazierski; **6.27** Dave Mazierski; **SEVEN: 7.1** Baylor College of Medicine/Peter Arnold, Inc.; **164** Lennart Nilsson/Bonnier Fakta; **7.2** Kam Yu; **7.3** CNRI/Science Photo Library; **7.4** Dave Mazierski; **7.5** Martyn Lengden and Dave Mazierski; **7.6** Dave Mazierski; **170** Martyn Lengden; **7.7** Dave Mazierski; **7.8** Kam Yu; **7.3A** Martyn Lengden; **7.3B** Kam Yu; **174** Richard Ustinich/The Image Bank; **7.10** Kam Yu; **7.11** Kam Yu; **7.12** Kam Yu; **7.13** Kam Yu; **7.14** Kam Yu; **7.15** Kam Yu; **7.16** Kam Yu; **7.17** Photographs courtesy of Dr. Gilla Kaplan, Cohen Simon Lab, Rockefeller University, New York, N.Y.; **7.18** Dave Mazierski; **7.19** Martyn Lengden; **7.20** Martin Bond/Science Photo Library; **7.21** Dave Mazierski; **182** Left, Prof. Luc Montagnier, Institut Pasteur/CNRI/Science Photo Library; Top right, Francine Gervais/McGill University; Centre right, Dr. Emil Skamene/McGill University; Bottom right, Dr. Paul Jolicoeur/Clinical Research Institute of Montreal;

Image Bank; **16.28** Kathleen Norris Cook/The Image Bank; **16.29** Bob Rose Productions/Miller Comstock; **16.30** Superstock/FourByFive Inc.; **16.31** Margo Stahl; **16.32** W. Belsey/Miller Comstock; **382** Left, W. Heck/University of Manitoba; Right, Jim Brandenburg/Minden Pictures; **384** Martyn Lengden; **SEVENTEEN: 17.1** Jessie Parker/First Light; **386** K. Wothe/The Image Bank; **17.2** Nick Bergkessel, The National Audubon Society Collection/Photo Researchers; **17.3** Martyn Lengden; **17.4** Martyn Lengden; **17.5** Margo Stahl; **17.6A** J.R. Wicinsons/Animals Animals/Earth Scenes; **17.6B** Greg Vaughn/Tom Stack & Assoc.; **17.7** Leef Snyder, The National Audubon Society Collection/Photo Researchers; **17.8** Martyn Lengden; **17.9** Martyn Lengden; **17.10** Martyn Lengden; **17.11** Martyn Lengden; **17.12** Martyn Lengden; **17.13** Martyn Lengden; **17.14** Martyn Lengden; **17.15** Martyn Lengden; **17.16** G.C. Kelley, The National Audubon Society Collection/Photo Researchers; **17.17** Martyn Lengden; **17.18** Martyn Lengden; **17.19** Martyn Lengden; **17.20** Martyn Lengden; **17.21** Martyn Lengden; **17.22** Marty Stouffer/Animals Animals/Earth Scenes; **17.23** Hans Pfletschinger/Peter Arnold Inc.; **17.24** Top, Cath Ellis, Dept. of Zoology, University of Hull; Bottom, Robert Hermes, The National Audubon Society Collection/Photo Researcher; **17.25** Henry Ausloos/Animals Animals/Earth Scenes; **17.26** Dr. E.R. Degginger; **408** Left, P. Wallick/Miller Comstock; Right, Dr. Austin Reed; **17.27** Barbara Von Hoffman/Tom Stack & Associates; **17.28** Margo Stahl; **17.29** Martyn Lengden; **413** Top left, Victor Last; Bottom left, Richard Kolar/Animals Animals/Earth Scenes; Right, Brian Milne/Animals Animals/Earth Scenes; **414** Top to bottom, A.E. Sirulnikoff/First Light; Doug Wechsler/Animals Animals/Earth Scenes; Doug Wechsler/Animals Animals/Earth Scenes; Dr. E.R. Degginger; **415** Left, LaFoto/H. Armstrong Roberts/Miller Comstock; Top center, Eric Hayes/Miller Comstock; Bottom center, J.D. Taylor/Miller Comstock; Top right, Barbara K. Deans/Canapress Photo Service; Bottom right, Elena Rooraid/Photo Edit; **EIGHTEEN: 18.1** John Foster/Masterfile; **420** Cabisco/Visuals Unlimited; **18.2** Courtesy of the Burndy Library, Cambridge, MA; **18.3** Kam Yu; **422** Margo Stahl; **18.4** Margo Stahl; **18.5** Bart Vallecoccia; **18.6** Bart Vallecoccia; **18.7** Kam Yu; **18.8** Kam Yu; Inset, David M. Phillips/Visuals Unlimited; **427** Kam Yu; **18.9** Dr. Hans Reichenbach, Gelleschaft für Biotechnologieche Forschung, Braunschweig; **18.10** Bart Vallecoccia; **18.11** George B. Chapman/Visuals Unlimited; **430** Martyn Lengden; **431** Top to bottom, Peter Parks/Oxford Scientific Films; Cabisco/Visuals Unlimited; Peter Parks/Oxford Scientific Films; **432** Top to bottom, Andrew Syred/Science Photo Library; Manfred Kage/Oxford Scientific Films; Eric Grave/Science Photo Library; CNRI/Science Photo Library; Roy Ficken, Memorial University; Bottom, S. Flegert/Visuals Unlimited; **18.12** Top right, R. Calentene/Visuals Unlimited; Bottom right, Cabisco/Visuals Unlimited; Left, Bart Vallecoccia; **434** Left, Onderstepoort Veterinary Research Institute; Right, Dale Wilson/First Light; **NINETEEN: 19.1** C. Reeves/FPG/Masterfile; **438** Phil Dodson/Photo Researchers; **19.2** Bart Vallecoccia; **19.3** CNRI/Science Photo Library; **19.4** Martyn Lengden; **441** Top to bottom, Dr. Tony Brain/Science Photo Library; Dr. Jeremy Burgess/Science Photo Library/Photo Researchers; J. L. Lepore/Photo Researchers; J. Forsdyke/Gene Cox/Science Photo Library/Photo Researchers; G. Eikerts; Bottom, CNRI/Science Photo Library; **442** Kam Yu; Inset, Dr. Jeremy Burgess/Science Photo Library; **19.5A** L. West/Photo Researchers; **19.5B** Ken Brate/Photo Researchers; **19.6** Martyn Lengden; **445** Left to right, Andrew Syred/Science Photo Library; Lawrence Gould/Oxford Scientific Films; Sinclair Stammers/Science Photo Library; Biophoto Assoc./Photo Researchers; **19.7** Left, Kam Yu; Right, G.I. Bernard/Oxford Scientific Films; **19.8A** Harold Taylor/ABIPP/Oxford Scientific Films; **19.8B** Gregory K. Scott/Photo Researchers; **447** Martyn Lengden; **19.9** Top left, Kjell B. Sandved/Photo Researchers; Top right, Scott Camazine/Photo Researchers; Bottom left, Biophoto Assoc./Photo Researchers; Bottom right, Kjell B. Sandved/Photo Researchers; **19.10** Bart Vallecoccia; Inset, Thomas Kitchin/First Light; **19.11A** Kathie Atkinson/Oxford Scientific Films; **19.11B** S.W. Carter/Photo Researchers; **19.12** Kam Yu; **19.13** Bart Vallecoccia; Inset, Ted Russell/Image Bank; **19.14** Bart Vallecoccia; **454** Top to bottom, G.I. Bernard/Oxford Scientific Films; Thomas W. Martin,APSA/Photo Researchers; S. Homer/First Light; G. Eikerts; Robert Young Pelton/First Light; **455** Left, Geoff Kidd/Oxford Scientific Films; Right, Virginia P. Weinland/Photo Researchers; Top right, Kam Yu; **456** Left, courtesy Dr. Faye Murrin; Right, G.I. Bernard/Oxford Scientific Films; **19.15** Kam Yu; **19.16A,B,C,D,E,G** Mike Sumner/University of Manitoba; **19.16F** Biophoto Assoc./Photo Researchers; **19.16H** Dr. Jeremy Burgess/Science Photo Library; **19.17** Left, Bart Vallecoccia; Right, London Scientific Films/Oxford Scientific Films; **19.18** Left, P. Dayanandan/Photo Researchers; Right, Bart Vallecoccia; **19.19** Left, Kam Yu; Right, Mike Sumner/University of Manitoba; **19.20** Left, Doug Wilson/First Light; Right, Kam Yu; **19.21** Top inset, M. I. Walker/Photo Researchers; Bottom inset, Biophoto Assoc./Photo Researchers; Bart Vallecoccia; **19.22** Bart Vallecoccia; **19.23A** Jerome Wexler/Photo Researchers; **19.23B** G.A MacLean/Oxford Scientific Films; **19.23C** Kjell B. Sandved/Oxford Scientific Films; **19.24** Sherman Hines/Masterfile; **TWENTY: 20.1** Richard Packwood/Oxford Scientific Films; **468** Bob Semple; **20.2** Bart Vallecoccia; **20.3** Bart Vallecoccia; **20.4** Kam Yu; **20.5** Kam Yu; **20.6** Kam Yu; **472** Margo Stahl; **20.7** Bart Vallecoccia; **20.8A** J.A.L. Cooke/Oxford Scientific Films; **20.8B** Peter Parks/Oxford Scientific Films; **474** Margo Stahl; **20.9** Peter Parks/Oxford Scientific Films; **476** Kam Yu; **20.10** Rudie Kuiter/Oxford Scientific Films; **20.11** David Thompson/Oxford Scientific Films; **20.12** Bart Vallecoccia; **20.13** Kathie Atkinson/Oxford Scientific Films; **20.14** Martin Dohrn/Science Photo Library; **480** Margo Stahl; **481** Kam Yu; **482** Kam Yu; **20.15** Kjell B. Sandved/Photo Researchers; **20.16A** Howard Hall/Oxford Scientific Films; **20.16B** Gregory Dimijian/Photo Researchers; **20.17** Top, Jason Puddifoot/First Light; Center, Bottom, G.I. Bernard/Oxford Scientific Films; **20.18** Top left, Peter Parks/Oxford Scientific Films; Bottom left, Norbert Wu/Oxford Scientific Films; Right, Douglas Faulkner/Photo Researchers; **20.19** Bob Semple; **20.20** Kam Yu; **20.21** Dave Mazierski; **20.22** Bart Valleccoccia; **20.23** Left,Dr. Jeremy Burgess/Science Photo Library; Right, Kam Yu; **20.24** Ken M. Highfill/Photo Researchers; **20.25** Martyn Lengden; **20.26** Bob Semple; **20.27A** Tom McHugh/Photo Researchers; **20.27B** Michael Fogden/Oxford Scientific Films; **20.28A** Michael Fogden/Oxford Scientific Films; **20.28B** Phil Devries/Oxford Scientific Films; **20.28C** Michael Fogden/Oxford Scientific Films; **20.28D** Bob Semple; **20.29** Fredrik Ehrenstrom/Oxford Scientific Films; **20.30** Left, Colin Milkins/Oxford Scientific Films; Right, Bart Vallecoccia; **490** Margo Stahl; **491** Top left, courtesy Dr. Vladys Steele; Bottom left, courtesy Dr. William Threlfall; Right, Freeman Patterson/Masterfile; **20.331** Martyn Lengden; **20.32** Sinclair Stammers/Oxford Scientific Films; **20.33** Top, Peter Parks/Oxford Scientific Films; Bottom, G.I. Bernard/Oxford Scientific Films; **20.34** John Paling/Oxford Scientific Films; **20.35** Norbert Wu/Oxford Scientific Films; **20.36A** Tom McHugh/Photo Researchers; **20.36B** John Paling/Oxford Scientific Films; **20.36C** Max Gibbs/Oxford Scientific Films; **20.36D** Bob Semple; **20.37** Philippe Summ/Photo Researchers; **20.38** Bart Vallecoccia; **20.39** Peter Scoones/Planet Earth Pictures; **20.40A** J.A.L. Cooke/Oxford Scientific Films; **20.40B** Joseph T. Collins/Photo Researchers; **20.40C** G.I. Bernard/Oxford Scientific Films; **20.40D** Stephen Dalton/Oxford Scientific Films; **20.41** Bart Vallecoccia; **20.42A** Michael Fogden/Oxford Scientific Films; **20.42B** Richard Packwood/Oxford Scientific Films; **20.42C** Guiliano Colliva/Image Bank; **20.43** James L. Amos/Photo Researchers; Inset, Kam Yu; **20.44** Margo Stahl; **20.45** Kathie Atkinson/Oxford Scientific Films; **20.46** Bart Vallecoccia; **503** Margo Stahl; **TWENTY-ONE: 21.1** Carl Haycock/First Light; **508** A.B. Dowsett/Science Photo Library/Masterfile; **21.2A** Dave Mazierski; **21.2B** Secchi, Lecaque, Roussell, Uclaf, CNRI/Science Photo Library; **21.3** Dave Mazierski; **21.4** Dave Mazierski; **21.5** Martyn Lengden: **21.6** Dave Mazierski; **21.7** Dave Mazierski; **21.8** Kam Yu; **21.9** Martyn Lengden; **521** Left, Kam Yu; Right, Martyn